Common Laplace Transform Pairs

$$u(t) \leftrightarrow \frac{1}{s}$$

$$u(t) - u(t - a) \leftrightarrow \frac{1 - e^{-as}}{s}, \qquad a > 0$$

$$t^n \leftrightarrow \frac{n!}{s^{n+1}}, \qquad n = 1, 2, 3, \ldots$$

$$\delta(t) \leftrightarrow 1$$

$$\delta(t - c) \leftrightarrow e^{-cs}, \qquad c > 0$$

$$e^{-bt} \leftrightarrow \frac{1}{s + b}, \qquad b \text{ real or complex}$$

$$t^n e^{-bt} \leftrightarrow \frac{n!}{(s + b)^{n+1}}, \qquad n = 1, 2, 3, \ldots$$

$$\cos \omega t \leftrightarrow \frac{s}{s^2 + \omega^2}$$

$$\sin \omega t \leftrightarrow \frac{\omega}{s^2 + \omega^2}$$

$$\cos^2 \omega t \leftrightarrow \frac{s^2 + 2\omega^2}{s(s^2 + 4\omega^2)}$$

$$\sin^2 \omega t \leftrightarrow \frac{2\omega^2}{s(s^2 + 4\omega^2)}$$

$$e^{-bt} \cos \omega t \leftrightarrow \frac{s + b}{(s + b)^2 + \omega^2}$$

$$e^{-bt} \sin \omega t \leftrightarrow \frac{\omega}{(s + b)^2 + \omega^2}$$

$$t \cos \omega t \leftrightarrow \frac{s^2 - \omega^2}{(s^2 + \omega^2)^2}$$

$$t \sin \omega t \leftrightarrow \frac{2\omega s}{(s^2 + \omega^2)^2}$$

Introduction to Signals and Systems

EDWARD KAMEN

University of Pittsburgh

INTRODUCTION TO SIGNALS AND SYSTEMS

Macmillan Publishing Company

New York

Collier Macmillan Publishers

London

Macmillan Publishing Company
866 Third Avenue, New York, New York 10022

Collier Macmillan Canada, Inc.

LIBRARY OF CONGRESS CATALOGING-IN-PUBLICATION DATA

Kamen, Edward W.
  Introduction to signals and systems.

  Includes index.
  1. Signal theory (Telecommunication)
2. System analysis. I. Title.
TK5102.5.K316 1987   621.38'043   86-16175
ISBN 0-02-362950-9

Printing: 1 2 3 4 5 6 7 8   Year: 7 8 9 0 1 2 3 4 5 6

ISBN 0-02-362950-9

To my mother and father, Betty and Stan

PREFACE

The purpose of this book is to present an introductory, yet comprehensive, treatment of signals and systems, with a strong emphasis on computing using programs written in Basic.

The material in this book has been taught over a period of several years to juniors in the Department of Electrical Engineering at both the Georgia Institute of Technology and the University of Florida. The potential audience for this book is actually much broader than electrical engineers. In particular, the book should be usable by students and professionals in mechanical engineering, chemical engineering, industrial engineering, and mathematics. The book can be followed without the reader having had a course in electrical circuits, although electrical circuits do provide a rich source of examples of systems, and many such examples are given. The background needed for reading the book consists of the usual freshman/sophomore courses in calculus, physics, and elementary differential equations.

The presentation is geared to students who are being exposed to the basic concepts of signals and systems for the first time. The level of the presentation is introductory, although the scope of the work is quite broad. There are over 250 illustrative examples and 280 homework problems. Most of the results are derived from basic facts. The derivation of results is fairly precise from a mathematical standpoint, yet mathematical rigor is not overemphasized.

A significant effort has been made to find simple ways to develop the theory of signals and systems. Much emphasis has been placed on the meaning and significance of the various concepts and how these concepts fit together to give an overall picture.

A major feature of the text is the ''side-by-side'' development of continuous-time and discrete-time systems. For every concept or mathematical model given in the continuous-time case, the corresponding discrete-time version is also studied. Similarities and differences in the continuous-time and discrete-time cases are pointed out throughout the book.

Instead of treating the continuous-time and discrete-time cases in parallel, by selecting appropriate sections, an instructor could cover continuous-time systems first, then discrete-time systems. In fact, there is enough material in the book for a one-semester course on the continuous-time case and a one-semester course on the discrete-time case.

There is a good deal of emphasis on the generation of discrete-time models for the study of continuous-time systems. In particular, several methods are given for discretizing in time a given continuous-time model. Examples include the discretization in time of differential-equation models, convolution models, transfer function models, and state models. Sampling is also discussed in some detail.

There is a strong emphasis on computing, as evidenced by the inclusion of 14 computer programs written in Basic. The programs have been written for the IBM PC; however, with minor changes they should run on any of the standard personal computers that accept programs in Basic. As noted in Appendix A, the programs are available on a diskette.

The computer programs go hand in hand with the theory developed in the book. For instance, most of the results involving discrete-time systems have been implemented in program form. Examples include the solution of difference equations, convolution, and the fast Fourier transform. The programs are primarily for instructional purposes rather than for serious applications, although they can be used for the latter purpose.

The book begins with time-domain models of continuous-time and discrete-time systems. These models, which are studied in Chapters 1 to 4, include the input/output operator, the input/output differential-equation or difference-equation model, and the convolution representation. The Laplace transform and the z-transform are then studied in Chapters 5 to 7, while the continuous-time and discrete-time Fourier transforms are studied in Chapters 9 to 11. The fundamental notion of frequency response is first developed in Chapter 8 by considering the steady-state response to sinusoidal inputs. Analog-to-digital converters and the design of digital filters are studied in Chapter 12. In Chapter 13 we present the state description of continuous-time and discrete-time systems.

The Laplace transform and z-transform are more ''algebraic'' than the analog and discrete Fourier transforms, which is probably the reason that many students seem to find the former much easier to use. This is one of the reasons why the Laplace and z transforms are considered before the analog and discrete Fourier transforms. Also, students in electrical engineering usually do see the Laplace transform in their first circuits course. Since we believe that it is best to start with the familiar first, this is another reason why we prefer to cover the Laplace and z transforms before the continuous-time and discrete-time Fourier transforms. However, an instructor could cover the chapters on the Fourier transform first.

Applications of the theory to communications, controls, and digital filtering are discussed in the book. The purpose of considering these applications is to illustrate the usefulness of the theory and to make the student aware of these major areas within engineering. Experience has shown that as a result of this introductory treatment, students are often motivated to take senior-level electives in these areas.

Although the emphasis in the book is on linear time-invariant systems, there is some material on time-varying and nonlinear systems. Since systems that are time-varying and/or nonlinear often arise in practice, we believe that some exposure to these classes of systems is appropriate in an introductory course. There is also some material on the multi-input multi-output case. Again, this case often arises in practice, so some attention devoted to it is worthwhile if time permits.

The main portion of the material in this book can be covered in a one-semester course or in a two-quarter sequence in the junior year. In a one-semester course, it is possible to cover most of the material in the unstarred sections of the first 11 chapters. A schedule for a one-semester course consisting of forty 50-minute lectures is given in the following table. This schedule has been found to work very well for a one-semester course offered during a 14- or 15-week period.

Schedule of Coverage for a One-Semester Course

Text Coverage	Number of 50-Minute Lectures
Chapter 1: 1.1, 1.2, 1.3, 1.4, 1.5, 1.6	4
Chapter 2: 2.1, 2.2, 2.3	3
Chapter 3: 3.1, 3.2	2
Chapter 4: 4.1, 4.2, 4.3, 4.4, 4.5	4
Chapter 5: All sections	5
Chapter 6: 6.1, 6.2, 6.3	2
Chapter 7: All sections	3
Chapter 8: 8.1, 8.4, 8.5, 8.6	3
8.8, 8.9	2
Chapter 9: All sections	6
Chapter 10: All sections	3
Chapter 11: 11.2, 11.3, 11.4, 11.5	3
Total	40

If the computational aspects are explored to their fullest possible extent, there is enough material for a course sequence spanning an entire academic year (two semesters or three quarters). However, the primary intended purpose of the book is for use in a one-semester course.

This book would not have been possible without the enthusiastic support of my students over the years. From their questions, I believe that I have been able to identify many of those aspects of the theory of signals and systems that are most difficult to understand for newcomers to the field. To a large extent, the book consists of the answers to questions asked by my students.

The final version of the book would not have been possible without the constructive comments of many colleagues. In particular, I would like to thank John O'Malley, Vernon Shaffer, and Kathy Ossman for having taught from an earlier draft and for making numerous suggestions on improving the work. I would also like to thank Leon Couch, Frank Lewis, and Tom Bullock for their comments. Thanks also go to my former students Hiep Vu and Robert Cooper, who helped write some of the programs. Finally, I wish to thank Mary Sue and Emma for their patience and understanding during the long period of time devoted to writing the book.

E. W. K.

CONTENTS

Introduction to Signals and Systems

Fundamental Concepts

MATHEMATICAL MODELS

To undertake a serious study of a given system, whether man-made or occurring in nature, we must have a mathematical model of the system. A *mathematical model* consists of a collection of equations describing the relationships between the signals appearing in the system. It is important to note that if a system is specified by a digital-computer or analog-computer simulation, then we have, in fact, a mathematical model: Simply write down the equations corresponding to the signal-flow diagram of the simulation.

A mathematical model of a system is usually an idealized representation of the system. In other words, many actual (physical) systems cannot be described exactly by a mathematical model. However, most (if not all) systems can be described sufficiently accurately by a model so that system behavior can be studied in terms of the model. Hence we can analyze the system by analyzing the model. In addition to system analysis, models are also indispensable in the design of new systems having various desirable operating characteristics. Thus mathematical models are used extensively in both system analysis and system design.

If a model of a system is to be useful, it must be tractable, and thus we should always attempt to construct the simplest possible model of the system under study. But the model must also be sufficiently accurate, which means that all primary characteristics (all first-order effects) must be included in the model. Usually, the more characteristics that we put into a model, the more complicated the model is, so there is a trade-off between simplicity of the model and accuracy of the model.

There are two basic types of mathematical models: input/output or external representations describing the relationship between the input and output signals of a system; and the state or internal model describing the relationship among the input, state, and output signals of a system. Input/output representations are studied in the first 12 chapters; the state model is considered in Chapter 13.

Five types of input/output representations are studied in this book:

1. The input/output operator.
2. The input/output differential equation or difference equation.
3. The convolution representation.
4. The transfer function representation.
5. The Fourier transform representation.

As will be shown, the convolution representation can be viewed as a special case of the input/output operator representation, and the Fourier transform representation can be viewed as a special case of the transfer function representation. Hence there are only three fundamentally different types of input/output representations that we shall study.

The first three representations listed above and the state model are referred to as *time-domain models* since these representations are given in terms of functions of time. The last two of the representations listed above are referred to as *frequency-domain models* since they are specified in terms of functions of a complex variable that is interpreted as a frequency variable. Both time-domain and frequency-domain models are used in system analysis and design. These different types of models are often used together in order to maximize our understanding of the behavior of the system under study.

In this chapter we begin with an introduction to continuous-time and discrete-time signals. Then in Section 1.4 we consider a general definition of a system based on the notion of the input/output operator. Examples of continuous-time and discrete-time systems are given in Section 1.5. In Section 1.6 we define the basic system properties of causality, finite dimensionality, linearity, and time invariance. Multi-input multi-output systems are considered in Section 1.7.

1.2

CONTINUOUS-TIME SIGNALS

A signal $x(t)$ is a *real-valued*, or *scalar-valued*, function of the time variable t. By "real valued" we mean that for any fixed value of the time variable t, the value of the signal at time t is a real number. When the time variable t takes its values from the set of real numbers, t is said to be a *continuous-time variable* and the signal $x(t)$ is said to be a *continuous-time signal* or an *analog signal*.

A common type of continuous-time signal is a voltage or current waveform in an *RLC* circuit or in an integrated circuit. Other common types of continuous-time signals are forces or torques applied to mechanical devices, angular positions or angular velocities of a rotor in a dc motor or a link in an industrial robot, and flow rates of liquids or gases in a chemical process.

It should be mentioned that the signals arising in some applications are functions of time and spatial coordinates. An example is the voltage $v(z, t)$ on a transmission line, where $v(z, t)$ is a function of spatial position z and time t. Another example is the irradiance $i(z_1, z_2, t)$ of a light beam in an optical system. In this case the signal $i(z_1, z_2, t)$ is a function of two spatial coordinates z_1, z_2 and time t. Signals that are functions of two or more variables are called *multidimensional signals*. In this book we restrict our attention to signals that are functions of a single variable (the time variable).

STEP AND RAMP FUNCTIONS   Two simple examples of continuous-time signals are the unit-step function $u(t)$ and the unit-ramp function $r(t)$. These functions are plotted in Figure 1.1.

The *unit-step function* $u(t)$ is defined mathematically by

$$u(t) = \begin{cases} 1, & t \geq 0 \\ 0, & t < 0. \end{cases}$$

Here "unit step" means that the amplitude of $u(t)$ is equal to 1 for $t \geq 0$. Note that we are following the convention that $u(0) = 1$. From a strict mathematical standpoint, $u(t)$ is not defined at $t = 0$. Nevertheless, we shall always take $u(0) = 1$. If K is an arbitrary nonzero number, $Ku(t)$ is the step function with amplitude K for $t \geq 0$.

For any continuous-time signal $x(t)$, the product $x(t)u(t)$ is equal to $x(t)$ for $t \geq 0$ and is equal to zero for $t < 0$. Thus multiplication of a signal $x(t)$ with $u(t)$ eliminates any nonzero values of $x(t)$ for $t < 0$.

The *unit-ramp function* $r(t)$ is defined mathematically by

$$r(t) = \begin{cases} t, & t \geq 0 \\ 0, & t < 0. \end{cases}$$

Note that for $t \geq 0$, the slope of $r(t)$ is 1. Thus $r(t)$ has "unit slope," which is

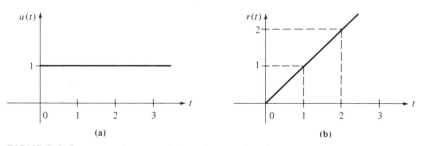

FIGURE 1.1.   (a) Unit-step and (b) unit-ramp functions.

the reason $r(t)$ is called the unit-ramp function. If K is an arbitrary nonzero scalar (real number), the ramp function $Kr(t)$ has slope K for $t \geq 0$.

The unit-ramp function $r(t)$ is equal to the integral of the unit-step function $u(t)$; that is,

$$r(t) = \int_{-\infty}^{t} u(\lambda) \, d\lambda.$$

Conversely, the first derivative of $r(t)$ with respect to t is equal to $u(t)$, except at $t = 0$, where the derivative of $r(t)$ is not defined.

THE IMPULSE   The *unit impulse* $\delta(t)$, also called the *delta function* or the *Dirac distribution*, is defined by

$$\delta(t) = 0, \qquad t \neq 0$$

$$\int_{-\epsilon}^{\epsilon} \delta(\lambda) \, d\lambda = 1 \qquad \text{for any real number } \epsilon > 0.$$

The first condition states that $\delta(t)$ is zero for all nonzero values of t, while the second condition states that the area under the impulse is 1, so $\delta(t)$ has unit area.

It is important to point out that the value $\delta(0)$ of $\delta(t)$ at $t = 0$ is not defined; in particular, $\delta(0)$ is not equal to infinity. The unit impulse is not actually a function. In mathematics, $\delta(t)$ is defined by a linear functional on a "space of test functions." We shall not consider this definition here.

Since $\delta(t)$ is not a function, it is not possible to generate an actual signal that has exactly the same properties as $\delta(t)$. However, $\delta(t)$ can be thought of as a pulse centered at the origin with amplitude A and time duration $1/A$, where A is a very large positive number. The pulse interpretation of $\delta(t)$ is displayed in Figure 1.2. It should be noted that the pulse shown does not converge to the unit impulse $\delta(t)$ as A goes to infinity.

For any real number K, $K\delta(t)$ is the impulse with area K. It is defined by

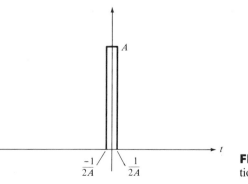

FIGURE 1.2.   Pulse interpretation of $\delta(t)$.

$$K\delta(t) = 0, \qquad t \neq 0$$

$$\int_{-\epsilon}^{\epsilon} K\delta(\lambda) \, d\lambda = K \qquad \text{for any real number } \epsilon > 0.$$

The graphical representation of $K\delta(t)$ is shown in Figure 1.3. The notation "(K)" in the figure refers to the area of the impulse $K\delta(t)$.

The unit-step function $u(t)$ is equal to the integral of the unit impulse $\delta(t)$; more precisely, we have that

$$u(t) = \int_{-\infty}^{t} \delta(\lambda) \, d\lambda, \qquad \text{all } t \text{ except } t = 0.$$

To verify this relationship, first note that for $t < 0$,

$$\int_{-\infty}^{t} \delta(\lambda) \, d\lambda = 0 \qquad \text{since } \delta(t) = 0 \text{ for all } t < 0.$$

For $t > 0$,

$$\int_{-\infty}^{t} \delta(\lambda) \, d\lambda = \int_{-t}^{t} \delta(\lambda) \, d\lambda = 1 \qquad \text{since } \int_{-\epsilon}^{\epsilon} \delta(\lambda) \, d\lambda = 1 \text{ for any } \epsilon > 0.$$

TIME-SHIFTED SIGNALS Given a continuous-time signal $x(t)$, we will often need to consider a *time-shifted* version of $x(t)$: If t_1 is a positive real number, the signal $x(t - t_1)$ is $x(t)$ shifted to the right by t_1 seconds. If t_1 is negative, $x(t - t_1)$ is $x(t)$ shifted to the left by t_1 seconds. For instance, if $x(t)$ is the unit-step function $u(t)$ and $t_1 = 2$, $u(t - t_1)$ is the 2-second right shift of $u(t)$. If $t_1 = -2$, $u(t - t_1)$ is the 2-second left shift of $u(t)$. These shifted signals are plotted in Figure 1.4. To verify that $u(t - 2)$ is given by the plot in Figure 1.4a, evaluate $u(t - 2)$ for various values of t. For example, $u(t - 2) = u(-2) = 0$ when $t = 0$, $u(t - 2) = u(-1) = 0$ when $t = 1$, $u(t - 2) = u(0) = 1$ when $t = 2$, and so on.

For any fixed positive or negative real number t_1, the t_1-second shift $K\delta(t - t_1)$ of the impulse $K\delta(t)$ is equal to the impulse with area K located at the point $t = t_1$; in other words,

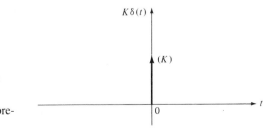

FIGURE 1.3. Graphical representation of the impulse $K\delta(t)$.

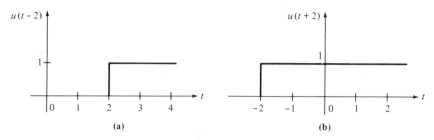

FIGURE 1.4. Two-second shifts of $u(t)$: (a) right shift; (b) left shift.

$$K\delta(t - t_1) = 0, \qquad t \neq t_1$$

$$\int_{t_1-\epsilon}^{t_1+\epsilon} K\delta(\lambda - t_1)\, d\lambda = K, \qquad \text{any } \epsilon > 0.$$

CONTINUOUS AND PIECEWISE-CONTINUOUS SIGNALS

A continuous-time signal $x(t)$ is said to be *discontinuous* at a fixed point t_1 if $x(t_1^-) \neq x(t_1^+)$, where $t_1 - t_1^-$ and $t_1^+ - t_1$ are infinitesimal positive numbers. Roughly speaking, a signal $x(t)$ is discontinuous at a point t_1 if the value of $x(t)$ "jumps" as t goes through the point t_1.

A signal $x(t)$ is *continuous* at the point t_1 if $x(t_1^-) = x(t_1) = x(t_1^+)$. If a signal $x(t)$ is continuous at all points t, we say that $x(t)$ is a *continuous signal*. The reader should note that we are using the term "continuous" in two different ways; that is, we have the notion of a continuous-time signal and we have the notion of a continuous-time signal that is continuous (as a function of t). This dual use of "continuous" should be clear from the context.

Many continuous-time signals of interest in engineering are continuous. An example is the ramp function $Kr(t)$. Another example of a continuous signal is the triangular pulse function displayed in Figure 1.5.

As indicated in the figure, the triangular pulse is equal to $(2t/\tau) + 1$ for $-\tau/2 \leq t \leq 0$ and is equal to $(-2t/\tau) + 1$ for $0 \leq t \leq \tau/2$.

There are also many continuous-time signals of interest in engineering that are not continuous at all points t. An example is the step function $Ku(t)$, which is discontinuous at the point $t = 0$ (assuming that $K \neq 0$). Another example of

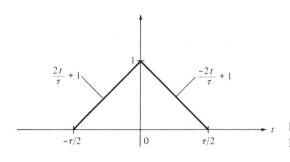

FIGURE 1.5. Triangular pulse function.

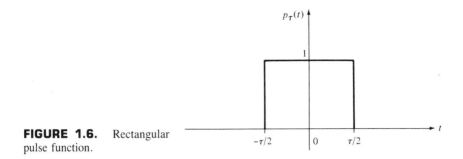

FIGURE 1.6. Rectangular pulse function.

a signal that is not continuous everywhere is the rectangular pulse function $p_\tau(t)$ defined by

$$p_\tau(t) = \begin{cases} 1, & \dfrac{-\tau}{2} \leq t < \dfrac{\tau}{2} \\[3mm] 0, & t < \dfrac{-\tau}{2}, t \geq \dfrac{\tau}{2}. \end{cases}$$

Here τ is a fixed positive number equal to the time duration of the pulse. The rectangular pulse function $p_\tau(t)$ is displayed in Figure 1.6. It is obvious from Figure 1.6 that $p_\tau(t)$ is continuous at all t except $t = -\tau/2$ and $t = \tau/2$.

Note that we can write $p_\tau(t)$ in the form

$$p_\tau(t) = u\left(t + \frac{\tau}{2}\right) - u\left(t - \frac{\tau}{2}\right).$$

Note also that the triangular pulse function shown in Figure 1.5 is equal to $(1 - 2|t|/\tau)p_\tau(t)$, where $|t|$ is the absolute value of t defined by $|t| = t$ when $t > 0$, $|t| = -t$ when $t < 0$.

A continuous-time signal $x(t)$ is said to be *piecewise continuous* if it is continuous at all t except at a finite or countably infinite collection of points t_i, $i = 1, 2, 3, \ldots$. Examples of piecewise-continuous functions are the step function $Ku(t)$ and the rectangular pulse function $p_\tau(t)$. Another example of a piecewise-continuous signal is the pulse train shown in Figure 1.7. This signal is continuous at all t except at $t = 0, \pm 1, \pm 2, \ldots$.

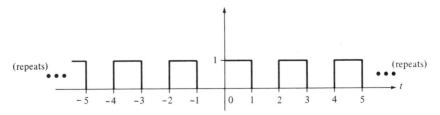

FIGURE 1.7. Signal that is discontinuous at $t = 0, \pm 1, \pm 2, \ldots$.

DERIVATIVE OF A CONTINUOUS-TIME SIGNAL   A continuous-time
signal $x(t)$ is said to be *differentiable* at a fixed point t_1 if

$$\frac{x(t_1 + h) - x(t_1)}{h}$$

has a limit as $h \to 0$, where h approaches zero from above ($h > 0$) and from
below ($h < 0$). If the limit exists, $x(t)$ has a *derivative* at the point t_1 defined
by

$$\left. \frac{dx(t)}{dt} \right|_{t=t_1} = \lim_{h \to 0} \frac{x(t_1 + h) - x(t_1)}{h}.$$

This definition of the derivative of $x(t)$ is sometimes called the *ordinary derivative* of $x(t)$.

   To be differentiable at a point t_1, it is necessary (but not sufficient in general)
that the signal $x(t)$ be continuous at t_1. Hence continuous-time signals that are
not continuous at all points cannot be differentiable at all points. In particular,
piecewise-continuous signals are not differentiable at all points. However, piecewise-continuous signals may have a derivative in the generalized sense. Suppose
that $x(t)$ is differentiable at all t except $t = t_1$. Then the generalized derivative
of $x(t)$ is defined to be

$$\frac{dx(t)}{dt} + [x(t_1^+) - x(t_1^-)]\delta(t - t_1),$$

where $dx(t)/dt$ is the ordinary derivative of $x(t)$ at all t except $t = t_1$, and
$\delta(t - t_1)$ is the unit impulse concentrated at the point $t = t_1$. Thus the generalized derivative of a signal at a point of discontinuity t_1 is equal to an impulse
located at t_1 and with area equal to the amount the function "jumps" at the
point t_1.

EXAMPLE 1.1.   Let $x(t)$ be the step function $Ku(t)$. We know that the
ordinary derivative of $Ku(t)$ is equal to zero at all t except at $t = 0$. Therefore,
the generalized derivative of $Ku(t)$ is equal to

$$K[u(0^+) - u(0^-)]\delta(t - 0) = K\delta(t).$$

Taking $K = 1$, we have that the generalized derivative of the unit-step
function $u(t)$ is equal to the unit impulse $\delta(t)$.

EXAMPLE 1.2.   Consider the piecewise-continuous signal $x(t)$ defined by

$$x(t) = \begin{cases} 2t + 1, & 0 \le t < 1 \\ 1, & 1 \le t \le 2 \\ -t + 3, & 2 \le t \le 3 \\ 0, & \text{all other } t. \end{cases}$$

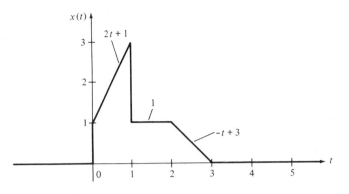

FIGURE 1.8. Signal in Example 1.2.

The signal $x(t)$ is plotted in Figure 1.8. From the plot we see that $x(t)$ is continuous at all t except at $t = 0, 1$. The ordinary derivative of $x(t)$ at all t (except $t = 0, 1, 2, 3$) is equal to

$$2[u(t) - u(t - 1)] - [u(t - 2) - u(t - 3)].$$

The generalized derivative of $x(t)$ is equal to

$$2[u(t) - u(t - 1)] - [u(t - 2) - u(t - 3)] + [x(0^+) - x(0^-)]\delta(t)$$
$$+ [x(1^+) - x(1^-)]\delta(t - 1),$$

which simplifies to

$$2[u(t) - u(t - 1)] - [u(t - 2) - u(t - 3)] + \delta(t) - 2\delta(t - 1).$$

The generalized derivative of $x(t)$ is displayed in Figure 1.9.

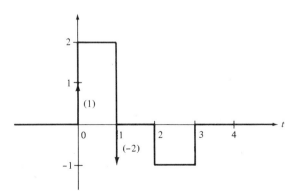

FIGURE 1.9. Generalized derivative of the signal in Example 1.2.

SIGNALS DEFINED INTERVAL BY INTERVAL   Continuous-time signals are often defined interval by interval. For example, suppose that $x(t)$ is given by

$$x(t) = \begin{cases} x_1(t), & t_1 \leq t < t_2 \\ x_2(t), & t_2 \leq t < t_3 \\ x_3(t), & t \geq t_3, \end{cases}$$

where $x_1(t)$, $x_2(t)$, and $x_3(t)$ are arbitrary functions of t. Any such signal can be expressed analytically in terms of the unit-step function $u(t)$ and time shifts of $u(t)$. For example, for the above signal we have that

$$x(t) = x_1(t)[u(t - t_1) - u(t - t_2)] + x_2(t)[u(t - t_2) - u(t - t_3)]$$
$$+ x_3(t)u(t - t_3), \qquad t \geq t_1. \tag{1.1}$$

Rearranging terms in (1.1), we can write $x(t)$ in the form

$$x(t) = x_1(t)u(t - t_1) + [x_2(t) - x_1(t)]u(t - t_2)$$
$$+ [x_3(t) - x_2(t)]u(t - t_3), \qquad t \geq t_1. \tag{1.2}$$

From (1.2) we see that we have expressed $x(t)$ in the form

$$x(t) = f_1(t)u(t - t_1) + f_2(t)u(t - t_2) + f_3(t)u(t - t_3), \qquad t \geq t_1, \tag{1.3}$$

where

$$f_1(t) = x_1(t)$$

$$f_2(t) = x_2(t) - x_1(t)$$

$$f_3(t) = x_3(t) - x_2(t).$$

Conversely, suppose that $x(t)$ is given in the form (1.3) for some functions $f_1(t), f_2(t), f_3(t)$. Then

$$x(t) = \begin{cases} f_1(t), & t_1 \leq t < t_2 \\ f_1(t) + f_2(t), & t_2 \leq t < t_3 \\ f_1(t) + f_2(t) + f_3(t), & t \geq t_3. \end{cases} \tag{1.4}$$

The signal $x(t)$ given by (1.4) can also be written in the form

$$x(t) = f_1(t)[u(t - t_1) - u(t - t_2)] + [f_1(t) + f_2(t)]$$
$$\times [u(t - t_2) - u(t - t_3)] + (f_1(t) + f_2(t)$$
$$+ f_3(t))u(t - t_3), \qquad t \geq t_1.$$

EXAMPLE 1.3. For the signal $x(t)$ in Example 1.2, we have that

$$x(t) = (2t + 1)[u(t) - u(t - 1)] + (1)[u(t - 1) - u(t - 2)] + (-t + 3)[u(t - 2) - u(t - 3)].$$

Writing $x(t)$ in the form (1.3), we get

$$x(t) = (2t + 1)u(t) - 2tu(t - 1) + (-t + 2)u(t - 2) + (t - 3)u(t - 3).$$

1.3

DISCRETE-TIME SIGNALS

The time variable t is said to be a *discrete-time variable* if t takes on only the discrete values $t = kT$, where T is a fixed positive real number and where k ranges over the set of integers (i.e., $k = \ldots, -2, -1, 0, 1, 2, \ldots$). If $T = 1$, the discrete-time variable $t = k$ takes on integer values only. A *discrete-time signal* $x(t)$ is a signal that is a function of the discrete-time variable $t = kT$. A discrete-time signal $x(t)$ is defined for only a discrete set of time points. For instance, the signal $x(t) = t$, $t = 0, T, 2T, \ldots, x(t) = 0, t = -T, -2T$, $\ldots$, is a discrete-time signal. This signal is a discrete-time version of the unit-ramp function $r(t)$. A plot of $x(t)$ versus the discrete-time variable $t = kT$ is given in Figure 1.10.

Discrete-time signals arise naturally in many areas of business, science, and engineering. In applications to business, the discrete-time variable may be the month, quarter, or year of a specified period of time. In Section 1.5 we give an example where the discrete-time variable is the month of a loan-repayment period.

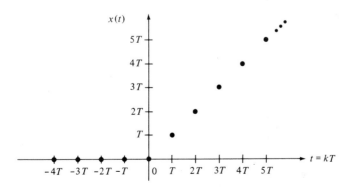

FIGURE 1.10. Example of a discrete-time signal.

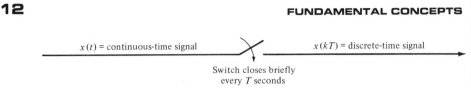

$x(t)$ = continuous-time signal $x(kT)$ = discrete-time signal

Switch closes briefly
every T seconds

FIGURE 1.11. Sampling process.

SAMPLING One of the most common ways in which discrete-time signals arise is in sampling continuous-time signals: Suppose that we apply a continuous-time signal $x(t)$ to an electronic switch that is closed briefly every T seconds. If the amount of time during which the switch is closed is much smaller than T, we can view the output of the switch as a discrete-time signal that is a function of the discrete time variable $t = kT$, where $k = \ldots, -2, -1, 0, 1, 2, \ldots$. The resulting discrete-time signal is called the *sampled version* of the original continuous-time signal and T is called the *sampling interval*. The sampling process is illustrated in Figure 1.11. We shall study sampling in depth in Chapters 10 and 12.

We can generate a large class of discrete-time signals by sampling continuous-time signals. For example, the discrete-time signal plotted in Figure 1.10 can be viewed as the sampled unit-ramp function. If we sample the unit-step function $u(t)$, we get the sampled unit-step function plotted in Figure 1.12.

DISCRETE-TIME RECTANGULAR PULSE Let N be a positive integer. An important example of a discrete-time signal is the *discrete-time rectangular pulse function* $p_{2NT}(t)$, defined by

$$p_{2NT}(t) = \begin{cases} 1, & t = -NT, -NT + T, \ldots, NT - T, NT \\ 0, & \text{all other } k. \end{cases}$$

The discrete-time rectangular pulse is displayed in Figure 1.13.

UNIT PULSE It should be noted that there is no sampled version of the unit impulse $\delta(t)$ since $\delta(0)$ is not defined. However, there is a discrete-time signal that is the "discrete-time counterpart" of the unit impulse. This is the *unit-pulse function* $\Delta(t)$, defined by

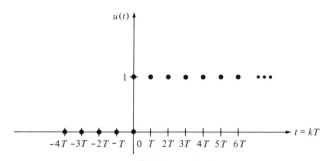

FIGURE 1.12. Sampled unit-step function.

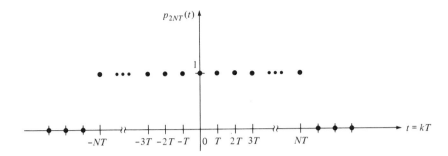

FIGURE 1.13.　Discrete-time rectangular pulse.

$$\Delta(t) = \begin{cases} 1, & t = 0 \\ 0, & t = kT, \quad k \neq 0. \end{cases}$$

The unit-pulse function is plotted in Figure 1.14.

DIGITAL SIGNALS　Let $\{a_1, a_2, \ldots, a_N\}$ be a set of N real numbers. A *digital signal* $x(t)$ is a discrete-time signal whose values belong to the finite set $\{a_1, a_2, \ldots, a_N\}$; that is, at each time point $t = kT$, $x(t) = a_i$ for some i, where $1 \leq i \leq N$. So a digital signal can have only a finite number of different values.

A sampled continuous-time signal is not necessarily a digital signal. For example, the sampled unit-ramp function is not a digital signal since $r(t)$ takes on an infinite range of values when $t = kT$, $k = \ldots, -2, -1, 0, 1, 2, \ldots$.

A *binary signal* is a digital signal whose values are equal to 1 or 0; that is, $x(t) = 0$ or 1 for $t = kT$, $k = \ldots, -2, -1, 0, 1, 2, \ldots$. The sampled unit-step function and the unit-pulse function are both examples of binary signals.

TIME-SHIFTED SIGNALS　Given a discrete-time signal $x(t)$, $t = kT$, and an integer k_1, the discrete-time signal $x(t - k_1T)$ is equal to a k_1T-second shift of $x(t)$. If k_1 is a positive integer, $x(t - k_1T)$ is a k_1T-second right shift of $x(t)$, and if k_1 is a negative integer, $x(t - k_1T)$ is a k_1T-second left shift of $x(t)$. For example, $p_{2T}(t - 2T)$ is the 2T-second right shift of the discrete-time rectangular

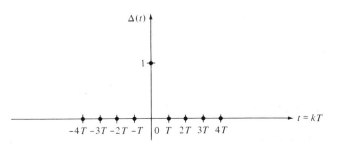

FIGURE 1.14.　Unit-pulse function.

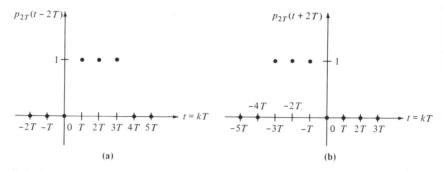

FIGURE 1.15. $2T$-second shifts of $p_{2T}(t)$: (a) right shift; (b) left shift.

pulse $p_{2T}(t)$, and $p_{2T}(t + 2T)$ is the $2T$-second left shift of $p_{2T}(t)$. The shifted signals are plotted in Figure 1.15.

DISCRETE-TIME SIGNALS DEFINED INTERVAL BY INTERVAL

As in the case of continuous-time signals, discrete-time signals are sometimes defined interval by interval. For instance, $x(kT)$ may be specified by

$$x(kT) = \begin{cases} x_1(kT), & k_1 \le k < k_2 \\ x_2(kT), & k_2 \le k < k_3 \\ x_3(kT), & k \ge k_3, \end{cases}$$

where $x_1(kT)$, $x_2(kT)$, and $x_3(kT)$ are arbitrary discrete-time signals. Using the sampled unit-step function $u(kT)$ and time shifts of $u(kT)$, we can write $x(kT)$ in the form

$$\begin{aligned} x(kT) = \; & x_1(kT)[u(kT - k_1T) - u(kT - k_2T)] \\ & + x_2(kT)[u(kT - k_2T) - u(kT - k_3T)] \\ & + x_3(kT)u(kT - k_3T), \quad k \ge k_1. \end{aligned} \quad (1.5)$$

Combining terms in (1.5), we have that

$$\begin{aligned} x(kT) = \; & x_1(kT)u(kT - k_1T) + [x_2(kT) - x_1(kT)]u(kT - k_2T) \\ & + [x_3(kT) - x_2(kT)]u(kT - k_3T), \quad k \ge k_1. \end{aligned} \quad (1.6)$$

Note that the expressions (1.5) and (1.6) are simply discrete-time versions of the corresponding expressions in the continuous-time case [see equations (1.1) and (1.2)].

1.4

GENERAL DEFINITION OF A SYSTEM

A common way of viewing a system is in terms of a "black box" with terminals, as illustrated in Figure 1.16. In the figure, $x_1(t), x_2(t), \ldots, x_p(t)$ are the signals

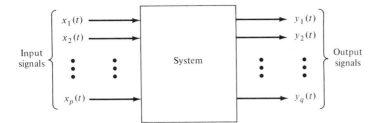

FIGURE 1.16. System with p inputs and q outputs.

applied to the p input terminals of the system and $y_1(t)$, $y_2(t)$, . . ., $y_q(t)$ are the resulting responses appearing at the q output terminals of the system. In general, p is not equal to q; in other words, the number of input terminals may not equal the number of output terminals. When $p = q = 1$, the system is a single-input single-output system, sometimes referred to as a *scalar system*. We should note that the input and output terminals shown in Figure 1.16 do not include "ground" connections. For example, if the $x_i(t)$ and the $y_i(t)$ are voltages relative to ground, we do not count the ground as a terminal.

The input signals $x_1(t)$, $x_2(t)$, . . ., $x_p(t)$ and the output signals $y_1(t)$, $y_2(t)$, . . ., $y_q(t)$ are real-valued functions of the time variable t; that is, for any fixed value of t, $x_i(t)$ and $y_i(t)$ are real numbers. A system may be subjected to different types of inputs. Common examples are control inputs, reference inputs, and disturbance inputs (such as noise). Certain types of input signals, such as disturbance inputs, may not be directly measurable. In contrast, the output signals of a system are usually assumed to be measurable using sensing devices.

When the time variable t takes its values from the set of real numbers, the system is said to be a *continuous-time system* or an *analog system*. When t takes on the discrete values $t = kT$, $k = \ldots, -2, -1, 0, 1, 2, \ldots$, the system is said to be a *discrete-time system*. A digital system is a discrete-time system whose inputs and outputs are digital signals; that is, the input and output signals can take on only a finite number of different values (e.g., 0 or 1).

INPUT/OUTPUT OPERATOR We shall represent (i.e., model) the system illustrated in Figure 1.16 by an input/output operator as follows. First, we say that the system has zero initial energy or is at rest at time t_0 if there is no energy stored in the system at time t_0. For instance, if the system is an interconnection of resistors, capacitors, and inductors, zero initial energy at time t_0 means that the initial voltages on the capacitors at time t_0 are all zero and the initial currents in the inductors at time t_0 are all zero.

If there is no initial energy in the system at time t_0, the system is said to be in the *zero state* at time t_0. (The notion of state is defined in Chapter 13.) If a system has zero initial energy at time t_0, the outputs $y_i(t)$ are all zero for $t > t_0$ whenever the inputs $x_i(t)$ are all zero for $t > t_0$.

Now suppose that the system under study is a single-input single-output system, so that the input $x(t)$ and output $y(t)$ are real-valued (scalar-valued) functions of the time variable t. Given an arbitrary input signal $x(t)$ with $x(t) = 0$

for all $t < t_0$ and with no initial energy in the system at time t_0, the resulting output response $y(t)$ can be expressed in the form

$$y(t) = (Fx)(t), \qquad t > t_0, \qquad (1.7)$$

where F is an operator or map that acts on the input signal x to produce the output signal y, and t_0 is the initial time. The output response $y(t)$ given by (1.7) is sometimes called the *zero-state response* since it is the response when the system is in the zero state (no initial energy) prior to the application of the input $x(t)$. The operator F, which is fixed for a given system, is called the *input/output operator* of the system. The input/output operator F describes how the system operates on input signals to produce output signals. In other words, F is a mathematical representation of the "flow" of signals through the system.

It is important to note that the output $y(t)$ at a fixed time t depends in general on the entire input signal x, not just on the value $x(t)$ of x at time t. Thus the input/output operator F acts on the input function x rather than the value $x(t)$ of x at time t. This is why the input/output relationship is written in the form

$$y(t) = (Fx)(t)$$

rather than in the form

$$y(t) = F(x(t)).$$

The input/output operator representation (1.7) and the representations to be given later are sufficiently general to include many different types of systems, such as electrical systems, mechanical systems, chemical systems, and combinations of different types of systems, such as electromechanical systems (e.g., an electric car). The concepts and techniques given in this book are applicable to these different types of systems. In particular, the system inputs and outputs may be voltages, currents, forces, angular positions of mechanical links, flow rates of liquids or gases, and so on.

The problem of explicitly determining the input/output operator F for a given system is called the *modeling* or *identification problem*. In some cases F is determined from input/output data consisting of the measured responses to a particular collection of known input signals (test signals). We shall consider this approach to system identification at various points in the book. In many applications the input/output operator F is determined by applying the laws of physics. Examples are given in the next section.

1.5

EXAMPLES OF SYSTEMS

In this section we give four examples of a system. The first three examples are continuous-time systems, while the fourth example is a discrete-time system.

To keep the discussion at an introductory level, in this section we restrict our attention to simple examples of systems. More complicated examples are considered later.

RC CIRCUIT   Consider the *RC* circuit shown in Figure 1.17. The *RC* circuit can be viewed as a single-input single-output continuous-time system with input *x(t)* equal to the current *i(t)* into the parallel connection and with output *y(t)* equal to the voltage $v_c(t)$ across the capacitor. By Kirchhoff's current law, we have that

$$i_c(t) + i_R(t) = i(t), \tag{1.8}$$

where $i_c(t)$ is the current in the capacitor and $i_R(t)$ is the current in the resistor. Now

$$i_c(t) = C \frac{dy(t)}{dt} \tag{1.9}$$

and

$$i_R(t) = \frac{1}{R} y(t). \tag{1.10}$$

Inserting (1.9) and (1.10) into (1.8), we get the following linear differential equation:

$$C \frac{dy(t)}{dt} + \frac{1}{R} y(t) = x(t). \tag{1.11}$$

The differential equation (1.11) is called the *input/output differential equation* of the circuit. It provides an implicit relationship between the input *x(t)* and the output *y(t)*. An explicit expression for *y(t)* in terms of *x(t)* is the input/output relationship $y(t) = (Fx)(t)$, where *F* is the input/output operator of the circuit. To compute *F* for this circuit, we must solve the input/output differential equation (1.11) for an arbitrary input *x(t)* applied for $t \geq t_0$. The steps are as follows.

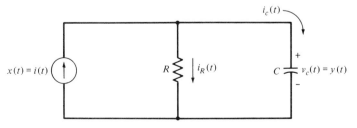

FIGURE 1.17   *RC* circuit.

First, it is easy to see that if $x(t) = 0$ for all $t \geq t_0$, the solution $y(t)$ to (1.11) is given by

$$y(t) = \exp\left[\frac{-1}{RC}(t - t_0)\right]y(t_0), \qquad t \geq t_0, \tag{1.12}$$

where $y(t_0)$ is the initial voltage on the capacitor at time t_0. To show that (1.12) is the solution to (1.11) when $x(t) = 0$, $t \geq t_0$, simply verify that (1.11) is satisfied when the expression (1.12) for $y(t)$ is inserted into (1.11). We invite the reader to check this.

Now suppose that the input $x(t)$ is not necessarily zero for $t \geq t_0$. The solution to (1.11) can be computed by first multiplying both sides of (1.11) by an integrating factor equal to $(1/C)\exp[(1/RC)t]$. This gives

$$\exp\left(\frac{1}{RC}t\right)\left[\frac{dy(t)}{dt} + \frac{1}{RC}y(t)\right] = \frac{1}{C}\exp\left(\frac{1}{RC}t\right)x(t). \tag{1.13}$$

The left-hand side of (1.13) is equal to the derivative with respect to t of $\exp[(1/RC)t]y(t)$, and thus

$$\frac{d}{dt}\left[\exp\left(\frac{1}{RC}t\right)y(t)\right] = \frac{1}{C}\exp\left(\frac{1}{RC}t\right)x(t).$$

This equation is in the form

$$\frac{dv(t)}{dt} = q(t), \tag{1.14}$$

where

$$v(t) = \exp\left(\frac{1}{RC}t\right)y(t) \tag{1.15}$$

$$q(t) = \frac{1}{C}\exp\left(\frac{1}{RC}t\right)x(t). \tag{1.16}$$

Integrating both sides of (1.14) with respect to t and using the Fundamental Theorem of Calculus, we have

$$v(t) = v(t_0) + \int_{t_0}^{t} q(\lambda)\,d\lambda, \qquad t \geq t_0. \tag{1.17}$$

Inserting the expressions (1.15) and (1.16) into (1.17), we obtain

$$\exp \left(\frac{1}{RC} t \right) y(t) = \exp \left(\frac{1}{RC} t_0 \right) y(t_0)$$
$$+ \int_{t_0}^{t} \frac{1}{C} \exp \left(\frac{1}{RC} \lambda \right) x(\lambda) \, d\lambda, \qquad t \geq t_0. \tag{1.18}$$

Finally, multiplying both sides of (1.18) by $\exp [-(1/RC)t]$, we have

$$y(t) = \exp \left[\frac{-1}{RC} (t - t_0) \right] y(t_0)$$
$$+ \int_{t_0}^{t} \frac{1}{C} \exp \left[\frac{-1}{RC} (t - \lambda) \right] x(\lambda) \, d\lambda, \qquad t \geq t_0. \tag{1.19}$$

The expression (1.19) for $y(t)$ is the complete output response of the RC circuit resulting from initial voltage $y(t_0) = v_c(t_0)$ and arbitrary input $x(t) = i(t)$ applied for $t \geq t_0$.

We should note that if the input $x(t)$ contains an impulse $K\delta(t - t_0)$, it is necessary to take the initial condition at time t_0^-, where $t_0 - t_0^-$ is an infinitesimal positive number. The reason for this is that an impulsive input can instantaneously change the value of the output. If the initial condition is taken at time t_0^-, the expression (1.19) for the output becomes

$$y(t) = \exp \left[\frac{-1}{RC} (t - t_0) \right] y(t_0^-)$$
$$+ \int_{t_0^-}^{t} \frac{1}{C} \exp \left[\frac{-1}{RC} (t - \lambda) \right] x(\lambda) \, d\lambda, \qquad t \geq t_0.$$

The RC circuit is at rest at time t_0 if and only if $y(t_0) = 0$, in which case the output response is

$$y(t) = \int_{t_0}^{t} \frac{1}{C} \exp \left[\frac{-1}{RC} (t - \lambda) \right] x(\lambda) \, d\lambda, \qquad t \geq t_0. \tag{1.20}$$

From (1.20) we have that the input/output operator F of the RC circuit is an *integral operator* given by

$$y(t) = (Fx)(t) = \int_{t_0}^{t} \frac{1}{C} \exp \left[\frac{-1}{RC} (t - \lambda) \right] x(\lambda) \, d\lambda, \qquad t \geq t_0.$$

In a subsequent chapter we shall see that the input/output operator F of a system specified by a linear input/output differential equation with constant coefficients can be determined without having to solve the input/output differential equation.

Note that by the input/output relationship (1.20), the output $y(t)$ at time t depends on the input $x(\lambda)$ for $t_0 \leq \lambda \leq t$. In other words, the value of the output at time t depends in general on the values of the input signal over the time interval from the initial time t_0 to the present time t.

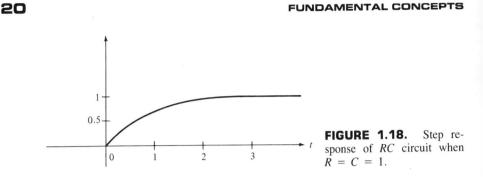

FIGURE 1.18. Step response of RC circuit when $R = C = 1$.

Using (1.20), we can compute the output response resulting from the application of any input current $x(t)$. For example, suppose that we take the initial time t_0 to be zero and we take the input $x(t)$ to be the unit-step function $u(t)$. Then, from (1.20),

$$y(t) = \int_0^t \frac{1}{C} \exp\left[\frac{-1}{RC}(t - \lambda)\right] d\lambda$$

$$= R \exp\left[\frac{-1}{RC}(t - \lambda)\right]_{\lambda=0}^{\lambda=t}$$

$$= R\left[1 - \exp\left(\frac{-1}{RC}t\right)\right], \qquad t \geq 0. \qquad (1.21)$$

The output response $y(t)$ given by (1.21) is called the *step response* since $y(t)$ is the output when the input is the unit-step function with the system at rest prior to the application of $u(t)$. If at $t = 0$ we switched on a constant current source of amplitude 1 [so that $x(t) = u(t)$], the resulting voltage across the capacitor would be given by (1.21). For the case $R = 1$ and $C = 1$, the step response is as plotted in Figure 1.18.

CAR ON A LEVEL SURFACE Consider an automobile on a horizontal surface, as illustrated in Figure 1.19. As indicated, the output $y(t)$ is the position of the car at time t relative to some reference, and the input $x(t)$ is the drive or braking force applied to the car at time t. From Newton's second law of motion, we have that $y(t)$ and $x(t)$ are related by the following second-order linear differential equation:

$$M \frac{d^2y(t)}{dt^2} + k_f \frac{dy(t)}{dt} = x(t), \qquad (1.22)$$

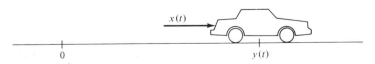

FIGURE 1.19. Car with drive or braking force $x(t)$.

where M is the mass of the car and k_f is the coefficient of friction. Note that k_f will change if there is a significant change in the road surface, for example, in going from a paved to an unpaved surface.

As in the RC circuit example, to compute the input/output operator F of the car, we must solve the input/output differential equation (1.22). To simplify the notation, in the following development we denote the derivative $dy(t)/dt$ by $\dot{y}(t)$.

Let us first determine the output response $y(t)$ when $x(t) = 0$ for $t \geq t_0$. With $y(t_0) = $ initial position of the car at initial time t_0 and with $\dot{y}(t_0) = $ initial velocity of the car, if the input $x(t)$ is zero for $t \geq t_0$, the output is

$$y(t) = y(t_0) + \frac{M}{k_f} \dot{y}(t_0) \left\{ 1 - \exp\left[\frac{-k_f}{M}(t - t_0) \right] \right\}, \qquad t \geq t_0. \quad (1.23)$$

The expression (1.23) is valid as long as $k_f \neq 0$. To check that (1.23) is the output when the input is zero, simply insert (1.23) into (1.22) and verify that (1.22) is satisfied.

From (1.23) we see that if the initial velocity $\dot{y}(t_0)$ is zero, then $y(t) = y(t_0)$ for all $t \geq t_0$. In other words, if the car does not have any initial velocity and no input is applied, it remains at its initial position $y(t_0)$.

Letting $t \to \infty$ in (1.23), we have that the car stops at the position.

$$y(\infty) = y(t_0) + \frac{M}{k_f} \dot{y}(t_0). \quad (1.24)$$

Note that it takes an infinite amount of time before the car stops, whereas an actual car would stop in a finite amount of time. This discrepancy is a result of the fact that our model of the car is an idealized model. Despite this, our model is sufficiently accurate to be useful in studying the ''dynamics'' of the car.

Also note that by measuring the stopping position $y(\infty)$, we can solve (1.24) for k_f. Hence we can identify (i.e., determine) the friction coefficient k_f in the system representation (1.22).

Now let us compute the output response $y(t)$ resulting from an arbitrary input $x(t)$ with the system at rest at time t_0. In this example, ''at rest'' means that the initial position $y(t_0)$ and initial velocity $\dot{y}(t_0)$ are both zero.

Let $v(t) = dy(t)/dt$, so that $v(t)$ is the velocity of the car at time t. Rewriting (1.22) in terms of $v(t)$, we obtain

$$M \frac{dv(t)}{dt} + k_f v(t) = x(t). \quad (1.25)$$

The first-order differential equation (1.25) is called the *velocity model* of the car. Assuming that $v(t_0) = 0$ and using the integrating-factor method given in the RC circuit example, we have that the solution $v(t)$ to (1.25) is given by

$$v(t) = \int_{t_0}^{t} \frac{1}{M} \exp\left[\frac{-k_f}{M}(t - \lambda)\right] x(\lambda) \, d\lambda, \qquad t \geq t_0. \qquad \text{(1.26)}$$

Now since $v(t) = dy(t)/dt$,

$$y(t) = y(t_0) + \int_{t_0}^{t} v(\tau) \, d\tau. \qquad \text{(1.27)}$$

Taking $y(t_0) = 0$ and inserting (1.26) into (1.27), we get

$$y(t) = \int_{t_0}^{t} \left\{ \int_{t_0}^{\tau} \frac{1}{M} \exp\left[\frac{-k_f}{M}(\tau - \lambda)\right] x(\lambda) \, d\lambda \right\} d\tau. \qquad \text{(1.28)}$$

This expression for $y(t)$ can be simplified as follows.

Again letting $u(t)$ denote the unit-step function, we can write (1.28) in the form

$$y(t) = \int_{t_0}^{t} \left\{ \int_{t_0}^{t} \frac{1}{M} \exp\left[\frac{-k_f}{M}(\tau - \lambda)\right] u(\tau - \lambda) x(\lambda) \, d\lambda \right\} d\tau.$$

Interchanging the integrals gives

$$y(t) = \int_{t_0}^{t} \frac{1}{M} \exp\left(\frac{k_f}{M}\lambda\right) \left[\int_{t_0}^{t} \exp\left(\frac{-k_f}{M}\tau\right) u(\tau - \lambda) \, d\tau \right] x(\lambda) \, d\lambda.$$

Since $u(\tau - \lambda) = 0$ for $\tau < \lambda$ and $u(\tau - \lambda) = 1$ for $\tau \geq \lambda$,

$$y(t) = \int_{t_0}^{t} \frac{1}{M} \exp\left(\frac{k_f}{M}\lambda\right) \left[\int_{\lambda}^{t} \exp\left(\frac{-k_f}{M}\tau\right) d\tau \right] x(\lambda) \, d\lambda.$$

Performing the integration with respect to the τ variable, we have

$$y(t) = \int_{t_0}^{t} \frac{-1}{k_f} \exp\left(\frac{k_f}{M}\lambda\right) \left[\exp\left(\frac{-k_f}{M}t\right) - \exp\left(\frac{-k_f}{M}\lambda\right) \right] x(\lambda) \, d\lambda$$

$$y(t) = \int_{t_0}^{t} \frac{1}{k_f} \left\{ 1 - \exp\left[\frac{-k_f}{M}(t - \lambda)\right] \right\} x(\lambda) \, d\lambda, \qquad t \geq t_0. \qquad \text{(1.29)}$$

The expression (1.29) is the output response resulting from the application of an input force $x(t)$ with the system at rest at initial time t_0 [i.e., $y(t_0) = 0$ and $\dot{y}(t_0) = 0$]. Hence (1.29) is the expression for the input/output operator F of the car. Note that F is also an integral operator for this system, as was the case in the RC circuit example. In Chapter 4 we show that the input/output operator F is an integral operator for a large class of continuous-time systems.

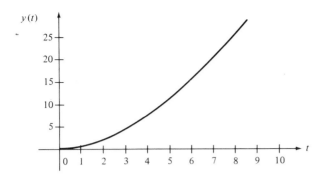

FIGURE 1.20. Step response of car with $M = 1$ and $k_f = 0.1$.

From the input/output relationship (1.29), we can compute the output $y(t)$ resulting from the application of any input force $x(t)$ applied to the car. For example, if $t_0 = 0$ and $x(t) = u(t)$, we have

$$
y(t) = \int_0^t \frac{1}{k_f} \left\{ 1 - \exp\left[\frac{-k_f}{M}(t - \lambda)\right] \right\} d\lambda,
$$

$$
= \frac{1}{k_f} \left[\lambda - \frac{M}{k_f} \exp\left[\frac{-k_f}{M}(t - \lambda)\right]\right] \Big\}_{\lambda=0}^{\lambda=t}
$$

$$
= \frac{1}{k_f} \left[t - \frac{M}{k_f} + \frac{M}{k_f} \exp\left(\frac{-k_f}{M}t\right)\right], \qquad t \geq 0. \tag{1.30}
$$

The output $y(t)$ given by (1.30) is the step response of the car. Note that $y(t) \to \infty$ as $t \to \infty$; in other words, the application of a constant positive force will move the car to infinity. With the normalized mass $M = 1$ and with $k_f = 0.1$, the step response is as plotted in Figure 1.20.

If we differentiate both sides of the expression (1.30) for the step response, we have that the velocity $v(t)$ of the car in response to a step input is given by

$$
v(t) = \frac{dy(t)}{dt} = \frac{1}{k_f} \left[1 - \exp\left(\frac{-k_f}{M}t\right)\right], \qquad t \geq 0. \tag{1.31}
$$

From (1.31) we see that $v(t) \to 1/k_f$ as $t \to \infty$. This result says that the velocity of the car will approach the "steady-state value" $1/k_f$ if a constant force of amplitude 1 is applied to the car.

SIMPLE PENDULUM Consider a pendulum of length L and mass M as illustrated in Figure 1.21. Here the input $x(t)$ is the force applied to the mass M tangential to the direction of motion of the mass, and $Mg \sin \theta(t)$ is the force due to gravity tangential to the motion. The output $y(t)$ is defined to be the angle $\theta(t)$ between the pendulum and the vertical position.

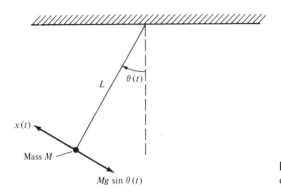

FIGURE 1.21. Simple pendulum.

From the laws of mechanics, we have the following second-order differential equation relating the input and output:

$$I \frac{d^2\theta(t)}{dt^2} + MgL \sin \theta(t) = Lx(t), \tag{1.32}$$

where g is the gravity constant and I is the moment of inertia given by $I = M(L^2)$.

As a result of the term $\sin \theta(t)$, the input/output differential equation (1.32) is a nonlinear differential equation. Due to the nonlinearity, it is not possible to derive an explicit expression for $y(t)$ in terms of $x(t)$ as we were able to do in the preceding two examples. Thus, in this example, there is no analytical expression for the input/output operator F. The system does have an input/output operator F, but it is not possible to express it in an analytical form.

It is interesting to observe that if the magnitude $|\theta(t)|$ of the angle $\theta(t)$ is small, so that $\sin \theta(t)$ is approximately equal to $\theta(t)$, we can approximate the nonlinear differential equation (1.32) by the linear differential equation

$$I \frac{d^2\theta(t)}{dt^2} + MgL\theta(t) = Lx(t). \tag{1.33}$$

The representation (1.33) is called a *small-signal model* of the given system since it is a good approximation to the given system if $|\theta(t)|$ is small. It is possible to derive an expression for the input/output operator F of the small-signal model (see Problem 1.12).

DISCRETE-TIME SYSTEM   The repayment of a bank loan can be modeled as a discrete-time system in the following manner. With k equal to the month of the repayment period, define the output $y(k)$ to be the balance of the loan after the kth month, and define the input $x(k)$ to be the amount of the loan payment in the kth month. Here k is the discrete-time variable that takes on the integer values $k = 0, 1, 2, \ldots$. The initial condition $y(0)$ is the amount of the loan. Usually, the loan payments $x(k)$ are constant; that is, $x(k) = c$, $k = 1, 2, 3, \ldots$, where c is a constant. We shall allow $x(k)$ to vary from one value of k to the next.

The repayment of the loan is described by the difference equation

$$y(k + 1) - \left(1 + \frac{I}{12}\right) y(k) = -x(k + 1), \qquad k = 0, 1, 2, \ldots, \qquad (1.34)$$

where I is the yearly interest rate in decimal form. For example, if the yearly interest rate were 10 percent, I would equal 0.1. The term $(I/12)y(k)$ in (1.34) is the interest on the loan in the $(k + 1)$th month.

Equation (1.34) is a first-order linear difference equation. It is the input/output difference equation of the system consisting of the loan-repayment process.

It is interesting to observe that by allowing the interest rate I to be a variable function of k, the representation (1.34) describes a variable-rate loan for which the interest rate may change from month to month.

We can compute the output $y(k)$ by solving (1.34) recursively as follows. First, let us rewrite (1.34) in the form

$$y(k + 1) = \left(1 + \frac{I}{12}\right) y(k) - x(k + 1). \qquad (1.35)$$

Here we are assuming that I is constant (independent of k). Now inserting $k = 0$ in (1.35), we have

$$y(1) = \left(1 + \frac{I}{12}\right) y(0) - x(1). \qquad (1.36)$$

Inserting $k = 1$ in (1.35), we obtain

$$y(2) = \left(1 + \frac{I}{12}\right) y(1) - x(2). \qquad (1.37)$$

Taking $k = 2$, we get

$$y(3) = \left(1 + \frac{I}{12}\right) y(2) - x(3). \qquad (1.38)$$

Continuing in this manner, we can compute $y(k)$ for any finite range of integer values of k.

From (1.36)–(1.38) we see that the next value of the output is computed from the present value of the output plus an input term. This is why the process is called a *recursion*. In this example, the recursion is a first-order recursion.

```
10 REM Loan balance program
20 REM Program computes loan balance y(k)
25 REM with y(0) = amount of loan,
30 REM I = yearly interest rate in decimal form,
40 REM and x(k) = C = monthly loan payment.
45 CLS
50 DIM Y(360)
60 PRINT:INPUT "Amount of loan";Y(0)
70 PRINT:INPUT "Yearly interest rate";I
80 PRINT:INPUT "Monthly loan payment";C
90 PRINT:PRINT "k              y(k) (in dollars)":PRINT
100 FOR K=1 TO 360
110 Y(K) = (1 + (I/12))*Y(K-1) - C
120 IF Y(K) < -C THEN END
130 PRINT K;TAB(12);.01*INT(100*Y(K))
140 NEXT K
```

FIGURE 1.22.   Loan-balance program.

A program for carrying out the recursion defined by (1.35) is given in Figure 1.22. As an example, suppose that we take out a $6,000 loan for a new car with the yearly interest rate equal to 12 percent (so $I = 0.12$). With monthly payments equal to $200, the output (loan balance) $y(k)$ is shown in Table 1.1.

TABLE 1.1.   Loan balance with $200 monthly payments.

k	y(k)	k	y(k)
1	$5859.99	19	$3086.47
2	5718.59	20	2917.33
3	5575.78	21	2746.51
4	5431.54	22	2573.97
5	5285.85	23	2399.71
6	5138.71	24	2223.71
7	4990.1	25	2045.95
8	4840	26	1866.41
9	4688.4	27	1685.07
10	4535.29	28	1501.92
11	4380.64	29	1316.94
12	4224.44	30	1130.11
13	4066.69	31	941.41
14	3907.36	32	750.83
15	3746.43	33	558.33
16	3583.89	34	363.92
17	3419.73	35	167.56
18	3253.93	36	-30.77

TABLE 1.2. Loan balance with $300 monthly payments.

k	y(k)	k	y(k)
1	$5759.99	13	$2685.76
2	5517.59	14	2412.61
3	5272.77	15	2136.74
4	5025.5	16	1858.11
5	4775.75	17	1576.69
6	4523.51	18	1292.46
7	4268.75	19	1005.38
8	4011.43	20	715.43
9	3751.55	21	422.59
10	3489.06	22	126.81
11	3223.95	23	− 171.92
12	2956.19		

When the monthly payments are $300, the output $y(k)$ is as displayed in Table 1.2. Note that in the first case, it takes 36 months to pay off the loan, whereas in the latter case, the loan is paid off in 23 months.

Usually, in taking out a loan the number of months in the payoff period is specified and then the monthly payments are determined. It is possible to solve for the monthly payments using the representation (1.34) [or (1.35)], but we shall not consider this here (see Problem 1.15).

We now derive an expression for the loan balance $y(k)$ in terms of the loan amount $y(0)$ and the loan payments $x(k)$. Inserting the expression (1.36) for $y(1)$ into (1.37), we obtain

$$y(2) = \left(1 + \frac{I}{12}\right)\left[\left(1 + \frac{I}{12}\right)y(0) - x(1)\right] - x(2).$$

$$= \left(1 + \frac{I}{12}\right)^2 y(0) - \left(1 + \frac{I}{12}\right)x(1) - x(2). \tag{1.39}$$

Inserting (1.39) into (1.38), we have

$$y(3) = \left(1 + \frac{I}{12}\right)^3 y(0) - \left(1 + \frac{I}{12}\right)^2 x(1) - \left(1 + \frac{I}{12}\right)x(2) - x(3). \tag{1.40}$$

From the pattern in (1.36), (1.39), and (1.40), we see that for an arbitrary integer value of $k \geq 1$,

$$y(k) = \left(1 + \frac{I}{12}\right)^k y(0) - \sum_{i=1}^{k}\left(1 + \frac{I}{12}\right)^{k-i} x(i), \qquad k = 1, 2, \ldots . \tag{1.41}$$

The expression (1.41) for $y(k)$ is the loan balance for $k \geq 1$ starting from loan amount $y(0)$ and with loan payments $x(k)$, $k \geq 1$.

It follows from (1.41) that if the loan payments $x(k)$ are not sufficiently large, the loan balance $y(k)$ will grow as k increases. In particular, if $x(k) = 0$ for all $k \geq 1$, then

$$y(k) = \left(1 + \frac{I}{12} \right)^k y(0), \qquad k \geq 1.$$

Clearly, $y(k) \to \infty$ as $k \to \infty$ since $1 + (I/12) > 1$ when $I > 0$.

Let us now compute the input/output operator F of this system. First, the system is at rest at initial time $k_0 = 0$ if and only if $y(0) = 0$. If $y(0) = 0$, from (1.41) we have

$$y(k) = -\sum_{i=1}^{k} \left(1 + \frac{I}{12} \right)^{k-i} x(i), \qquad k = 1, 2, \ldots . \qquad (1.42)$$

Since the input/output operator F is the expression of $y(k)$ in terms of $x(k)$ with the system at rest prior to the application of $x(k)$, from (1.42) we have that

$$y(k) = (Fx)(k) = -\sum_{i=1}^{k} \left(1 + \frac{I}{12} \right)^{k-i} x(i), \qquad k = 1, 2, \ldots .$$

It is interesting that in this example, the input/output operator F does not have any "physical significance" since F relates $y(k)$ and $x(k)$ with the loan amount $y(0)$ equal to zero. However, the operator F is still a valid representation of the loan-repayment system.

1.6

BASIC SYSTEM PROPERTIES

The extent to which a system can be studied using analytical techniques depends on the properties of the system. Two of the most fundamental properties are linearity and time invariance. We shall see in this book that there exists an extensive analytical theory for the study of systems possessing the properties of linearity and time invariance. These two properties and other basic properties are defined in this section.

Throughout this section we consider single-input single-output systems specified by the input/output relationship

$$y(t) = (Fx)(t). \qquad (1.43)$$

Recall that $y(t)$ is the output response of the system resulting from input $x(t)$ with no initial energy in the system prior to the application of the input. Here the time variable t may take on real values or only the discrete values $t = kT$,

$k = \ldots, -2, -1, 0, 1, 2, \ldots$; that is, the system may be continuous time or discrete time.

CAUSALITY   The system given by (1.43) is said to be *causal* or *nonanticipatory* if for any time t_1 and any input $x(t)$ with $x(t) = 0$ for all $t < t_1$, the response $y(t) = (Fx)(t)$ is zero for all $t < t_1$. Thus, in a causal system, it is not possible to get an output before an input is applied to the system. A system is said to be *noncausal* or *anticipatory* if it is not causal.

> **EXAMPLE 1.4.**   Suppose that the system is a continuous-time system given by
>
> $$y(t) = (Fx)(t) = x(t + 1).$$
>
> This system is noncausal since the value $y(t)$ of the output at time t depends on the value $x(t + 1)$ of the input at time $t + 1$. Noncausality can also be seen by considering the response of the system to a 1-second input pulse shown in Figure 1.23a. From the relationship $y(t) = x(t + 1)$, we have that the output $y(t)$ resulting from the input pulse is the pulse shown in Figure 1.23b. Since the output pulse appears before the input pulse is applied, the system is noncausal. The system with the input/output relationship $y(t) = x(t + 1)$ is called an *ideal predictor*. Most physicists would argue that ideal predictors do not exist, at least not in this universe.

> **EXAMPLE 1.5.**   Consider the system with input/output relationship
>
> $$y(t) = (Fx)(t) = x(t - 1).$$
>
> This system is causal since the value of the output at time t depends on only the value of the input at time $t - 1$. If we apply the pulse shown in Figure 1.24a to this system, the output is the pulse shown in Figure 1.24b. From Figure 1.24 we see that the system delays the input pulse by 1 second. In fact, the system delays all inputs by 1 second; in other words, the system is an *ideal time delay*. There are a number of techniques for realizing time delays. For instance, the time delay between the record and playback heads of a tape recorder can be used to realize time delays with delays on the order of milliseconds.

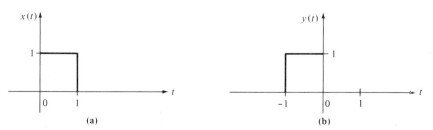

FIGURE 1.23.   (a) Input and (b) output pulse of system in Example 1.4.

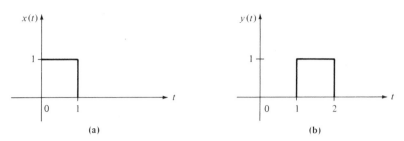

FIGURE 1.24.   (a) Input and (b) output pulse of system in Example 1.5.

EXAMPLE 1.6.   Consider the *RC* circuit that was studied in Section 1.5. If we take the initial time t_0 to be zero, by (1.20) the input/output relationship for the circuit is

$$y(t) = \int_0^t \frac{1}{C} \exp\left[\frac{-1}{RC}(t - \lambda)\right] x(\lambda) \, d\lambda, \qquad t \geq 0. \qquad (1.44)$$

Now suppose that $x(t) = 0$ for all $t < t_1$, where t_1 is an arbitrary positive real number. Then $x(\lambda) = 0$ for all $\lambda < t_1$ and the integral in (1.44) is zero when $t < t_1$. Hence $y(t) = 0$ for all $t < t_1$, so the *RC* circuit is causal.

Memoryless Systems and Systems with Memory: A causal system $y(t) = (Fx)(t)$ is *memoryless* if for any time t_1, the value of the output at time t_1 depends only on the value of the input at time t_1. In this case we write the input/output relationship in the form $y(t) = F(x(t))$.

EXAMPLE 1.7.   Suppose that $y(t) = Kx(t)$, where K is a fixed real number. At any time t_1, $y(t_1) = Kx(t_1)$, and thus $y(t_1)$ depends only on the value of the input at time t_1. Hence the system is memoryless. Since an ideal amplifier or attenuator can be represented by the input/output relationship $y(t) = Kx(t)$, we see that these devices are memoryless.

A causal system that is not memoryless is said to have *memory*. A system has memory if the output at time t_1 depends in general on the values of the input $x(t)$ for some range of values of t up to $t = t_1$.

EXAMPLE 1.8.   Again consider the *RC* circuit with the input/output relationship (1.44). From (1.44) we see that the output $y(t)$ of the *RC* circuit at time t depends on the values of the input $x(\lambda)$ for $0 \leq \lambda \leq t$. Thus the circuit does have memory.

FINITE DIMENSIONALITY   Now suppose that we have a continuous-time system given by $y(t) = (Fx)(t)$. For any nonnegative integer i, we shall let $y^{(i)}(t)$ and $x^{(i)}(t)$ denote the ith derivative of the output $y(t)$ and input $x(t)$. When $i = 0$, $y^{(i)}(t) = y(t)$ and $x^{(i)}(t) = x(t)$. The system $y(t) = (Fx)(t)$ is said to be *finite dimensional* or *lumped* if for some nonnegative integer n, the nth derivative of the output $y(t)$ at time t is equal to a function of $y^{(i)}(t)$ and $x^{(i)}(t)$ at time t for $0 \le i \le n - 1$. The nth derivative of the output at time t may also depend on the ith derivative of the input at time t for $i \ge n$. In mathematical terms, the system is finite dimensional if for some nonnegative integers n and m, $y^{(n)}(t)$ can be written in the form

$$y^{(n)}(t) = f(y(t), y^{(1)}(t), \ldots, y^{(n-1)}(t), x(t), x^{(1)}(t), \ldots, x^{(m)}(t), t), \quad (1.45)$$

where f is a function of the variables $y(t)$, $y^{(1)}(t)$, $\ldots$, $y^{(n-1)}(t)$, $x(t)$, $x^{(1)}(t)$, $\ldots$, $x^{(m)}(t)$ and time t. The integer n is called the *order* of the differential equation (1.45). The integer n is also referred to as the *order* or *dimension* of the system with the input/output relationship (1.45).

A continuous-time system is *infinite dimensional* if it is not finite dimensional. In other words, a system is infinite dimensional if and only if it is not possible to express the nth derivative of the output in the form (1.45) for some nonnegative integer n.

EXAMPLE 1.9. Consider the pendulum given by the input/output differential equation (1.32). We can rewrite (1.32) in the form

$$\frac{d^2\theta(t)}{dt^2} = \frac{-MgL}{I} \sin \theta(t) + \frac{L}{I} x(t). \quad (1.46)$$

Clearly, (1.46) is in the form (1.45), so the system is finite dimensional. The dimension of the system is 2.

EXAMPLE 1.10. Consider the continuous-time system given by

$$\frac{dy(t)}{dt} + ay(t - 1) = x(t), \quad (1.47)$$

where a is an arbitrary nonzero constant. Due to the delay term $y(t - 1)$ in (1.47), it is not possible to write (1.47) in the form (1.45) for any positive integer n. Thus the system is infinite dimensional. The system defined by (1.47) is an example of a *system with time delays*. Systems with time delays are always infinite dimensional. The input/output differential equation (1.47) of the system is called a *delay differential equation*. Delay differential equations are much more complicated than ordinary differential equations. In particular, it is seldom possible to express the solution of a delay differential equation in analytical form.

Finite-Dimensional Discrete-Time Systems: Now suppose that the given system is a discrete-time system with the input/output relationship $y(kT) = (Fx)(kT)$, where k is an integer variable. The system is said to be *finite-dimensional* or *recursive* if for some nonnegative integers n and m, the nT-second left shift of $y(kT)$ can be written in the form

$$y(kT + nT) = f(y(kT), y(kT + T), \ldots, y(kT + nT - T), x(kT),$$
$$x(kT + T), \ldots, x(kT + mT), kT). \tag{1.48}$$

The integer n is called the *order* or *dimension* of the system. A discrete-time system that is not finite dimensional is said to be *infinite dimensional* or *nonrecursive*. (In Chapter 3 we explain the meaning of the terms "recursive" and nonrecursive.")

EXAMPLE 1.11. Consider the loan-repayment system that was studied in Section 1.5. The system is given by the input/output difference equation

$$y(k + 1) - \left(1 + \frac{I}{12}\right) y(k) = -x(k + 1). \tag{1.49}$$

Equation (1.49) can be written in the form (1.48) with $n = 1$, $m = 1$, and $T = 1$, and thus the system is finite dimensional with dimension 1.

LINEARITY The system defined by $y(t) = (Fx)(t)$ is said to be *additive* if

$$[F(x + v)](t) = (Fx)(t) + (Fv)(t) \tag{1.50}$$

for any two inputs x and v. In other words, the system is additive if the output response resulting from the input $x(t) + v(t)$ is equal to the output response resulting from $x(t)$ plus the output response resulting from $v(t)$. The system is said to be *homogeneous* if

$$[F(ax)](t) = a(Fx)(t) \tag{1.51}$$

for any input x and any scalar a. By (1.51) the system is homogeneous if the output response resulting from the input $ax(t)$ is equal to a times the output response resulting from $x(t)$.

The system $y(t) = (Fx)(t)$ is *linear* if it is both additive and homogeneous; that is, both (1.50) and (1.51) are satisfied for all inputs $x(t)$, $v(t)$, and all scalars a. It is easy to see that the conditions (1.50) and (1.51) are equivalent to the single condition

$$[F(ax + bv)](t) = a(Fx)(t) + b(Fv)(t) \tag{1.52}$$

for all inputs $x(t)$, $v(t)$ and all scalars a, b. Hence the given system is linear if and only if condition (1.52) is satisfied. By (1.52), the system is linear if for any inputs $x(t)$ and $v(t)$ and any scalars a and b, the output response resulting from the input $ax(t) + bv(t)$ is equal to a times the response to $x(t)$ plus b times the response to $v(t)$. A system that is not linear is said to be *nonlinear*.

In mathematics, (1.52) is the condition for linearity of the operator F. So linearity of a system is equivalent to requiring that the input/output operator of the system be a linear operator.

Linearity is an extremely important property. If a system is linear, we can apply the vast collection of existing results on linear operators in the study of system behavior and structure. In contrast, the analytical theory of nonlinear systems is very limited in scope. In practice, we often attempt to approximate a given nonlinear system by a linear system so that analytical techniques for linear systems can then be applied. A widely used approximation method is based on linearization, which we consider in Section 2.5.

A very common type of nonlinear system is a circuit containing diodes, as shown in the following example.

EXAMPLE 1.12. Consider the circuit with the ideal diode shown in Figure 1.25. Here the output $y(t)$ is the voltage across the resistor with resistance R_2. The ideal diode is a short circuit when the current $i(t)$ is positive and it is an open circuit when $i(t)$ is negative. Thus the input/output relationship $y(t) = (Fx)(t)$ of the circuit is given by

$$y(t) = \begin{cases} \dfrac{R_2}{R_1 + R_2}\, x(t), & \text{when } x(t) \geq 0 \\ \\ 0 & \text{when } x(t) \leq 0. \end{cases} \tag{1.53}$$

Now suppose that the input $x(t)$ is the unit-step function $u(t)$. Then, from (1.53), the resulting response is

$$y(t) = \frac{R_2}{R_1 + R_2}\, u(t). \tag{1.54}$$

If we multiply the unit-step input by the scalar -1, so that the input is $-u(t)$, by (1.53) the resulting response is zero for all $t \geq 0$. But this is not equal to -1 times the response to $u(t)$ given by (1.54). Hence the system is not homogeneous, and thus it is not linear. It is also easy to see that the circuit is not additive.

Nonlinearity may also be a result of the presence of signal multipliers. This is illustrated by the following example.

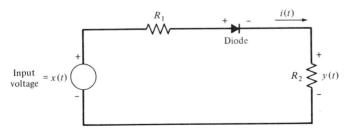

FIGURE 1.25. Resistive circuit with an ideal diode.

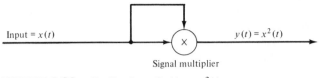

Input = $x(t)$ $\times$ $y(t) = x^2(t)$

Signal multiplier

FIGURE 1.26.   Realization of $y(t) = x^2(t)$.

EXAMPLE 1.13.   Consider the continuous-time system with the input/output relationship

$$y(t) = (Fx)(t) = x^2(t). \tag{1.55}$$

This system can be realized using a signal multiplier as shown in Figure 1.26. The signal multiplier in Figure 1.26 can be built (approximately) using operational amplifiers and diodes (see Truxal [1972] for the details).

The system defined by (1.55) is sometimes called a *square-law device*. Note that the system is memoryless. Now given a scalar a and an input $x(t)$, by (1.55) the response to $ax(t)$ is $a^2x^2(t)$. But a times the response to $x(t)$ is equal to $ax^2(t)$, which is not equal to $a^2x^2(t)$ in general. Thus the system is not homogeneous, and the system is not linear.

Another way in which nonlinearity arises is in systems containing devices that go into "saturation" when signal levels become too large, as in the following example.

EXAMPLE 1.14.   Consider an ideal amplifier with the input/output relationship $y(t) = Kx(t)$, where K is a fixed positive real number. A plot of the output $y(t)$ versus the input $x(t)$ is given in Figure 1.27. The ideal amplifier is clearly linear, but this is not the case for an actual (nonideal) amplifier, since the output $y(t)$ will not equal $Kx(t)$ for arbitrarily large input signals. In a nonideal amplifier, the output versus input characteristic may be as shown in Figure 1.28. From the figure we see that $y(t) = Kx(t)$ only when the

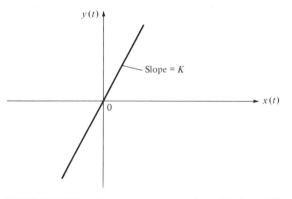

FIGURE 1.27.   Output versus input in an ideal amplifier.

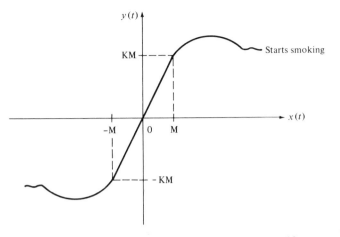

FIGURE 1.28. Output versus input in a nonideal amplifier.

magnitude $|x(t)|$ of the input is less than M. The nonideal amplifier is not homogeneous since the response to $ax(t)$ is not equal to a times the response to $x(t)$ unless $|ax(t)| < M$. We can view the nonideal amplifier as a linear system only if we can guarantee that the magnitude of the input applied to the amplifier will never exceed M.

Although nonlinear systems are very common, many systems arising in practice can be modeled as linear systems. A large class of linear systems is defined in the next example.

EXAMPLE 1.15. Consider a single-input single-output continuous-time system with the input/output relationship

$$y(t) = (Fx)(t) = \int_{t_0}^{t} h(t - \lambda)x(\lambda)\, d\lambda, \qquad t \geq t_0, \qquad (1.56)$$

where $h(t)$ is an arbitrary real-valued function of t with $h(t) = 0$ for all $t < 0$. In Chapter 4 we will see that $h(t)$ is the impulse response of the system defined by (1.56). Note that if we take $h(t) = (1/C) \exp [(-1/RC)t]$ for $t \geq 0$, we get the RC circuit considered in Section 1.5, or if we take

$$h(t) = \frac{1}{k_f} \left[1 - \exp \left(\frac{-k_f}{M} t \right) \right] \qquad \text{for } t \geq 0,$$

we get the automobile considered in Section 1.5. Now suppose that the input is $ax(t) + bv(t)$, where $x(t)$, $v(t)$, a, and b are arbitrary. By (1.56) the resulting output response is

$$y(t) = \int_{t_0}^{t} h(t - \lambda)[ax(\lambda) + bv(\lambda)]\, d\lambda.$$

Using the linearity property of integration, we obtain

$$y(t) = a \int_{t_0}^{t} h(t - \lambda)x(\lambda) \, d\lambda + b \int_{t_0}^{t} h(t - \lambda)v(\lambda) \, d\lambda.$$

Therefore,

$$y(t) = a(Fx)(t) + b(Fv)(t),$$

which proves that the system is linear. Since the input/output operator of the RC circuit and the automobile can be expressed in the form (1.56), we see that both these systems are linear.

Linear Finite-Dimensional Systems: Suppose that the system under study is a finite-dimensional continuous-time system with the input/output differential equation

$$y^{(n)}(t) = f(y(t), y^{(1)}(t), \ldots, y^{(n-1)}(t), x(t), x^{(1)}(t), \ldots, x^{(m)}(t), t). \quad (1.57)$$

The system defined by (1.57) can be shown to be linear if and only if (1.57) is a linear differential equation, which is the case if and only if there exist scalar functions $a_0(t), a_1(t), \ldots, a_{n-1}(t), b_0(t), b_1(t), \ldots, b_m(t)$ such that

$$f(y(t), \ldots, y^{(n-1)}(t), x(t), \ldots, x^{(m)}(t), t)$$
$$= -\sum_{i=0}^{n-1} a_i(t)y^{(i)}(t) + \sum_{i=0}^{m} b_i(t)x^{(i)}(t).$$

Thus the system defined by (1.57) is linear if and only if the nth derivative of the output can be expressed in the form

$$y^{(n)}(t) = -\sum_{i=0}^{n-1} a_i(t)y^{(i)}(t) + \sum_{i=0}^{m} b_i(t)x^{(i)}(t). \quad (1.58)$$

EXAMPLE 1.16. The input/output differential equation (1.46) of the pendulum is nonlinear as a result of the term $-(MgL/I) \sin \theta(t)$. Thus the pendulum is a nonlinear system.

Now consider a finite-dimensional discrete-time system with the input/output difference equation

$$y(kT + nT) = f(y(kT), y(kT + T), \ldots, y(kT + nT - T), x(kT), \quad (1.59)$$
$$x(kT + T), \ldots, x(kT + mT), kT).$$

It can be shown that the system defined by (1.59) is linear if and only (1.59) is a linear difference equation, which is the case if and only if (1.59) can be written in the form

$$y(kT + nT) = -\sum_{i=0}^{n-1} a_i(k)y(kT + iT) + \sum_{i=0}^{m} b_i(k)x(kT + iT). \quad (1.60)$$

EXAMPLE 1.17. The loan-repayment system with the input/output difference equation (1.49) is linear since (1.49) can be written in the form (1.60) and thus is a linear difference equation.

TIME INVARIANCE Given a real number t_1 and a signal $x(t)$, recall that $x(t - t_1)$ is equal to $x(t)$ shifted to the right by t_1 seconds when $t_1 > 0$, and that $x(t - t_1)$ is equal to $x(t)$ shifted to the left by t_1 seconds when $t_1 < 0$. Now suppose that we have a continuous-time or discrete-time system with the input/output relationship $y(t) = (Fx)(t)$. The system is said to be *time invariant* or *constant* if for any input $x(t)$ and any t_1 (t_1 real in the continuous-time case, $t_1 = k_1T$ where k_1 is integer in the discrete-time case),

$$y(t - t_1) = (F\bar{x})(t), \quad (1.61)$$

where $\bar{x}(t) = x(t - t_1)$. By (1.61), the system is time invariant if for any t_1 and any input $x(t)$ that produces response $y(t)$, the response to the time-shifted input $\bar{x}(t) = x(t - t_1)$ is equal to the time shift $y(t - t_1)$ of $y(t)$. Therefore, in a time-invariant system the response to a left or right shift of the input $x(t)$ is equal to a corresponding left or right shift in the response $y(t)$. In a time-invariant system, there are no changes in system structure as a function of time t. The system is *time varying* or *time variant* if it is not time invariant.

EXAMPLE 1.18. Suppose that $y(t) = tx(t)$. It is easy to see that this system is memoryless and linear. Now for any t_1, we have

$$y(t - t_1) = (Fx)(t - t_1) = (t - t_1)x(t - t_1).$$

But setting $\bar{x}(t) = x(t - t_1)$, by definition of F we have

$$(F\bar{x})(t) = t\bar{x}(t) = tx(t - t_1),$$

which does not equal $(t - t_1)x(t - t_1)$ in general. Hence $y(t - t_1)$ is not equal to $(F\bar{x})(t)$ in general, and thus the system is time varying. Note that this system could be viewed as an ideal amplifier with time-varying gain t.

Linear Time-Invariant Finite-Dimensional Systems: Suppose that we have a linear finite-dimensional continuous-time system specified by the input/output differential equation (1.58). The system can be shown to be time invariant if and only if the coefficients of the differential equation are constant (independent of t); that is,

$$a_i(t) = a_i \quad \text{and} \quad b_i(t) = b_i \qquad \text{for all } i \text{ and all real numbers } t,$$

where the a_i and b_i are constants.

The input/output differential equations of the RC circuit and the automobile studied in the preceding section have constant coefficients, and thus both these systems are time invariant.

If the system is a linear finite-dimensional discrete-time system with the input/output difference equation (1.60), it can be shown that the system is time invariant if and only if the coefficients of the difference equation are constant; that is,

$$a_i(k) = a_i \quad \text{and} \quad b_i(k) = b_i \quad \text{for all } i \text{ and all integers } k.$$

EXAMPLE 1.19. The input/output difference equation (1.49) of the loan-repayment system has constant coefficients if the interest rate I is constant, and thus in this case the system is time invariant. If the interest rate I varies as a function of k (a variable-rate loan), one of the coefficients of the input/output difference equation is time varying, and thus the system is time varying.

★1.7

MULTI-INPUT MULTI-OUTPUT SYSTEMS

It is often the case that the system under consideration has more than one input terminal and/or more than one output terminal. All of the concepts introduced in this chapter generalize to systems with p inputs and q outputs (i.e., p input terminals and q output terminals), where p and q are arbitrary positive integers. We shall illustrate this by considering a two-input two-output system.

Given a two-input two-output system, we shall let $x_1(t)$ and $x_2(t)$ denote the input signals applied to the first and second input terminals, respectively, and we shall let $y_1(t)$ and $y_2(t)$ denote the resulting output responses at the first and second output terminals, respectively. The system under study may be continuous time or discrete time.

If the system has no initial energy before the application of the input signals, we have that

$$\begin{aligned} y_1(t) &= [F_1(x_1, x_2)](t) \\ y_2(t) &= [F_2(x_1, x_2)](t). \end{aligned} \tag{1.62}$$

In other words, the output $y_1(t)$ at the first output terminal is a function F_1 of the input signals x_1 and x_2, and the output $y_2(t)$ at the second output terminal is a function F_2 of the input signals x_1 and x_2. The relationship (1.62) is the input/output operator representation of the two-input two-output system.

The system defined by (1.62) is linear if for any inputs $x_1(t)$, $v_1(t)$, $x_2(t)$, $v_2(t)$ and any scalars a and b, we have

$$F_1(ax_1 + bv_1, ax_2 + bv_2) = aF_1(x_1, x_2) + bF_1(v_1, v_2) \qquad (1.63)$$

and

$$F_2(ax_1 + bv_1, ax_2 + bv_2) = aF_2(x_1, x_2) + bF_2(v_1, v_2). \qquad (1.64)$$

By the first condition (1.63), the response at the first output terminal resulting from $ax_1 + bv_1$ applied to the first input terminal and $ax_2 + bv_2$ applied to the second input terminal is equal to a times $F_1(x_1, x_2)$ plus b times $F_1(v_1, v_2)$. The second condition (1.64) has a corresponding interpretation. Here $F_1(x_1, x_2)$ is the response at the first output terminal resulting from x_1 applied to the first input terminal and x_2 applied to the second input terminal, $F_1(v_1, v_2)$ is the response at the first output terminal resulting from v_1 applied to the first input terminal and v_2 applied to the second input terminal.

The conditions (1.63) and (1.64) for linearity are equivalent to requiring that the output responses $y_1(t)$, $y_2(t)$ resulting from the inputs $x_1(t)$, $x_2(t)$ are expressible in the form

$$\begin{aligned} y_1(t) &= (F_{11}x_1)(t) + (F_{12}x_2)(t) \\ y_2(t) &= (F_{21}x_1)(t) + (F_{22}x_2)(t), \end{aligned} \qquad (1.65)$$

where F_{11}, F_{12}, F_{21}, and F_{22} are *linear operators*. The operator F_{ij}, for $i = 1$, 2 and $j = 1$, 2, can be viewed as the input/output operator from the jth input terminal to the ith output terminal. It follows from (1.65) that the given two-input two-output system can be viewed as an interconnection of four scalar (single-input single-output) subsystems with the input/output operators F_{11}, F_{12}, F_{21}, F_{22}. The interconnection is shown in Figure 1.29.

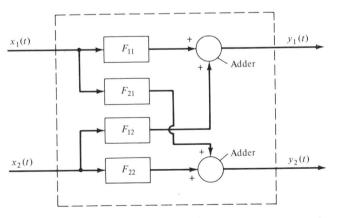

FIGURE 1.29. Realization of two-input two-output system in terms of scalar subsystems.

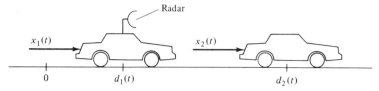

FIGURE 1.30.   Two-car system.

The two-input two-output linear system defined by (1.65) is finite dimensional if each of the scalar subsystems is finite dimensional, and the system is time invariant if each of the scalar subsystems is time invariant.

> **EXAMPLE 1.20.**   Consider two cars moving along a level surface as shown in Figure 1.30. We assume that the mass of both cars is equal to M and that the friction coefficient of both cars is equal to k_f. As illustrated, $d_1(t)$ is the position of the first car at time t, $d_2(t)$ is the position of the second car at time t, $x_1(t)$ is the drive or braking force applied to the first car, and $x_2(t)$ is the drive or braking force applied to the second car. The first car also has radar, which gives a measurement of the distance
>
> $$w(t) = d_2(t) - d_1(t)$$
>
> between the two cars at time t. We define the inputs of the two-car system to be $x_1(t)$ and $x_2(t)$. The outputs are defined to be
>
> $$y_1(t) = v_1(t) = \dot{d}_1(t) = \text{velocity of the first car}$$
>
> and
>
> $$y_2(t) = w(t).$$
>
> Both these outputs are measurable at the first car: $y_1(t)$ is the output of the speedometer of the first car and $y_2(t)$ is the output of the radar on the first car. We could define additional outputs involving the second car, but we shall not do so since we want to limit our attention to the two-input two-output case. Now we want to show that the input/output relationship of the two-car system can be expressed in the form (1.65). First, from the analysis of the single-car system given in Section 1.5, when $v_1(t_0) = 0$ we have that
>
> $$y_1(t) = v_1(t) = \int_{t_0}^{t} \frac{1}{M} \exp\left[\frac{-k_f}{M}(t - \lambda)\right] x_1(\lambda)\, d\lambda. \qquad (1.66)$$
>
> Comparing (1.66) with the first part of (1.65), we have that
>
> $$(F_{11}x_1)(t) = \int_{t_0}^{t} \frac{1}{M} \exp\left[\frac{-k_f}{M}(t - \lambda)\right] x_1(\lambda)\, d\lambda \qquad (1.67)$$

and

$$(F_{12}x_2)(t) = 0. \tag{1.68}$$

Again from the analysis of the single-car system, when the initial conditions are all zero we have that

$$d_i(t) = \int_{t_0}^{t} \frac{1}{k_f} \left\{ 1 - \exp\left[\frac{-k_f}{M} (t - \lambda) \right] \right\} x_i(\lambda) \, d\lambda, \qquad i = 1, 2. \tag{1.69}$$

Since

$$y_2(t) = -d_1(t) + d_2(t),$$

using (1.69) we obtain

$$
\begin{aligned}
y_2(t) = &- \int_{t_0}^{t} \frac{1}{k_f} \left\{ 1 - \exp\left[\frac{-k_f}{M} (t - \lambda) \right] \right\} x_1(\lambda) \, d\lambda \\
&+ \int_{t_0}^{t} \frac{1}{k_f} \left\{ 1 - \exp\left[\frac{-k_f}{M} (t - \lambda) \right] \right\} x_2(\lambda) \, d\lambda.
\end{aligned}
\tag{1.70}
$$

Comparing (1.70) and the second part of (1.65), we get

$$(F_{21}x_1)(t) = - \int_{t_0}^{t} \frac{1}{k_f} \left\{ 1 - \exp\left[\frac{-k_f}{M} (t - \lambda) \right] \right\} x_1(\lambda) \, d\lambda \tag{1.71}$$

and

$$(F_{22}x_2)(t) = \int_{t_0}^{t} \frac{1}{k_f} \left\{ 1 - \exp\left[\frac{-k_f}{M} (t - \lambda) \right] \right\} x_2(\lambda) \, d\lambda. \tag{1.72}$$

The input/output operators F_{11}, F_{12}, F_{21}, and F_{22} of the scalar subsystems are completely specified by (1.67), (1.68), (1.71), and (1.72). The scalar subsystems defined by these input/output operators are linear, time invariant, and finite dimensional. Thus the two car-system is linear, time invariant, and finite dimensional.

PROBLEMS

Chapter 1

1.1 Consider the continuous-time signals displayed in Figure P1.1. Show that each of these signals is equal to a sum of rectangular pulses $p_\tau(t)$ and/or triangular pulses $(1 - 2|t|/\tau)p_\tau(t)$.

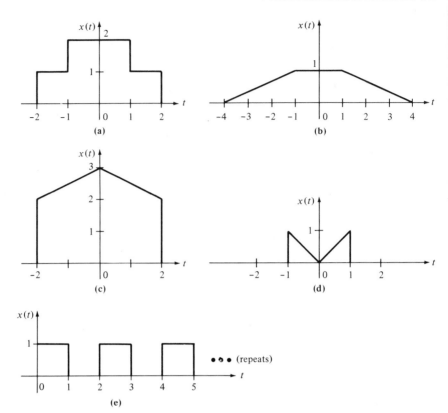

FIGURE P1.1

1.2   Sketch the following continuous-time signals.
(a) $x(t) = u(t + 1) - 2u(t - 1) + u(t - 3)$.
(b) $x(t) = (t + 1)u(t - 1) - tu(t) - u(t - 2)$.
(c) $x(t) = e^{-t}u(t) + e^{-t}[\exp(2t - 4) - 1]u(t - 2) - e^{t-4}u(t - 4)$.
(d) $x(t) = \cos t[u\left(t + \dfrac{\pi}{2}\right) - 2u(t - \pi)] + (\cos t)u\left(t - \dfrac{3\pi}{2}\right)$.

1.3   For each of the signals in Problem 1.2, give the generalized derivative in analytical form, and then sketch the generalized derivative.

1.4   Express each of the signals in Problem 1.2 in the form

$$x(t) = x_1(t)[u(t - t_1) - u(t - t_2)] + x_2(t)[u(t - t_2) - u(t - t_3)] + \cdots .$$

Give the signals $x_1(t)$, $x_2(t)$, ... in the simplest possible analytical form.

1.5   Consider the continuous-time signals shown in Figure P1.5. Express each signal in the form

$$x(t) = f_1(t)u(t - t_1) + f_2(t)u(t - t_2) + \cdots .$$

Give the functions $f_1(t)$, $f_2(t)$, ... in the simplest possible analytical form.

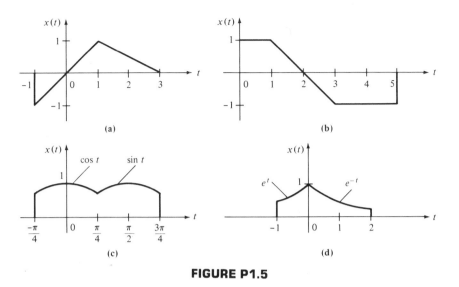

FIGURE P1.5

1.6 For each of the signals in Problem 1.5:
(a) Give the generalized derivative in analytical form.
(b) Sketch the generalized derivative.

1.7 Given a continuous-time signal $x(t)$ and a constant c, consider the signal $x(t)u(t - c)$.
(a) Show that there exists a signal $v(t)$ such that

$$x(t)u(t - c) = v(t - c)u(t - c).$$

Express $v(t)$ in terms of $x(t)$.
(b) Determine the simplest possible analytical form for $v(t)$ when:
(i) $x(t) = e^{-2t}$ and $c = 3$.
(ii) $x(t) = t^2 - t + 1$ and $c = 2$.
(iii) $x(t) = \sin 2t$ and $c = \dfrac{\pi}{4}$.

1.8 Sketch the following discrete-time signals.
(a) $x(k) = u(k) - 2u(k - 1) + u(k - 4)$.
(b) $x(k) = (k + 2)u(k + 2) - 2u(k) - ku(k - 4)$.
(c) $x(k) = \Delta(k + 1) - \Delta(k) + u(k + 1) - u(k - 2)$.
(d) $x(k) = e^{0.2k}u(k + 1) + u(k) - 2e^{0.1k}u(k - 3) - (1 - e^{0.1k})^2u(k - 5)$.

1.9 Express each of the discrete-time signals in Problem 1.8 in the form

$$x(k) = x_1(k)[u(k - k_1) - u(k - k_2)]$$
$$+ x_2(k)[u(k - k_2) - u(k - k_3)] + \cdots.$$

Give the signals $x_1(k)$, $x_2(k)$, . . . in the simplest possible analytical form.

1.10 Consider the *RL* circuit shown in Figure P1.10.

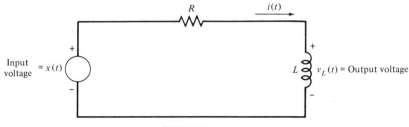

FIGURE P1.10

(a) Determine the input/output differential equation of the circuit.
(b) Let $q(t) = v_L(t) - x(t)$. *Using your result in part (a), show that*

$$\frac{dq(t)}{dt} + \frac{R}{L} q(t) = \frac{-R}{L} \dot{x}(t).$$

(c) *By first solving the differential equation in part (b)*, determine the input/output operator F of the *RL* circuit. Take the initial time t_0 to be zero.
(d) *Using your result in part (c)*, compute the step response of the *RL* circuit.
(e) Compute $v_L(t)$ for all $t > 0$ when $v_L(0) = v_0$ and $x(t) = 1$ for $t \geq 0$.
(f) Compute $v_L(t)$ for all $t > 0$ when $v_L(0) = 0$ and $x(t) = 1 + (R/L)t$ for $t \geq 0$.

1.11 An automobile on an inclined surface is illustrated in Figure P1.11. The velocity model of the car is given by the input/output differential equation

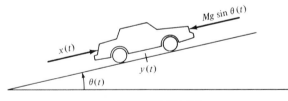

FIGURE P1.11

$$M \frac{dv(t)}{dt} + k_f v(t) = x(t) - Mg \sin \theta(t).$$

Here $x(t)$ is the drive or braking force applied to the car, $Mg \sin \theta(t)$ is the force on the car due to gravity, g is the gravity constant, $\theta(t)$ is the angle of the inclined surface, and $v(t) = dy(t)/dt$ is the velocity of the car.
(a) Determine the input/output operator F, where $v(t) = (Fx)(t)$. Take the initial time t_0 to be zero.
(b) Suppose that $\theta(t)$ is equal to a constant θ for all $t \geq 0$. Using your result in part (a), compute the step response of the velocity model.

(c) Repeat part (b) with $\theta(t)$ (in degrees) equal to $tu(t) - (t - 10)u(t - 10)$.

(d) Suppose that $v(0) = v_0$, $\theta(t)$ is a constant θ and $x(t) = 0$ for all $t \geq 0$. Compute $v(t)$ for $t > 0$.

1.12 Again consider the simple pendulum given by the small-signal model

$$\frac{d^2\theta(t)}{dt^2} + \frac{MgL}{I}\theta(t) = \frac{L}{I}x(t).$$

Recall that $\theta(t)$ is the angle of the pendulum from the vertical reference and $x(t)$ is the force applied to the pendulum tangential to the motion. Let $q_1(t)$ be the solution to

$$\frac{dq_1(t)}{dt} + j\sqrt{\frac{MgL}{I}}\,q_1(t) = \frac{L}{I}x(t),$$

where $j = \sqrt{-1}$. Let $q_2(t)$ be the solution to

$$\frac{dq_2(t)}{dt} - j\sqrt{\frac{MgL}{I}}\,q_2(t) = q_1(t).$$

In all of the following parts, take the initial time t_0 to be zero.

(a) Show that

$$\frac{d^2q_2(t)}{dt^2} + \frac{MgL}{I}q_2(t) = \frac{L}{I}x(t).$$

(b) Define the operator F_1 by $q_1(t) = (F_1x)(t)$. Determine F_1.

(c) Define the operator F_2 by $q_2(t) = (F_2q_1)(t)$. Determine F_2.

(d) Using the results in parts (a), (b), and (c), determine the input/output operator F for the small-signal model of the pendulum.

(e) Using your result in part (d), compute the step response of the pendulum.

1.13 A tank filled with water is shown in Figure P1.13. Here $h(t)$ is the water level at time t, $x(t)$ is the flow rate (in gal/sec) of the water coming into the tank, and $y(t)$ is the flow rate (in gal/sec) of the water coming out of the tank. Letting R denote the valve resistance, we have that $y(t) = Rh(t)$. Assume that the tank is cylindrical with the area of the base equal to A. Then the volume of water in the tank at time t is equal to $Ah(t)$.

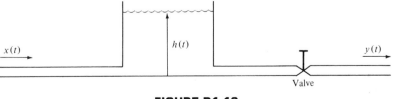

FIGURE P1.13

(a) Determine the input/output differential equation of the system.

(b) Determine the input/output operator F of the system. Take the initial time t_0 to be zero.

1.14 Consider the circuit with two ideal diodes shown in Figure P1.14. Determine the input/output operator F of the circuit when

(a) $R_1 = R_2$.

(b) $R_1 \neq R_2$.

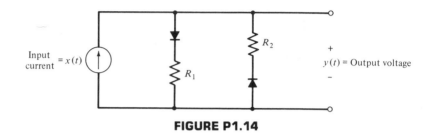

FIGURE P1.14

1.15 Again consider the loan-balance system with the input/output difference equation

$$y(k + 1) - \left(1 + \frac{I}{12}\right) y(k) = -x(k + 1).$$

Recall that $y(0)$ is the amount of the loan, $y(k)$ is the loan balance at the end of the kth month, $x(k)$ is the loan payment in the kth month, and I is the interest rate in decimal form. We assume that the monthly payments $x(k)$ for $k \geq 1$ are equal to a constant c. Suppose that the number of months in the repayment period is N. Derive an expression for the monthly payments c in terms of $y(0)$, N, and I.

HINT: Use the relationship

$$\sum_{i=1}^{N} a^{N-i} = \frac{1 - a^N}{1 - a}, \qquad a \neq 1.$$

1.16 A savings account in a bank can be modeled by the input/output difference equation

$$y(k + 1) - \left(1 + \frac{I}{4}\right) y(k) = x(k + 1),$$

where $y(k)$ is the amount in the account at the end of the kth quarter, $x(k)$ is the amount deposited in the kth quarter, and I is the yearly interest rate in decimal form.

(a) Suppose that $I = 10$ percent. Compute $y(k)$ for $k = 1, 2, 3, 4$ when $y(0) = 1000$ and $x(k) = 1000$ for $k \geq 1$.

(b) Determine the input/output operator F of the system.

(c) Suppose that $x(k) = c$ for $k \geq 1$ and $y(0) = 0$. Given an integer N, suppose that we want to have an amount $y(N)$ in the savings account at the end of the Nth quarter. Using your result in part (b), derive an expression for N in terms of $y(N)$, c, and I.

(d) Suppose that you set up an IRA (individual retirement account) with $y(0) = \$2000$, $I = 10$ percent and $x(k) = \$500$, $k \geq 1$ ($k =$ quarter). How many years will it take to amass $\$1,000,000$ in the account?

1.17 Determine whether the following continuous-time systems are causal or noncausal, have memory or are memoryless. Justify your answers. In the following parts, $x(t)$ is an arbitrary input and $y(t)$ is the zero-state response to $x(t)$.

(a) $y(t) = x(t) + 1$.

(b) $y(t) = |x(t)| = \begin{cases} x(t) & \text{when } x(t) \geq 0 \\ -x(t) & \text{when } x(t) < 0. \end{cases}$

(c) $\dfrac{dy(t)}{dt} = y(t)x(t)$.

(d) $y(t) = \displaystyle\int_{t-1}^{t} x(\lambda)\, d\lambda$.

(e) $y(t) = \sin x(t)$.

(f) For any $x(t)$, $y(t) = \begin{cases} 1, & t \geq 0 \\ 0, & t < 0. \end{cases}$

(g) $y(t) = e^{-t}x(t)$.

(h) $y(t) = \begin{cases} x(t) & \text{when } |x(t)| \leq 10 \\ 10 & \text{when } |x(t)| > 10. \end{cases}$

(i) $y(t) = \displaystyle\int_{-\infty}^{t} (t - \lambda)x(\lambda)\, d\lambda$.

1.18 For each of the systems in Problem 1.17, determine whether the system is linear or nonlinear. Justify your answers.

1.19 For each of the systems in Problem 1.17, determine whether the system is time invariant or time varying. Justify your answers.

1.20 A continuous-time system is said to have a *dead zone* if the output response $y(t)$ is zero for any input $x(t)$ with $|x(t)| < A$, where A is a constant called the *threshold*. An example is a dc motor (see Example 2.3) that is unable to supply any torque until the input voltage exceeds a threshold value. Show that any system with a dead zone is nonlinear.

1.21 Determine whether the following discrete-time systems are causal or noncausal, have memory or are memoryless, are linear or nonlinear, are time invariant or time varying. Justify your answers. In the following parts, $x(kT)$ is an arbitrary input and $y(kT)$ is the zero-state response to $x(kT)$.

(a) $y(kT) = x(kT) + x(kT - T)$.

(b) $y(kT) = x(kT) + x(kT + T)$.

(c) $y(kT + T) = y(kT)x(kT)$.

(d) $y(kT) = u(kT)x(kT)$.

(e) $y(kT + T) + ky(kT) = x(kT)$.

(f) $y(k) = \sum\limits_{i=-\infty}^{k} (0.5)^{k-i} x(i).$

(g) $y(k) = \sum\limits_{i=-\infty}^{k} (0.5)^{k} x(i).$

1.22   Again consider the car on an inclined surface (see Figure P1.11). The velocity model of the car is

$$M \frac{dv(t)}{dt} + k_f v(t) = x(t) - Mg \sin \theta(t).$$

In all of the following parts, the output of the system is the velocity $v(t)$ of the car.

(a) With $\theta(t) = 0$ for all t and with the input equal to $x(t)$, determine whether the car system is linear and time invariant. Justify your answers.

(b) Now suppose that $\theta(t) \neq 0$. With the input equal to $x(t)$, determine whether the system is linear and time invariant. Justify your answers.

(c) We can make the system into a two-input system by taking the first input to be $x(t)$ and the second input to be $\sin \theta(t)$. With this definition of the inputs, is the system linear and time invariant? Justify your answers.

Continuous-Time Systems Defined by an Input/Output Differential Equation

In this chapter we study finite-dimensional continuous-time systems specified in terms of input/output differential equations and integrator realizations. Linear input/output differential equations with constant coefficients are considered first. We show that such equations can be realized (or simulated) using integrators, which in turn are realizable by operational amplifiers. Linear input/output differential equations with time-varying coefficients are studied in Section 2.4. In Section 2.5 we consider systems given by a nonlinear input/output differential equation. Using the method of linearization, we show that nonlinear differential equations can be studied in terms of a linear differential equation, which describes system behavior in response to a small perturbation from some nominal operation. In Section 2.6 we consider multi-input multi-output systems given by a collection of input/output differential equations.

2.1

LINEAR INPUT/OUTPUT DIFFERENTIAL EQUATIONS WITH CONSTANT COEFFICIENTS

Suppose that we have a single-input single-output continuous-time system given by the input/output differential equation

$$\frac{d^n y(t)}{dt^n} + \sum_{i=0}^{n-1} a_i \frac{d^i y(t)}{dt^i} = \sum_{i=0}^{m} b_i \frac{d^i x(t)}{dt^i}. \tag{2.1}$$

In (2.1), $x(t)$ is the input applied to the system for $t \geq 0$ and $y(t)$ is the output response for $t > 0$ resulting from $x(t)$ and initial conditions at time $t = 0$ (defined later). Note that we are taking the initial time t_0 to be zero. The coefficients a_0, $a_1, \ldots, a_{n-1}$ and $b_0, b_1, \ldots, b_m$ are real constants.

Recall from Chapter 1 that since (2.1) is a linear differential equation with constant coefficients, the system defined by (2.1) is finite dimensional, linear, and time invariant. The integer n in (2.1) is the order or dimension of the system. Finally, any system given by (2.1) can be shown to be causal.

The class of linear time-invariant continuous-time systems that can be described by a differential equation of the form (2.1) is very large. For example, RLC circuits can be modeled by an equation of the form (2.1). The input/output differential equation for an RLC circuit can be determined by applying Kirchhoff's laws. We refer the reader to a textbook in circuits for the details; for example, see Van Valkenburg [1974], Hayt and Kemmerly [1978], or Nilsson [1983].

INITIAL CONDITIONS   To solve (2.1) for $t > 0$, we need the n initial conditions

$$y(0), y^{(1)}(0), \ldots, y^{(n-1)}(0), \tag{2.2}$$

where $y^{(i)}(t)$ is the ith derivative of $y(t)$. To solve (2.1) for $t > 0$, we could also take the initial conditions at time $t = 0^-$, where 0^- is an infinitesimal negative number. In this case, the n initial conditions are

$$y(0^-), y^{(1)}(0^-), \ldots, y^{(n-1)}(0^-). \tag{2.3}$$

If the mth derivative of the input $x(t)$ contains an impulse $k\delta(t)$ or a derivative of an impulse, to solve (2.1) for $t > 0$ it is necessary to take the initial time to be 0^-. The reason for this is that if the term

$$b_m \frac{d^m x(t)}{dt^m}$$

in (2.1) contains an impulse at $t = 0$, the values of the output $y(t)$ and its derivatives up to order $n - 1$ may change instantaneously at time $t = 0$. So initial conditions must be taken just prior to time $t = 0$.

Using solution techniques for ordinary linear differential equations, we can solve (2.1) for the output response $y(t)$ resulting from the initial conditions (2.2) or (2.3) and an input $x(t)$. In Chapter 5 we show that (2.1) can be solved using the Laplace transform.

FIRST-ORDER CASE   In the first-order case ($n = 1$), it is possible to express the solution to (2.1) in a useful general form: Suppose that the system is given by the first-order differential equation

$$\frac{dy(t)}{dt} + ay(t) = bx(t), \tag{2.4}$$

where a and b are arbitrary constants. Equation (2.4) can be solved using the integrating-factor method as illustrated in the RC circuit example in Chapter 1. The result is that the output response $y(t)$ resulting from initial condition $y(0)$ and input $x(t)$ is given by

$$y(t) = e^{-at}y(0) + \int_0^t \exp\left[-a(t - \lambda)\right]bx(\lambda)\, d\lambda, \qquad t \geq 0. \tag{2.5}$$

If the initial time is taken to be 0^-, the output response is given by

$$y(t) = e^{-at}y(0^-) + \int_{0^-}^t \exp\left[-a(t - \lambda)\right]bx(\lambda)\, d\lambda, \qquad t \geq 0. \tag{2.6}$$

Note that if we set $a = 1/RC$ and $b = 1/C$, (2.5) or (2.6) gives the output response of the RC circuit studied in Chapter 1.

We can generalize (2.4) by adding a term involving the derivative of the input. Suppose that the system is given by the input/output differential equation

$$\frac{dy(t)}{dt} + ay(t) = b_1 \frac{dx(t)}{dt} + b_0 x(t). \tag{2.7}$$

Equation (2.7) is still a first-order differential equation, but its solution is more complicated than the solution of (2.4) due to the $dx(t)/dt$ term. It is possible to solve (2.7) without having to differentiate the input $x(t)$. We will show this by first reducing (2.7) to an equation of the form (2.4). Let

$$q(t) = y(t) - b_1 x(t), \tag{2.8}$$

so that

$$y(t) = q(t) + b_1 x(t). \tag{2.9}$$

Differentiating both sides of (2.8), we have

$$\frac{dq(t)}{dt} = \frac{dy(t)}{dt} - b_1 \frac{dx(t)}{dt}.$$

Using the expression (2.7) for $dy(t)/dt$, we get

$$\frac{dq(t)}{dt} = -ay(t) + b_1 \frac{dx(t)}{dt} + b_0 x(t) - b_1 \frac{dx(t)}{dt}$$

$$= -ay(t) + b_0 x(t). \tag{2.10}$$

Inserting the expression (2.9) for $y(t)$ into (2.10), we get

$$\frac{dq(t)}{dt} = -a(q(t) + b_1 x(t)) + b_0 x(t)$$

$$= -aq(t) + (b_0 - ab_1)x(t). \tag{2.11}$$

Now (2.11) is in the form of (2.4), so we can solve for $q(t)$ using (2.5). This gives

$$q(t) = e^{-at}q(0) + \int_0^t \exp\left[-a(t - \lambda)\right](b_0 - ab_1)x(\lambda)\, d\lambda, \qquad t \geq 0. \tag{2.12}$$

Then, inserting the expression (2.12) for $q(t)$ into (2.9) and using the relationship $q(0) = y(0) - b_1 x(0)$, we have

$$y(t) = e^{-at}[y(0) - b_1 x(0)]$$

$$+ \int_0^t \exp\left[-a(t - \lambda)\right](b_0 - ab_1)x(\lambda)\, d\lambda + b_1 x(t), \qquad t \geq 0. \tag{2.13}$$

The expression (2.13) for $y(t)$ is the complete solution of (2.7) with initial condition $y(0)$ and with input $x(t)$ applied for $t \geq 0$.

If the derivative of the input $x(t)$ contains an impulse $k\delta(t)$, we must take the initial time to be 0^-, in which case the output response is

$$y(t) = e^{-at}[y(0^-) - b_1 x(0^-)]$$

$$+ \int_{0^-}^t \exp\left[-a(t - \lambda)\right](b_0 - ab_1)x(\lambda)\, d\lambda + b_1 x(t), \qquad t \geq 0.$$

If $x(0^-) = 0$, the output response $y(t)$ is

$$y(t) = e^{-at}y(0^-) + \int_{0^-}^t \exp\left[-a(t - \lambda)\right](b_0 - ab_1)x(\lambda)\, d\lambda$$

$$+ b_1 x(t), \qquad t \geq 0. \tag{2.14}$$

Note that if the input $x(t)$ is the unit-step function $u(t)$, the derivative of $x(t)$ is the unit impulse $\delta(t)$, and we must use (2.14) for the output response. Evaluating (2.14) with $x(t) = u(t)$, we get

$$y(t) = e^{-at}y(0^-) + \int_0^t \exp\left[-a(t - \lambda)\right](b_0 - ab_1)\, d\lambda + b_1, \qquad t \geq 0.$$

$$= e^{-at}y(0^-) + \left(\frac{b_0}{a} - b_1\right)[1 - e^{-at}] + b_1, \qquad t \geq 0.$$

$$= \left[y(0^-) - \frac{b_0}{a} + b_1\right]e^{-at} + \frac{b_0}{a}, \qquad t \geq 0. \tag{2.15}$$

From (2.15) we see that there is an instantaneous change of amount b_1 in the value of the output $y(t)$ in going from 0^- to 0^+ seconds. This is due to the $b_1[dx(t)/dt]$ term in the input/output differential equation (2.7) of the system.

SOLUTION OF SECOND-ORDER AND HIGHER-ORDER EQUATIONS By working with matrix equations, we can generalize the form of the solution (2.5) in the first-order case to the nth-order case. The matrix-equation formulation is constructed by defining a state model for the system. The details are carried out in Chapter 13.

2.2

THE OPERATIONAL AMPLIFIER AND BASIC SYSTEM COMPONENTS

Any finite-dimensional linear time-invariant continuous-time system given by the input/output differential equation (2.1) with $m \leq n$ can be realized (or simulated) via an interconnection of integrators, adders, subtracters, and scalar multipliers. These components are in turn realizable from interconnections of resistors, capacitors, and operational amplifiers. In this section we first define the (ideal) operational amplifier, and then we show that it can be utilized to realize integrators and basic system operations.

OPERATIONAL AMPLIFIER A schematic diagram of an operational amplifier (denoted by ''op amp'') in a common type of configuration is shown in Figure 2.1. The input applied to the op amp is the voltage $v_{in}(t)$, while the output is the voltage $v_{out}(t)$. The op amp contains transistors, and thus it is an *active device*, in contrast to a *passive device*, which does not require a dc power source for operation.

In an ideal op amp, the voltage $v_0(t)$ and the current $i_0(t)$ in Figure 2.1 are both zero. Note that $i_0(t) = 0$ implies that the input impedance of the op amp is infinite.

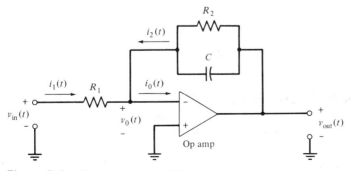

Figure 2.1. Op amp with an RC feedback loop.

Since both $v_0(t)$ and $i_0(t)$ are zero, from the wiring diagram in Figure 2.1 we have that

$$i_1(t) + i_2(t) = 0, \tag{2.16}$$

$$v_{\text{in}}(t) = R_1 i_1(t), \tag{2.17}$$

$$C \frac{dv_{\text{out}}(t)}{dt} + \frac{1}{R_2} v_{\text{out}}(t) = i_2(t). \tag{2.18}$$

Inserting (2.17) and (2.18) into (2.16), we have

$$\frac{1}{R_1} v_{\text{in}}(t) + C \frac{dv_{\text{out}}(t)}{dt} + \frac{1}{R_2} v_{\text{out}}(t) = 0$$

or

$$C \frac{dv_{\text{out}}(t)}{dt} = -\frac{1}{R_2} v_{\text{out}}(t) - \frac{1}{R_1} v_{\text{in}}(t). \tag{2.19}$$

Equation (2.19) is the input/output differential equation of the op-amp circuit in Figure 2.1.

If $C = 0$, from (2.19) we have

$$v_{\text{out}}(t) = -\frac{R_2}{R_1} v_{\text{in}}(t). \tag{2.20}$$

Therefore, if $C = 0$ and $R_2 = R_1$, the op-amp circuit is a *signal inverter*.

If $R_2 = \infty$ (an open circuit), from (2.19) we have

$$C \frac{dv_{\text{out}}(t)}{dt} = -\frac{1}{R_1} v_{\text{in}}(t). \tag{2.21}$$

Integrating both sides of (2.21) with the initial time equal to t_0, we get

$$v_{\text{out}}(t) = v_{\text{out}}(t_0) + \int_{t_0}^{t} \frac{-1}{R_1 C} v_{\text{in}}(\lambda) \, d\lambda, \qquad t \geq t_0. \tag{2.22}$$

We shall see below that the device with the input/output relationship (2.22) can be used to realize an integrator.

BASIC SYSTEM COMPONENTS   By interconnecting op-amp circuits with various values of R_1, R_2, and C, we can realize a number of basic system components. We begin with the integrator.

The Integrator.   A key element in the theory and practice of systems engineering is the integrator, which is illustrated in Figure 2.2. As shown, the

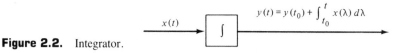

Figure 2.2. Integrator.

output $y(t)$ of the integrator at time t is equal to the initial value $y(t_0)$ plus the integral of the input $x(\lambda)$ from $\lambda = t_0$ to t. Hence the device is an "integrator." In mathematical terms, we have

$$y(t) = y(t_0) + \int_{t_0}^{t} x(\lambda) \, d\lambda, \qquad t \geq t_0. \qquad (2.23)$$

Differentiating both sides of the input/output relationship (2.23), we have

$$\frac{dy(t)}{dt} = x(t). \qquad (2.24)$$

Equation (2.24) is the input/output differential equation of the integrator. From (2.24), we see that if the input to the integrator is the derivative of a signal $v(t)$, the resulting output is $v(t)$. This makes sense since integration "undoes" differentiation.

The integrator is at rest at time t_0 if and only if $y(t_0) = 0$, in which case the output response is given by

$$y(t) = (Fx)(t) = \int_{t_0}^{t} x(\lambda) \, d\lambda, \qquad t \geq t_0.$$

The integrator can be realized using two op amps, as shown in Figure 2.3. Let us verify that the circuit shown is a realization of the integrator. First, by (2.20),

$$v_2(t) = -\frac{R_1}{R_1} v_{in}(t) = -v_{in}(t). \qquad (2.25)$$

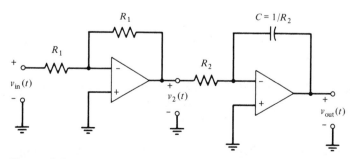

Figure 2.3. Op-amp realization of the integrator.

Thus the first stage of the circuit is a signal inverter. We should note that in writing (2.25), we are assuming that the second op amp does not load the output of the first op amp.

Now, by (2.22),

$$v_{out}(t) = v_{out}(t_0) + \int_{t_0}^{t} \frac{-1}{R_2 C} v_2(\lambda) \, d\lambda, \qquad t \geq t_0$$

$$= v_{out}(t_0) - \int_{t_0}^{t} v_2(\lambda) \, d\lambda, \qquad t \geq t_0. \qquad (2.26)$$

Inserting (2.25) into (2.26), we get

$$v_{out}(t) = v_{out}(t_0) + \int_{t_0}^{t} v_{in}(\lambda) \, d\lambda, \qquad t \geq t_0. \qquad (2.27)$$

Equation (2.27) is the input/output relationship of the integrator, and hence the circuit in Figure 2.3 is a realization of the integrator.

Adders, Subtracters, and Scalar Multipliers.   Op amps can also be used to realize adders, subtracters, and scalar multipliers. These operations are illustrated in Figure 2.4. Op-amp realizations of these operations are given in Figure 2.5. When using adders and subtracters, we shall sometimes combine them as illustrated in Figure 2.6.

Finally, we note that by eliminating redundant inverters, it is possible to minimize the number of op amps required to realize an interconnection of integrators, adders, subtracters, and scalar multipliers. For example, a cascade connection of a scalar multiplier and an integrator can be realized using only two op amps.

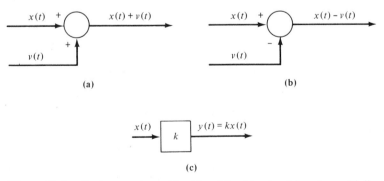

Figure 2.4.   Basic operations: (a) adder; (b) subtracter; (c) scalar multiplier.

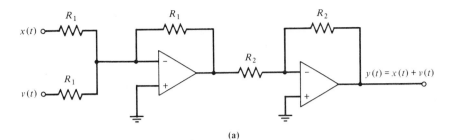

(a)

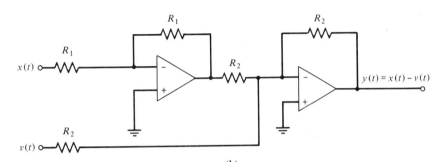

(b)

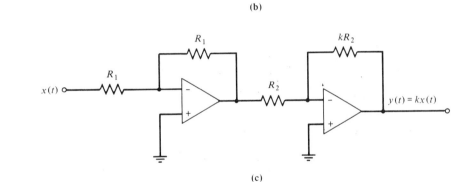

(c)

Figure 2.5. Op-amp realizations: (a) adder; (b) subtracter; (c) scalar multiplier.

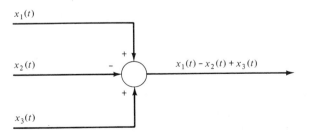

Figure 2.6. Combined adder and subtracter.

INTEGRATOR REALIZATIONS

In this section we show that any system given by the nth-order input/output differential equation (2.1) with $m \leq n$ can be realized by an interconnection of n integrators and combinations of adders, subtracters, and scalar multipliers. Such a realization is sometimes called an *analog-computer simulation* since it can be constructed from op amps, the basic building block of the analog computer.

From the results in this section, we will see that there is a one-to-one correspondence between nth-order differential equations and integrator interconnections. This correspondence gives a certain ''concreteness'' to differential-equation models.

We begin with the $n = 1$ case.

FIRST-ORDER CASE Suppose that our system has the input/output differential equation

$$\frac{dy(t)}{dt} + ay(t) = b_1 \frac{dx(t)}{dt} + b_0 x(t). \qquad (2.28)$$

We claim that this system is realized by the interconnection shown in Figure 2.7. To say that this interconnection realizes (or simulates) the system defined by (2.28), we mean that the input/output differential equation of the system shown in Figure 2.7 is the same as (2.28). We shall verify that this is the case.

Denoting the output of the integrator in Figure 2.7 by $q(t)$, we have that the input to the integrator is equal to $dq(t)/dt$. The input to the integrator is equal to the output of the subtracter, which in turn is equal to

$$-ay(t) + b_0 x(t).$$

Thus

$$\frac{dq(t)}{dt} = -ay(t) + b_0 x(t). \qquad (2.29)$$

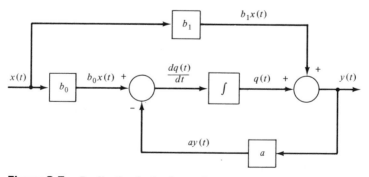

Figure 2.7. Realization in the first-order case.

In addition, $y(t)$ is the output of the adder, and thus

$$y(t) = q(t) + b_1 x(t). \tag{2.30}$$

Differentiating both sides of (2.30) and using (2.29), we get

$$\frac{dy(t)}{dt} = -ay(t) + b_0 x(t) + b_1 \frac{dx(t)}{dt}.$$

This is the same as (2.28), and hence the interconnection in Figure 2.7 is a realization of the given system.

Since the components comprising the interconnection in Figure 2.7 can be realized using op amps, the system with input/output differential equation (2.28) can be realized using op amps. In the case when $a > 0$, $b_1 < 0$, and $b_0 - b_1 a > 0$, a realization containing two op amps is given in Figure 2.8. We shall verify that this circuit is a realization. First, we have that

$$y(t) = -v(t) + b_1 x(t).$$

Differentiating both sides gives

$$\frac{dy(t)}{dt} = -\frac{dv(t)}{dt} + b_1 \frac{dx(t)}{dt}.$$

Now, using (2.19), we have

$$\frac{dv(t)}{dt} = -av(t) - (b_0 - b_1 a)x(t).$$

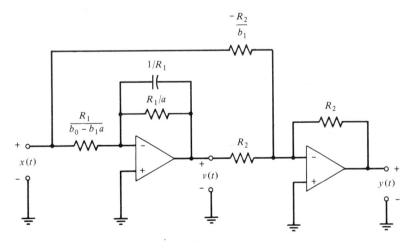

Figure 2.8.   Op-amp realization of first-order system.

Inserting this into the expression above for $dy(t)/dt$, we get

$$\frac{dy(t)}{dt} = av(t) + (b_0 - b_1 a)x(t) + b_1 \frac{dx(t)}{dt}.$$

Now

$$av(t) = -ay(t) + ab_1 x(t),$$

and thus

$$\frac{dy(t)}{dt} = -ay(t) + b_0 x(t) + b_1 \frac{dx(t)}{dt}.$$

This is equivalent to (2.28), and thus the circuit in Figure 2.8 is a realization.

SECOND-ORDER CASE   Now consider the system given by the second-order input/output differential equation

$$\frac{d^2y(t)}{dt^2} + a_1 \frac{dy(t)}{dt} + a_0 y(t) = b_2 \frac{d^2x(t)}{dt^2} + b_1 \frac{dx(t)}{dt} + b_0 x(t). \quad (2.31)$$

We claim that this system is realized by the interconnection in Figure 2.9. To verify that this is a realization, let $q_1(t)$ denote the output of the first integrator and let $q_2(t)$ denote the output of the second integrator. From the interconnection in Figure 2.9, we have

$$\frac{dq_1(t)}{dt} = -a_0 y(t) + b_0 x(t) \quad (2.32)$$

$$\frac{dq_2(t)}{dt} = -a_1 y(t) + q_1(t) + b_1 x(t) \quad (2.33)$$

$$y(t) = q_2(t) + b_2 x(t). \quad (2.34)$$

Differentiating both sides of (2.34) and using (2.33), we get

$$\frac{dy(t)}{dt} = -a_1 y(t) + q_1(t) + b_1 x(t) + b_2 \frac{dx(t)}{dt}. \quad (2.35)$$

Differentiating both sides of (2.35) and using (2.32), we have

$$\frac{d^2y(t)}{dt^2} = -a_1 \frac{dy(t)}{dt} - a_0 y(t) + b_0 x(t) + b_1 \frac{dx(t)}{dt} + b_2 \frac{d^2x(t)}{dt^2}.$$

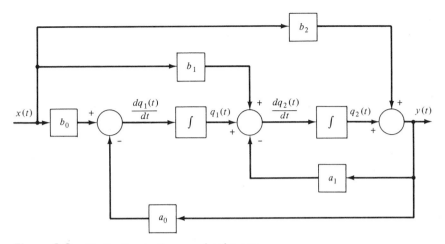

Figure 2.9. Realization in the second-order case.

This is the same as (2.31), so the interconnection in Figure 2.9 is a realization of the system with the input/output differential equation (2.31).

The integrator realization in Figure 2.9 can be used to simulate any system defined by a second-order input/output differential equation. Two examples follow.

EXAMPLE 2.1. Consider the automobile given by the input/output differential equation

$$M \frac{d^2y(t)}{dt^2} + k_f \frac{dy(t)}{dt} = x(t). \qquad (2.36)$$

Clearly, we can write (2.36) in the form (2.31) by taking $a_1 = k_f/M$, $a_0 = 0$, $b_2 = 0$, $b_1 = 0$, and $b_0 = 1/M$. Thus, with these values of the scalars, the interconnection in Figure 2.9 is a simulation of the car. The simulation is shown in Figure 2.10.

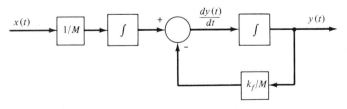

Figure 2.10. Integrator simulation of the automobile.

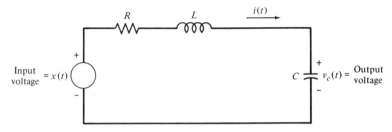

Figure 2.11.   Series *RLC* circuit.

EXAMPLE 2.2.   Consider the series *RLC* circuit shown in Figure 2.11. We shall first derive the input/output differential equation of the circuit. Summing the voltages around the loop, we have

$$L \frac{di(t)}{dt} + Ri(t) + v_c(t) = x(t).$$

Also, the current $i(t)$ in the loop is given by

$$i(t) = C \frac{dv_c(t)}{dt} .$$

Inserting the expression for $i(t)$ into the first equation, we obtain the input/output differential equation

$$LC \frac{d^2v_c(t)}{dt^2} + RC \frac{dv_c(t)}{dt} + v_c(t) = x(t).$$

Clearly, this can be written in the form (2.31) with

$$a_1 = \frac{RC}{LC} = \frac{R}{L}, \qquad a_0 = \frac{1}{LC}, \qquad b_2 = b_1 = 0, \qquad b_0 = \frac{1}{LC}.$$

With these values of the scalars, the interconnection in Figure 2.9 is an integrator realization of the *RLC* circuit. This realization is given in Figure

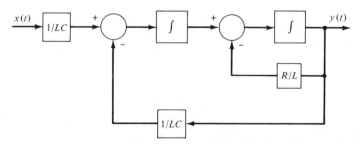

Figure 2.12.   Integrator realization of *RLC* circuit.

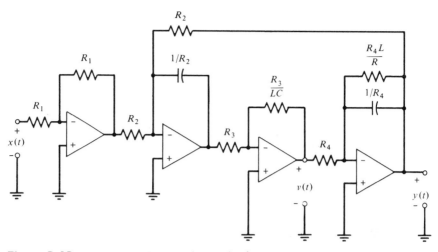

Figure 2.13. Op-amp realization of *RLC* circuit.

2.12. Realizing the components using op amps, we obtain the op-amp synthesis of the *RLC* circuit shown in Figure 2.13.

To verify that the circuit in Figure 2.13 is a realization, first observe that

$$\frac{1}{R_4}\frac{dy(t)}{dt} = -\frac{R}{R_4 L}y(t) - \frac{1}{R_4}v(t).$$

Multiplying both sides by R_4 and differentiating with respect to t, we get

$$\frac{d^2y(t)}{dt^2} = -\frac{R}{L}\frac{dy(t)}{dt} - \frac{dv(t)}{dt}.$$

Now we also have that

$$\frac{dv(t)}{dt} = \frac{1}{LC}[y(t) - x(t)].$$

Inserting this into the expression above for $d^2y(t)/dt^2$, we obtain the input/output differential equation of the *RLC* circuit.

The circuit in Figure 2.13 is called an *active-network realization* of the *RLC* circuit since the realization is constructed from op amps that are active-circuit elements.

We should point out that the realization of (2.31) given in Figure 2.9 is only one among an infinite number of realizations of (2.31) (with each realization containing two integrators). For example, a second realization is given in Figure 2.14. The input/output differential equation of this system is identical to the input/output differential equation of the system shown in Figure 2.9. (*Exercise:* Verify this.)

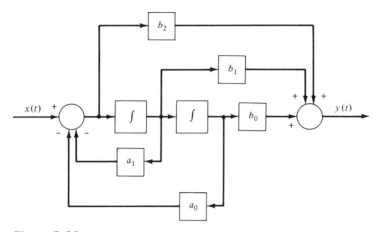

Figure 2.14. Second realization in the second-order case.

nTH-ORDER CASE Now consider the nth-order case with $m \leq n$. The input/output differential equation (2.1) can be written as

$$\frac{d^n y(t)}{dt^n} + \sum_{i=0}^{n-1} a_i \frac{d^i y(t)}{dt^i} = \sum_{i=0}^{n} b_i \frac{d^i x(t)}{dt^i} . \qquad (2.37)$$

Note that in going from (2.1) to (2.37), we have replaced m by n. This does not result in any loss of generality since we are assuming that $m \leq n$. If $m = n - r$ for some nonnegative integer r, the coefficients $b_n, b_{n-1}, \ldots, b_{n-r+1}$ in (2.37) are all zero.

The form of the realization in Figure 2.9 generalizes to the nth-order case given by (2.37). The integrator realization is given in Figure 2.15. In other words, the input/output differential equation of the realization shown in Figure 2.15 is the same as (2.37).

The form of the realization in Figure 2.14 also generalizes to the nth-order case given by (2.37). The realization is shown in Figure 2.16.

Again, it can be shown that the input/output differential equation of the system in Figure 2.16 is the same as (2.37). Thus, even though the diagrams of the systems in Figures 2.15 and 2.16 are quite different, these two systems have the same input/output differential equation. Thus both realizations have exactly the same input/output behavior. Which realization is used in practice depends on the application. There are realization forms other than those given in Figures 2.15 and 2.16 that are also used. One of these is the cascade realization, which we consider in Chapter 6.

Since the realizations shown in Figures 2.15 and 2.16 can be built using only resistors, capacitors, and op amps, we see that finite-dimensional linear time-invariant continuous-time systems can be realized without having to use inductors. Inductorless realizations are very desirable since actual inductors do not behave very much like ideal inductors over a wide range of operating characteristics (e.g., frequency range). This does not mean that inductors are obsolete since there are situations when integrator realizations cannot be employed. An

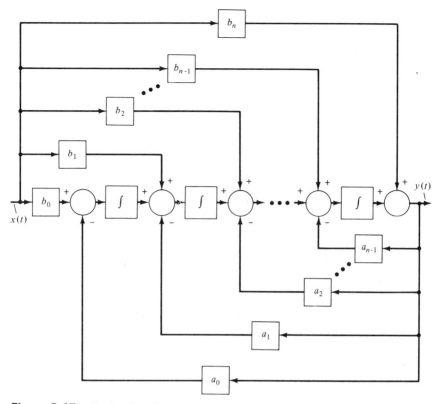

Figure 2.15. Realization of nth-order differential equation.

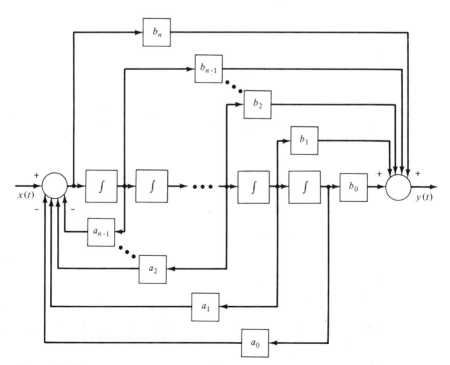

Figure 2.16. Second realization in the nth-order case.

example is in applications where the power requirements are too high for integrator-circuit implementations.

EXAMPLE 2.3. A field-controlled dc (direct current) motor is illustrated in Figure 2.17. The input to the motor is the voltage $v_f(t)$ applied to the field circuit and the output is the angle $\theta(t)$ of the motor shaft. The torque $T(t)$ developed by the motor is related to the angle $\theta(t)$ by the differential equation

$$J \frac{d^2\theta(t)}{dt^2} + f \frac{d\theta(t)}{dt} = T(t),$$

where J is the moment of inertia of the motor and load and f is the viscous friction coefficient of the motor and load. In addition, $T(t) = ki_f(t)$, where k is a constant and $i_f(t)$ is the current in the field circuit. Inserting this expression for $T(t)$ into the above differential equation, we get

$$J \frac{d^2\theta(t)}{dt^2} + f \frac{d\theta(t)}{dt} = ki_f(t).$$

Now summing voltages in the field circuit, we have

$$L_f \frac{di_f(t)}{dt} + R_f i_f(t) = v_f(t).$$

Multiplying both sides by k gives

$$L_f \frac{dki_f(t)}{dt} + R_f ki_f(t) = kv_f(t).$$

Finally, setting $ki_f(t) = J\ddot{\theta}(t) + f\dot{\theta}(t)$ in this equation, we obtain the following input/output differential equation of the dc motor

$$L_f J \frac{d^3\theta(t)}{dt^3} + (L_f f + R_f J) \frac{d^2\theta(t)}{dt^2} + R_f f \frac{d\theta(t)}{dt} = kv_f(t).$$

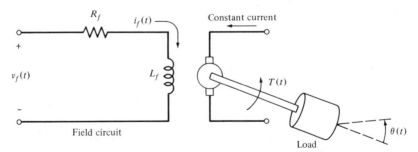

Figure 2.17. Field-controlled dc motor with load.

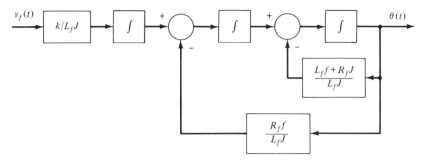

Figure 2.18. First realization of *dc* motor.

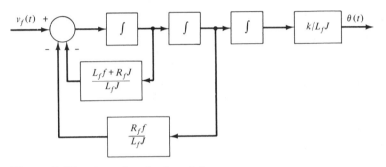

Figure 2.19. Second realization of *dc* motor.

In this example, n is equal to 3 and the integrator realization with the form in Figure 2.15 is shown in Figure 2.18. The realization with the form in Figure 2.16 is given in Figure 2.19. Comparing Figures 2.18 and 2.19, we see that in this case the two realizations are very similar. In fact, from the rules of block-diagram algebra (to be discussed in Chapter 6), it is easy to see that the two realizations are equivalent.

REALIZATIONS CONTAINING DIFFERENTIATORS It is important to note that we were able to realize the input/output differential equations given above without having to use differentiators. A differentiator is a device with the input/output relationship

$$y(t) = \frac{dx(t)}{dt},$$

where $x(t)$ and $y(t)$ have the same dimensions (e.g., both are voltages). A differentiator can be realized using op-amps (see Problem 2.8).

It is very desirable to avoid using differentiators whenever possible because differentiation can greatly amplify any high-frequency noise components. For example, suppose that the input is given by

$$x(t) = x_1(t) + 10^{-6} \cos 10^6 t,$$

where $10^{-6} \cos 10^6 t$ is a noise component with frequency 10^6 rad/sec. Although the magnitude of the noise component is very small, after differentiation it is equal to $\sin 10^6 t$, which may be large in comparison to the signal component of the derivative of $x(t)$.

If the input $x(t)$ and the output $y(t)$ have the same dimensions and the number of differentiations of $x(t)$ exceeds the number of differentiations of $y(t)$ in the input/output differential equation [i.e., $m > n$ in the notation of (2.1)], then in order to realize the system it is necessary to use differentiators. We leave the details of this case to the interested reader.

INPUT/OUTPUT DIFFERENTIAL EQUATION OF INTEGRATOR INTERCONNECTIONS

Any interconnection of n integrators and combinations of adders, subtracters, and scalar multipliers can be modeled by an nth-order input/output differential equation of the form (2.1) or (2.37). In particular, any analog-computer simulation of a finite-dimensional linear time-invariant continuous-time system can be described by a linear differential equation with constant coefficients. The computation of the differential equation can be carried out by first labeling the outputs of the integrators in the simulation and writing a first-order differential equation for each integrator. By combining these equations with the equation for the output, we will obtain the input/output differential equation. The procedure is illustrated by the following example.

EXAMPLE 2.4. Consider the system given by the interconnection in Figure 2.20, where we have labeled the outputs of the integrators $q_1(t)$ and $q_2(t)$. From the interconnection, we have

$$\frac{dq_1(t)}{dt} = -2q_1(t) + 4x(t) \tag{2.38}$$

$$\frac{dq_2(t)}{dt} = -2q_2(t) - 3q_1(t) + x(t) \tag{2.39}$$

$$y(t) = q_1(t) + q_2(t). \tag{2.40}$$

Now we can combine (2.38), (2.39), and (2.40) to get the input/output differential equation of the given system. Since we want a differential equation relating $x(t)$ and $y(t)$, we need to combine these equations in such a way that we eliminate $q_1(t)$ and $q_2(t)$. Adding (2.38) and (2.39) and using (2.40), we get

$$\frac{dq_1(t)}{dt} + \frac{dq_2(t)}{dt} = \frac{dy(t)}{dt} = -2[q_1(t) + q_2(t)] - 3q_1(t) + 5x(t).$$

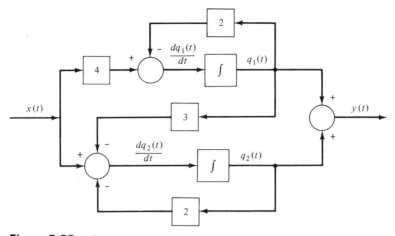

Figure 2.20. System with two integrators.

Thus

$$\frac{dy(t)}{dt} = -2y(t) - 3q_1(t) + 5x(t). \tag{2.41}$$

Differentiating both sides of (2.41), we have

$$\frac{d^2y(t)}{dt^2} = -2\frac{dy(t)}{dt} - 3\frac{dq_1(t)}{dt} + 5\frac{dx(t)}{dt},$$

and inserting (2.38), we get

$$\frac{d^2y(t)}{dt^2} = -2\frac{dy(t)}{dt} - 3[-2q_1(t) + 4x(t)] + 5\frac{dx(t)}{dt}. \tag{2.42}$$

Finally, solving (2.41) for $q_1(t)$ and inserting into (2.42), we have

$$\frac{d^2y(t)}{dt^2} = -2\frac{dy(t)}{dt} - \frac{6}{3}\left[\frac{dy(t)}{dt} + 2y(t) - 5x(t)\right] - 12x(t) + 5\frac{dx(t)}{dt}$$

$$= -4\frac{dy(t)}{dt} - 4y(t) - 2x(t) + 5\frac{dx(t)}{dt}. \tag{2.43}$$

Equation (2.43) is the input/output differential equation of the system in Figure 2.20.

The process of determining the input/output differential equation of an interconnection of integrators is greatly simplified by using the Laplace transform. We consider this approach in Chapter 6.

LINEAR INPUT/OUTPUT DIFFERENTIAL EQUATIONS WITH TIME-VARYING COEFFICIENTS

We can generalize the class of linear systems under study to include time-varying systems by allowing the coefficients of the input/output differential equation to vary with time. For example, consider the first-order case given by

$$\frac{dy(t)}{dt} + a(t)y(t) = b(t)x(t). \tag{2.44}$$

In (2.44), the coefficient $a(t)$ and/or the coefficient $b(t)$ may vary with t. If both $a(t)$ and $b(t)$ are constant (i.e., do not vary with t), (2.44) is a first-order differential equation with constant coefficients, which we have already studied in detail.

Systems given by an input/output differential equation with time-varying coefficients arise in many applications. Examples include circuits with time-varying components. For instance, resistors may have a time-varying resistance due to heating effects. Both capacitors and inductors may be time varying. A capacitor with a time-varying capacitance is illustrated in Figure 2.21.

The time variance $C(t)$ of the capacitance is a result of changing the position of the dielectric. To determine the current–voltage relationship in this case, first observe that the charge $q(t)$ on the capacitor plates is given by

$$q(t) = C(t)v_c(t), \tag{2.45}$$

where $v_c(t)$ is the voltage across the capacitor. The current $i_c(t)$ into the capacitor is equal to the derivative of the charge $q(t)$. Thus, taking the derivative of both sides of (2.45), we get

$$i_c(t) = C(t)\frac{dv_c(t)}{dt} + \dot{C}(t)v_c(t), \tag{2.46}$$

where $\dot{C}(t) = dC(t)/dt$. Note that if $C(t)$ is constant, so that $\dot{C}(t) = 0$, (2.46) reduces to the current–voltage relationship of the time-invariant capacitor.

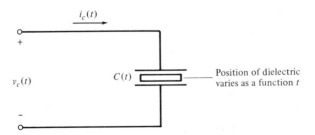

Figure 2.21.   Time-varying capacitance.

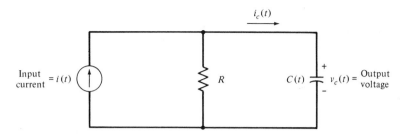

Figure 2.22. *RC* circuit with time-varying capacitor.

Now we shall reconsider the *RC* circuit studied in Chapter 1 with the capacitor given by the relationship (2.46). The circuit is shown in Figure 2.22. From Kirchhoff's current law we have

$$i_c(t) + \frac{1}{R} v_c(t) = i(t). \tag{2.47}$$

Inserting the expression (2.46) for $i_c(t)$ into (2.47), we have

$$C(t) \frac{dv_c(t)}{dt} + \left[\frac{1}{R} + \dot{C}(t) \right] v_c(t) = i(t). \tag{2.48}$$

Assuming that $C(t) > 0$ for all t, we can divide both sides of (2.48) by $C(t)$. This gives

$$\frac{dv_c(t)}{dt} + \frac{(1/R) + \dot{C}(t)}{C(t)} v_c(t) = \frac{1}{C(t)} i(t). \tag{2.49}$$

Equation (2.49) is the input/output differential equation of the *RC* circuit with time-varying capacitor. Note that (2.49) is a first-order differential equation with both coefficients equal to functions of t [assuming that $C(t)$ varies with t].

Let us return to the first-order input/output differential equation given by (2.44). Again using the integrating-factor method, we have that the solution $y(t)$ to (2.44) with initial condition $y(0)$ and input $x(t)$ is given by

$$y(t) = \exp \left[\int_0^t -a(\tau) \, d\tau \right] y(0)$$
$$+ \int_0^t \exp \left[\int_\lambda^t -a(\tau) \, d\tau \right] b(\lambda) x(\lambda) \, d\lambda, \qquad t \geq 0. \tag{2.50}$$

When both coefficients $a(t)$ and $b(t)$ are constant, (2.50) reduces to the expression (2.5) for the solution of the first-order differential equation (2.4) with constant coefficients. To see this, note that if the coefficient $a(t)$ is equal to a constant a, then

$$\int_\lambda^t -a(\tau)\, d\tau = \int_\lambda^t -a\, d\tau = -a \Big]_{\tau=\lambda}^{\tau=t} = -a(t - \lambda). \qquad (2.51)$$

Inserting (2.51) into (2.50) and setting $b(t)$ equal to a constant b, we get (2.5).

Unlike the constant-coefficient case, second-order and higher-order differential equations with time-varying coefficients cannot be solved exactly in general. In such cases, solutions must be obtained using numerical-solution techniques.

INTEGRATOR REALIZATIONS Differential equations with time-varying coefficients can also be realized by using integrators, except that in this case we need to employ time-varying scalar multipliers. A time-varying scalar multiplier is illustrated in Figure 2.23. The realization of a time-varying scalar multiplier may be difficult to achieve unless the time variation $k(t)$ is simple [e.g., $k(t)$ is sinusoidal]. A time-varying scalar multiplier may be realized using an amplifier whose gain can be varied as a function of time.

Now consider the system with the first-order input/output differential equation (2.44). This system has the integrator realization shown in Figure 2.24. The output of the subtracter shown in the figure is equal to

$$-a(t)y(t) + b(t)x(t).$$

Since the input to the integrator is $dy(t)/dt$, we see that the system in Figure 2.24 is a realization of (2.44).

Now consider the second-order case

$$\frac{d^2y(t)}{dt^2} + a_1(t)\frac{dy(t)}{dt} + a_0(t)y(t) = b_1(t)\frac{dx(t)}{dt} + b_0(t)x(t). \qquad (2.52)$$

Here all four of the coefficients $a_1(t)$, $a_0(t)$, $b_1(t)$, $b_0(t)$ may be time varying. We claim that this equation has the realization shown in Figure 2.25. To verify that the interconnection shown is a realization, first note that

$$\frac{dy(t)}{dt} = -a_1(t)y(t) + q(t) + b_1(t)x(t) \qquad (2.53)$$

$$\frac{dq(t)}{dt} = -[a_0(t) - \dot{a}_1(t)]y(t) + [b_0(t) - \dot{b}_1(t)]x(t). \qquad (2.54)$$

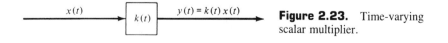

Figure 2.23. Time-varying scalar multiplier.

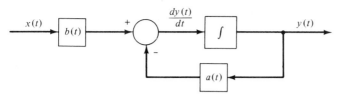

Figure 2.24. Realization of first-order differential equation with time-varying coefficients.

Differentiating (2.53) and using (2.54), we have

$$\frac{d^2y(t)}{dt^2} = -a_1(t)\frac{dy(t)}{dt} - \dot{a}_1(t)y(t) - [a_0(t) - \dot{a}_1(t)]y(t)$$
$$+ [b_0(t) - \dot{b}_1(t)]x(t) + b_1(t)\frac{dx(t)}{dt} + \dot{b}_1(t)x(t)$$

$$= -a_1(t)\frac{dy(t)}{dt} - a_0(t)y(t) + b_0(t)x(t) + b_1(t)\frac{dx(t)}{dt}. \quad (2.55)$$

Equation (2.55) is equivalent to (2.52), and thus the interconnection in Figure 2.25 is a realization.

Note that due to the presence of the derivatives $\dot{a}_1(t)$ and $\dot{b}_1(t)$ in Figure 2.25, this realization is a nontrivial generalization of the realization in the time-invariant case (see Figure 2.9). Of course, the realization in Figure 2.25 does reduce to the realization in the time-invariant case when $\dot{a}_1(t)$ and $\dot{b}_1(t)$ are both zero.

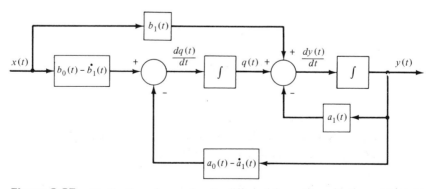

Figure 2.25. Realization of second-order differential equation with time-varying coefficients.

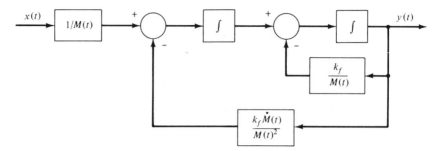

Figure 2.26.   Realization of the car with time-varying mass $M(t)$.

EXAMPLE 2.5.   Again consider the automobile studied in Section 1.5. Due to fuel consumption, the mass M of the car actually varies with time t. If the percent change in the mass is small, the variance in M can be neglected. But in some physical systems, the change in mass cannot be neglected. An example is a rocket or spacecraft, where much of the mass is fuel. In such cases it is usually not possible to neglect the variation in mass due to fuel consumption. If we allow the mass M of the car to vary with t, the input/output differential equation becomes

$$\frac{d^2y(t)}{dt^2} + \frac{k_f}{M(t)}\frac{dy(t)}{dt} = \frac{1}{M(t)}x(t). \tag{2.56}$$

Equation (2.56) is a special case of (2.52). In particular, if we take $a_1(t) = k_f/M(t)$, $a_0(t) = 0$, $b_1(t) = 0$, and $b_0(t) = 1/M(t)$, then (2.52) is identical to (2.56). Thus, with these expressions for the coefficients, the realization in Figure 2.25 is a realization of the car with a time-varying mass $M(t)$. The realization is shown in Figure 2.26. Note that the realization shown here is more complicated than the realization in the constant-mass case (see Figure 2.10). Clearly, the realization in Figure 2.26 does reduce to the realization in the constant-mass case.

Linear nth-order input/output differential equations with time-varying coefficients can also be realized using integrators and time-varying scalar multipliers. The gains in the time-varying scalar multipliers are functions of the coefficients and the derivatives of the coefficients of the input/output differential equation. We can derive a general expression for the gains in a realization, but omit the details.

★**2.5**

NONLINEAR INPUT/OUTPUT DIFFERENTIAL EQUATIONS AND LINEARIZATION

Many nonlinear systems arising in applications are specified by a nonlinear input/output differential equation. We shall first consider a nonlinear time-invariant system given by the first-order input/output differential equation

$$\frac{dy(t)}{dt} = f(y(t), x(t)), \qquad t > t_0. \tag{2.57}$$

Here $f(y(t), x(t))$ is a function of the output $y(t)$ at time t and the input $x(t)$ at time t.

Recall from Chapter 1 that the system defined by (2.57) is linear if and only if (2.57) is a linear differential equation. The differential equation (2.57) is linear if and only if there exist constants a and b such that

$$f(y(t), x(t)) = -ay(t) + bx(t). \tag{2.58}$$

In this case, the input/output differential equation (2.57) becomes

$$\frac{dy(t)}{dt} = -ay(t) + bx(t).$$

This equation was studied in Section 2.1.

Let us now suppose that the system defined by (2.57) is nonlinear so that the function f cannot be expressed in the form (2.58). In general, due to the nonlinearity it is not possible to derive an analytical expression for the output response $y(t)$ resulting from an initial condition $y(t_0)$ and an input $x(t)$. However, it is sometimes possible to build a realization of the system using an integrator and signal multipliers, as shown in the following example.

EXAMPLE 2.6. Consider the system with the nonlinear input/output differential equation

$$\frac{dy(t)}{dt} = -y^2(t) + x(t). \tag{2.59}$$

In this example the function $f(y(t), x(t))$ is given by

$$f(y(t), x(t)) = -y^2(t) + x(t).$$

There is no general expression for the solution $y(t)$ to (2.59) resulting from an arbitrary input $x(t)$. Nevertheless, the system does have the realization shown in Figure 2.27. By realizing the signal multiplier shown here (e.g.,

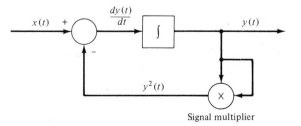

Figure 2.27. Realization of a nonlinear input/output differential equation.

using diodes and op amps), we could build the system and use this as a simulation of the given system. In particular, we could determine the response of the given system to an input $x(t)$ by applying $x(t)$ to the simulation and measuring the resulting output response $y(t)$.

LINEARIZATION   A common way of treating nonlinear systems is to approximate the system by a linear model that is valid for some range of operating values. One procedure for doing this is based on an approximation of system behavior with respect to some nominal operation. This technique is called *linearization*. We shall first describe the linearization procedure for a system given by the first-order nonlinear differential equation (2.57).

FIRST-ORDER CASE   Let $y_{nom}(t)$ denote the solution to (2.57) when the input $x(t)$ is equal to a given function $x_{nom}(t)$, called the *nominal input function*, and where the initial condition $y_{nom}(0)$ is some given real number y_0, called the *nominal initial condition*. In other words, we have that

$$\frac{dy_{nom}(t)}{dt} = f(y_{nom}(t), x_{nom}(t)), \qquad t > 0, \qquad (2.60)$$

with initial condition $y_{nom}(0) = y_0$.

The function $y_{nom}(t)$ is called the *nominal output response* resulting from the nominal input function $x_{nom}(t)$ and nominal initial condition y_0. The nominal response $y_{nom}(t)$ is often computed using a numerical solution technique for nonlinear differential equations. The function $y_{nom}(t)$ must be computed before we can apply the linearization procedure.

Now the idea is to linearize the system with respect to the nominal behavior given by the nominal functions $x_{nom}(t)$ and $y_{nom}(t)$. Suppose that the new input $x(t)$ is given by

$$x(t) = x_{nom}(t) + \Delta x(t), \qquad (2.61)$$

and the new initial condition is

$$y(0) = y_0 + \Delta y_0. \qquad (2.62)$$

In (2.61) and (2.62), we are assuming that the magnitudes of $\Delta x(t)$ and Δy_0 are small in comparison with the magnitudes of $x_{nom}(t)$ and $y_{nom}(0)$. The input $\Delta x(t)$ and initial condition Δy_0 are referred to as *perturbations* in the nominal input $x_{nom}(t)$ and nominal initial condition y_0, respectively.

The output response $y(t)$ resulting from the input $x(t) = x_{nom}(t) + \Delta x(t)$ and initial condition $y(0) = y_0 + \Delta y_0$ can be written in the form

$$y(t) = y_{nom}(t) + \Delta y(t). \qquad (2.63)$$

In (2.63), $\Delta y(t)$ is the perturbation in the nominal output response $y_{nom}(t)$ resulting from the perturbations $\Delta x(t)$ and Δy_0 in the nominal input and initial condition. The output $y(t)$ is the solution to (2.57) with input $x(t)$ given by (2.61) and with initial condition $y(0)$ given by (2.62). Hence

$$\frac{dy(t)}{dt} = \frac{dy_{nom}(t)}{dt} + \frac{d\,\Delta y(t)}{dt} = f(y_{nom}(t) + \Delta y(t), x_{nom}(t) + \Delta x(t)). \quad (2.64)$$

Under certain conditions on the function f, we can expand

$$f(y_{nom}(t) + \Delta y(t), x_{nom}(t) + \Delta x(t))$$

into a Taylor series about the nominal functions $y_{nom}(t)$ and $x_{nom}(t)$. The result is

$$\begin{aligned} f(y_{nom}(t) &+ \Delta y(t), x_{nom}(t) + \Delta x(t)) \\ &= f(y_{nom}(t), x_{nom}(t)) + c(t)\,\Delta y(t) + d(t)\,\Delta x(t) \\ &+ \text{higher-order terms involving } (\Delta y(t))^2, \Delta y(t)\,\Delta x(t), (\Delta x(t))^2, \text{ etc.} \quad (2.65) \end{aligned}$$

In equation (2.65), $c(t)$ and $d(t)$ are the partial derivatives of f with respect to x and y, with the partial derivatives evaluated at the nominal functions. In mathematical terms, we have

$$c(t) = \left.\frac{\partial f}{\partial y}\right|_{\substack{y(t)=y_{nom}(t) \\ x(t)=x_{nom}(t)}} \quad (2.66)$$

$$d(t) = \left.\frac{\partial f}{\partial x}\right|_{\substack{y(t)=y_{nom}(t) \\ x(t)=x_{nom}(t)}} \quad (2.67)$$

If the magnitudes of $\Delta y(t)$ and $\Delta x(t)$ are suitably small, we can neglect the higher-order terms in (2.65), which gives

$$\begin{aligned} f(y_{nom}(t) + \Delta y(t), x_{nom}(t) + \Delta x(t)) &= f(y_{nom}(t), x_{nom}(t)) \\ &+ c(t)\,\Delta y(t) + d(t)\,\Delta x(t). \quad (2.68) \end{aligned}$$

Now inserting (2.68) into (2.64), we have

$$\frac{dy_{nom}(t)}{dt} + \frac{d\,\Delta y(t)}{dt} = f(y_{nom}(t), x_{nom}(t)) + c(t)\,\Delta y(t) + d(t)\,\Delta x(t). \quad (2.69)$$

Using (2.60), we have that (2.69) simplifies to

$$\frac{d\,\Delta y(t)}{dt} = c(t)\,\Delta y(t) + d(t)\,\Delta x(t). \quad (2.70)$$

The linear differential equation (2.70) is called the *linearized equation* with respect to the nominal functions $y_{nom}(t)$ and $x_{nom}(t)$. The representation (2.70) describes the perturbation in system behavior due to the perturbation $\Delta x(t)$ in the nominal input $x_{nom}(t)$ and the perturbation Δy_0 in the nominal initial condition $y_{nom}(0)$. The input/output differential equation (2.70) is an accurate model of system behavior as long as the magnitudes of $\Delta y(t)$ and $\Delta x(t)$ are small in comparison with the magnitudes of $x_{nom}(t)$ and $y_{nom}(t)$.

Since the differential equation (2.70) is linear, it can be studied using linear techniques. This is illustrated in the following example.

EXAMPLE 2.7. Consider a high-speed vehicle on a horizontal surface. The vehicle is described by the input/output differential equation

$$M \frac{d^2y(t)}{dt^2} + k_f \frac{dy(t)}{dt} + k_d \left[\frac{dy(t)}{dt} \right]^2 = x(t). \tag{2.71}$$

In (2.71) $y(t)$ is the position of the vehicle at time t, $x(t)$ is the drive or braking force at time t, M is the mass of the vehicle, k_f is the coefficient of friction, and k_d is the drag coefficient due to air resistance. Note that (2.71) is the same as the input/output differential equation of the car in Example 2.1, except that (2.71) contains an additional term due to air resistance. Also note that the air resistance term depends on the square of velocity, and thus (2.71) is a nonlinear differential equation. We shall linearize the system with respect to the nominal operation consisting of the vehicle moving at a constant velocity v_0. We shall be concerned only with the velocity and not the position, and thus we shall define the output to be the velocity $v(t)$ of the vehicle. The vehicle is then described by the velocity model given by

$$\frac{dv(t)}{dt} + \frac{k_f}{M} v(t) + \frac{k_d}{M} v^2(t) = \frac{1}{M} x(t). \tag{2.72}$$

We shall linearize the velocity model with respect to a constant nominal velocity v_0; that is, we want

$$v_{nom}(t) = v_0 \qquad \text{for all } t \geq 0. \tag{2.73}$$

The nominal output given by (2.73) is produced by applying the nominal input function

$$x_{nom}(t) = (k_f + k_d v_0)v_0, \qquad t \geq 0. \tag{2.74}$$

with the nominal initial condition

$$v_{nom}(0) = v_0.$$

To verify this, simply check that the differential equation (2.72) is satisfied with $v_{nom}(t)$ and $x_{nom}(t)$ given by (2.73) and (2.74), respectively.

Now in this example,

$$f(v(t), x(t)) = -\frac{k_f}{M} v(t) - \frac{k_d}{M} v^2(t) + \frac{1}{M} x(t).$$

Then

$$\frac{\partial f}{\partial v} = -\frac{k_f}{M} - 2 \frac{k_d}{M} v(t) \tag{2.75}$$

$$\frac{\partial f}{\partial x} = \frac{1}{M}. \tag{2.76}$$

Evaluating the partial derivatives (2.75) and (2.76) at the nominal functions given by (2.73) and (2.74), from (2.66) and (2.67), we have

$$c(t) = -\frac{1}{M} (k_f + 2k_d v_0)$$

$$d(t) = \frac{1}{M}.$$

Note that in this example, both $c(t)$ and $d(t)$ are constant. The linearized equation is then

$$\frac{d \, \Delta v(t)}{dt} = -\frac{1}{M} (k_f + 2k_d v_0) \, \Delta v(t) + \frac{1}{M} \Delta x(t). \tag{2.77}$$

The differential equation (2.77) describes the perturbed behavior of the vehicle resulting from some small perturbation $\Delta x(t)$ in the nominal input $x_{\text{nom}}(t)$ given by (2.74) and/or some small perturbation Δv_0 in the nominal velocity v_0. Since (2.77) is a first-order linear differential equation with constant coefficients, we can solve (2.77) using the general form derived in Section 2.1. For example, suppose that $\Delta x(t)$ is a small step change in the drive force applied to the vehicle; that is,

$$\Delta x(t) = \epsilon u(t),$$

where $|\epsilon|$ is a small number. Taking $\Delta v_0 = 0$, we have that the resulting solution $\Delta v(t)$ to the linearized equation (2.77) is

$$
\begin{aligned}
\Delta v(t) &= \int_0^t \exp\left[\frac{-1}{M}(k_f + 2k_d v_0)(t - \lambda)\right] \frac{1}{M} \epsilon \, d\lambda \\
&= \frac{\epsilon}{k_f + 2k_d v_0} \exp\left[\frac{-1}{M}(k_f + 2k_d v_0)(t - \lambda)\right]_{\lambda=0}^{\lambda=t} \\
&= \frac{\epsilon}{k_f + 2k_d v_0} \left\{ 1 - \exp\left[\frac{-1}{M}(k_f + 2k_d v_0)t\right] \right\}, \qquad t \geq 0. \tag{2.78}
\end{aligned}
$$

The expression (2.78) is the change in the velocity from the nominal v_0 resulting from the perturbation $\epsilon u(t)$ in the force applied to the vehicle. The total velocity $v(t)$ of the vehicle resulting from the force $x(t) = x_{\text{nom}}(t) + \epsilon u(t)$ is given by

$$v(t) = v_0 + \Delta v(t)$$

$$= v_0 + \frac{\epsilon}{k_f + 2k_d v_0} \left\{ 1 - \exp\left[\frac{-1}{M} (k_f + 2k_d v_0)t \right] \right\}, \quad t \geq 0.$$

SECOND-ORDER CASE Now suppose that the system under study is given by the second-order input/output differential equation

$$\frac{d^2 y(t)}{dt^2} = f\left(y(t), \frac{dy(t)}{dt}, x(t), \frac{dx(t)}{dt} \right). \tag{2.79}$$

The system defined by (2.79) is linear if and only if there exist constants a_1, a_0, b_1, b_0 such that

$$f\left(y(t), \frac{dy(t)}{dt}, x(t), \frac{dx(t)}{dt} \right) = -a_0 y(t) - a_1 \frac{dy(t)}{dt} + b_0 x(t) + b_1 \frac{dx(t)}{dt}. \tag{2.80}$$

Let us suppose that the system is nonlinear, so that f cannot be written in the form (2.80). We shall linearize the system with respect to a given nominal input function $x_{\text{nom}}(t)$ and nominal initial conditions $y_{\text{nom}}(0)$, $\dot{y}_{\text{nom}}(0)$. The resulting nominal output function will be denoted by $y_{\text{nom}}(t)$. As in the first-order case, the nominal response $y_{\text{nom}}(t)$ is often computed using numerical techniques for solving nonlinear differential equations. Since the nominal output $y_{\text{nom}}(t)$ satisfies the differential equation (2.79) when the input $x(t)$ is equal to the nominal input $x_{\text{nom}}(t)$, we have that

$$\frac{d^2 y_{\text{nom}}(t)}{dt^2} = f\left(y_{\text{nom}}(t), \frac{dy_{\text{nom}}(t)}{dt}, x_{\text{nom}}(t), \frac{dx_{\text{nom}}(t)}{dt} \right). \tag{2.81}$$

Now let $\Delta x(t)$ be a perturbation in the nominal input function $x_{\text{nom}}(t)$, and let Δy_0 and $\Delta \dot{y}_0$ be perturbations in the nominal initial conditions $y_{\text{nom}}(0)$ and $\dot{y}_{\text{nom}}(0)$. The resulting output response $y(t)$ can be expressed in the form

$$y(t) = y_{\text{nom}}(t) + \Delta y(t), \tag{2.82}$$

where $\Delta y(t)$ is the perturbation in the nominal output $y_{\text{nom}}(t)$ resulting from the perturbations in the nominal input and nominal initial conditions.

We then have that

$$\frac{d^2 y(t)}{dt^2} = \frac{d^2 y_{nom}(t)}{dt^2} + \frac{d^2 \, \Delta y(t)}{dt^2}$$

$$= f\left(y_{nom}(t) + \Delta y(t), \frac{dy_{nom}(t)}{dt} + \frac{d \, \Delta y(t)}{dt}, \right. \tag{2.83}$$

$$\left. x_{nom}(t) + \Delta x(t), \frac{dx_{nom}(t)}{dt} + \frac{d \, \Delta x(t)}{dt}\right).$$

Under certain conditions on f, we can expand

$$f\left(y_{nom}(t) + \Delta y(t), \frac{dy_{nom}(t)}{dt} + \frac{d \, \Delta y(t)}{dt}, x_{nom}(t)\right.$$

$$\left. + \Delta x(t), \frac{dx_{nom}(t)}{dt} + \frac{d \, \Delta x(t)}{dt}\right)$$

into a Taylor series about $y_{nom}(t)$, $dy_{nom}(t)/dt$, $x_{nom}(t)$, $dx_{nom}(t)/dt$. This gives

$$f\left(y_{nom}(t) + \Delta y(t), \frac{dy_{nom}(t)}{dt} + \frac{d \, \Delta y(t)}{dt}, x_{nom}(t) + \Delta x(t), \frac{dx_{nom}(t)}{dt} + \frac{d \, \Delta x(t)}{dt}\right)$$

$$= f\left(y_{nom}(t), \frac{dy_{nom}(t)}{dt}, x_{nom}(t), \frac{dx_{nom}(t)}{dt}\right) + c_0(t) \, \Delta y(t) \tag{2.84}$$

$$+ c_1(t) \frac{d \, \Delta y(t)}{dt} + d_0(t) \, \Delta x(t) + d_1(t) \frac{d \, \Delta x(t)}{dt} + \text{higher-order terms},$$

where

$$c_0(t) = \left.\frac{\partial f}{\partial y}\right|_{\substack{y=y_{nom} \\ x=x_{nom}}}$$

$$c_1(t) = \left.\frac{\partial f}{\partial \dot{y}}\right|_{\substack{y=y_{nom} \\ x=x_{nom}}}$$

$$d_0(t) = \left.\frac{\partial f}{\partial x}\right|_{\substack{y=y_{nom} \\ x=x_{nom}}}$$

$$d_1(t) = \left.\frac{\partial f}{\partial \dot{x}}\right|_{\substack{y=y_{nom} \\ x=x_{nom}}}$$

If the perturbations are suitably small in magnitude, we can neglect the higher-order terms in (2.84). Then inserting (2.84) into (2.83) and using (2.81), we get

$$\frac{d^2 \, \Delta y(t)}{dt^2} = c_0(t) \, \Delta y(t) + c_1(t) \frac{d \, \Delta y(t)}{dt} \tag{2.85}$$

$$+ d_0(t) \, \Delta x(t) + d_1(t) \frac{d \, \Delta x(t)}{dt}.$$

Equation (2.85) is the linearized input/output differential equation with respect to the nominal input function $x_{nom}(t)$ and nominal output $y_{nom}(t)$.

EXAMPLE 2.8.   Consider the simple pendulum that was studied in Section 1.5 (see Figure 1.21). Recall that the input/output differential equation of the pendulum is

$$I \frac{d^2\theta(t)}{dt^2} + MgL \sin \theta(t) = Lx(t). \tag{2.86}$$

In this example the output $y(t)$ is the angle $\theta(t)$ between the pendulum and the vertical reference. We can write (2.86) in the form

$$\frac{d^2\theta(t)}{dt^2} = f(\theta(t), x(t)) = -\frac{MgL}{I} \sin \theta(t) + \frac{L}{I} x(t). \tag{2.87}$$

Let us take the nominal input function $x_{nom}(t)$ to be the zero function and the nominal initial conditions to be zero also. With this input and initial conditions, the solution $\theta_{nom}(t)$ to (2.87) is the zero solution; that is,

$$\theta_{nom}(t) = 0 \qquad \text{for all } t \geq 0.$$

In other words, the nominal behavior is the pendulum resting in the vertical position. Now consider the input

$$x(t) = x_{nom}(t) + \Delta x(t) = \Delta x(t)$$

and the initial conditions

$$\theta(0) = \Delta\theta_0$$

$$\dot{\theta}(0) = \Delta\dot{\theta}_0.$$

Now since $f(\theta(t), x(t))$ does not depend on $\dot{\theta}(t)$ and $\dot{x}(t)$, we have that

$$\frac{\partial f}{\partial \dot{\theta}} = 0 \qquad \text{and} \qquad \frac{\partial f}{\partial \dot{x}} = 0.$$

In addition,

$$c_0(t) = \frac{\partial f}{\partial \theta}\bigg|_{\substack{\theta=0 \\ x=0}} = -\frac{MgL}{I} \cos \theta(t)\bigg|_{\substack{\theta=0 \\ x=0}} = -\frac{MgL}{I}$$

$$d_0(t) = \frac{\partial f}{\partial x}\bigg|_{\substack{\theta=0 \\ x=0}} = \frac{L}{I}.$$

Hence the linearized equation is

$$\frac{d^2 \,\Delta\theta(t)}{dt^2} = -\frac{MgL}{I}\,\Delta\theta(t) + \frac{L}{I}\,\Delta x(t). \tag{2.88}$$

Note that (2.88) is the same as the small-signal model given in Section 1.5.

Now suppose that we perturb the nominal initial position by amount α, where $|\alpha|$ is a small number, but we do not perturb the nominal input force $x_{nom}(t)$ or the nominal initial velocity $\dot\theta_0$ (both of which are zero). This corresponds to someone moving the pendulum to the angular position α at time $t = 0$ and then releasing the pendulum without imparting any initial velocity to it. The resulting perturbed response $\Delta\theta(t)$ is given by

$$\Delta\theta(t) = \alpha \cos\left(\sqrt{\frac{MgL}{I}}\,t\right), \qquad t \geq 0. \tag{2.89}$$

To verify that (2.89) is the perturbed response, check that (2.89) satisfies the linearized equation (2.88) when $\Delta x(t) = 0$.

From (2.89), we see that when the pendulum is released at the initial angle α, it oscillates about the vertical position at a frequency of $\sqrt{MgL/I}$ rad/sec. It should be stressed that this result is valid only if the magnitude $|\alpha|$ of the initial angle is small.

Let's now change the nominal conditions: Suppose that the nominal output $\theta_{nom}(t)$ is

$$\theta_{nom}(t) = \frac{\pi}{2}\text{ rad}, \qquad t \geq 0. \tag{2.90}$$

The nominal input force $x_{nom}(t)$ that produces the nominal output (2.90) is given by

$$x_{nom}(t) = Mg, \qquad t \geq 0. \tag{2.91}$$

To show that $x_{nom}(t)$ given by (2.91) does produce the nominal output given by (2.90), simply insert (2.90) and (2.91) into the differential equation (2.86) and check that (2.86) is satisfied. Note that the nominal input force given by (2.91) holds the pendulum in a horizontal position (90° from the vertical resting position).

Evaluating the partial derivatives of f with respect to the nominal functions (2.90) and (2.91), we obtain

$$c_0(t) = \frac{\partial f}{\partial \theta}\bigg|_{\substack{\theta=\pi/2 \\ x=Mg}} = -\frac{MgL}{I}\cos\theta(t)\bigg|_{\substack{\theta=\pi/2 \\ x=Mg}} = 0$$

$$d_0(t) = \frac{\partial f}{\partial x}\bigg|_{\substack{\theta=\pi/2 \\ x=Mg}} = \frac{L}{I}.$$

The resulting linearized equation is

$$\frac{d^2 \, \Delta\theta(t)}{dt^2} = \frac{L}{I} \, \Delta x(t).$$ (2.92)

Equation (2.92) is the linearized equation with respect to the horizontal nominal position, while (2.88) is the linearized equation with respect to the vertical nominal position. Since (2.92) and (2.88) differ significantly, we would expect that the behavior of the pendulum resulting from a small perturbation in the nominal vertical position is quite different from the behavior of the pendulum resulting from a small perturbation in the nominal horizontal position. That this is the case is seen from the following example.

In the linearized model (2.92), let us take $\Delta x(t) = 0$, $\Delta\theta_0 = 0$, and $\Delta\dot{\theta}_0 = \beta$, where $|\beta|$ is a small number. This corresponds to giving the pendulum a small initial velocity (say, by hitting it with an object) at time $t = 0$ when the pendulum is in the nominal horizontal position. The resulting solution of (2.92) is

$$\Delta\theta(t) = \beta t, \qquad t \geq 0.$$ (2.93)

[Insert (2.93) into (2.92) to verify that it is the solution.] From (2.93) we see that a small perturbation β in the nominal initial velocity causes the pendulum to "fly off" from the horizontal position. In contrast, with $\Delta x(t) = 0$, $\Delta\theta_0 = 0$, and $\Delta\dot{\theta}_0 = \beta$, the solution of the linearized equation (2.88) is

$$\Delta\theta(t) = \beta \sqrt{\frac{I}{MgL}} \, \sin\left(\sqrt{\frac{MgL}{I}} \, t\right), \qquad t \geq 0.$$ (2.94)

From (2.94) we have that the pendulum oscillates with frequency $\sqrt{MgL/I}$ in response to a small perturbation β in the nominal initial velocity when the pendulum is in the vertical resting position.

Now suppose that the pendulum is attached to a pivot so that the pendulum can move through a $360°$ circle. In this case we can take the nominal output $\theta_{nom}(t)$ to be

$$\theta_{nom}(t) = \pi \text{ rad}, \qquad t \geq 0.$$

The linearization of the pendulum with respect to this nominal position is considered in the problems (see Problem 2.21).

nTH-ORDER CASE   Now consider a nonlinear system given by the nth-order input/output differential equation

$$y^{(n)}(t) = f(y(t), y^{(1)}(t), \ldots, y^{(n-1)}(t), x(t), x^{(1)}(t), \ldots, x^{(m)}(t)),$$ (2.95)

where $y^{(i)}(t)$ and $x^{(i)}(t)$ are the ith derivatives of $y(t)$ and $x(t)$, respectively. We shall linearize the system with respect to a nominal input function $x_{nom}(t)$ and nominal initial conditions $y^{(i)}_{nom}(0)$, $i = 0, 1, \ldots, n - 1$. The resulting nominal response is denoted by $y_{nom}(t)$.

Now consider the input

$$x(t) = x_{nom}(t) + \Delta x(t),$$

and initial conditions

$$y^{(i)}(0) = y^{(i)}_{nom}(0) + \Delta y_i, \qquad i = 0, 1, 2, \ldots, n - 1.$$

Writing the output $y(t)$ in the form

$$y(t) = y_{nom}(t) + \Delta y(t),$$

we have the linearized equation

$$\frac{d^n \Delta y(t)}{dt^n} = \sum_{i=0}^{n-1} c_i(t) \frac{d^i \Delta y(t)}{dt^i} + \sum_{i=0}^{m} d_i(t) \frac{d^i \Delta x(t)}{dt^i}, \qquad (2.96)$$

where

$$c_i(t) = \left. \frac{\partial f}{\partial y^{(i)}} \right|_{\substack{y = y_{nom}, \\ x = x_{nom}}} \qquad i = 0, 1, 2, \ldots, n - 1$$

$$d_i(t) = \left. \frac{\partial f}{\partial x^{(i)}} \right|_{\substack{y = y_{nom}, \\ x = x_{nom}}} \qquad i = 0, 1, 2, \ldots, m.$$

The linearized equation (2.96) can then be studied using techniques for linear differential equations with time-varying coefficients.

★2.6

MULTI-INPUT MULTI-OUTPUT SYSTEMS

Finite-dimensional continuous-time systems with more than one output terminal and/or more than one input terminal can be modeled by a collection of coupled input/output differential equations. For example, consider a two-input two-output finite-dimensional system with inputs $x_1(t)$, $x_2(t)$ and outputs $y_1(t)$, $y_2(t)$. If the system is linear and time invariant, it can be modeled by the input/output differential equations

$$\frac{d^n y_1(t)}{dt^n} + \sum_{i=0}^{n-1} a_{i1} \frac{d^i y_1(t)}{dt^i} + \sum_{i=0}^{n} a_{i2} \frac{d^i y_2(t)}{dt^i} = \sum_{i=0}^{n} b_{i1} \frac{d^i x_1(t)}{dt^i} + \sum_{i=0}^{n} b_{i2} \frac{d^i x_2(t)}{dt^i}$$

$$(2.97)$$

$$\frac{d^n y_2(t)}{dt^i} + \sum_{i=0}^{n-1} a_{i3} \frac{d^i y_2(t)}{dt^i} + \sum_{i=0}^{n} a_{i4} \frac{d^i y_1(t)}{dt^i} = \sum_{i=0}^{n} b_{i3} \frac{d^i x_2(t)}{dt^i} + \sum_{i=0}^{n} b_{i4} \frac{d^i x_1(t)}{dt^i}.$$

$$(2.98)$$

In (2.97) and (2.98), the coefficients a_{i1}, a_{i2}, a_{i3}, a_{i4}, b_{i1}, b_{i2}, b_{i3}, b_{i4} are constants.

EXAMPLE 2.9. Again consider the two-car system in Example 1.20. The system is illustrated in Figure 2.28. The inputs to the system are the forces $x_1(t)$, $x_2(t)$ applied to the two cars. The outputs are the velocity $v_1(t)$ of the first car and the distance

$$w(t) = d_2(t) - d_1(t) \tag{2.99}$$

between the two cars. In (2.99), $d_1(t)$ and $d_2(t)$ are the positions of the two cars at time t. We shall compute the input/output differential equations of the two-car system. First, we have the velocity model of the first car given by

$$\frac{dv_1(t)}{dt} + \frac{k_f}{M} v_1(t) = \frac{1}{M} x_1(t). \tag{2.100}$$

The differential equation (2.100) can be written in the form (2.97) by defining

$$a_{01} = \frac{k_f}{M}, \qquad a_{02} = 0, \qquad a_{12} = 0,$$

and

$$b_{01} = \frac{1}{M}, \qquad b_{10} = 0, \qquad b_{02} = 0, \qquad b_{12} = 0.$$

Hence (2.100) is the first part of the input/output differential-equation representation. Now differentiating both sides of (2.99), we obtain

$$\frac{dw(t)}{dt} = v_2(t) - v_1(t). \tag{2.101}$$

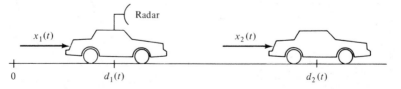

Figure 2.28. Two-car system.

The differential equation (2.101) is not in the form (2.98) due to the $v_2(t)$ term. However, we can eliminate $v_2(t)$ by first differentiating both sides of (2.101). This gives

$$\frac{d^2w(t)}{dt^2} = \frac{dv_2(t)}{dt} - \frac{dv_1(t)}{dt}. \tag{2.102}$$

From the velocity model for the second car, we have

$$\frac{dv_2(t)}{dt} = -\frac{k_f}{M} v_2(t) + \frac{1}{M} x_2(t). \tag{2.103}$$

Inserting (2.103) into (2.102), we get

$$\frac{d^2w(t)}{dt^2} = -\frac{k_f}{M} v_2(t) + \frac{1}{M} x_2(t) - \frac{dv_1(t)}{dt}. \tag{2.104}$$

Rewriting (2.101), we have

$$v_2(t) = \frac{dw(t)}{dt} + v_1(t). \tag{2.105}$$

Finally, inserting (2.105) into (2.104), we obtain

$$\frac{d^2w(t)}{dt^2} = -\frac{k_f}{M} \left[\frac{dw(t)}{dt} + v_1(t) \right] + \frac{1}{M} x_2(t) - \frac{dv_1(t)}{dt}$$

$$= -\frac{k_f}{M} \frac{dw(t)}{dt} - \frac{k_f}{M} v_1(t) - \frac{dv_1(t)}{dt} + \frac{1}{M} x_2(t). \tag{2.106}$$

The differential equation (2.106) can be written in the form (2.98) by defining

$$a_{03} = 0, \qquad a_{13} = \frac{k_f}{M}, \qquad a_{04} = \frac{k_f}{M}, \qquad a_{14} = 1, \qquad a_{24} = 0,$$

and

$$b_{03} = \frac{1}{M}, \qquad b_{ij} = 0 \qquad \text{for all other } i, j.$$

Thus (2.106) is the second part of the input/output differential-equation representation of the two-car system.

PROBLEMS

Chapter 2

2.1   For the *RLC* circuit in Figure P2.1, find the input/output differential equation.

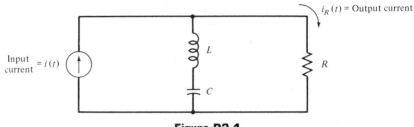

Figure P2.1

2.2   Find the input/output differential equation relating the input $i(t)$ to the output $v_C(t)$ for the circuit in Figure P2.2.

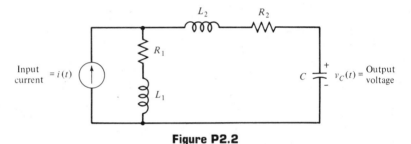

Figure P2.2

2.3   For the circuit in Figure P2.3, determine the input/output differential equation when:
(a) The output $y(t) = i_R(t)$.
(b) The output $y(t) = i_L(t)$.
(c) The output $y(t) = v_L(t)$.
(d) The output $y(t) = v_C(t)$.

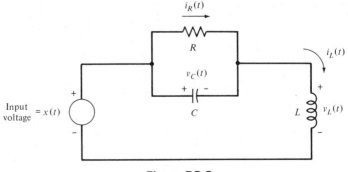

Figure P2.3

2.4   For the series *RLC* circuit in Figure P2.4, find the input/output differential equation when:

(a) $y(t) = v_R(t)$.
(b) $y(t) = i(t)$.
(c) $y(t) = v_L(t)$.
(d) $y(t) = v_C(t)$.

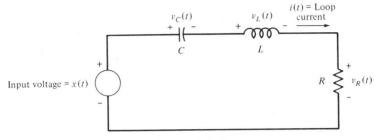

Figure P2.4

2.5 For the circuit in Figure P2.5, find the input/output differential equation when:
(a) $y(t) = i_1(t)$.
(b) $y(t) = i_2(t)$.
(c) $y(t) = i_3(t)$.
(d) $y(t) = v_C(t)$.

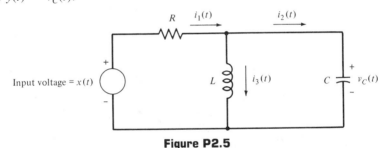

Figure P2.5

2.6 A mass M sits on top of a vibration absorber as illustrated in Figure P2.6. As shown in Figure P2.6, a force $x(t)$ (e.g., a vibrational force) is applied to the mass M, whose base is located at position $y(t)$. The force applied to the mass resulting from the springs is equal to $-Ky(t)$ and the force resulting from the dashpot is $-D\dot{y}(t)$. You may neglect the mass of the platform on which the mass sits. Derive the input/output differential equation of the system.

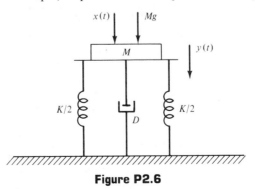

Figure P2.6

2.7 Find the input/output differential equation for each of the op-amp circuits shown in Figure P2.7.

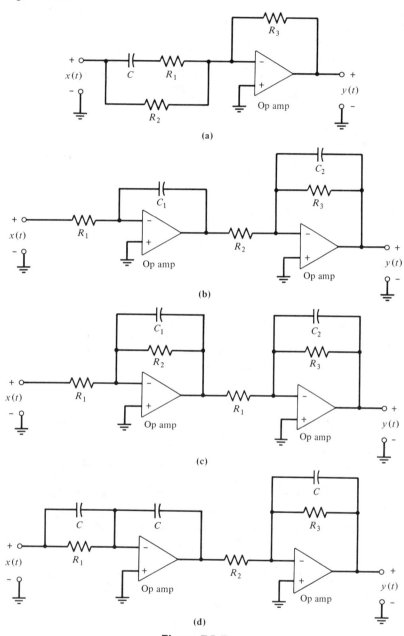

Figure P2.7

2.8 Using two op amps and resistors and capacitors (no inductors), realize each of the following input/output differential equations.

(a) $y(t) = b \dfrac{dx(t)}{dt}, \quad b > 0.$

(b) $y(t) = ax(t) + b\dfrac{dx(t)}{dt}, \quad a > 0 \text{ and } b > 0.$

(c) $y(t) = ax(t) + b\displaystyle\int_{-\infty}^{t} x(\lambda)\, d\lambda, \quad a > 0 \text{ and } b > 0.$

(d) $\dfrac{dy(t)}{dt} = -ay(t) + b\dfrac{dx(t)}{dt}, \quad a > 0 \text{ and } b > 0.$

2.9   Using two integrators and combinations of adders, subtracters, and scalar multipliers (no differentiators), realize each of the following input/output differential equations.

(a) $\dfrac{d^2y(t)}{dt^2} + y(t) = \dfrac{d^2x(t)}{dt^2}.$

(b) $\dfrac{d^2y(t)}{dt^2} = x(t) + \dfrac{d^2x(t)}{dt^2}.$

2.10   Using integrators, adders, subtracters, and scalar multipliers (no differentiators), realize the system with the input/output equation

$$\frac{d^2y(t)}{dt^2} + y(t) = x(t) + \frac{dx(t)}{dt} + \int_{-\infty}^{t} x(\lambda)\, d\lambda.$$

2.11   Using integrators, adders, subtracters, and scalar multipliers (no differentiators), realize the *RLC* circuit in Figure P2.1.

2.12   Repeat Problem 2.11 for the circuit in Figure P2.2.

2.13   Repeat Problem 2.11 for the op-amp circuit in Figure P2.7a.

2.14   Find the input/output differential equation for the systems shown in Figure P2.14.

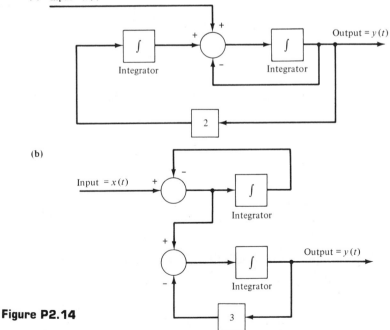

(a) Input = $x(t)$

(b)

Input = $x(t)$

Output = $y(t)$

Figure P2.14

2.15 Find the input/output differential equation for the system shown in Figure P2.15.

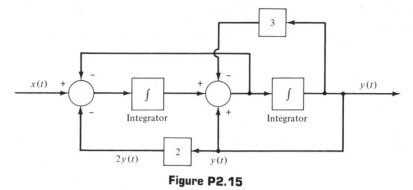

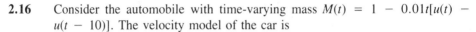

Figure P2.15

2.16 Consider the automobile with time-varying mass $M(t) = 1 - 0.01t[u(t) - u(t - 10)]$. The velocity model of the car is

$$M(t) \frac{dv(t)}{dt} + k_f v(t) = x(t),$$

where $v(t)$ is the velocity of the car, $x(t)$ is the drive or braking force, and k_f is the coefficient of friction. Suppose that $k_f = 0.1$. Compute the velocity $v(t)$ for all $t > 0$ when $v(0) = 0$ and $x(t) = 1$ for all $t \geq 0$. Your expression for $v(t)$ should be completely evaluated (i.e., all integrals should be evaluated).

2.17 Using *two* integrators, signal multipliers, adders, subtracters, and scalar multipliers (no differentiators), realize the following input/output differential equations.

(a) $\dfrac{d^2 y(t)}{dt^2} = \dfrac{dy(t)}{dt} y(t) + \dfrac{dx(t)}{dt}$.

(b) $\dfrac{d^2 y(t)}{dt^2} = \dfrac{dy(t)}{dt} \dfrac{dx(t)}{dt} - y(t)$.

2.18 Consider the continuous-time system shown in Figure P2.18. The output $f(y(t))$ of the saturating amplifier is given by

$$f(y(t)) = \begin{cases} 20y(t), & -2 < y(t) < 2 \\ 40, & y(t) > 2 \\ -40, & y(t) < -2. \end{cases}$$

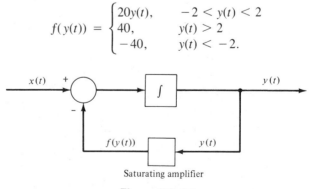

Saturating amplifier

Figure P2.18

(a) Compute the exact output response $y(t)$ for all $t > 0$ when $x(t) = 0$ for all $t \geq 0$ and $y(0) = 5$.

(b) Compute the exact output response $y(t)$ for all $t > 0$ when $x(t) = -t$ for all $t \geq 0$ and $y(0) = 0$.

2.19 Consider the nonlinear continuous-time system given by the input/output differential equation

$$\frac{dy(t)}{dt} + y^2(t) + ay(t) = bx(t),$$

where a and b are arbitrary constants.

(a) Suppose that $x(t) = 0$ for $t \geq 0$, and let $v(t) = 1/y(t)$. Show that $v(t)$ satisfies the differential equation

$$\frac{dv(t)}{dt} - av(t) = 1, \qquad t \geq 0.$$

(b) With $v(0)$ equal to an arbitrary constant, derive an expression for the solution $v(t)$ to the differential equation $\dot{v}(t) - av(t) = 1$.

(c) Using your result in part (b), derive an expression for the solution $y(t)$ to the differential equation

$$\frac{dy(t)}{dt} + y^2(t) + ay(t) = 0,$$

with initial condition $y(0)$.

(d) We continue to assume that $x(t) = 0$ for all $t \geq 0$. Determine all values of $y(0)$ such that $y(t) = \infty$ for some finite value of t (i.e., $0 < t < \infty$). Such values of t are called *finite escape times*.

2.20 Consider the nonlinear continuous-time system given by the input/output differential equation

$$\frac{dy(t)}{dt} + x(t)y(t) = x(t).$$

(a) Compute the nominal output response $y_{nom}(t)$ when the nominal input is $x_{nom}(t) = 1$ for all $t \geq 0$, with nominal initial condition $y_{nom}(0) = 0$.

(b) Compute the linearized equation with respect to the nominal functions in part (a).

(c) Repeat parts (a) and (b) with $x_{nom}(t) = 0$, $t \geq 0$, and $y_{nom}(0) = 1$.

2.21 Again consider the simple pendulum that was studied in Example 2.8.

(a) Compute the linearized equation with respect to the nominal functions $x_{nom}(t) = 0$ for all $t \geq 0$, $\theta_{nom}(t) = \pi$ radians for all $t \geq 0$.

(b) Using your result in part (a), determine the approximate response $\theta(t)$ when $\theta(0) = \pi + \alpha$, where $|\alpha|$ is a small number. We are assuming that $\dot{\theta}(0) = 0$ and $x(t) = 0$ for all $t \geq 0$.

(c) Using your result in part (a), compute the approximate response $\theta(t)$ when $\theta(0) = \pi$ and $\dot{\theta}(0) = \beta$, where $|\beta|$ is a small number. We are still assuming that $x(t) = 0$ for $t \geq 0$.

(d) Compare your results in parts (b) and (c) with the corresponding results in the cases $\theta_{nom}(t) = 0$ for $t \geq 0$ and $\theta_{nom}(t) = \pi/2$ for $t \geq 0$.

2.22 A continuous-time system is specified by the input/output differential equation

$$\frac{dy(t)}{dt} = -y^2(t) + x(t).$$

(a) Compute the linearized equation with respect to the nominal behavior $x_{nom}(t) = 0$ for all $t \geq 0$, and $y_{nom}(0) = 0$.

(b) Compute the nominal output response $y_{nom}(t)$ when $x_{nom}(t) = 0$ for all $t \geq 1$ and $y_{nom}(1) = 1$.

(c) Compute the linearized equation with respect to the nominal functions in part (b).

2.23 A continuous-time system has the input/output differential equation

$$\frac{d^2y(t)}{dt^2} + \left[\frac{dy(t)}{dt}\right]^2 + y(t) = y(t)x(t).$$

(a) Compute the linearized equation with respect to the nominal behavior $x_{nom}(t) = 0$ for all $t \geq 0$, $y_{nom}(0) = 0$.

(b) Using your result in part (a), find the approximate output response $y(t)$ when $x(t) = A, t \geq 0$, and $y(0) = y_0$. We are assuming that $|A|$ and $|y_0|$ are small, and that $\dot{y}(0) = 0$.

2.24 Consider the two-car system given by

$$\dot{v}_1(t) + \frac{k_f}{M} v_1(t) = \frac{1}{M} x_1(t)$$

$$\dot{v}(t) + \frac{k_f}{M} v_2(t) = \frac{1}{M} x_2(t)$$

$$\dot{w}(t) = v_2(t) - v_1(t).$$

Here $v_1(t)(v_2(t))$ is the velocity of the first (second) car, $x_1(t)(x_2(t))$ is the drive or braking force applied to the first (second) car, and $w(t)$ is the distance between the cars.

(a) Compute $v_1(t)$, $v_2(t)$, and $w(t)$ for *all* $t > 0$ when $v_1(0) = v_2(0) = 55$, and $w(0) = 100$. Assume that $x_1(t) = x_2(t) = 0$ for all $t \geq 0$.

(b) Now suppose that the first car is hit by a gust of wind given by the force

$$f(t) = \begin{cases} c, & 0 \leq t \leq 1 \\ 0, & \text{all other } t. \end{cases}$$

Take the model of the first car to be

$$\dot{v}_1(t) + \frac{k_f}{M} v_1(t) = \frac{1}{M} [x_1(t) - f(t)].$$

Compute $v_1(t)$, $v_2(t)$, and $w(t)$ for all $t > 0$ when $v_1(0) = v_2(0) = 55$, $w(0) = 100$, and $x_1(t) = x_2(t) = 0$ for all $t \geq 0$.

(c) In your results for part (b), for what range of values of c will the first car crash into the second car? Give all values of c for which this is the case.

Discrete-Time Systems Defined by an Input/Output Difference Equation

In this chapter we give the discrete-time counterpart to the input/output differential-equation representation studied in Chapter 2. We begin by considering discrete-time systems specified by a linear input/output difference equation with constant coefficients. It is shown that input/output difference equations can be solved very easily using recursion. A computer program is given for carrying out the recursion. The program is actually a *software realization* of the discrete-time system.

In Section 3.2 we show that discrete-time systems modeled by a linear input/output difference equation with constant coefficients can be realized from interconnections of unit delayers, adders, subtracters, and scalar multipliers. Such a realization is called a *hardware realization* of the discrete-time system. Results are then given on discrete-time systems whose input/output difference equation is nonlinear or has time-varying coefficients. In Section 3.4 we consider the discretization in time of input/output differential equations. By discretizing in time, we can study continuous-time systems using discrete-time representations.

3.1

LINEAR INPUT/OUTPUT DIFFERENCE EQUATIONS WITH CONSTANT COEFFICIENTS

Consider the single-input single-output discrete-time system given by the input/output difference equation

$$y(kT + nT) + \sum_{i=0}^{n-1} a_i y(kT + iT) = \sum_{i=0}^{m} b_i x(kT + iT). \qquad (3.1)$$

In (3.1), T is a fixed positive real number, k is a variable that takes its values from the set of integers, and the coefficients $a_0, a_1, \ldots, a_{n-1}, b_0, b_1, \ldots, b_m$ are constants. The discrete-time signal $x(kT)$ is the input applied to the system and the discrete-time signal $y(kT)$ is the output response of the system resulting from input $x(kT)$ with the n initial conditions $y(-T), y(-2T), \ldots, y(-nT)$. Note that we are taking the initial time to be zero.

The integer n in (3.1) is the order of the input/output difference equation. If the coefficient b_m in (3.1) is nonzero, the system defined by (3.1) is causal if and only if the integer m in (3.1) is less than or equal to the order n; that is, $m \leq n$. To see this, if we set $k = 0$ in (3.1) and solve for $y(nT)$, we have

$$y(nT) = -a_0 y(0) - a_1 y(T) - \cdots$$
$$- a_{n-1} y(nT - T) + b_0 x(0) + b_1 x(T) \qquad (3.2)$$
$$+ \cdots + b_m x(mT).$$

From (3.2) we see that the output $y(nT)$ at time nT depends on the value $x(mT)$ of the input at time mT. Hence, if $m > n$, the output $y(nT)$ at time nT depends on the value of the input at future time mT, and therefore the system is noncausal. From here on we shall always assume that $m \leq n$, so that the system is causal.

We can rewrite the input/output difference equation (3.1) in the form

$$y(kT + nT) = - \sum_{i=0}^{n-1} a_i y(kT + iT) + \sum_{i=0}^{m} b_i x(kT + iT). \qquad (3.3)$$

From (3.3) we see that $y(kT + nT)$ is a function of $y(kT + iT)$ for $i = 0, 1, \ldots, n - 1$ and a function of $x(kT + iT)$ for $i = 0, 1, \ldots, m$. Thus, by the definition of finite dimensionality given in Section 1.6, we have that the system defined by (3.1) [or (3.3)] is finite dimensional with order or dimension equal to n. Further, since (3.1) is a linear difference equation with constant coefficients, the system defined by (3.1) is linear and time invariant.

There is another form of the input/output difference equation (3.1) that is sometimes used. By shifting both sides of (3.1) to the right by nT seconds, we obtain

$$y(kT) + \sum_{i=0}^{n-1} a_i y(kT + iT - nT) = \sum_{i=0}^{m} b_i x(kT + iT - nT). \quad (3.4)$$

Some authors prefer to work with the form (3.4) rather than the form (3.1). We shall use the form (3.1) [or (3.3)] for the most part, although later we will need the form (3.4).

SOLUTION BY RECURSION   Unlike linear input/output differential equations, linear input/output difference equations can be solved by a direct numerical procedure. More precisely, the output $y(kT)$ for some finite range of integer values of k can be computed recursively as follows.

First, we rewrite (3.1) in the form (3.3). Setting $k = -n$ in (3.3), we have

$$y(0) = - \sum_{i=0}^{n-1} a_i y(-nT + iT) + \sum_{i=0}^{m} b_i x(-nT + iT),$$

or

$$y(0) = -a_0 y(-nT) - a_1 y(-nT + T) - \cdots$$
$$- a_{n-1} y(-T) + b_0 x(-nT) + b_1 x(-nT + T)$$
$$+ \cdots + b_m x(-nT + mT).$$

So the output $y(0)$ at time 0 is a linear combination of $y(-nT)$, $y(-nT + T)$, $\ldots$, $y(-T)$ and $x(-nT)$, $x(-nT + T)$, $\ldots$, $x(-nT + mT)$.

Setting $k = -n + 1$ in (3.3), we have

$$y(T) = - \sum_{i=0}^{n-1} a_i y(-nT + T + iT) + \sum_{i=0}^{m} b_i x(-nT + T + iT)$$

or

$$y(T) = -a_0 y(-nT + T) - a_1 y(-nT + 2T)$$
$$- \cdots - a_{n-1} y(0) + b_0 x(-nT + T)$$
$$+ b_1 x(-nT + 2T) + \cdots + b_m x(-nT + mT + T).$$

Hence the output $y(T)$ at time T is a linear combination of $y(-nT + T)$, $y(-nT + 2T)$, $\ldots$, $y(0)$ and $x(-nT + T)$, $x(-nT + 2T)$, $\ldots$, $x(-nT + T + mT)$.

If we continue, it is clear that the next value of the output is a linear combination of the n past values of the output and $m + 1$ values of the input. At each step of the computation, it is necessary to store only the n past values of the output (plus, of course, the input values). This process is called an nth-order recursion. Here the term *recursion* refers to the property that the next value of the output is computed from the n previous values of the output (plus input values). The discrete-time system given by (3.3) [or (3.1)] is sometimes called a *recursive discrete-time system* or a *recursive digital filter* since its output can be computed recursively.

It is important to note that the recursive structure described above is a result of the assumption that the system is finite dimensional. If the system is infinite dimensional, the output response cannot be computed recursively. Thus infinite-dimensional discrete-time systems are sometimes referred to as *nonrecursive systems* or *nonrecursive digital filters*.

From the expression (3.3) for $y(kT + nT)$, we see that if $m = n$ the computation of $y(kT)$ for each integer value of k requires, in general, $2n + 1$ additions and $2n + 1$ multiplications. So the "computational complexity" of recursion is directly proportional to the order n of the recursion. In particular, note that the number of computations required to compute $y(kT)$ does not depend on k. When we get to the convolution model (Chapter 4), we shall see that the number of computations required to compute $y(kT)$ does depend on k.

EXAMPLE 3.1.   Suppose that the discrete-time system is given by the second-order difference equation

$$y(kT + 2T) - 1.5y(kT + T) + y(kT) = 2x(kT). \qquad (3.5)$$

Let us take the input $x(kT)$ to be the discrete-time unit step function $u(kT)$. We shall take the initial conditions to be $y(-T) = 1$, $y(-2T) = 2$. Now rewriting (3.5) in the form

$$y(kT + 2T) = 1.5y(kT + T) - y(kT) + 2x(kT), \qquad (3.6)$$

we shall compute the output values $y(0)$, $y(T)$, $y(2T)$, $y(3T)$ by solving (3.6) recursively. First, setting $k = -2$ in (3.6), we have

$$y(0) = 1.5y(-T) - y(-2T) + 2x(-2T)$$
$$= (1.5)(1) - 2 + 2(0) = -0.5.$$

Setting $k = -1$ in (3.6), we get

$$y(T) = 1.5y(0) - y(-T) + 2x(-T)$$
$$= (1.5)(-0.5) - 1 + 2(0) = -1.75.$$

Continuing, we have

$$y(2T) = 1.5y(T) - y(0) + 2x(0)$$
$$= 1.5(-1.75) + 0.5 + 2(1) = -0.125$$
$$y(3T) = 1.5y(2T) - y(T) + 2x(T)$$
$$= 1.5(-0.125) + 1.75 + 2(1) = 3.5625.$$

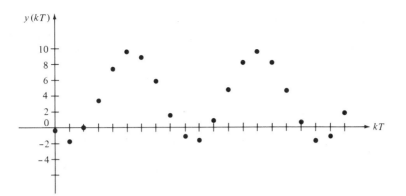

Figure 3.1.   Plot of output response in Example 3.1.

In solving (3.1) or (3.3) recursively, we can begin the process of computing the output $y(kT)$ at any desired initial time point. In the above development, the first value of the output that was computed was $y(0)$. If the first value of the output that we want to compute is the value $y(qT)$ at time qT, we can begin the recursion process by setting $k = -n + q$ in (3.3). For example, if the first value we want to compute is $y(nT)$, then $q = n$ and we set $k = 0$ in (3.3) to start the recursion. In this case, the initial output values are $y(nT - T)$, $y(nT - 2T), \ldots, y(0)$.

A program for solving the nth-order difference equation (3.1) is given in Figure A.3 of Appendix A. The program computes the output values $y(0)$, $y(T)$, $\ldots, y(QT)$ with the initial conditions $y(-T)$, $y(-2T), \ldots, y(-nT)$ and initial input values $x(-T)$, $x(-2T), \ldots, x(-nT)$. The input $x(kT)$ for $k = 0, 1, 2, \ldots$ may be given value by value or may be given in function form on line 320 of the program. As an example, the second-order difference equation in Example 3.1 was solved using this program with the number of time points equal to 20. A plot of the resulting output response $y(kT)$ is given in Figure 3.1.

ANALYTICAL EXPRESSIONS FOR SOLUTIONS   The recursion process will yield the values of $y(kT)$ for any finite range of integer values of k. If we want an analytical expression for the solution $y(k)$ to (3.1) or (3.3) that is valid for all integers $k \geq 0$, we must use techniques for solving linear difference equations. These techniques resemble methods for solving linear differential equations. Here we shall derive the solution to the general first-order linear difference equation

$$y(kT + T) = -ay(kT) + b_0 x(kT) + b_1 x(kT + T). \qquad (3.7)$$

Setting $k = -1$ in (3.7), we have

$$y(0) = -ay(-T) + b_0 x(-T) + b_1 x(0). \qquad (3.8)$$

Setting $k = 0$ in (3.7) gives

$$y(T) = -ay(0) + b_0x(0) + b_1x(T). \qquad (3.9)$$

Inserting the expression (3.8) for $y(0)$ into (3.9), we have

$$y(T) = a^2y(-T) - ab_0x(-T) - ab_1x(0) + b_0x(0) + b_1x(T).$$
$$= a^2y(-T) - ab_0x(-T) + (b_0 - ab_1)x(0) + b_1x(T). \qquad (3.10)$$

Setting $k = 1$ in (3.7) gives

$$y(2T) = -ay(T) + b_0x(T) + b_1x(2T). \qquad (3.11)$$

Inserting the expression (3.10) for $y(T)$ into (3.11), we have

$$y(2T) = -a^3y(-T) + a^2b_0x(-T)$$
$$- a(b_0 - ab_1)x(0) - ab_1x(T) + b_0x(T) + b_1x(2T)$$
$$= -a^3y(-T) + a^2b_0x(-T)$$
$$- a(b_0 - ab_1)x(0) + (b_0 - ab_1)x(T) + b_1x(2T). \qquad (3.12)$$

From the pattern in (3.10) and (3.12), we have that

$$y(kT) = (-a)^{k+1}y(-T) + (-a)^kb_0x(-T)$$
$$+ \sum_{i=0}^{k-1} [(-a)^{k-1-i}(b_0 - ab_1)x(iT)] + b_1x(kT), \qquad k \geq 1. \qquad (3.13)$$

Equations (3.8) and (3.13) give the complete output response $y(kT)$ for $k \geq 0$ resulting from initial condition $y(-T)$, initial input $x(-T)$, and input $x(kT)$ applied for $k = 0, 1, 2, \ldots$. This solution is the discrete-time counterpart to the solution (2.13) of the general first-order linear input/output differential equation (2.7). It is also worth pointing out that the solution (1.42) of the loan-repayment system in Chapter 1 is a special case of (3.13).

Analytical expressions for the solution of second-order and higher-order linear constant-coefficient difference equations can be derived in terms of matrix equations. As in the continuous-time case, the matrix formulation is constructed by defining a state model for the given system. We refer the reader to Chapter 13. In Chapter 7 we show that solutions to nth-order difference equations can be computed by applying the z-transform.

EXAMPLE 3.2. Consider the problem of raising hamsters starting from a single adult male/female pair. We assume that two male/female pairs are born every three months to each pair of adults. We also assume that male/female pairs remain permanently paired and that the hamsters do not die during the time period under consideration here. A newborn male and female hamster can produce offspring in six months. Now let $y(k)$ denote the number of male/female pairs at the end of the kth three-month period. We assume that $y(0) = 1$, so that $y(1) = 3$. For $k \geq 2$, $y(k) - y(k - 1)$ is the number of hamsters born at the end of the kth period. This must be equal to $2y(k - 2)$ [rather than $2y(k - 1)$] since newborn hamsters cannot have offspring for six months. Therefore, the process is described by the difference equation

$$y(k) - y(k - 1) = 2y(k - 2).$$

Replacing k by $k + 2$ in both sides, we get

$$y(k + 2) - y(k + 1) = 2y(k)$$

or

$$y(k + 2) = y(k + 1) + 2y(k).$$

Note that this process does not have an input. Solving this difference equation with $y(0) = 1$, $y(1) = 3$, we obtain the values $y(2) = 5$, $y(3) = 11$, $y(4) = 21$, $y(5) = 43$, ... We can also express $y(k)$ in the analytical form

$$y(k) = \tfrac{1}{3}[4(2^k) - (-1)^k], \qquad k \geq 2.$$

Solutions of this form can be generated by using z-transform techniques discussed in Chapter 7. To verify that the solution is correct, insert it into the difference equation and verify that the difference equation is satisfied. We invite the reader to check this.

MULTI-INPUT MULTI-OUTPUT DISCRETE-TIME SYSTEMS Multi-input multi-output linear time-invariant finite-dimensional discrete-time systems can be represented by a collection of coupled linear input/output difference equations with constant coefficients. These equations can be solved numerically by using a matrix version of the recursion process. We shall consider multi-input multi-output discrete-time systems from the state viewpoint in Chapter 13.

3.2

UNIT-DELAYER REALIZATIONS

In this section we show that any discrete-time system given by the input/output difference equation (3.1) can be realized by interconnecting unit delayers and discrete-time versions of adders, subtracters, and scalar multipliers. We begin with the definition of the unit delayer.

UNIT DELAYER A basic building block of discrete-time systems is the unit delayer, which is illustrated in Figure 3.2. As indicated, the response $y(kT)$ of the unit delayer to input $x(kT)$ is given by

$$y(kT) = x(kT - T), \qquad k = k_0, k_0 + 1, \ldots \qquad (3.14)$$

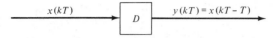

Figure 3.2. Unit delayer.

where k_0 is the initial time. From (3.14) we see that the unit delayer holds each value of the input signal for T seconds, and then outputs the input value at the next time point. The output $y(kT)$ of the unit delayer is a T-second right shift of the input signal $x(kT)$. For example, suppose that the input $x(kT)$ is the pulse shown in Figure 3.3a. From (3.14) we have that $y(kT) = 0$ for all integers $k < 0$ since $x(kT - T)$ is zero for all $k < 0$. Setting $k = 0, 1, 2, 3$ in (3.14), we have

$$y(0) = x(-T) = 0$$
$$y(T) = x(0) = 1$$
$$y(2T) = x(T) = 1$$
$$y(3T) = x(2T) = 1.$$

Finally, for all integer values of $k \geq 4$, we have that $y(kT) = 0$ since $x(kT - T)$ is zero for all $k \geq 4$. The output $y(kT)$ is plotted in Figure 3.3b. From Figure 3.3 we see that $y(kT)$ is a T-second right shift of the input $x(kT)$. Replacing k by $k + 1$ in both sides of (3.14), we have

$$y(kT + T) = x(kT). \tag{3.15}$$

Equation (3.15) is the input/output difference equation of the unit delayer. Note that (3.15) is a linear first-order difference equation. A discrete-time (or digital) device with the input/output relationship (3.15) can be realized using a *shift register*.

UNIT-DELAYER REALIZATIONS Using the unit delayer as a basic "building block," we can construct realizations of any discrete-time system modeled by a linear constant-coefficient input/output difference equation. Let us begin with the first-order case.

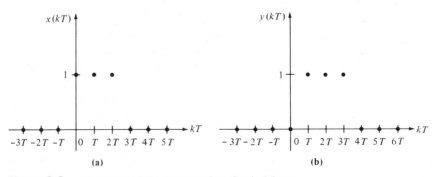

Figure 3.3. (a) Input and (b) output pulse of unit delayer.

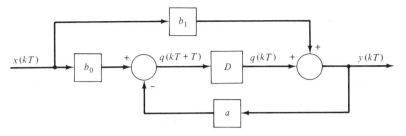

Figure 3.4. Realization of first-order input/output difference equation.

Consider the system defined by the first-order difference equation

$$y(kT + T) + ay(kT) = b_1x(kT + T) + b_0x(kT). \qquad (3.16)$$

The system defined by (3.16) has the unit-delayer realization shown in Figure 3.4. We shall verify that the interconnection in Figure 3.4 is a realization. First, with the output of the unit delayer denoted by $q(kT)$, we have that the input to the delayer is $q(kT + T)$. But $q(kT + T)$ is also the output of the subtracter, and thus

$$q(kT + T) = -ay(kT) + b_0x(kT). \qquad (3.17)$$

The output $y(kT)$ is given by

$$y(kT) = q(kT) + b_1x(kT). \qquad (3.18)$$

Replacing k by $k + 1$ in (3.18), we have

$$y(kT + T) = q(kT + T) + b_1x(kT + T). \qquad (3.19)$$

Inserting the expression (3.17) for $q(kT + T)$ into (3.19), we have

$$y(kT + T) = -ay(kT) + b_0x(kT) + b_1x(kT + T). \qquad (3.20)$$

Equation (3.20) can be written in the form (3.16), and thus the interconnection in Figure 3.4 is a realization of (3.16).

Now consider the discrete-time system given by the second-order input/output difference equation

$$y(kT + 2T) + a_1y(kT + T) + a_0y(kT) \qquad (3.21)$$
$$= b_2x(kT + 2T) + b_1x(kT + T) + b_0x(kT).$$

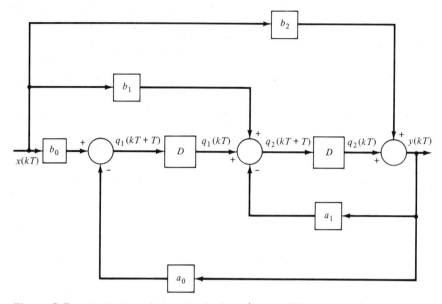

Figure 3.5. Realization of second-order input/output difference equation.

We claim that (3.21) is realized by the interconnection shown in Figure 3.5. Let us check that the system in Figure 3.5 is a realization. From Figure 3.5 we have

$$q_1(kT + T) = -a_0 y(kT) + b_0 x(kT) \tag{3.22}$$

$$q_2(kT + T) = -a_1 y(kT) + q_1(kT) + b_1 x(kT). \tag{3.23}$$

$$y(kT) = q_2(kT) + b_2 x(kT). \tag{3.24}$$

Replacing k by $k + 1$ in (3.23) and using the expression (3.22) for $q_1(kT + T)$, we get

$$q_2(kT + 2T) = -a_1 y(kT + T) + q_1(kT + T) + b_1 x(kT + T)$$

$$= -a_1 y(kT + T) + [-a_0 y(kT) + b_0 x(kT)] + b_1 x(kT + T). \tag{3.25}$$

Replacing k by $k + 2$ in (3.24) and using the expression (3.25) for $q_2(kT + 2T)$, we get

$$y(kT + 2T) = q_2(kT + 2T) + b_2 x(kT + 2T)$$

$$= -a_1 y(kT + T) - a_0 y(kT) + b_0 x(kT)$$
$$+ b_1 x(kT + T) + b_2 x(kT + 2T). \tag{3.26}$$

Equation (3.26) can be written in the form (3.21), and hence the interconnection in Figure 3.5 is a realization of (3.21).

It is interesting to note that the realizations in Figures 3.4 and 3.5 have exactly the same form as the realizations of the first-order and second-order input/output differential equations given in Chapter 2 (see Figures 2.7 and 2.9). The only difference in the realizations is that in the discrete-time case, the basic building block is the unit delayer; whereas in the continuous-time case, the basic building block is the integrator.

This correspondence between unit-delayer realizations of discrete-time systems and integrator realizations of continuous-time systems extends to the nth-order case. Consider the discrete-time system given by the nth-order input/output difference equation

$$y(kT + nT) + \sum_{i=0}^{n-1} a_i y(kT + iT) = \sum_{i=0}^{n} b_i x(kT + iT). \qquad (3.27)$$

The system specified by (3.27) can be realized by an interconnection of n unit delayers and combinations of adders, subtracters, and scalar multipliers. In particular, if the integrators in the realization shown in Figure 2.15 are replaced by unit delayers and the continuous-time signals are replaced by discrete-time signals, the resulting interconnection is a realization of (3.27). This realization is shown in Figure 3.6.

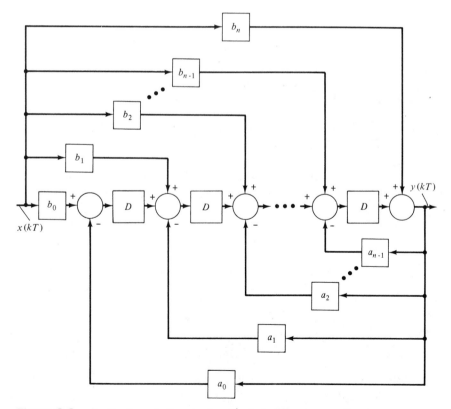

Figure 3.6. Realization of nth-order input/output difference equation.

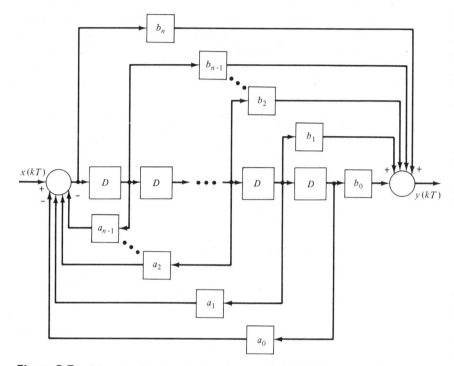

Figure 3.7. Second realization of nth-order input/output difference equation.

If the integrators in the realization shown in Figure 2.16 are replaced by unit delayers, we obtain another realization of (3.27). This second realization is shown in Figure 3.7.

INPUT/OUTPUT DIFFERENCE EQUATION OF UNIT-DELAYER IN-
TERCONNECTIONS There is a converse to the result above that any linear constant-coefficient input/output difference equation can be realized by an interconnection of unit delayers: Any linear time-invariant discrete-time system consisting of an interconnection of n unit delayers and combinations of adders, subtracters, and scalar multipliers can be modeled by an nth-order constant-coefficient input/output difference equation of the form (3.1) or (3.27). Thus there is a one-to-one correspondence between linear constant-coefficient difference equations and linear time-invariant discrete-time systems consisting of an interconnection of unit delayers, adders, subtracters, and scalar multipliers.

The input/output difference equation of a given interconnection of unit delayers can be determined by first denoting the outputs of the unit delayers by $q_1(kT)$, $q_2(kT)$, . . . , $q_n(kT)$. The input $q_i(kT + T)$ of the ith unit delayer can then be expressed as a linear combination of $q_1(kT + jT)$, $q_2(kT + jT)$, . . . , $q_n(kT + jT)$, and the input $x(kT)$, where $j = 0, 1$. The output $y(kT)$ can also be expressed as a linear combination of $q_1(kT + jT)$, $q_2(kT + jT)$, . . . , $q_n(kT + jT)$, and $x(kT)$, where again $j = 0, 1$. These equations can then be combined, which yields the input/output difference equation of the given interconnection. The process is analogous to the procedure we used to determine the input/output differential equation for an interconnection of integrators (see Example 2.4). In Chapter 7 we show that the z-transform can be employed to simplify the process of determining the input/output difference equation for a given interconnection.

★3.3

LINEAR TIME-VARYING DISCRETE-TIME SYSTEMS AND NONLINEAR SYSTEMS

We can generalize the representation studied in Section 3.1 to include linear time-varying discrete-time systems by allowing the coefficients of the input/output difference equation to be functions of the discrete-time index k. In the first-order time-varying case, the input/output difference equation has the general form

$$y(kT + T) + a(k)y(kT) = b_1(k)x(kT + T) + b_0x(kT). \qquad (3.28)$$

In Equation (3.28) the coefficients $a(k)$, $b_1(k)$, and $b_0(k)$ are in general functions of the integer variable k. The discrete-time system defined by (3.28) is time invariant if and only if the coefficients $a(k)$, $b_1(k)$, and $b_0(k)$ are constant (i.e., do not vary with k).

EXAMPLE 3.3. Again consider the loan-repayment process defined in Section 1.5. We assume that the loan is a variable-rate loan, so that the interest rate $I(k)$ varies as a function of k, where k is the month of the repayment period. The input/output difference equation of the loan-repayment process is

$$y(k + 1) - \left[1 + \frac{I(k)}{12}\right] y(k) = -x(k + 1), \qquad (3.29)$$

where $x(k)$ is the loan payment in the kth month and $y(k)$ is the loan balance in the kth month. Clearly, the difference equation (3.29) is a special case of (3.28), where

$$a(k) = -1 - \frac{I(k)}{12}$$

$$b_1(k) = -1$$

$$b_0(k) = 0.$$

In the general nth-order time-varying coefficient case, the input/output difference equation is given by

$$y(kT + nT) + \sum_{i=0}^{n-1} a_i(k)y(kT + iT) = \sum_{i=0}^{m} b_i(k)x(kT + iT), \qquad (3.30)$$

where $m \leq n$. The system defined by (3.30) is time invariant if and only if the coefficients $a_0(k)$, $a_1(k)$, . . . , $a_{n-1}(k)$ are constant and the coefficients $b_0(k)$, $b_1(k)$, . . . , $b_m(k)$ are constant.

SOLUTION BY RECURSION   Even though the coefficients of (3.30) are in general time varying, the difference equation (3.30) can be solved by recursion for any finite range of integer values of k. The recursive procedure is a straightforward generalization of the recursive procedure for solving a difference equation with constant coefficients. In particular, if we are given the initial values $y(-T)$, $y(-2T)$, . . . , $y(-nT)$, we can start the recursion by setting $k = -n$ in (3.30). The program in Figure A.3 for the constant-coefficient case can be modified to work for the time-varying case by allowing the coefficients to be functions of k. The coefficients could be inputed in function form or could be inputed value by value.

UNIT-DELAYER REALIZATIONS   Time-varying discrete-time systems given by the input/output difference equation (3.30) can be realized from an interconnection of unit delayers, adders, subtracters, and time-varying scalar multipliers. For example, the system defined by the first-order equation (3.28) has the realization shown in Figure 3.8. From the figure we have

$$q(kT + T) = -a(k)y(kT) + b_0(k)x(kT) \tag{3.31}$$

$$y(kT) = q(kT) + b_1(k - 1)x(kT). \tag{3.32}$$

Replacing k by $k + 1$ in (3.32) and using the expression (3.31) for $q(kT + T)$, we have

$$y(kT + T) = -a(k)y(kT) + b_0(k)x(kT) + b_1(k)x(kT + T). \tag{3.33}$$

Equation (3.33) can be written in the form (3.28), and thus the interconnection in Figure 3.8 is a realization of (3.28).

Note the right shift $b_1(k - 1)$ of the coefficient $b_1(k)$ in the second forward path of the interconnection in Figure 3.8. If all the coefficients are constant, the realization in Figure 3.8 reduces to the realization in Figure 3.4 of the first-order constant-coefficient case.

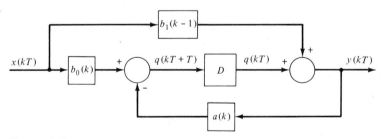

Figure 3.8.   Realization of first-order difference equation with time-varying coefficients.

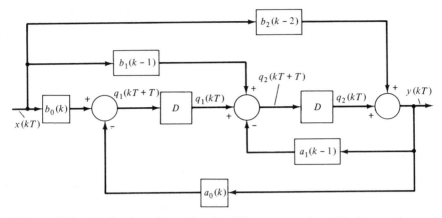

Figure 3.9. Realization of second-order difference equation with time-varying coefficients.

Now consider the second-order case given by

$$y(kT + 2T) + \sum_{i=0}^{1} a_i(k)y(kT + iT) = \sum_{i=0}^{2} b_i(k)x(kT + iT). \quad (3.34)$$

We claim that (3.34) has the unit-delayer realization shown in Figure 3.9. We shall verify that the diagram in Figure 3.9 is a realization. From the figure we have

$$q_1(kT + T) = -a_0(k)y(kT) + b_0(k)x(kT) \quad (3.35)$$

$$q_2(kT + T) = -a_1(k - 1)y(kT) + q_1(kT) + b_1(k - 1)x(kT) \quad (3.36)$$

$$y(kT) = q_2(kT) + b_2(k - 2)x(kT). \quad (3.37)$$

Replacing k by $k + 1$ in (3.36) and k by $k + 2$ in (3.37) and using (3.35), we obtain (3.34). Hence the interconnection in Figure 3.9 is a realization of (3.34).

There is a definite pattern in the realizations shown in Figures 3.8 and 3.9. From this one may guess that a realization in the nth-order case is given by the interconnection shown in Figure 3.10. The proof that this is a realization is omitted.

NONLINEAR INPUT/OUTPUT DIFFERENCE EQUATIONS   Nonlinear finite-dimensional discrete-time systems may be modeled by a nonlinear input/output difference equation. In the first-order time-invariant case, a nonlinear input/output difference equation has the general form

$$y(kT + T) = f(y(kT), x(kT), x(kT + T)), \quad (3.38)$$

where $f(y(kT), x(kT), x(kT + T))$ is a function of the output at time kT and the input at times kT and $kT + T$. In Section 3.4 we show that a nonlinear

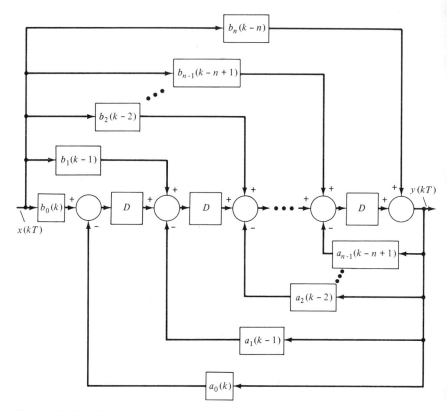

Figure 3.10. Realization of nth-order difference equation with time-varying coefficients.

equation of the form (3.38) may arise as a result of discretizing in time a nonlinear first-order input/output differential equation.

In the second-order time-invariant case, a nonlinear input/output difference equation has the general form

$$y(kT + 2T) = f(y(kT), y(kT + T), x(kT), x(kT + T), x(kT + 2T)),$$
(3.39)

where f is now a function of the output at times kT and $kT + T$ and a function of the input at times kT, $kT + T$, and $kT + 2T$.

First-order and second-order nonlinear input/output difference equations of the form (3.38) and (3.39), and higher-order nonlinear difference equations, can be solved by recursion for any finite range of integer values of k. If an explicit expression is given for the function f in the input/output difference equation, the recursion procedure can be programed as we did in the linear case. We illustrate this in Section 3.4 when we consider a time discretization of a nonlinear differential equation.

EXAMPLE 3.4. The following nonlinear difference equation arises in genetics (see Luenberger [1979]):

$$y(k + 1) = \frac{y(k)}{1 + y(k)}.$$

Note that in this case there is no input. This equation can be solved very easily using recursion. For example, suppose that $y(0) = 1$. Then

$$y(1) = \frac{1}{1 + 1} = \frac{1}{2}$$

$$y(2) = \frac{\frac{1}{2}}{1 + \frac{1}{2}} = \frac{1}{3}$$

$$y(3) = \frac{\frac{1}{3}}{1 + \frac{1}{3}} = \frac{1}{4}$$

REALIZATION OF NONLINEAR DISCRETE-TIME SYSTEMS Nonlinear discrete-time systems can sometimes be realized by an interconnection of unit delayers and signal multipliers as shown in the following example.

EXAMPLE 3.5. Consider the discrete-time system given by the nonlinear input/output difference equation

$$y(kT + T) = -y(kT)x(kT) + x(kT). \tag{3.40}$$

The system defined by (3.40) can be realized using a unit delayer and a signal multiplier. The realization is shown in Figure 3.11.

★3.4

DISCRETIZATION OF DIFFERENTIAL EQUATIONS

As an application of the difference-equation framework, in this section we show that an input/output differential equation can be discretized in time, resulting

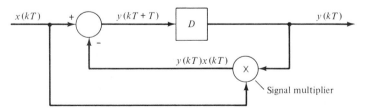

Figure 3.11. Realization of a nonlinear discrete-time system.

in a difference equation which can then be solved by recursion. This discretization in time actually yields a discrete-time simulation of the continuous-time system defined by the given input/output differential equation. We begin with the first-order case.

FIRST-ORDER CASE   Consider the time-invariant continuous-time system with the first-order input/output differential equation

$$\frac{dy(t)}{dt} = f(y(t), x(t)). \tag{3.41}$$

Here $f(y(t), x(t))$ is an arbitrary function of $y(t)$ and $x(t)$. The system defined by (3.41) may be linear, in which case there exist constants a and b such that

$$f(y(t), x(t)) = -ay(t) + bx(t).$$

We do not assume linearity in the first part of the following development.

Given a fixed positive number T, let us set $t = kT$ in (3.41), where k takes on integer values only. This gives

$$\frac{dy(kT)}{dt} = f(y(kT), x(kT)). \tag{3.42}$$

We compute an approximation to (3.42) given in terms of a first-order difference equation. First, we can approximate the derivative in (3.42) by

$$\frac{dy(kT)}{dt} \approx \frac{y(kT + T) - y(kT)}{T}, \tag{3.43}$$

where the symbol "$\approx$" means "approximately equal to." If T is suitably small, the approximation (3.43) to the derivative $dy(kT)/dt$ will be accurate. This approximation is sometimes called the *Euler approximation* of the derivative.

Inserting the approximation (3.43) into (3.42), we get

$$\frac{y(kT + T) - y(kT)}{T} \approx f(y(kT), x(kT)). \tag{3.44}$$

Replacing the approximation symbol in (3.44) by equality and rearranging terms, we obtain an approximation to (3.42) given by the first-order difference equation

$$y(kT + T) = y(kT) + Tf(y(kT), x(kT)). \tag{3.45}$$

Equation (3.45) is a difference-equation approximation to the given differential equation (3.41). We shall call this the Euler approximation since it is based on the Euler approximation of the derivative.

To compute the discrete values $y(kT)$ of the solution $y(t)$ to (3.41), we can solve (3.45). If the function f is given in an explicit form, we can solve (3.45) by recursion. The procedure is illustrated by the following example.

EXAMPLE 3.6. Consider the high-speed vehicle moving on a level surface that was first studied in Example 2.7. The velocity model of the vehicle is given by the first-order nonlinear input/output differential equation

$$\frac{dv(t)}{dt} = -\frac{k_f}{M} v(t) - \frac{k_d}{M} v^2(t) + \frac{1}{M} x(t). \qquad (3.46)$$

In (3.46), k_f is the coefficient of friction, k_d is the drag coefficient due to air resistance, M is the mass of the vehicle, $v(t)$ is the velocity of the vehicle at time t, and $x(t)$ is the drive or braking force applied to the vehicle at time t. Setting $t = kT$ in (3.46) and using the approximation (3.43) for the derivative, we obtain

$$v(kT + T) = v(kT) + T\left[-\frac{k_f}{M} v(kT) - \frac{k_d}{M} v^2(kT) + \frac{1}{M} x(kT) \right]$$

$$= \left(1 - T\frac{k_f}{M}\right) v(kT) - \frac{Tk_d}{M} v^2(kT) + \frac{T}{M} x(kT). \qquad (3.47)$$

The nonlinear difference equation (3.47) can be solved recursively, which will give approximate values $v(kT)$ of the velocity of the vehicle resulting from initial velocity $v(0)$ and the application of an input force $x(t)$ for $t \geq 0$. As an example, let us take $M = 1$, $k_f = 0.1$, $v(0) = 0$, and $x(t)$ equal to the step function $u(t)$. The resulting solution $v(kT)$ to (3.47) is plotted in Figure 3.12 for $T = 1$ and three different values of the drag coefficient k_d.

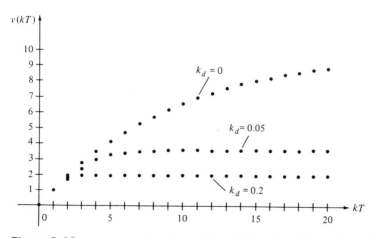

Figure 3.12. Approximation of velocity with $T = 1$ and three values of k_d.

From the plot we see that the steady-state value of the velocity in response to a step input depends on the magnitude of the drag coefficient k_d. Since there is no analytical expression for the exact solution $v(t)$, it is not possible to compare directly the approximations in Figure 3.12 with the exact solution. In the linear case such a comparison is possible. The details follow.

LINEAR CASE Now suppose that the input/output differential equation (3.41) is linear so that

$$\frac{dy(t)}{dt} = -ay(t) + bx(t). \tag{3.48}$$

Setting $t = kT$ in (3.48) and using the approximation (3.43) for the derivative, we obtain

$$y(kT + T) = (1 - aT)y(kT) + Tbx(kT). \tag{3.49}$$

Let us solve (3.49) with arbitrary initial condition $y(0)$ and with input $x(kT)$ equal to zero for all integers $k \geq 0$. We claim that the solution is

$$y(kT) = (1 - aT)^k y(0), \qquad k = 0, 1, 2, \ldots . \tag{3.50}$$

To check that (3.50) is the solution, insert the expression (3.50) for $y(kT)$ into (3.49) with $x(kT) = 0$. This gives

$$(1 - aT)^{k+1} y(0) = (1 - aT)(1 - aT)^k y(0)$$
$$= (1 - aT)^{k+1} y(0).$$

Hence (3.49) is satisfied, which shows that (3.50) is the solution.

The expression (3.50) for $y(kT)$ gives approximate values of the solution $y(t)$ to (3.48) with arbitrary initial condition $y(0)$ and with zero input. Let us compare (3.50) with the exact values of $y(kT)$. First, from the results in Section 2.1 we have that the exact solution $y(t)$ to (3.48) with initial condition $y(0)$ and with zero input is given by

$$y(t) = e^{-at} y(0), \qquad t \geq 0. \tag{3.51}$$

Taking $t = kT$ in (3.51), we have the following exact expression for $y(kT)$:

$$y(kT) = e^{-akT} y(0), \qquad k = 0, 1, 2, \ldots . \tag{3.52}$$

Using the property that

$$e^{ab} = (e^a)^b \qquad \text{for any real numbers } a, b,$$

we can rewrite (3.52) in the form

$$y(kT) = (e^{-aT})^k y(0), \qquad k = 0, 1, 2, \ldots . \tag{3.53}$$

Further, using the expansion for the exponential e^{-aT}, from (3.53) we have the following exact expression for $y(kT)$:

$$y(kT) = \left(1 - aT + \frac{a^2T^2}{2} - \frac{a^3T^3}{6} + \cdots\right)^k y(0), \qquad k = 0, 1, 2, \ldots.$$

(3.54)

Comparing (3.50) and (3.54), we see that our approximation (3.50) is accurate if $1 - aT$ is a good approximation to the exponential e^{-aT}. This will be the case if the magnitude of aT is much less than 1, in which case the magnitudes of the powers of aT will be much smaller than the quantity $1 - aT$.

SECOND-ORDER CASE The discretization technique for first-order differential equations described above can be generalized to second-order and higher-order differential equations. In the second-order case we can use the approximations

$$\frac{dy(kT)}{dt} \approx \frac{y(kT + T) - y(kT)}{T}$$

(3.55)

$$\frac{d^2y(kT)}{dt^2} \approx \frac{\dfrac{dy(kT + T)}{dt} - \dfrac{dy(kT)}{dt}}{T}.$$

(3.56)

By combining (3.55) and (3.56), we have the following approximation to the second derivative:

$$\frac{d^2y(kT)}{dt^2} \approx \frac{y(kT + 2T) - 2y(kT + T) + y(kT)}{T^2}.$$

(3.57)

The approximation (3.57) is called the Euler approximation to the second derivative.

Now consider a linear time-invariant continuous-time system with the second-order input/output differential equation

$$\frac{d^2y(t)}{dt^2} + a_1 \frac{dy(t)}{dt} + a_0 y(t) = b_1 \frac{dx(t)}{dt} + b_0 x(t).$$

(3.58)

Replacing t by kT in (3.58) and using the approximations (3.55) and (3.57), we obtain the following difference-equation approximation to (3.58):

$$\frac{y(kT + 2T) - 2y(kT + T) + y(kT)}{T^2} + a_1 \frac{y(kT + T) - y(kT)}{T} + a_0 y(kT)$$

$$= b_1 \frac{x(kT + T) - x(kT)}{T} + b_0 x(kT).$$

(3.59)

Combining terms, we can write (3.59) in the form

$$y(kT + 2T) + (a_1T - 2)y(kT + T) + (1 - a_1T + a_0T^2)y(kT) \quad (3.60)$$
$$= b_1Tx(kT + T) + (b_0T^2 - b_1T)x(kT).$$

Equation (3.60) is a difference-equation approximation of the input/output differential equation (3.58). Note that (3.60) can be viewed as the input/output difference equation of a linear time-invariant discrete-time system. The discrete-time system defined by (3.60) is a discrete-time simulation of the given continuous-time system [specified by the input/output differential equation (3.58)].

We can obtain an approximate solution to (3.58) by solving (3.60). In particular, suppose that the initial values for (3.58) are $y(0)$ and $\dot{y}(0)$. Then using the approximation

$$\dot{y}(0) \approx \frac{y(T) - y(0)}{T},$$

we can solve for $y(T)$, which gives

$$y(T) = y(0) + T\dot{y}(0). \quad (3.61)$$

Now with the initial values $y(0)$ and $y(T)$, we can solve (3.60) by recursion. The process is started by setting $k = 0$ in (3.60). The computations can be carried out by using the recursion program in Figure A.3. We shall apply the discretization procedure to the *RLC* circuit in Figure 3.13.

EXAMPLE 3.7.　Consider the series *RLC* circuit shown in Figure 3.13. As indicated, the input $x(t)$ is the voltage applied to the circuit and the output $y(t)$ is the voltage $v_c(t)$ across the capacitor. In Example 2.2, we showed that the input/output differential equation of the circuit is

$$LC\frac{d^2v_c(t)}{dt^2} + RC\frac{dv_c(t)}{dt} + v_c(t) = x(t). \quad (3.62)$$

Equation (3.62) is a second-order differential equation that can be written in the form (3.58) with

$$a_1 = \frac{RC}{LC} = \frac{R}{L}, \qquad a_0 = \frac{1}{LC}. \quad (3.63)$$

$$b_1 = 0, \qquad b_0 = \frac{1}{LC}. \quad (3.64)$$

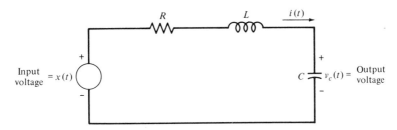

Figure 3.13. Series *RLC* circuit.

Inserting (3.63) and (3.64) into the discretized equation (3.60), we have

$$
v_c(kT + 2T) + \left(\frac{RT}{L} - 2\right) v_c(kT + T)
$$

$$
+ \left(1 - \frac{RT}{L} + \frac{T^2}{LC}\right) v_c(kT) = \frac{T^2}{LC} x(kT). \tag{3.65}
$$

Equation (3.65) is the difference-equation approximation of the *RLC* circuit. Using this discretization, we shall compute the voltage $v_c(t)$ across the capacitor when $R = 2$, $L = C = 1$, $v_c(0) = 1$, $\dot{v}_c(0) = -1$, and $x(t) = (\sin t)u(t)$. From (3.61) we have

$$
v_c(T) = v_c(0) + T\dot{v}_c(0) = 1 - T.
$$

Now we can solve (3.65) using the program in Figure A.3. In running the program, set

$$
x(kT) = \sin(kT + 2T)
$$

$$
y(-T) = v_c(T) = 1 - T, \qquad x(-T) = \sin T,
$$

$$
y(-2T) = v_c(0) = 1, \qquad x(-2T) = \sin 0 = 0.
$$

With the initial conditions set up in this way, the output $y(kT)$ of the program will be equal to $v_c(kT + 2T)$. The solution of the difference-equation approximation (3.65) is plotted in Figure 3.14 for $T = 0.2$. Using solution techniques for differential equations (or the Laplace transform method given in Chapter 5), we have that the exact solution to the differential equation (3.62) is

$$
v_c(t) = 0.5 [(3 + t) e^{-t} - \cos t], \qquad t \geq 0.
$$

The exact solution is also plotted in Figure 3.14. From the plots we see that there is a significant error in the approximation. The smaller we take the

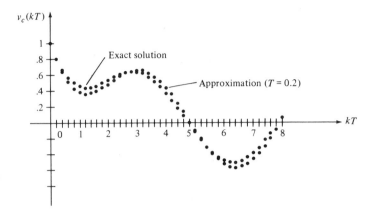

Figure 3.14.   Exact and approximate output responses in Example 3.7.

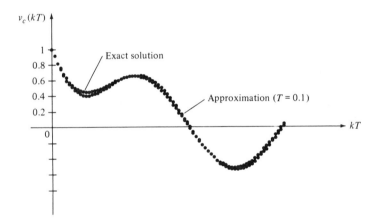

Figure 3.15.   Exact and approximate output responses in Example 3.7.

discretization interval T, the closer the approximation should be to the exact solution. For example, if we take $T = 0.1$, we get the approximation displayed in Figure 3.15. The exact solution is also shown in Figure 3.15. Clearly, the approximation with $T = 0.1$ shown in Figure 3.15 is closer to the exact solution than the approximation with $T = 0.2$ shown in Figure 3.14.

SECOND DISCRETIZATION TECHNIQUE   For continuous-time systems given by a linear input/output differential equation, there are a number of discretization techniques that are much more accurate (for a given value of T) than the above technique based on the Euler approximation of derivatives. One of these is based on an approximation to the integral expression of the solution to the input/output differential equation. We shall carry out the details for the first-order case.

Again suppose that we are given the first-order input/output differential equation (3.48). Using the integrating-factor method (see the *RC* circuit example in Section 1.5), we have that the general solution $y(t)$ to (3.48) with initial condition $y(t_0)$ and arbitrary input $x(t)$ applied for $t > t_0$ is given by

$$y(t) = \exp\left[-a(t - t_0)\right]y(t_0) + \int_{t_0}^{t} b \exp\left[-a(t - \lambda)\right]x(\lambda)\, d\lambda. \quad (3.66)$$

Now fix $T > 0$. Then setting $t_0 = kT$ and $t = kT + T$ in (3.66), where $k = 0, 1, 2, \ldots$, we have

$$
\begin{aligned}
y(kT + T) &= \exp\left[-a(kT + T - kT)\right]y(kT) \\
&\quad + \int_{kT}^{kT+T} b \exp\left[-a(kT + T - \lambda)\right]x(\lambda)\, d\lambda \\
&= e^{-aT}y(kT) \\
&\quad + \int_{kT}^{kT+T} b \exp\left[-a(kT + T - \lambda)\right]x(\lambda)\, d\lambda.
\end{aligned}
\quad (3.67)
$$

Except for the second term on the right-hand side of (3.67), this equation looks like the input/output difference equation of a linear time-invariant discrete-time system. We can simplify (3.67) by assuming that the input $x(t)$ is approximately constant over each T-second interval $kT \le t < kT + T$; that is,

$$x(t) \approx x(kT) \qquad \text{for } kT \le t < kT + T. \quad (3.68)$$

For example, the signal $x(t)$ in Figure 3.16 is exactly equal to a constant over each T-second interval $kT \le t < kT + T$.

If the input signal $x(t)$ is a continuous function of t (see Section 1.2), the assumption (3.68) will be valid for a suitably small value of T.

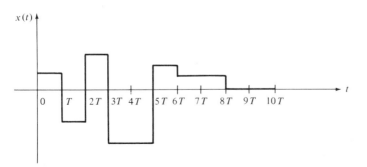

Figure 3.16. Signal that is constant over each interval $kT \le t < kT + T$.

Now by the assumption (3.68), we can rewrite the integral in (3.67) in the form

$$\int_{kT}^{kT+T} b \exp[-a(kT + T - \lambda)]x(\lambda)\, d\lambda$$

$$= \left\{ \int_{kT}^{kT+T} b \exp[-a(kT + T - \lambda)]d\lambda \right\} x(kT)$$

$$= b \exp[-a(kT + T)] \left[\int_{kT}^{kT+T} \exp(a\lambda)\, d\lambda \right] x(kT)$$

$$= b \exp[-a(kT + T)] \left[\frac{1}{a} e^{a\lambda} \right]_{\lambda=kT}^{\lambda=kT+T} x(kT)$$

$$= \frac{b}{a}(1 - e^{-aT})x(kT).$$

Now define

$$a_d = -e^{-aT} \quad \text{and} \quad b_d = \frac{b}{a}(1 - e^{-aT}),$$

where d stands for "discrete." Then we can write (3.67) in the form

$$y(kT + T) = -a_d y(kT) + b_d x(kT). \tag{3.69}$$

Equation (3.69) is another difference-equation approximation of the first-order differential equation (3.48). We shall call this the *integral approximation* since it is based on the approximation of the integral expression for the solution to the input/output differential equation.

The discretization (3.69) is, in general, much more accurate (for a fixed value of T) than the Euler discretization defined by (3.49). In fact, if the input $x(t)$ is constant over each T-second interval $kT \le t < kT + T$ (as is the signal in Figure 3.16), the integral approximation gives exact values of the output response; whereas the Euler approximation is usually not exact. This is illustrated by the following example.

EXAMPLE 3.8. Again consider the automobile with velocity model

$$M \frac{dv(t)}{dt} + k_f v(t) = x(t). \tag{3.70}$$

Taking $M = 1$ and $k_f = 0.1$, we have that the Euler approximation of (3.70) is given by

$$v(kT + T) = (1 - 0.1T)v(kT) + Tx(kT). \tag{3.71}$$

The integral approximation of (3.70) is given by

$$v(kT + T) = e^{-0.1T}v(kT) + 10(1 - e^{-0.1T})x(kT). \qquad (3.72)$$

Clearly, these discrete-time approximations are not identical. Let us assume that $v(-T) = x(-T) = 0$. Then from (3.8) and (3.13), the solution to (3.71) is

$$v(0) = 0$$

$$v(kT) = \sum_{i=0}^{k-1} (1 - 0.1T)^{k-i-1}Tx(iT), \qquad k \geq 1.$$

and the solution to (3.72) is

$$v(0) = 0$$

$$v(kT) = \sum_{i=0}^{k-1} [10(1 - e^{-0.1T})] \exp[-0.1(k - i - 1)T]x(iT), \qquad k \geq 1.$$

Now let us take $T = 2$ and $x(t) = 10$ for $0 \leq t < 10$, $x(t) = 0$ for $t \geq 10$. This input is constant over each 2-second interval $2k \leq t < 2k + 2$, and thus the integral approximation gives exact values of the output response. The response is displayed in Figure 3.17. Also plotted in Figure 3.17 is the approximate solution resulting from the Euler approximation with $T = 2$. The solutions were computed using the program shown in Figure A.3.

The integral-approximation method extends to second-order and higher-order input/output differential equations. The approximation can be constructed in terms of a state representation of the given system. This is studied in Chapter 13.

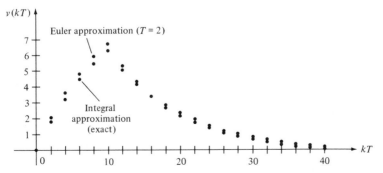

Figure 3.17. Integral and Euler approximations of output response in Example 3.8.

PROBLEMS

Chapter 3

3.1   A discrete-time system is given by the input/output difference equation

$$y(kT + 2T) + 1.5y(kT + T) + 0.5y(kT) = x(kT + 2T) - x(kT).$$

(a) Compute $y(kT)$ for $k = 0, 1, 2, 3$ when $y(-2T) = -1$, $y(-T) = 2$, and $x(kT) = 0$, $k = -2, -1, 0, 1, 2, \ldots$.
(b) Compute $y(kT)$ for $k = 0, 1, 2, 3$ when $y(-2T) = y(-T) = 0$ and $x(kT) = 1$ for $k = -2, -1, 0, 1, 2, \ldots$.
(c) Compute $y(kT)$ for $k = 0, 1, 2, 3$ when $y(-2T) = -1$, $y(-T) = 2$, and $x(kT) = 1$ for $k = -2, -1, 0, 1, 2, \ldots$.
(d) Compute $y(kT)$ for $k = 3, 4, 5$ when $y(T) = 2$, $y(2T) = 3$, and $x(kT) = \sin(\pi k/2)$ for $k = 0, 1, 2, \ldots$.
(e) Compute $y(kT)$ for $k = -1, 0, 1$ when $y(-3T) = -2$, $y(-2T) = 4$, and $x(kT) = (0.5)^{k-1}u(kT - T)$ for all k.

3.2   Consider the discrete-time signal $x(k)$ where $x(k) = 0$ for $k = 0, -1, -2, \ldots$, $x(1) = 1$, $x(2) = 1$, $x(3) = 2$, $x(4) = 3$, $x(5) = 5$, $x(6) = 8$, $x(7) = 13$, etc. This signal is called the *Fibonacci sequence*. Show that there are constants a_0 and a_1 such that the Fibonacci sequence $x(k)$ is the solution to the difference equation

$$x(k + 2) + a_1 x(k + 1) + a_0 x(k) = \Delta(k + 1)$$

with initial data $x(-2) = x(-1) = 0$. Here $\Delta(k + 1)$ is the unit pulse located at $k = -1$ [i.e., $\Delta(k + 1) = 0$ for all k except $k = -1$, and $\Delta(0) = 1$]. Determine a_0 and a_1.

3.3   When the unit pulse $\Delta(kT)$ is applied to a particular linear time-invariant discrete-time system, the resulting zero-state response is $y(kT) = (-1)^k$, $k = 0, 1, 2, \ldots$. Compute the input/output difference equation of the system.

   HINT: Observe that $y(kT + 2T) = y(kT)$ for $k = 0, 1, 2, \ldots$.

3.4   Find a unit-delayer realization of the discrete-time system with the input/output difference equation

$$y(kT + 2T) = x(kT - 2T) + x(kT).$$

3.5   Find a unit-delayer realization of the discrete-time system with the input/output difference equation

$$y(kT + T) + y(kT) - y(kT - T) = x(kT - 2T).$$

3.6   Find the input/output difference equation of the discrete-time system shown in Figure P3.6.

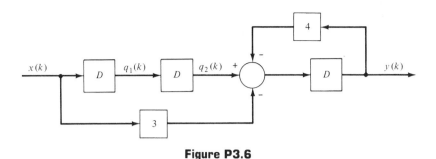

Figure P3.6

3.7 For the system in Figure P3.6, compute $y(1)$, $y(2)$, and $y(3)$ when
(a) $q_1(0) = -1$, $q_2(0) = 1$, $y(0) = 3$, and $x(0) = 2$, $x(1) = -2$, $x(2) = 4$.
(b) $y(-2) = -3$, $y(-1) = -1$, $y(0) = 2$, and $x(-2) = 2$, $x(-1) = -2$, $x(0) = 4$, $x(1) = 3$, $x(2) = -1$.

3.8 Find the input/output difference equations of the discrete-time systems shown in Figure P3.8.

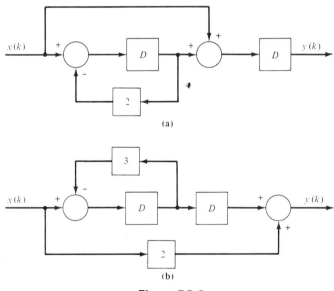

(a)

(b)

Figure P3.8

3.9 Consider the discrete-time model for a savings account given by the difference equation

$$y(k + 1) - \left[1 + \frac{I(k)}{4}\right] y(k) = x(k + 1).$$

Here k is the quarter, $y(k)$ is the amount in the account at the end of the kth quarter, $x(k)$ is the amount deposited in the kth quarter, and $I(k)$ is the interest rate in the kth quarter.

(a) Suppose that $I(k) = 0.1 + 0.01k[u(k) - u(k - 5)]$, so the interest rate is increasing by 1 percent each quarter. Compute $y(k)$ for $k = 1, 2, 3, 4$ when $y(0) = 1000$ and $x(k) = 1000$ for $k = 1, 2, 3, \ldots$.

(b) Repeat part (a) with $I(k) = 0.1 - 0.01k[u(k) - u(k - 5)]$, so that now the interest rate is decreasing by 1 percent each quarter.

3.10 Using two unit delayers, signal multipliers, adders, subtracters, and scalar multipliers, realize the following input/output difference equations.

(a) $y(kT + 2T) = y(kT + T)x(kT + T) + y(kT) + x(kT)$.

(b) $y(kT + 2T) = y(kT + T)y(kT) + x(kT + T)$.

3.11 For each of the systems in Problem 3.10, compute the output response $y(kT)$ for $k = 3, 4, 5$ when $y(T) = -1$, $y(2T) = 3$, and $x(kT) = \tan(\pi k/5)[u(kT) - u(kT - 5T)]$.

3.12 Consider the nonlinear discrete-time system given by the input/output difference equation

$$y(k + 1)y(k) + ay(k + 1) - by(k) = cx(k),$$

where a, b, and c are arbitrary constants.

(a) Suppose that $x(k) = 0$ for $k = 0, 1, 2, \ldots$, and let $v(k) = (b - y(k))/y(k)$. Show that for some constants g, h, $v(k)$ satisfies the difference equation

$$v(k + 1) - gv(k) = h, \qquad k = 0, 1, 2, \ldots .$$

(b) Using your result in part (a), derive an explicit analytical expression for the solution $y(k)$ to the difference equation

$$y(k + 1)y(k) + ay(k + 1) - by(k) = 0,$$

with arbitrary initial condition $y(0)$.

(c) With $x(k) = 0$ for $k = 0, 1, 2, \ldots$, determine all values of $y(0)$ such that $y(k) = \infty$ for some finite value of k. This is the discrete-time version of the notion of finite escape times.

3.13 A continuous-time system is given by the input/output differential equation

$$\frac{d^2y(t)}{dt^2} + a\frac{dy(t)}{dt} = bx(t).$$

Recall that this is the input/output differential equation of a car on a level surface if we take $a = k_f/M$ and $b = 1/M$.

(a) Determine a discretization of the system by using the Euler approximation. Let the discretization interval T be arbitrary.

(b) With $a = 0.1$, $b = 1$, and $T = 1$, use the discretization in part (a) and the program in Figure A.3 to compute $y(k)$ for $k = 1, 2, 3, \ldots , 50$ when $y(0) = 0$, $\dot{y}(0) = 0$, and $x(t) = \sin 0.1\pi t$ for $t \geq 0$.

(c) Compare your results in part (b) with the exact values of the response.

(d) Repeat parts (b) and (c) with $T = 0.5$.

3.14 Consider the connection shown in Figure P3.14. System 1 is a continuous-time system with the input/output differential equation $\dot{y}_1(t) + y_1(t) = x(t)$, and system 2 is a discrete-time system with the input/output difference equation $y(kT + T) + 3y(kT) = 2y_1(kT)$. Compute the exact output values $y(T)$, $y(2T)$, $y(3T)$ when

(a) $x(t) = 0$ for all $t \geq 0$ and $y_1(0) = 1$, $y(0) = 1$.
(b) $x(t) = 1$ for all $t \geq 0$ and $y_1(0) = 1$, $y(0) = 1$.
(c) $x(t) = e^{-t}$ for all $t \geq 0$ and $y_1(0) = -2$, $y(0) = -1$.

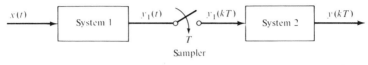

Figure P3.14

3.15 A continuous-time system is given by the input/output differential equation

$$\frac{dy(t)}{dt} + ay(t) = b_0x(t) + b_1\frac{dx(t)}{dt}.$$

If we define $q(t) = y(t) - b_1x(t)$, as shown in Section 2.1, we have that

$$\frac{dq(t)}{dt} + aq(t) = (b_0 - ab_1)x(t).$$

(a) By first discretizing the equation for $q(t)$ using the Euler approximation, compute a discretization of the given system. Let the discretization interval T be arbitrary.
(b) Repeat part (a) using the integral approximation of the equation for $q(t)$.

3.16 In this problem we compute a discretization of the system in Problem 3.13 by approximating integrals. Consider the connection shown in Figure P3.16. As illustrated, the input to system 2 is the output $y_1(t)$ of system 1. Here system 1 has input/output differential equation $\dot{y}_1(t) = x(t)$ (i.e., system 1 is an integrator), and the input/output differential equation of system 2 is $\dot{y}(t) + ay(t) = by_1(t)$.

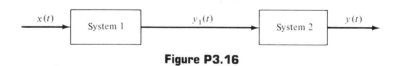

Figure P3.16

(a) Show that the input/output differential equation of the connection in Figure P3.16 is

$$\frac{d^2y(t)}{dt^2} + a\frac{dy(t)}{dt} = bx(t).$$

(b) By using integral approximations, discretize the two systems in Figure P3.16.

(c) Using your results in part (b), compute a discretization of the system with input/output differential equation

$$\frac{d^2y(t)}{dt^2} + a\,\frac{dy(t)}{dt} = bx(t).$$

(d) Using the discretization in part (c) and the program in Figure A.3, compute $y(k)$ for $k = 1, 2, 3, \ldots, 50$ when $a = 0.1$, $b = 1$, $T = 1$, $y(0) = 0$, $\dot{y}(0) = 0$, and $x(t) = \sin 0.1\pi t$ for $t \geq 0$. Compare your results with those obtained in parts (b) and (c) of Problem 3.13.

3.17   Consider the pendulum given by the input/output differential equation

$$I\,\frac{d^2\theta(t)}{dt^2} + MgL \sin \theta(t) = Lx(t).$$

(a) Discretize the system by using the Euler approximation. Let T be arbitrary.   **NOTE:** Do *not* approximate the system by the small-signal model.

(b) Let $I = M = L = 1$, $g = 9.8$. Using the discretization in part (a) with $T = 0.04$, compute $\theta(0.04k)$ for $k = 1, 2, 3, \ldots, 50$ when $\theta(0) = \pi/2$ (radians), $\dot{\theta}(0) = 0$, and $x(t) = 0$ for all $t \geq 0$.

(c) Repeat part (b) with $T = 0.02$. Are the results significantly different?

(d) Now consider the small-signal model given by

$$\frac{d^2\theta(t)}{dt^2} + 9.8\theta(t) = x(t).$$

When $\theta(0) = \pi/2$, $\dot{\theta}(0) = 0$, and $x(t) = 0$ for $t \geq 0$, the exact solution to the small-signal model is $\theta(t) = (\pi/2) \cos 3.13t$. Compare these values with the values of $\theta(0.02k)$ obtained in part (c). (The values should be plotted.) What do you conclude from this? In particular, is the small-signal model accurate when $\theta(t)$ is large, as is the case here?

3.18   Consider a mass M sitting on top of the vibration absorber described in Problem 2.6 (see Figure P2.6). Let $M = 1$, $D = 4$, $K = 1$, and $g = 9.8$.

(a) Discretize the system by using the Euler approximation. Let T be arbitrary.

(b) Using the discretization in part (a) with $T = 0.1$, compute $y(0.1k)$ for $k = 1, 2, \ldots, 50$ when $y(0) = 9.8$, $\dot{y}(0) = 0$, and $x(t) = 10 \sin 2t$, $t \geq 0$. This input corresponds to a vibrational force.

(c) Repeat part (b) with $x(t) = 10 \sin 4t$, $t \geq 0$.

(d) If the absorber is absorbing the vibration, $y(t)$ should converge to the value 9.8 as $t \to \infty$. By modifying the constant D and repeating part (c), see if you can improve the performance of the absorber.

Convolution
Representation

In Chapters 2 and 3 we studied time-domain models of continuous-time and discrete-time systems given in terms of input/output differential equations and input/output difference equations. In this chapter we consider another type of time-domain model based on the convolution operation. The convolution representation is actually a special case of the input/output operator representation that we defined in Chapter 1.

We begin in Section 4.1 with the definition of the convolution of two discrete-time signals. A computer program is given for computing the convolution. In Section 4.2 we develop the convolution representation for the class of linear time-invariant discrete-time systems. The convolution of continuous-time signals and the convolution representation of linear time-invariant continuous-time systems are studied in Sections 4.3 and 4.4. In Section 4.5 we show that the continuous-time convolution representation can be discretized in time, which results in a discrete-time simulation of the given continuous-time system. Although the convolution representation does not apply to time-varying systems, in Section 4.6 we show that there is an input/output relationship for linear time-varying systems which reduces to the convolution relationship in the time-invariant case.

CONVOLUTION OF DISCRETE-TIME SIGNALS

Given two discrete-time signals $x(kT)$ and $v(kT)$, the convolution of $x(kT)$ and $v(kT)$, denoted by $(x * v)(kT)$, is defined by

$$(x * v)(kT) = \sum_{i=-\infty}^{\infty} x(iT)v(kT - iT). \qquad (4.1)$$

The summation on the right side of (4.1) is called the *convolution sum*.

It should be noted that there are discrete-time signals $x(kT)$, $v(kT)$ for which the convolution sum does not exist; that is, there are discrete-time signals that cannot be convolved. As shown below, the convolution sum always exists when $x(kT)$ and $v(kT)$ are both zero for all integers $k < 0$.

If $x(kT)$ and $v(kT)$ are zero for all integers $k < 0$, then $x(iT) = 0$ for all integers $i < 0$ and $v(kT - iT) = 0$ for all integers $k - i < 0$ (or $k < i$). Thus the summation on i in (4.1) may be taken from $i = 0$ to $i = k$, and the convolution operation is given by

$$(x * v)(kT) = \begin{cases} 0, & k = -1, -2, \ldots \\ \sum_{i=0}^{k} x(iT)v(kT - iT), & k = 0, 1, 2, \ldots \end{cases} \qquad (4.2)$$

Since the summation in (4.2) is over a finite range of integers ($i = 0$ to $i = k$), the convolution sum exists. Hence any two signals that are zero for all integers $k < 0$ can be convolved.

To compute the convolution (4.1) or (4.2), we must first change the discrete-time index k to i in the signals $x(kT)$ and $v(kT)$. The resulting signals $x(iT)$ and $v(iT)$ are then functions of the discrete-time variable iT. The next step is to determine $v(kT - iT)$ and then form the product $x(iT)v(kT - iT)$. The signal $v(kT - iT)$ is a *folded and shifted version* of the signal $v(iT)$. More precisely, $v(-iT)$ is $v(iT)$ folded about the vertical axis, and $v(kT - iT)$ is $v(-iT)$ shifted by kT seconds. If $k > 0$, $v(kT - iT)$ is a kT-second right shift of $v(-iT)$. If the signal $v(kT)$ is given in analytical form, $v(kT - iT)$ can be formed by simply replacing k by $k - i$ in the expression for $v(kT)$. Once the product $x(iT)v(kT - iT)$ is formed, the value of the convolution $x * v$ at the point kT is computed by summing the values of $x(iT)v(kT - iT)$ as i ranges over the set of integers. The process is illustrated by the following example.

EXAMPLE 4.1. Suppose that $x(kT) = a^k u(kT)$ and $v(kT) = b^k u(kT)$, where $u(kT)$ is the discrete-time unit-step function and a and b are fixed nonzero real numbers. Since both $x(kT)$ and $v(kT)$ are zero for $k < 0$, the convolution $(x * v)(kT)$ is given by (4.2). Inserting $x(iT) = a^i u(iT)$ and $v(kT - iT) = b^{k-i}u(kT - iT)$ into (4.2), we have

$$(x * v)(kT) = \sum_{i=0}^{k} a^i u(iT) b^{k-i} u(kT - iT), \qquad k = 0, 1, 2, \ldots \quad (4.3)$$

Now $u(iT) = 1$ and $u(kT - iT) = 1$ for all integer values of i ranging from 0 to k, and thus (4.3) reduces to

$$(x * v)(kT) = \sum_{i=0}^{k} a^i b^{k-i} = b^k \sum_{i=0}^{k} \left(\frac{a}{b}\right)^i, \qquad k = 0, 1, 2, \ldots \quad (4.4)$$

If $a = b$,

$$\sum_{i=0}^{k} \left(\frac{a}{b}\right)^i = k + 1$$

and

$$(x * v)(kT) = b^k(k + 1) = a^k(k + 1), \qquad k = 0, 1, 2, \ldots.$$

If $a \neq b$, we have

$$\sum_{i=0}^{k} \left(\frac{a}{b}\right)^i = \frac{1 - (a/b)^{k+1}}{1 - (a/b)}. \qquad (4.5)$$

The relationship (4.5) can be verified by multiplying both sides of (4.5) by $1 - (a/b)$. Inserting (4.5) into (4.4), we obtain (assuming that $a \neq b$)

$$(x * v)(kT) = b^k \frac{1 - (a/b)^{k+1}}{1 - (a/b)} = \frac{b^{k+1} - a^{k+1}}{b - a}, \qquad k = 0, 1, 2, \ldots.$$

It is easy to generalize (4.2) to the case when $x(kT)$ and $v(kT)$ are not necessarily zero for all integers $k < 0$. In particular, suppose that $x(kT) = 0$ and $v(kT) = 0$ for all integers $k < -q$, where q is a fixed arbitrary positive integer. In this case the convolution operation (4.1) can be written in the form

$$(x * v)(kT) = \begin{cases} 0, & k < -q \\ \sum_{i=-q}^{k+q} x(iT)v(kT - iT), & k \geq -q. \end{cases} \qquad (4.6)$$

Note that the convolution sum in (4.6) is still finite, and thus the convolution $x * v$ exists.

DIRECT EVALUATION OF THE CONVOLUTION SUM

If the signals $x(kT)$ and $v(kT)$ are both zero for $k < -q$, the convolution $(x * v)(kT)$ can be computed for any finite range of values of k by directly evaluating the convolution sum in (4.6).

In particular, if $q = 0$, from (4.6) we have

$$(x * v)(0) = x(0)v(0)$$

$$(x * v)(T) = x(0)v(T) + x(T)v(0)$$

$$(x * v)(2T) = x(0)v(2T) + x(T)v(T) + x(2T)v(0)$$

$$\cdot$$
$$\cdot$$
$$\cdot$$

For signals $x(kT)$ and $v(kT)$ that are zero for $k < 0$, a program for computing $(x * v)(kT)$ for $k = 0, 1, 2, \ldots, Q$ is given in Figure A.4. The discrete-time signals $x(kT)$ and $v(kT)$ may be entered value by value or they may be given in function form on lines 210 and 280 of the program.

EXAMPLE 4.2. Suppose that $x(kT)$ and $v(kT)$ are equal to the pulse $p(kT)$ defined by

$$p(kT) = \begin{cases} 1, & 0 \le k \le 9, \\ 0, & \text{all other } k. \end{cases}$$

To run the program in Figure A.4, on line 210 we set

$$\text{DEF FNX(K)} = .5*(\text{SGN(K} + .5) - \text{SGN(K} - 9.5)),$$

and on line 280, we set

$$\text{DEF FNV(K)} = .5*(\text{SGN(K} + .5) - \text{SGN(K} - 9.5)).$$

In the expressions above, SGN is the "signum function," defined by

$$\text{SGN(A)} = \begin{cases} -1 & \text{if } A < 0, \\ 0 & \text{if } A = 0, \\ 1 & \text{if } A > 0. \end{cases}$$

With $x(kT)$ and $v(kT)$ as defined above, we ran the program with $Q = 25$. The results are plotted in Figure 4.1. Note that $(p * p)(kT)$ is a discrete-time triangular pulse.

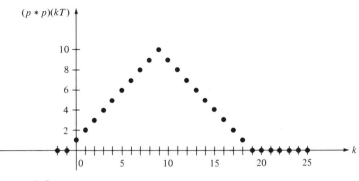

Figure 4.1. Plot of $(p * p)$ (kT).

The convolution operation $(x * v)(kT)$ satisfies a number of useful properties, which follow.

Associativity. For any signals $x(kT)$, $v(kT)$, and $w(kT)$, we have that

$$x * (v * w) = (x * v) * w,$$

so convolution satisfies the associativity property. The proof of associativity follows easily from the definition of convolution, and therefore the proof is omitted.

Commutativity. The convolution operation $(x * v)(kT)$ is commutative; that is,

$$(x * v)(kT) = (v * x)(kT) \tag{4.7}$$

or

$$\sum_{i=-\infty}^{\infty} x(iT)v(kT - iT) = \sum_{i=-\infty}^{\infty} v(iT)x(kT - iT). \tag{4.8}$$

To prove (4.8), let $\bar{i} = k - i$ in the definition (4.1) of convolution. Then $i = k - \bar{i}$ and (4.1) becomes

$$(x * v)(kT) = \sum_{\bar{i}=-\infty}^{\infty} x(kT - \bar{i}T)v(\bar{i}T). \tag{4.9}$$

Since the functions $x(kT)$ and $v(kT)$ are real valued and real numbers commute, (4.9) can be rewritten as

$$(x * v)(kT) = \sum_{\bar{i}=-\infty}^{\infty} v(\bar{i}T)x(kT - \bar{i}T). \tag{4.10}$$

Finally, since the index $\bar{i}$ of the summation in (4.10) can be changed to i, the right side of (4.10) is equal to the convolution $(v * x)(kT)$. Thus we have proved the commutativity property.

Distributivity with Addition.　The convolution operation is distributive with addition; that is, for any signals $x(kT)$, $v(kT)$, $w(kT)$,

$$x * (v + w) = x * v + x * w. \tag{4.11}$$

The relationship (4.11) follows directly from the definition of convolution. We omit the details.

Shift Property.　Given two signals $x(kT)$ and $v(kT)$ and a positive or negative integer q, let x_q and v_q denote the discrete-time signals defined by

$$x_q(kT) = x(kT - qT) \quad \text{and} \quad v_q(kT) = v(kT - qT).$$

When $q < 0$, x_q and v_q are qT-second left shifts of x and v, and when $q > 0$, x_q and v_q are qT-second right shifts of x and v.

Now we claim that

$$(x * v)(kT - qT) = (x_q * v)(kT) = (x * v_q)(kT). \tag{4.12}$$

By (4.12), the qT-second shift of the convolution $x * v$ is equal to the qT-second shift of x convolved with v, which in turn is equal to x convolved with the qT-second shift of v.

To prove (4.12), from (4.1) we have

$$(x * v)(kT - qT) = \sum_{i=-\infty}^{\infty} x(iT)v(kT - qT - iT). \tag{4.13}$$

By definition of $v_q(kT)$, we have

$$v(kT - qT - iT) = v_q(kT - iT). \tag{4.14}$$

Inserting (4.14) into (4.13), we obtain

$$(x * v)(kT - qT) = \sum_{i=-\infty}^{\infty} x(iT)v_q(kT - iT) = (x * v_q)(kT),$$

which proves the second part of (4.12). The first part of (4.12) is proved in a similar manner using the commutativity property. The details are omitted.

CONVOLUTION WITH THE UNIT PULSE
Let $\Delta(kT)$ denote the unit pulse located at $kT = 0$; that is, $\Delta(kT)$ is zero for all $k \neq 0$, and $\Delta(0) = 1$. For any signal $x(kT)$, we have

$$(x * \Delta)(kT) = x(kT). \tag{4.15}$$

By (4.15) the convolution of any discrete-time signal $x(kT)$ with the unit pulse $\Delta(kT)$ reproduces $x(kT)$. As a result of this property, the unit pulse $\Delta(kT)$ is said to be the *identity element* of the convolution operation. To prove (4.15), from (4.1) we have

$$(x * \Delta)(kT) = \sum_{i=-\infty}^{\infty} x(iT)\Delta(kT - iT). \tag{4.16}$$

Since $\Delta(kT - iT) = 0$ when $k \neq i$, and $\Delta(kT - iT) = 1$ when $k = i$, the summation in (4.16) reduces to $x(kT)$. Hence we have verified (4.15).

CONVOLUTION WITH THE SHIFTED UNIT PULSE   Given a positive or negative integer q, let $\Delta_q(kT)$ denote qT-second shift of the unit pulse $\Delta(kT)$; that is,

$$\Delta_q(kT) = \Delta(kT - qT).$$

Note that $\Delta_q(kT)$ is the unit pulse located at the point $kT = qT$. Now for any signal $x(kT)$, we have

$$(x * \Delta_q)(kT) = x(kT - qT). \tag{4.17}$$

The relationship (4.17) says that the convolution of $x(kT)$ with the shifted unit pulse $\Delta_q(kT)$ is equivalent to shifting $x(kT)$ by qT seconds. This characterization of the shift operation turns out to be rather useful. The proof of (4.17) follows from (4.15) and the shift property (4.12).

4.2

CONVOLUTION REPRESENTATION OF LINEAR TIME-INVARIANT DISCRETE-TIME SYSTEMS

Consider a single-input single-output discrete-time system given by the input/output relationship

$$y(kT) = (Fx)(kT), \tag{4.18}$$

where F is the input/output operator of the system. Recall from Chapter 1 that $y(kT)$ is the output response of the system resulting from input $x(kT)$ with no initial energy in the system prior to the application of $x(kT)$. Throughout this section we assume that the system is causal, linear, and time invariant. We do not require that the system be finite dimensional.

UNIT-PULSE RESPONSE   Let $h(kT)$ denote the response of the system when the input $x(kT)$ is equal to the unit pulse $\Delta(kT)$ with no initial energy in the system at time $kT = 0$. In terms of the input/output operator F, we have that

$$h(kT) = (F\Delta)(kT).$$

The response $h(kT)$ is called the *unit-pulse response* of the system.

It is sometimes possible to determine the unit-pulse response $h(kT)$ experimentally by applying the unit pulse $\Delta(kT)$ to the system and measuring the resulting output response. In this experiment the system must not have any initial energy prior to the application of the unit pulse. The unit-pulse response may also be computable from some known mathematical representation of the system. For example, suppose that the system is given by the input/output difference equation (3.1) studied in Section 3.1. Then $h(kT)$ can be calculated by solving (3.1) with $x(kT) = \Delta(kT)$ and with the initial conditions $y(iT) = 0$ for $i = -1, -2, \ldots, -n$.

EXAMPLE 4.3. Consider the finite-dimensional discrete-time system given by the input/output difference equation

$$y(kT + T) + ay(kT) = b_0x(kT) + b_1x(kT + T), \qquad (4.19)$$

where a, b_0, b_1 are arbitrary constants. We want to compute the unit-pulse response $h(kT)$ of this system. By definition, $h(kT)$ is equal to the solution of (4.19) with $x(kT) = \Delta(kT)$ and with initial condition $y(-T) = 0$. From the results in Section 3.1 [see (3.8) and (3.13)] we have that the solution $y(kT)$ of (4.19) with arbitrary input $x(kT)$ (where $x(-T) = 0$) and with $y(-T) = 0$ is given by

$$y(0) = b_1x(0)$$
$$\qquad (4.20)$$
$$y(kT) = \sum_{i=0}^{k-1} (-a)^{k-i-1}(b_0 - ab_1)x(iT) + b_1x(kT), \qquad k \geq 1.$$

Setting $x(kT) = \Delta(kT)$ in (4.20), we obtain

$$y(0) = h(0) = b_1$$
$$y(kT) = h(kT) = (-a)^{k-1}(b_0 - ab_1), \qquad k = 1, 2, \ldots.$$

Returning to the general case, since the system given by (4.18) is assumed to be causal, there can be no output response before an input is applied to the system. Thus it must be true that

$$h(kT) = 0 \qquad \text{for } k = -1, -2, \ldots,$$

since $\Delta(kT) = 0$ for $k = -1, -2, \ldots.$ So causality implies that the unit-pulse response $h(kT)$ must be zero for all integers $k < 0$. In addition, since the system is time invariant, for any positive integer i the output response resulting from the iT-second right shift of the unit pulse $\Delta(kT)$ must be equal to the

iT-second right shift of $h(kT)$; that is, the response to $\Delta(kT - iT)$ must be equal to $h(kT - iT)$. Letting $\Delta_i(kT) = \Delta(kT - iT)$, we have

$$h(kT - iT) = (F\Delta_i)(kT), \qquad i = 0, 1, 2, \ldots. \tag{4.21}$$

CONVOLUTION REPRESENTATION Now let $x(kT)$ be an arbitrary input with $x(kT) = 0$ for $k = -1, -2, \ldots.$ We claim that the output response $y(kT)$ resulting from input $x(kT)$ with zero initial energy at time $kT = 0$ is given by

$$y(kT) = (h * x)(kT) = \sum_{i=0}^{k} h(iT)x(kT - iT), \qquad k = 0, 1, 2, \ldots. \tag{4.22}$$

By commutativity of convolution, we can rewrite (4.22) in the form

$$y(kT) = (x * h)(kT) = \sum_{i=0}^{k} x(iT)h(kT - iT), \qquad k = 0, 1, 2, \ldots. \tag{4.23}$$

From the relationship (4.22) or (4.23), we have that the output response $y(kT)$ resulting from input $x(kT)$ with no initial energy at time $kT = 0$ is equal to the convolution of the unit-pulse response $h(kT)$ with the input $x(kT)$. Equation (4.22) [or (4.23)] is called the *convolution representation* of the system. This is a time-domain model since the components of (4.22) are functions of the discrete-time variable kT.

Comparing (4.18) and (4.22), we have that

$$(Fx)(kT) = (h * x)(kT), \qquad k = 0, 1, 2, \ldots. \tag{4.24}$$

From (4.24) we see that the input/output operator F of the system acts on the input $x(kT)$ by convolving the input with the unit-pulse response $h(kT)$. This type of input/output operator is called a *convolution operator*. By (4.24) we have the fundamental result that the input/output operator F of any causal linear time-invariant discrete-time system is a convolution operator.

An interesting consequence of the convolution representation (4.22) is the result that the system is completely determined by the unit-pulse response $h(kT)$. In particular, if $h(kT)$ is known, the output response resulting from any input $x(kT)$ can be computed by evaluating (4.22) or (4.23).

PROOF OF THE CONVOLUTION RELATIONSHIP For the sake of completeness, we shall prove the relationship (4.22). Let $x(kT)$ be an arbitrary input with $x(kT) = 0$ for $k = -1, -2, \ldots.$ For each nonnegative integer r, let v_r denote the discrete-time signal defined by

$$v_r = \sum_{i=0}^{r} x(iT)\Delta_i, \tag{4.25}$$

where again Δ_i is the unit pulse located at $kT = iT$. From (4.25), for any integer k ranging from 0 to r we have

$$v_r(kT) = \sum_{i=0}^{r} x(iT)\Delta_i(kT)$$

$$= \sum_{i=0}^{r} x(iT)\Delta(kT - iT)$$

$$= x(kT) \qquad \text{since } \Delta(kT - iT) = 0 \text{ when}$$
$$k \neq i \text{ and } \Delta(kT - iT) = 1 \text{ when } k = i.$$

Since $v_r(kT) = x(kT)$ for $0 \leq k \leq r$, by causality it must be true that the output responses resulting from v_r and x are equal at all time points kT, where $0 \leq k \leq r$. In mathematical terms we have

$$(Fv_r)(kT) = (Fx)(kT) \qquad \text{for } 0 \leq k \leq r. \tag{4.26}$$

Since v_r is equal to the finite sum in (4.25), by linearity of the given system we have

$$(Fv_r)(kT) = \sum_{i=0}^{r} x(iT)(F\Delta_i)(kT), \qquad 0 \leq k \leq r.$$

By (4.21)

$$(F\Delta_i)(kT) = h(kT - iT),$$

and thus

$$(Fv_r)(kT) = \sum_{i=0}^{r} x(iT)h(kT - iT), \qquad 0 \leq k \leq r. \tag{4.27}$$

Since the integer r is arbitrary, we can take $r = k$ in (4.26) and (4.27), which gives

$$(Fv_r)(kT) = (Fx)(kT) = \sum_{i=0}^{k} x(iT)h(kT - iT), \qquad k \geq 0. \tag{4.28}$$

The term on the far right side of (4.28) is equal to the convolution $(x * h)(kT)$, which in turn is equal to $(h * x)(kT)$ by commutativity of convolution. Hence we have

$$y(kT) = (Fx)(kT) = (h * x)(kT), \qquad k \geq 0,$$

and we have verified the convolution representation (4.22).

EXAMPLE 4.4. Suppose that the first few values of the unit-pulse response $h(kT)$ are $h(0) = -1$, $h(T) = 2$, $h(2T) = 1$, and that the values of $h(kT)$ for $k \geq 3$ are not known. We want to compute the output response $y(kT)$ resulting from input $x(kT) = (1/2)^k$, $k = 0, 1, 2, \ldots$, with zero initial energy in the system at time $kT = 0$. From (4.22) we have

$$y(0) = h(0)x(0) = (-1)(1) = -1$$

$$y(T) = h(0)x(T) + h(T)x(0) = (-1)(\tfrac{1}{2}) + (2)(1) = \tfrac{3}{2}$$

$$y(2T) = h(0)x(2T) + h(T)x(T) + h(2T)x(0)$$
$$= (-1)(\tfrac{1}{4}) + (2)(\tfrac{1}{2}) + (1)(1) = \tfrac{7}{4}$$

$$y(3T) = h(0)x(3T) + h(T)x(2T) + h(2T)x(T) + h(3T)x(0).$$

Note that $y(3T)$ cannot be calculated since we do not know the value of $h(3T)$. In fact, we cannot compute $y(kT)$ for any value of $k \geq 3$ since $h(kT)$ is unknown for $k \geq 3$.

From (4.22) we see that for each integer value of $k > 0$, the computation of $y(kT)$ requires in general $k + 1$ multiplications and $k + 1$ additions. Thus the computational complexity of the input/output convolution relationship increases with k. For instance, if we want to compute the output $y(kT)$ at time $kT = 1000T$, we will have to perform on the order of 1000 multiplications and 1000 additions in order to evaluate the convolution expression (4.22).

The dependency on k of the computational complexity of convolution is in contrast to the recursion process, where the computational complexity depends only on the order n of the recursion (see Section 3.1). The difference in the complexity of the two computational procedures is due to finite dimensionality. That is, the recursion process requires and exploits finite dimensionality of the given discrete-time system, whereas the convolution procedure does not require finite dimensionality and does not make effective use of finite dimensionality when the given system has this property.

Since the output response $y(kT)$ is equal to the discrete-time convolution $(h * x)(kT)$ given by (4.22), the response $y(kT)$ for any finite range of values of $k = 0, 1, \ldots, Q$ can be computed using the convolution program in Figure A.4. The procedure is illustrated by the following example.

EXAMPLE 4.5. Suppose that the unit-pulse response $h(kT)$ is equal to $\sin 0.5k$ for $k \geq 0$, and the input $x(kT)$ is equal to $\sin 0.1k$ for $k \geq 0$. Then in the program in Figure A.4, set

$$\text{DEF FNX(K)} = \text{SIN(.1*K)}$$

on line 210 and set

$$\text{DEF FNV(K)} = \text{SIN(.5*K)}$$

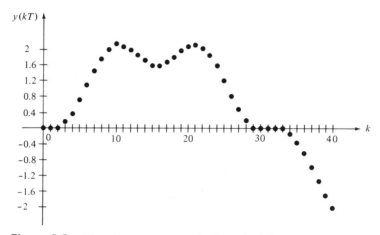

Figure 4.2. Plot of output response in Example 4.5.

on line 280. In running the program, one will observe a significant increase in the time required to compute $y(kT)$ as k becomes large. Running the program with $Q = 40$, we get the results shown in Figure 4.2.

REPRESENTATION IN TERMS OF THE STEP RESPONSE Again consider the discrete-time system defined by the input/output relationship (4.18). In addition to being able to express the output response $y(kT)$ in terms of the unit-pulse response $h(kT)$, it is also possible to express $y(kT)$ in terms of the system's step response. The *step response*, which we shall denote by $g(kT)$, is the response of the system when the input $x(kT)$ is equal to the discrete-time unit step function $u(kT)$ with no initial energy in the system at time $kT = 0$. In terms of the input/output operator F, we have

$$g(kT) = (Fu)(kT).$$

From the convolution representation (4.22), we have

$$g(kT) = (h * u)(kT) = \sum_{i=0}^{k} h(iT), \qquad (4.29)$$

and since the system is time invariant,

$$g(kT - T) = (h * u)(kT - T). \qquad (4.30)$$

Subtracting (4.30) from (4.29), we obtain

$$g(kT) - g(kT - T) = (h * u)(kT) - (h * u)(kT - T). \qquad (4.31)$$

Using the distributivity property (4.11) and shift property (4.12) of convolution, we get

$$g(kT) - g(kT - T) = [h * (u - u_1)](kT), \tag{4.32}$$

where $u_1(kT) = u(kT - T)$. But

$$u(kT) - u_1(kT) = \Delta(kT),$$

and thus (4.32) reduces to

$$g(kT) - g(kT - T) = (h * \Delta)(kT) = h(kT). \tag{4.33}$$

From (4.33) we see that the unit-pulse response $h(kT)$ of the system is equal to the step response $g(kT)$ minus the T-second right shift $g(kT - T)$ of the step response. Hence the unit-pulse response can be determined directly from the step response. Further, using the convolution representation (4.22), for any input $x(kT)$ we have

$$y(kT) = [(g - g_1) * x](kT), \tag{4.34}$$

where $g_1(kT) = g(kT - T)$. Since

$$g_1 * x = g * x_1,$$

where $x_1(kT) = x(kT - T)$, we can rewrite (4.34) in the form

$$y(kT) = [g * (x - x_1)] (kT). \tag{4.35}$$

The interesting feature of the two convolution representations (4.34) and (4.35) is that they are expressed in terms of the system's step response $g(kT)$. Either (4.34) or (4.35) may be used to compute the output response $y(kT)$ resulting from an arbitrary input $x(kT)$ with zero initial energy at time $kT = 0$.

4.3

CONVOLUTION OF CONTINUOUS-TIME SIGNALS

Given two continuous-time signals $x(t)$ and $v(t)$, the convolution of $x(t)$ and $v(t)$, denoted by $(x * v)(t)$, is defined by

$$(x * v)(t) = \int_{-\infty}^{\infty} x(\lambda)v(t - \lambda) \, d\lambda. \tag{4.36}$$

The integral in the right side of (4.36) is called the *convolution integral*. The convolution integral does not exist for some signals $x(t)$, $v(t)$; in other words, there are continuous-time signals that cannot be convolved. However, most continuous-time signals of interest in engineering can be convolved. In particular, as discussed below, most signals that are zero for $t < 0$ can be convolved. If $x(t)$ and $v(t)$ are both zero for all $t < 0$, then $x(\lambda) = 0$ for all $\lambda < 0$ and

$v(t - \lambda) = 0$ for all $t - \lambda < 0$ (or $t < \lambda$). In this case, the integration in (4.36) may be taken from $\lambda = 0$ to $\lambda = t$, and the convolution operation is given by

$$(x * v)(t) = \begin{cases} 0, & t < 0 \\ \int_0^t x(\lambda)v(t - \lambda) \, d\lambda, & t \geq 0. \end{cases} \qquad (4.37)$$

The integral in (4.37) exists for all $t > 0$ if the functions $x(t)$ and $v(t)$ are absolutely integrable for all $t > 0$; that is,

$$\int_0^t |x(\lambda)| \, d\lambda < \infty \quad \text{and} \quad \int_0^t |v(\lambda)| \, d\lambda < \infty \qquad \text{for all } t > 0. \quad (4.38)$$

Most functions of interest in engineering satisfy (4.38) and therefore can be convolved.

To compute the convolution (4.36) or (4.37), we must first determine $x(\lambda)$ and $v(t - \lambda)$. The signal $x(\lambda)$ is simply $x(t)$ with t replaced by λ, while $v(t - \lambda)$ is a folded and shifted version of $v(\lambda)$. A graphical representation of the folding and shifting operation is considered later in this section. If $v(t)$ is given in analytical form, $v(t - \lambda)$ can be formed simply by replacing t by $t - \lambda$ in the expression for $v(t)$. This is illustrated in the following example.

EXAMPLE 4.6. Suppose that $x(t) = e^{-at}u(t)$ and $v(t) = e^{-bt}u(t)$, where $u(t)$ is the unit-step function and a and b are fixed real numbers. Since $x(t)$ and $v(t)$ are zero for $t < 0$, the convolution $(x * v)(t)$ is given by (4.37). Inserting $x(\lambda) = e^{-a\lambda}u(\lambda)$ and $v(t - \lambda) = \exp[-b(t - \lambda)]u(t - \lambda)$ into (4.37), we obtain

$$(x * v)(t) = \int_0^t e^{-a\lambda}u(\lambda) \exp[-b(t - \lambda)]u(t - \lambda) \, d\lambda, \qquad t \geq 0. \tag{4.39}$$

Since $u(\lambda) = 1$ and $u(t - \lambda) = 1$ for any value of λ between 0 and t, (4.39) reduces to

$$(x * v)(t) = \int_0^t e^{-a\lambda} \exp[-b(t - \lambda)] \, d\lambda, \qquad t \geq 0$$

$$= e^{-bt} \int_0^t \exp[(b - a)\lambda] \, d\lambda, \qquad t \geq 0$$

If $a = b$,

$$(x * v)(t) = e^{-bt} \int_0^t (1) \, d\lambda = e^{-bt}t, \qquad t \geq 0.$$

If $a \neq b$,

$$
(x * v)(t) = e^{-bt} \frac{1}{b - a} \left\{ \exp \left[(b - a)\lambda \right] \right\}_{\lambda = 0}^{\lambda = t}
$$

$$
= \frac{e^{-bt}}{b - a} \left\{ \exp \left[(b - a)t \right] - 1 \right\}
$$

$$
= \frac{e^{-at} - e^{-bt}}{b - a}, \qquad t \geq 0.
$$

As in the discrete-time case, the convolution operation satisfies a number of useful properties.

Associativity. For any signals $x(t)$, $v(t)$, and $w(t)$, we have that

$$
(x * v) * w = x * (v * w),
$$

and thus the convolution operation is associative. The proof of associativity is omitted.

Commutativity. The convolution operation $(x * v)(t)$ is commutative; that is,

$$
(x * v)(t) = (v * x)(t). \tag{4.40}
$$

Writing out the definition of convolution, from (4.40) we have

$$
\int_{-\infty}^{\infty} x(\lambda)v(t - \lambda)d\lambda = \int_{-\infty}^{\infty} v(\lambda)x(t - \lambda) \, d\lambda. \tag{4.41}
$$

The property (4.41) can be proved by considering the change of variable $\bar{t} = t - \lambda$ in the convolution integral in (4.36). We invite the reader to work out the details.

Distributivity with Addition. Convolution is distributive with addition; that is, for any (convolvable) signals x, v, w,

$$
x * (v + w) = x * v + x * w. \tag{4.42}
$$

The proof of (4.42) is immediate from the definition of convolution.

Shift Property. Given two continuous-time signals $x(t)$ and $v(t)$, and a positive or negative real number c, let $x_c(t)$ and $v_c(t)$ denote the c-second shifts of $x(t)$ and $v(t)$ defined by

$$
x_c(t) = x(t - c) \qquad \text{and} \qquad v_c(t) = v(t - c).
$$

The signals $x_c(t)$ and $v_c(t)$ are c-second left shifts of $x(t)$ and $v(t)$ when $c < 0$, and are c-second right shifts of $x(t)$ and $v(t)$ when $c > 0$. Now we have

$$(x * v)(t - c) = (x_c * v)(t) = (x * v_c)(t). \tag{4.43}$$

By (4.43), the c-second shift of the convolution $x * v$ is equal to the c-second shift of x convolved with v, which in turn is equal to x convolved with the c-second shift of v. The proof of (4.43) follows directly from the definition of convolution. We omit the details.

Derivative Property. If the signal $x(t)$ has an ordinary first derivative $\dot{x}(t)$, the convolution $(x * v)(t)$ has an ordinary first derivative and

$$\frac{d}{dt} (x * v)(t) = (\dot{x} * v)(t). \tag{4.44}$$

To prove (4.44), by definition of convolution we have

$$\frac{d}{dt} (x * v)(t) = \frac{d}{dt} (v * x)(t) = \frac{d}{dt} \left[\int_{-\infty}^{\infty} v(\lambda)x(t - \lambda) \, d\lambda \right]$$

$$= \int_{-\infty}^{\infty} v(\lambda) \left[\frac{d}{dt} x(t - \lambda) \right] d\lambda. \tag{4.45}$$

But

$$\frac{d}{dt} x(t - \lambda) = \dot{x}(t - \lambda),$$

and thus the right side of (4.45) is equal to $v * \dot{x}$, which is equal to $\dot{x} * v$. Thus we have verified (4.44).

If both $x(t)$ and $v(t)$ are differentiable (i.e., have ordinary first derivatives), then using commutativity and applying the derivative property (4.44) twice, we have that $x * v$ has an ordinary second derivative given by

$$\frac{d^2}{dt^2} (x * v)(t) = (\dot{x} * \dot{v})(t).$$

This result shows that convolution is a *smoothing operation*; that is, the convolution $(x * v)(t)$ is, in general, a smoother function of t than either $x(t)$ or $v(t)$. Here smoothness is measured in terms of the number of times a function can be differentiated in the ordinary sense.

Integration Property. Let $x^{(-1)}(t)$ and $v^{(-1)}(t)$ denote the integrals of $x(t)$ and $v(t)$ defined by

$$x^{(-1)}(t) = \int_{-\infty}^{t} x(\lambda) \, d\lambda \text{ and } v^{(-1)}(t) = \int_{-\infty}^{t} v(\lambda) \, d\lambda.$$

Letting $(x * v)^{(-1)}$ denote the integral of $x * v$, we have

$$(x * v)^{(-1)} = x^{(-1)} * v = x * (v^{(-1)}). \qquad (4.46)$$

The relationship (4.46) follows easily from the definition of convolution. We leave the details to the reader.

CONVOLUTION WITH THE UNIT IMPULSE Let $\delta(t)$ denote the unit impulse located at the origin. Recall from Chapter 1 that $\delta(t) = 0$ for all $t \neq 0$ and

$$\int_{-\epsilon}^{\epsilon} \delta(\lambda) \, d\lambda = 1 \qquad \text{for any real number } \epsilon > 0. \qquad (4.47)$$

Now for any continuous-time signal $x(t)$, by definition of convolution we have

$$(x * \delta)(t) = (\delta * x)(t) = \int_{-\infty}^{\infty} \delta(\lambda) x(t - \lambda) \, d\lambda. \qquad (4.48)$$

Since $\delta(\lambda) = 0$ for all $\lambda \neq 0$, the integrand of the integral in (4.48) reduces to $\delta(\lambda) x(t)$, and thus

$$(x * \delta)(t) = \int_{-\infty}^{\infty} \delta(\lambda) x(t) \, d\lambda = x(t) \int_{-\infty}^{\infty} \delta(\lambda) \, d\lambda.$$

Then using (4.47) with $\epsilon = \infty$, we have

$$(x * \delta)(t) = x(t). \qquad (4.49)$$

By (4.49) we see that the convolution of any continuous-time signal $x(t)$ with the unit impulse $\delta(t)$ reproduces $x(t)$. Thus the unit impulse $\delta(t)$ is the identity element of the convolution operation.

CONVOLUTION WITH THE SHIFTED UNIT IMPULSE Given a positive or negative real number c, let $\delta_c(t)$ denote the unit impulse located at $t = c$; that is,

$$\delta_c(t) = \delta(t - c).$$

Given a signal $x(t)$, using (4.49) and the shift property (4.43), we have

$$(x * \delta_c)(t) = x(t - c). \qquad (4.50)$$

Hence convolving $x(t)$ with the shifted unit impulse $\delta_c(t)$ is equivalent to shifting $x(t)$ by c seconds.

CONVOLUTION OF SIGNALS DEFINED INTERVAL BY INTERVAL

Let us now suppose that $x(t)$ and/or $v(t)$ are defined interval by interval. For example, $x(t)$ could be defined by

$$x(t) = \begin{cases} 1, & 0 \leq t < 1 \\ -1, & 1 \leq t < 2 \\ 0, & \text{all other } t. \end{cases} \qquad (4.51)$$

To compute the convolution $x * v$ when x or v is defined interval by interval, it is often useful to graph the functions in the integrand of the convolution integral. This can help to determine the integrand and integration limits of the convolution integral. The steps of this graphical aid to computing the convolution integral are listed below. Here we are assuming that both $x(t)$ and $v(t)$ are zero for all $t < 0$. If $x(t)$ and $v(t)$ are not zero for all $t < 0$, we can use the shift property to reduce the problem to the case when $x(t)$ and $v(t)$ are zero for all $t < 0$.

Step 1. Graph $x(\lambda)$ and $v(-\lambda)$ as functions of λ. The function $v(-\lambda)$ is equal to the function $v(\lambda)$ folded about the vertical axis.

Step 2. Let $[0, a]$ denote the set of all t such that $0 \leq t \leq a$, where a is a positive number. For t equal to an arbitrary point of the interval $[0, a]$, graph $v(t - \lambda)$ and the product $x(\lambda)v(t - \lambda)$ as functions of λ. Note that $v(t - \lambda)$ is equal to $v(-\lambda)$ shifted to the right by t seconds. The function $v(t - \lambda)$ is a folded and shifted version of $v(\lambda)$. The value of a in the interval $[0, a]$ is the largest value of a for which the product $x(\lambda)v(t - \lambda)$ has the same analytical form [or the plot of $x(\lambda)v(t - \lambda)$ has the same form] for all values of $t \in [0, a]$.

Step 3. Integrate the product $x(\lambda)v(t - \lambda)$ as a function of λ with the limits of integration from $\lambda = 0$ to $\lambda = t$. The result is the convolution $(x * v)(t)$ for $t \in [0, a]$.

Step 4. For t equal to an arbitrary point of the interval $[a, b]$, graph $v(t - \lambda)$ and the product $x(\lambda)v(t - \lambda)$ as functions of λ. The value of b is the largest value of b for which the product $x(\lambda)v(t - \lambda)$ has the same analytical form for all values of $t \in [a, b]$.

Step 5. Integrate the product $x(\lambda)v(t - \lambda)$ as a function of λ with the limits of integration from $\lambda = 0$ to $\lambda = t$. The result is the convolution $(x * v)(t)$ for $t \in [a, b]$. Repeat Steps 4 and 5 as many times as necessary until $(x * v)(t)$ is computed for all $t > 0$.

The above procedure is illustrated by the following two examples.

EXAMPLE 4.7. Suppose that $x(t)$ is the signal given by (4.51), and that the signal $v(t)$ is the pulse

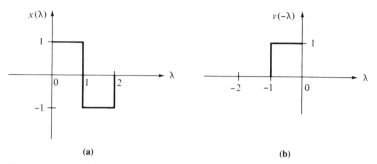

Figure 4.3.   Plots of (a) $x(\lambda)$ and (b) $v(-\lambda)$.

$$v(t) = \begin{cases} 1, & 0 \le t < 1 \\ 0, & \text{all other } t. \end{cases}$$

We shall follow the steps given above.

Step 1.   The functions $x(\lambda)$ and $v(-\lambda)$ are plotted in Figure 4.3.

Step 2.   For $t \in [0, 1]$, the plots of $x(\lambda)$, $v(t - \lambda)$ and the product $x(\lambda)v(t - \lambda)$ are given in Figure 4.4. For $t \in [1, 2]$, the form of the product $x(\lambda)v(t - \lambda)$ changes as shown in Figure 4.5c. Thus the value of a is 1.

Step 3.   Integrating the product $x(\lambda)v(t - \lambda)$ displayed in Figure 4.4c, for $0 \le t \le 1$ we have

$$(x * v)(t) = \int_0^t 1 \, d\lambda = t.$$

Step 4.   For $t \in [2, 3]$, the product $x(\lambda)v(t - \lambda)$ is plotted in Figure 4.6c. Comparing Figures 4.5c and 4.6c, we see that the form of the product $x(\lambda)v(t - \lambda)$ again changes from the interval $[1, 2]$ to the interval $[2, 3]$. Thus the value of b is 2.

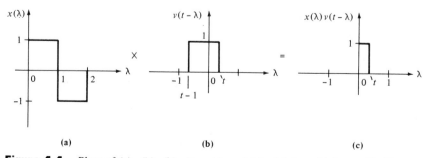

Figure 4.4.   Plots of (a) $x(\lambda)$, (b) $v(t - \lambda)$, and (c) $x(\lambda)v(t - \lambda)$ for $t \in [0, 1]$.

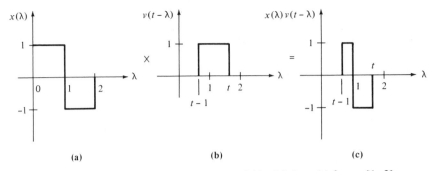

Figure 4.5 Plots of (a) $x(\lambda)$, (b) $v(t - \lambda)$, and (c) $x(\lambda)v(t - \lambda)$ for $t \in [1, 2]$.

Step 5. Integrating the product plotted in Figure 4.5c, for $1 \leq t \leq 2$ we have

$$(x * v)(t) = \int_{t-1}^{1} (1) \, d\lambda + \int_{1}^{t} (-1) \, d\lambda$$

$$= 1 - (t - 1) + (-1)(t - 1) = -2t + 3.$$

Repeating step 5 with $t \in [2, 3]$, from Figure 4.6c we have

$$(x * v)(t) = \int_{t-1}^{2} (-1) \, d\lambda$$

$$= (-1)[2 - (t - 1)] = t - 3, \qquad 2 \leq t \leq 3.$$

Finally, for $t \geq 3$, the product $x(\lambda)v(t - \lambda)$ is zero since there is no overlap between $x(\lambda)$ and $v(t - \lambda)$ when $t \geq 3$. Hence

$$(x * v)(t) = 0 \qquad \text{for all } t \geq 3.$$

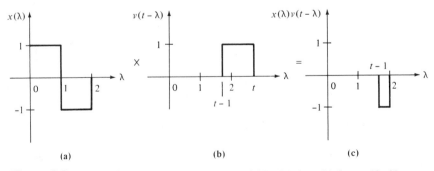

Figure 4.6. Plots of (a) $x(\lambda)$, (b) $v(t - \lambda)$, and (c) $x(\lambda)v(t - \lambda)$ for $t \in [2, 3]$.

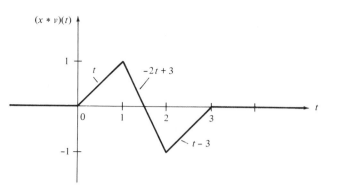

Figure 4.7. Sketch of $x * v$.

A sketch of the convolution $(x * v)(t)$ is shown in Figure 4.7. We can express the convolution $(x * v)(t)$ in analytical form as follows. From the plot in Figure 4.7 we have

$$(x * v)(t) = t[u(t) - u(t - 1)] + (-2t + 3)[u(t - 1) - u(t - 2)]$$
$$+ (t - 3)[u(t - 2) - u(t - 3)]$$
$$= tu(t) + (-3t + 3)u(t - 1) + (3t - 6)u(t - 2)$$
$$- (t - 3)u(t - 3).$$

EXAMPLE 4.8. Consider the signals $x(t)$ and $v(t)$ defined by

$$x(t) = \begin{cases} e^{-t}, & 0 \le t \le 4 \\ 0, & \text{all other } t \end{cases} \qquad v(t) = \begin{cases} e^t, & 0 \le t < 1 \\ e^{2-t}, & 1 \le t < 2 \\ 0, & \text{all other } t. \end{cases}$$

The signals $x(t)$ and $v(t)$ are plotted in Figure 4.8. It is usually easier to evaluate the convolution $(x * v)(t)$ by folding and shifting the "simpler" of

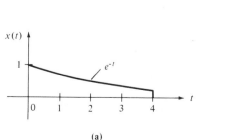

(a)

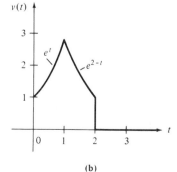

(b)

Figure 4.8. Plots of (a) $x(t)$ and (b) $v(t)$.

the two signals x and v. Here x is simpler than v, so we shall fold and shift x; that is, we shall compute the convolution $x * v$ using the expression

$$(x * v)(t) = \int_0^t v(\lambda) x(t - \lambda) \, d\lambda.$$

The functions $v(\lambda)$ and $x(-\lambda)$ are displayed in Figure 4.9. For $t \in [0, 1]$, the functions $x(t - \lambda)$ and $v(\lambda)x(t - \lambda)$ are plotted in Figure 4.10, and for $t \in [1, 2]$, these functions are plotted in Figure 4.11. Integrating the product $v(\lambda)x(t - \lambda)$ displayed in Figure 4.10b, for $t \in [0, 1]$ we have

$$\begin{aligned}
(x * v)(t) &= \int_0^t e^\lambda e^{-(t-\lambda)} \, d\lambda = e^{-t} \int_0^t e^{2\lambda} \, d\lambda \\
&= \frac{e^{-t}}{2} \left[e^{2\lambda} \right]_{\lambda=0}^{\lambda=t} = \frac{e^{-t}}{2} (e^{2t} - 1) \\
&= \tfrac{1}{2}(e^t - e^{-t}).
\end{aligned}$$

Integrating the product displayed in Figure 4.11b, for $t \in [1, 2]$ we obtain

$$\begin{aligned}
(x * v)(t) &= \int_0^1 e^\lambda e^{-(t-\lambda)} \, d\lambda + \int_1^t e^{2-\lambda} e^{-(t-\lambda)} \, d\lambda \\
&= \frac{e^{-t}}{2} \left[e^{2\lambda} \right]_{\lambda=0}^{\lambda=1} + e^2 e^{-t} \int_1^t (1) \, d\lambda \\
&= \frac{e^{-t}}{2} (e^2 - 1) + e^2 e^{-t}(t - 1) \\
&= \left[\frac{e^2 - 1}{2} + e^2(t - 1) \right] e^{-t}.
\end{aligned}$$

Continuing with the steps described above, for $t \in [2, 4]$ we have

$$\begin{aligned}
(x * v)(t) &= \int_0^1 e^\lambda \, e^{-(t-\lambda)} \, d\lambda + \int_1^2 e^{2-\lambda} e^{-(t-\lambda)} \, d\lambda \\
&= \frac{e^{-t}}{2} (e^2 - 1) + e^2 e^{-t} \int_1^2 (1) \, d\lambda \\
&= \frac{e^{-t}}{2} (e^2 - 1) + e^2 e^{-t} \\
&= \left(\frac{e^2 - 1}{2} + e^2 \right) e^{-t}.
\end{aligned}$$

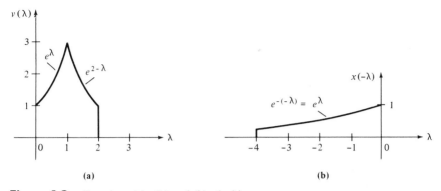

(a)   **(b)**

Figure 4.9.   Functions (a) $v(\lambda)$ and (b) $x(-\lambda)$.

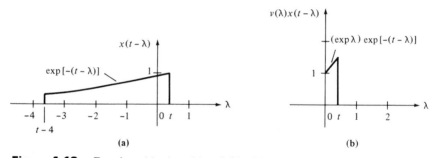

(a)   **(b)**

Figure 4.10.   Functions (a) $x(t-\lambda)$ and (b) $v(\lambda)x(t-\lambda)$ for $t \in [0, 1]$.

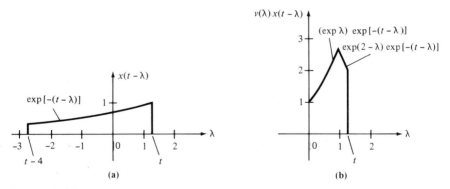

(a)   **(b)**

Figure 4.11.   Functions (a) $x(t-\lambda)$ and (b) $v(\lambda)x(t-\lambda)$ for $t \in [1, 2]$.

For $t \in [4, 5]$,

$$
\begin{aligned}
(x * v)(t) &= \int_{t-4}^{1} e^{\lambda} e^{-(t-\lambda)} \, d\lambda + \int_{1}^{2} e^{2-\lambda} e^{-(t-\lambda)} \, d\lambda \\
&= \frac{e^{-t}}{2} \left[e^{2\lambda} \right]_{\lambda=t-4}^{\lambda=1} + e^{2} e^{-t} \int_{1}^{2} (1) \, d\lambda \\
&= \frac{e^{-t}}{2} [e^{2} - e^{2(t-4)}] + e^{2} e^{-t} \\
&= \frac{1}{2} [3e^{2} - e^{2(t-4)}] e^{-t}.
\end{aligned}
$$

For $t \in [5, 6]$,

$$
\begin{aligned}
(x * v)(t) &= \int_{t-4}^{2} e^{2-\lambda} e^{-(t-\lambda)} \, d\lambda \\
&= e^{2} e^{-t} \int_{t-4}^{2} (1) \, d\lambda \\
&= e^{2} e^{-t} [2 - (t - 4)] \\
&= e^{2} (6 - t) e^{-t}.
\end{aligned}
$$

Finally, for $t \geq 6$, $(x * v)(t) = 0$ since the functions $v(\lambda)$ and $x(t - \lambda)$ do not overlap when $t \geq 6$.

4.4

CONVOLUTION REPRESENTATION OF LINEAR TIME-INVARIANT CONTINUOUS-TIME SYSTEMS

Consider a single-input single-output causal linear time-invariant continuous-time system given by the input/output relationship

$$
y(t) = (Fx)(t). \tag{4.52}
$$

The function $y(t)$ in (4.52) is the output response resulting from the input $x(t)$ with no initial energy in the system prior to the application of $x(t)$. The system defined by (4.52) may be infinite dimensional.

The convolution representation of the system defined by (4.52) is specified in terms of the impulse response of the system. We first define the impulse response, and then we give the convolution representation.

IMPULSE RESPONSE   The *impulse response* $h(t)$ of the system defined by (4.52) is the output response of the system when the input $x(t)$ is the unit impulse $\delta(t)$ with no initial energy in the system at time $t = 0^-$. In terms of the input/output operator F, we have

$$h(t) = (F\delta)(t).$$

The impulse response $h(t)$ could be determined experimentally by applying the unit impulse to the system and measuring the resulting output response. To carry out this experiment, it would be necessary to apply to the system a large-amplitude short-duration pulse, but in practice it is often not possible to apply such an input to a system. Later in this section, we show that the impulse response can be determined by measuring the output response resulting from a nonimpulsive input.

The impulse response $h(t)$ may be computable from some known mathematical representation of the given system. For instance, suppose that the system is modeled by the input/output differential equation (2.1) studied in Chapter 2. The impulse response $h(t)$ is equal to the solution to (2.1) with $x(t) = \delta(t)$ and with the initial conditions $y^{(i)}(0^-) = 0$ for $i = 0, 1, 2, \ldots, n - 1$.

EXAMPLE 4.9. Again consider the automobile on a level surface with the input/output differential equation

$$\frac{d^2y(t)}{dt^2} + \frac{k_f}{M}\frac{dy(t)}{dt} = \frac{1}{M}x(t). \tag{4.53}$$

Recall that M is the mass of the car, k_f is the friction coefficient, $y(t)$ is the position of the car at time t, and $x(t)$ is the drive or braking force applied to the car at time t. The value $h(t)$ of the impulse response of the car system at time t is the position $y(t)$ of the car at time t resulting from the impulsive force $\delta(t)$ applied to the car at time $t = 0$ with no initial energy in the system at time $t = 0^-$. Here no initial energy at time $t = 0^-$ means that both the position and velocity of the car are zero at time $t = 0^-$. The impulse response $h(t)$ is equal to the solution of (4.53) with input $x(t) = \delta(t)$ and with initial conditions $y(0^-) = 0$, $\dot{y}(0^-) = 0$. From the results in Section 1.5 [see (1.29)], we have that the solution of (4.53) with arbitrary input $x(t)$ and with zero initial conditions is given by

$$y(t) = \int_0^t \frac{1}{k_f}\left\{1 - \exp\left[\frac{-k_f}{M}(t - \lambda)\right]\right\} x(\lambda)\, d\lambda. \tag{4.54}$$

Setting $x(t) = \delta(t)$ in (4.54), we obtain

$$y(t) = \frac{1}{k_f}\left\{1 - \exp\left[\left(\frac{-k_f}{M}\right)t\right]\right\}, \qquad t \geq 0. \tag{4.55}$$

Thus the impulse response $h(t)$ of the car system is equal to $y(t)$ given by (4.55).

Again consider the system defined by (4.52). Since the system is assumed to be causal and $\delta(t) = 0$ for all $t < 0$, the impulse response $h(t)$ is zero for all $t < 0$. Further, by time invariance of the system, for any positive real number λ, the output response resulting from the λ-second right shift of the unit impulse $\delta(t)$ must be equal to the λ-second right shift of the impulse response $h(t)$. In other words, the response to the input $\delta(t - \lambda)$ is equal to $h(t - \lambda)$ for all real numbers $\lambda > 0$. Letting $\delta_\lambda(t) = \delta(t - \lambda)$, we have

$$h(t - \lambda) = (F\delta_\lambda)(t), \quad \text{all real numbers } \lambda > 0. \tag{4.56}$$

CONVOLUTION REPRESENTATION   We can express the output response resulting from an arbitrary input $x(t)$ in terms of the impulse response as follows. Suppose that $x(t) = 0$ for all $t < 0$. Then if there is no initial energy in the system at time $t = 0$, we claim that

$$y(t) = (h * x)(t) = \int_0^t h(\lambda)x(t - \lambda)\, d\lambda, \quad t \geq 0. \tag{4.57}$$

By commutativity of convolution, we can rewrite (4.57) in the form

$$y(t) = (x * h)(t) = \int_0^t x(\lambda)h(t - \lambda)\, d\lambda, \quad t \geq 0. \tag{4.58}$$

By (4.57) or (4.58), the output response $y(t)$ resulting from an arbitrary input $x(t)$ with no initial energy in the system prior to the application of $x(t)$ is equal to the convolution of the impulse response $h(t)$ with $x(t)$. The relationship (4.57) [or (4.58)] is called the *convolution representation* of the system. The convolution representation is a time-domain model of the system since the components of (4.57) are functions of the continuous-time variable t. Note that (4.57) is a natural continuous-time counterpart of the convolution representation (4.22) in the discrete-time case.

Comparing (4.52) and (4.57), we see that the input/output operator F of the system is given by

$$(Fx)(t) = (h * x)(t), \quad t \geq 0. \tag{4.59}$$

Hence, as in the discrete-time case, the input/output operator F of a linear time-invariant continuous-time system is a *convolution operator*; that is, F acts on an arbitrary input $x(t)$ by convolving it with the impulse response $h(t)$. Since the input/output operator F is completely specified in terms of the impulse response $h(t)$, the system is completely determined by $h(t)$.

PROOF OF THE CONVOLUTION RELATIONSHIP   We shall verify the convolution representation (4.57). Let $x(t)$ be an arbitrary input with $x(t) = 0$ for all $t < 0$. For each positive real number τ, let v_τ denote the continuous-time signal defined by

$$v_\tau(t) = \begin{cases} \int_0^\tau x(\lambda)\delta_\lambda(t)\,d\lambda, & 0 \le t \le \tau \\ \\ 0, & \text{all other } t. \end{cases} \qquad (4.60)$$

By definition of $\delta_\lambda(t)$, we have

$$\begin{aligned} v_\tau(t) &= \int_0^\tau x(\lambda)\delta(t - \lambda)\,d\lambda, & 0 \le t \le \tau \\ &= x(t), & 0 \le t \le \tau. \end{aligned}$$

Since $v_\tau(t) = x(t)$ for $0 \le t \le \tau$, by causality the output response resulting from v_τ is equal to the output response resulting from x for $0 \le t \le \tau$; that is,

$$(Fv_\tau)(t) = (Fx)(t) \qquad \text{for } 0 \le t \le \tau. \qquad (4.61)$$

Since v_τ is equal to the finite integral in (4.60), when we apply the input/output operator F to v_τ, we obtain

$$(Fv_\tau)(t) = \int_0^\tau x(\lambda)(F\delta_\lambda)(t)\,d\lambda, \qquad 0 \le t \le \tau. \qquad (4.62)$$

Actually, in order to move the operator F inside the integral as we did in (4.62), in addition to being linear, the operator F must be "well behaved" (e.g., F maps continuous functions into continuous functions). This condition is usually satisfied, so we shall not consider it.

Now by (4.56),

$$(F\delta_\lambda)(t) = h(t - \lambda), \qquad \text{all } \lambda > 0,$$

and thus (4.62) becomes

$$(Fv_\tau)(t) = \int_0^\tau x(\lambda)h(t - \lambda)\,d\lambda, \qquad 0 \le t \le \tau. \qquad (4.63)$$

Since the real number τ is arbitrary, we can take $\tau = t$ in (4.61) and (4.63), which gives

$$(Fv_\tau)(t) = (Fx)(t) = \int_0^t x(\lambda)h(t - \lambda)\,d\lambda, \qquad t \ge 0. \qquad (4.64)$$

The term on the far right side of (4.64) is equal to the convolution $(x * h)(t)$. By commutativity of convolution, $(x * h)(t) = (h * x)(t)$, and thus from (4.64),

$$y(t) = (Fx)(t) = (h * x)(t).$$

We have therefore verified the convolution relationship (4.57).

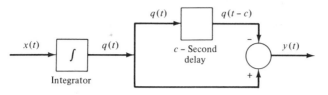

Figure 4.12. Realization of system with $h(t) = u(t) - u(t - c)$.

EXAMPLE 4.10. Given a fixed positive real number c, consider the system with impulse response

$$h(t) = u(t) - u(t - c).$$

We claim that this system can be realized using one integrator and the c-second time delay shown in Figure 4.12. To verify that $h(t) = u(t) - u(t - c)$ is the impulse response of the system shown in the figure, we must show that the output response $y(t)$ of the system is equal to $u(t) - u(t - c)$ when the input $x(t)$ is the unit impulse $\delta(t)$ with no initial energy in the system at time $t = 0^-$. For the system in the figure, no initial energy means that the output $q(t)$ of the integrator is zero when $t = 0^-$ and there is nothing stored in the delay line at time $t = 0^-$. The latter condition is equivalent to requiring that $q(t) = 0$ for $-c \le t < 0$. By definition of the integrator, if $x(t) = \delta(t)$, the output $q(t)$ of the integrator is the integral of the input, and thus $q(t) = u(t)$. Then the output of the delayor is $u(t - c)$ and

$$y(t) = u(t) - u(t - c),$$

which gives the desired result. Now if the system has zero initial energy at time $t = 0$, by the convolution representation (4.58), the output response $y(t)$ resulting from an arbitrary input $x(t)$ with $x(t) = 0$ for $t < 0$ is given by

$$y(t) = (x * h)(t) = \int_0^t x(\lambda)h(t - \lambda)\, d\lambda. \tag{4.65}$$

Since $h(t) = 1$ for $0 \le t < c$ and $h(t) = 0$ for all other t, we have

$$h(t - \lambda) = \begin{cases} 1, & 0 \le t - \lambda < c \\ 0, & \text{all other } t - \lambda. \end{cases}$$

Now $0 \le t - \lambda < c$ is equivalent to $-t \le -\lambda < c - t$, which is equivalent to $t \ge \lambda > t - c$, so we have

$$h(t - \lambda) = \begin{cases} 1, & t - c < \lambda \le t \\ 0, & \text{all other } \lambda. \end{cases} \tag{4.66}$$

Inserting (4.66) into (4.65), we get

$$y(t) = \int_{t-c}^{t} x(\lambda) \, d\lambda, \tag{4.67}$$

The expression (4.67) for the output response $y(t)$ shows that at time t, the system processes an input $x(t)$ by integrating the input over the past c-second interval. Thus the system is a *finite-time integrator*.

REPRESENTATION IN TERMS OF THE STEP RESPONSE   Again consider the system given by the input/output relationship (4.52). Let $g(t)$ denote the output response of the system when the input $x(t)$ is the unit-step function $u(t)$ with no initial energy in the system at time $t = 0$. The response $g(t)$ is called the *step response* of the system. In terms of the input/output operator F of the system, we have

$$(Fu)(t) = g(t).$$

From the convolution representation (4.57), we also have

$$g(t) = (h * u)(t). \tag{4.68}$$

Differentiating both sides of (4.68) and using the derivative property of convolution, we get

$$\dot{g}(t) = (\dot{h} * u)(t) = (h * \dot{u})(t). \tag{4.69}$$

Since $\dot{u}(t) = \delta(t)$ and $h(t) = (h * \delta)(t)$, from (4.69) we get

$$\dot{g}(t) = h(t). \tag{4.70}$$

Hence the impulse response $h(t)$ is equal to the derivative of the step response $g(t)$. Integrating both sides of (4.70), we get

$$g(t) = \int_{0}^{t} h(\lambda) \, d\lambda, \tag{4.71}$$

and thus the step response is equal to the integral of the impulse response.

Now let $x(t)$ be an arbitrary input with $x(t) = 0$ for $t < 0$. Inserting (4.70) into the convolution relationship (4.57), we obtain

$$y(t) = (\dot{g} * x)(t). \tag{4.72}$$

By the derivative property of convolution, $\dot{g} * x = g * \dot{x}$, and thus we can rewrite (4.72) in the form

$$y(t) = (g * \dot{x})(t). \tag{4.73}$$

The relationships (4.72) and (4.73) are two additional convolution representations of the system, both of which are given in terms of the step response $g(t)$ of the system. Either (4.72) or (4.73) may be used to compute the output response resulting from an arbitrary input $x(t)$ with no initial energy in the system at time $t = 0$.

4.5

NUMERICAL CONVOLUTION

Consider the causal linear time-invariant continuous-time system given by the convolution relationship

$$y(t) = (h * x)(t) = \int_0^t h(\lambda)x(t - \lambda) \, d\lambda. \tag{4.74}$$

In the first part of this section we do not assume that the input $x(t)$ is necessarily zero for $t < 0$, and thus we must take the upper limit of the integral in (4.74) to be infinity. Therefore,

$$y(t) = \int_0^\infty h(\lambda)x(t - \lambda) \, d\lambda. \tag{4.75}$$

In this section we give a numerical procedure for computing the convolution integral in (4.75).

Let T be a fixed positive real number. We can discretize the convolution integral in (4.75) by setting $t = kT$, where k is an integer variable. This gives

$$y(kT) = \int_0^\infty h(\lambda)x(kT - \lambda) \, d\lambda. \tag{4.76}$$

We can evaluate the convolution integral in (4.76) by breaking the integral up into a sum of integrals over T-second intervals; that is, we have

$$y(kT) = \int_0^T h(\lambda)x(kT - \lambda) \, d\lambda + \int_T^{2T} h(\lambda)x(kT - \lambda) \, d\lambda + \cdots$$
$$+ \int_{iT}^{iT+T} h(\lambda)x(kT - \lambda) \, d\lambda + \cdots. \tag{4.77}$$

Using the summation symbol, we can write (4.77) in the form

$$y(kT) = \sum_{i=0}^\infty \int_{iT}^{iT+T} h(\lambda)x(kT - \lambda) \, d\lambda. \tag{4.78}$$

Now if T is taken suitably small, for each positive integer i we can approximate $h(\lambda)$ and $x(kT - \lambda)$ on the interval $iT \leq \lambda < iT + T$ by

$$h(\lambda) = h(iT), \qquad iT \leq \lambda < iT + T \qquad (4.79)$$
$$x(kT - \lambda) = x(kT - iT), \qquad iT \leq \lambda < iT + T.$$

Inserting the approximations (4.79) into (4.78), we have the following approximation of $y(kT)$:

$$y(kT) = \sum_{i=0}^{\infty} \int_{iT}^{iT+T} h(iT)x(kT - iT) \, d\lambda. \qquad (4.80)$$

Since $h(iT)x(kT - iT)$ is independent of the variable λ of integration, we can move this product outside of the integrand of the integrals in (4.80). This gives

$$y(kT) = \sum_{i=0}^{\infty} \left[\int_{iT}^{iT+T} (1) \, d\lambda \right] h(iT)x(kT - iT)$$

$$= \sum_{i=0}^{\infty} \left[\lambda \Big|_{\lambda=iT}^{\lambda=iT+T} \right] h(iT)x(kT - iT)$$

$$= \sum_{i=0}^{\infty} Th(iT)x(kT - iT). \qquad (4.81)$$

The expression (4.81) is an approximation of the output response $y(kT)$ at the time points $t = kT$, where k takes on integer values. In general, the approximation will be more accurate the smaller T is.

It is important to observe that (4.81) can be viewed as the convolution representation of a linear time-invariant discrete-time system with unit-pulse response equal to $Th(kT)$. This discrete-time system can be viewed as a discrete-time simulation of the given continuous-time system. Thus we can study any linear time-invariant continuous-time system with impulse response $h(t)$ in terms of the linear time-invariant discrete-time system with unit-pulse response $Th(kT)$. The discrete-time simulation defined by (4.81) is not the same as the discrete-time simulations constructed in Section 3.4 by discretizing the input/output differential equation of the given continuous-time system. Hence we have a second approach to the discretization in time of continuous-time systems.

Since the expression (4.81) for $y(kT)$ is given in terms of a convolution sum, $y(kT)$ can be computed simply by evaluating the sum. Hence we have a numerical procedure for computing the convolution integral in the continuous-time case. In particular, suppose that the input $x(t)$ applied to the continuous-time system is zero for all $t < 0$. Then $x(kT) = 0$ for $k = -1, -2, \ldots$, and the approximation (4.81) for $y(kT)$ becomes

$$y(kT) = \sum_{i=0}^{k} Th(iT)x(kT - iT), \qquad k = 0, 1, 2, \ldots. \qquad (4.82)$$

A program for evaluating the finite sum in (4.82) is given in Figure A.5. The program computes the output values $y(kT)$ for $k = 0, 1, 2, \ldots, Q$. The discretized impulse response $h(kT)$ and the discretized input $x(kT)$ may be entered value by value, or they may be given in function form on lines 250 and 320 of the program. The procedure is illustrated by the following example.

EXAMPLE 4.11. Again consider the car with the input/output differential equation (4.53). In Example 4.9 we showed that the impulse response of the car is

$$h(t) = \frac{1}{k_f}\left[1 - \exp\left(-\frac{k_f}{M}t\right)\right], \qquad t \geq 0. \tag{4.83}$$

Let T be a fixed positive real number. Then the unit-pulse response of the discrete-time simulation of the car is

$$Th(kT) = \frac{T}{k_f}\left[1 - \exp\left(-\frac{k_f T}{M}k\right)\right], \qquad k = 0, 1, 2, \ldots.$$

The response $y(kT)$ of the car to an arbitrary input force $x(t)$ applied for $t \geq 0$ with zero initial position and zero initial velocity is then given (approximately) by

$$y(kT) = \sum_{i=0}^{k} \frac{T}{k_f}\left[1 - \exp\left(-\frac{k_f T}{M}i\right)\right] x(kT - iT), \qquad k \geq 0. \tag{4.84}$$

For a specific example, let us set $M = 1$, $k_f = 0.1$, and take the input $x(t)$ to be the force shown in Figure 4.13.

Before running the program in Figure A.5, on line 250 we set

$$\text{DEF FNH(K)} = 10*(1 - \text{EXP}(-.1*T*K)),$$

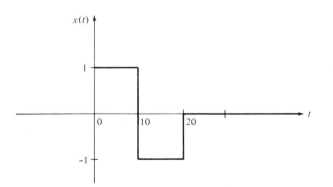

Figure 4.13. Force $x(t)$ applied to car in Example 4.11.

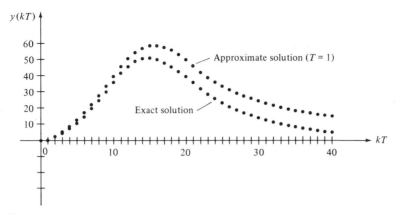

Figure 4.14. Exact and approximate solutions of output response in Example 4.11.

and on line 320, we set

$$DEF\ FNX(K) = .5*SGN(K*T + .5) - SGN(K*T - 10.5) + .5*SGN(K*T - 20.5).$$

The program was then run with $T = 1$ and $Q = 40$. The results are plotted in Figure 4.14. Evaluating the convolution relationship (4.57), we have that the exact solution is

$$y(t) = \begin{cases} 100(0.1t - 1 + e^{-0.1t}), & 0 \le t < 10 \\ -100(0.1t - 3 + (2e - 1)e^{-0.1t}), & 10 \le t < 20 \\ 100(1 - 2e + e^2)e^{-0.1t}, & t > 20. \end{cases}$$

The exact solution is also plotted in Figure 4.14. Note that the approximate solution does differ from the exact solution. To get a more accurate approximation, we can choose a smaller discretization interval T. For instance, if we take $T = 0.5$, we get the results shown in Figure 4.15. Clearly, this approximation is closer to the exact solution than the approximation with $T = 1$.

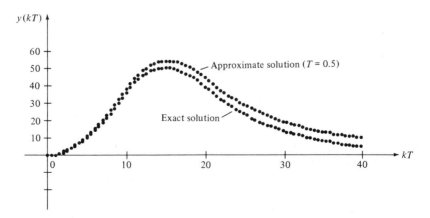

Figure 4.15. Exact and approximate solutions of output in Example 4.11.

★**4.6**

LINEAR TIME-VARYING SYSTEMS

If the system under study is time varying, it is not possible to express the output response as the convolution of the input with the impulse response (or unit-pulse response). However, in the time-varying case there still is a useful expression for the output response that reduces to the convolution relationship in the time-invariant case. In this section we describe this input/output relationship for linear time-varying systems. We begin with the discrete-time case.

DISCRETE-TIME TIME-VARYING SYSTEMS   Consider a causal linear single-input single-output discrete-time system with the input/output relationship

$$y(kT) = (Fx)(kT). \tag{4.85}$$

The system may be time varying and it may be infinite dimensional.

For each fixed integer i, let $h(kT, iT)$ denote the response of the system when the input $x(kT)$ is equal to the unit pulse $\Delta(kT - iT)$ located at $kT = iT$. We are assuming that there is no initial energy in the system before the application of $\Delta(kT - iT)$. In mathematical terms we have

$$h(kT, iT) = (F\Delta_i)(kT), \qquad k \geq i, \tag{4.86}$$

where $\Delta_i(kT) = \Delta(kT - iT)$. Note that by causality,

$$h(kT, iT) = 0, \qquad k < i.$$

Also note that

$$h(kT, 0) = h(kT),$$

where $h(kT)$ is the unit-pulse response defined in Section 4.2. The collection of responses $h(kT, iT)$ is called the *family of unit-pulse responses* of the system.

Now let $x(kT)$ be an arbitrary input with $x(kT) = 0$ for $k = -1, -2, \ldots$. Then the output response $y(kT)$ resulting from $x(kT)$ with no initial energy at time $kT = 0$ is given by

$$y(kT) = \sum_{i=0}^{k} x(iT)h(kT, iT), \qquad k \geq 0. \tag{4.87}$$

Equation (4.87) is a generalization of the convolution relationship (4.23) for linear time-invariant systems. Note that (4.87) and (4.23) are identical if

$$h(kT, iT) = h(kT - iT) \qquad \text{for all } k \geq i. \tag{4.88}$$

The condition (4.88) holds if and only if the given system is time invariant. Hence (4.87) reduces to the convolution relationship in the time-invariant case. The proof of (4.87) is similar to the proof of the convolution relationship (4.23) which we gave in Section 4.2. We leave the verification to the reader.

EXAMPLE 4.12. Consider the linear time-varying discrete-time system given by the input/output difference equation

$$y(kT + T) + a(k)y(kT) = b(k)x(kT). \tag{4.89}$$

In (4.89) the coefficients $a(k)$, $b(k)$ are arbitrary functions of k. The unit-pulse responses $h(kT, iT)$ are the solutions of (4.89) with $x(kT) = \Delta(kT - iT)$ and with $y(iT) = 0$. Solving (4.89) by recursion, we obtain:

$$h(kT, iT)$$
$$= \begin{cases} 0, & k \le i \\ b(i), & k = i + 1 \\ (-1)^{k-i-1}a(k - 1)a(k - 2) \cdots a(i + 1)b(i), & k \ge i + 2. \end{cases} \tag{4.90}$$

Inserting (4.90) into (4.87), we obtain the input/output relationship:

$$y(kT)$$
$$= \begin{cases} 0, & k = 0 \\ x(0)b(0), & k = 1 \\ \displaystyle\sum_{i=0}^{k-2} x(iT)(-1)^{k-i-1}a(k - 1)a(k - 2) \cdots a(i + 1)b(i), & k \ge 2. \end{cases}$$

CONTINUOUS-TIME CASE Now suppose that the system under consideration is a causal linear single-input single-output continuous-time system with the input/output relationship

$$y(t) = (Fx)(t). \tag{4.91}$$

For each fixed real number λ, let $h(t, \lambda)$ denote the output response resulting from the unit impulse $x(t) = \delta(t - \lambda)$ located at $t = \lambda$. The system is assumed to be at rest prior to the application of $\delta(t - \lambda)$. In mathematical terms we have

$$h(t, \lambda) = (F\delta_\lambda)(t),$$

where $\delta_\lambda(t) = \delta(t - \lambda)$. By causality

$$h(t, \lambda) = 0, \qquad t < \lambda.$$

Note that

$$h(t, 0) = h(t),$$

where $h(t)$ is the impulse response defined in Section 4.4. The collection of responses $h(t, \lambda)$ is called the *family of impulse responses* of the system.

Now if $x(t)$ is an arbitrary input with $x(t) = 0$ for $t < 0$, the response resulting from $x(t)$ is

$$y(t) = \int_0^t x(\lambda)h(t, \lambda) \, d\lambda, \qquad t \geq 0. \tag{4.92}$$

The input/output relationship (4.92) is identical to the convolution relationship (4.58) if

$$h(t, \lambda) = h(t - \lambda) \qquad \text{for all } t \geq \lambda. \tag{4.93}$$

The condition (4.93) is satisfied if and only if the given system is time invariant, and thus (4.92) reduces to the convolution expression in the time-invariant case.

EXAMPLE 4.3.   Consider the linear time-varying continuous-time system given by the input/output differential equation

$$\dot{y}(t) + a(t)y(t) = b(t)x(t), \tag{4.94}$$

where $a(t)$ and $b(t)$ are arbitrary functions of t. The impulse responses $h(t, \lambda)$ are the solutions of (4.94) with $x(t) = \delta(t - \lambda)$ and with $y(\lambda^-) = 0$. From the theory of differential equations with time-varying coefficients, we have

$$h(t, \lambda) = \exp\left[-\int_\lambda^t a(\beta) \, d\beta \right] b(\lambda). \tag{4.95}$$

Inserting (4.95) into (4.92), we obtain the input/output relationship

$$y(t) = \int_0^t x(\lambda) \exp\left[-\int_\lambda^t a(\beta) \, d\beta \right] b(\lambda) \, d\lambda, \qquad t \geq 0.$$

Although we were able to give $h(t, \lambda)$ in analytical form in the above example, in general there is no closed-form analytical expression for the impulse responses in the time-varying case. In particular, this is usually the case for linear time-varying systems given by a second-order or higher-order input/output differential equation with time-varying coefficients.

PROBLEMS

Chapter 4

4.1   For the discrete-time signals $x(kT)$ and $v(kT)$ given in each of the following parts, compute the convolution $(x * v)(kT)$ for *all* integers $k \geq 0$.

(a) $x(0) = 1$, $x(T) = -2$, $x(2T) = 5$, $x(3T) = 3$, $x(kT) = 0$ for all other integers k; $v(0) = 4$, $v(T) = -7$, $v(2T) = 2$, $v(kT) = 0$ for all other integers k.

(b) $x(kT) = 2^k$ for all integers $k \le 3$, including $k = -1, -2, -3, \ldots$, $x(kT) = 0$ for all integers $k > 3$; $v(0) = 2$, $v(T) = -3$, $v(2T) = 0$, $v(3T) = 6$, $v(kT) = 0$ for all other integers k.

(c) $x(kT) = 1/k$ for $k = 2, 3, 4, 5$, $x(kT) = 0$ for all other integers k; $v(2T) = -2$, $v(3T) = -5$, $v(kT) = 0$ for all other integers k.

(d) $x(kT) = u(kT)$, $v(kT) = u(kT)$, where $u(kT)$ is the discrete-time step function.

(e) $x(kT) = u(kT)$, $v(kT) = \ln k$ for all integers $k \ge 1$, $v(kT) = 0$ for all integers $k < 1$.

(f) $x(kT) = \Delta(kT) - \Delta(kT - 2T)$, where $\Delta(kT)$ is the unit pulse concentrated at $kT = 0$; $v(kT) = \cos(\pi k/3)$ for all integers $k \ge 0$, $v(kT) = 0$ for all integers $k < 0$.

4.2 A discrete-time system is given by the input/output difference equation

$$y(kT + T) = 0.5y(kT) + x(kT) + 2x(kT + T).$$

(a) Compute the unit-pulse response $h(kT)$ for all integers $k \ge 0$.

(b) Compute the step response $g(kT)$ for all integers $k \ge 0$.

4.3 A linear time-invariant discrete-time system has the unit-pulse response

$$h(kT) = (-1)^k \qquad \text{for all integers } k \ge 0.$$

(a) Compute the step response $g(kT)$ for all integers $k \ge 0$.

(b) Compute the output response $y(kT)$ for all integers $k \ge 0$ when the input is $x(kT) = u(kT) - u(kT - 5T)$ with zero initial energy in the system prior to the application of the input.

4.4 A linear time-invariant discrete-time system has the step response

$$g(kT) = \frac{2^k}{k + 1} \left[\sin \frac{\pi k}{2} \right] u(kT).$$

(a) Compute the unit-pulse response $h(kT)$ for all integers $k \ge 0$.

(b) Compute the output response $y(kT)$ for all integers $k \ge 0$ when $x(0) = 2$, $x(T) = -1$, $x(2T) = 4$, $x(kT) = 0$ for all other integers k. Assume that there is no initial energy in the system at time $kT = 0$.

4.5 Let $r(kT)$ denote the discrete-time unit-ramp function defined by $r(kT) = kTu(kT)$.

(a) Express the unit-pulse function $\Delta(kT)$ in terms of $r(kT)$ and time shifts of $r(kT)$.

(b) Let $h(kT)$ denote the unit-pulse response of a linear time-invariant discrete-time system. Using your result in part (a), derive an expression for $h(kT)$ in terms of the response of the system to $r(kT)$.

4.6 Two linear time-invariant discrete-time systems are connected as shown in Figure P4.6. System 1 has unit-pulse response $h_1(kT)$ and system 2 has unit-pulse response $h_2(kT)$. Derive an expression for the unit-pulse response of the overall system. Express your answer in terms of $h_1(kT)$ and $h_2(kT)$.

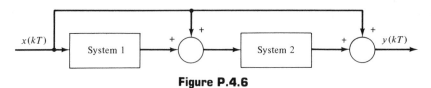

Figure P.4.6

4.7 For the continuous-time signals $x(t)$ and $v(t)$ shown in Figure P4.7, compute the convolution $(x * v)(t)$ for all $t \geq 0$.

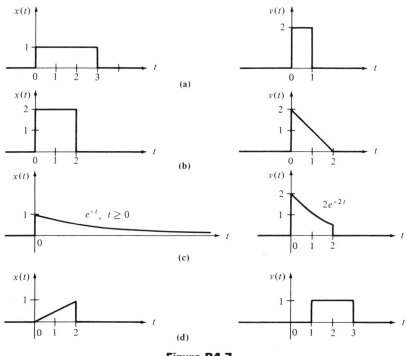

Figure P4.7

4.8 Compute the impulse response $h(t)$ of the small-signal model of the pendulum given by the input/output differential equation

$$\frac{d^2\theta(t)}{dt^2} + \frac{MgL}{I}\theta(t) = \frac{L}{I}x(t).$$

Here M, g, L, and I are arbitrary positive numbers.

HINT: Use the results derived in Problem 1.12.

4.9 A linear time-invariant continuous-time system has impulse response

$$h(t) = e^{-t} + \sin t, \qquad t \geq 0.$$

(a) Compute the step response $g(t)$ for all $t \geq 0$.
(b) Compute the output response $y(t)$ for all $t \geq 0$ when the input is $u(t) - u(t - 2)$ with no initial energy in the system at time $t = 0$.

4.10 A linear time-invariant continuous-time system has impulse response $h(t) = (\sin t)u(t - 2)$. Compute the output response $y(t)$ for all $t \geq 0$ when $x(t) = u(t) - u(t - 1)$ with no initial energy in the system.

4.11 A linear time-invariant continuous-time system has impulse response $h(t)$ displayed in Figure P4.11a. Find the output response $y(t)$ for $4 \leq t \leq 6$ only resulting from the input shown in Figure P4.11b. Assume that there is no initial energy.

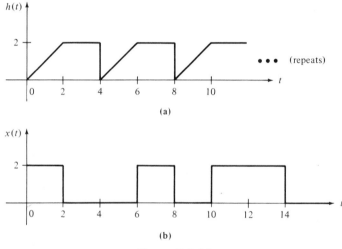

(a)

(b)

Figure P4.11

4.12 A linear time-invariant continuous-time system has impulse response $h(t)$ shown in Figure P4.12. The input $x(t) = 2e^{-t}[u(t) - u(t - 1)]$ is applied to the system with no initial energy at time $t = 0$. Compute the resulting output response $y(t)$ for $0 \leq t \leq 2$ only.

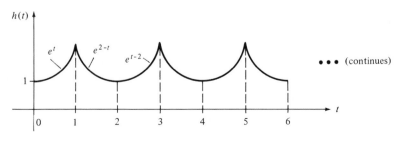

Figure P4.12

4.13 A linear time-invariant continuous-time system has impulse response $h(t)$ given by

$$h(t) = \sin t[u(t) - u(t - \pi)] + \sin t[u(t - 2\pi) - u(t - 3\pi)].$$

The input

$$x(t) = 4[u(t) - u(t - \pi)] - [u(t - \pi) - u(t - 2\pi)]$$
$$+ t[u(t - 3\pi) - u(t - 4\pi)]$$

is applied to the system with no initial energy. By folding and shifting $x(t)$, compute the resulting output response $y(t)$ for $2\pi \le t \le 3\pi$.

4.14 A linear time-invariant continuous-time system has impulse response $h(t)$. The output response of the system resulting from an input $x(t)$ is equal to $y(t)$, where again there is no initial energy in the system prior to the application of $x(t)$. Derive an expression for the response to the following inputs. Express your answer in terms of $y(t)$ and/or $x(t)$.

(a) $\dfrac{dx(t)}{dt}$.

(b) $\dfrac{d^2x(t)}{dt^2}$.

(c) $\displaystyle\int_{-\infty}^{t} x(\lambda)\, d\lambda$.

(d) $(x * x)(t)$.

4.15 A linear time-invariant continuous-time system has step response $g(t)$. The system receives the input

$$x(t) = \sum_{i=0}^{N-1} c_i[u(t - t_i) - u(t - t_{i+1})],$$

where $t_0 < t_1 < \cdots < t_N$, N is a positive integer, and the c_i are constants. Express the resulting output in terms of $g(t)$ and time shifts of $g(t)$. Assume that there is no initial energy in the system prior to the application of the input.

4.16 Consider the linear time-invariant continuous-time system with the step response $g(t)$ displayed in Figure P4.16a. Using your result in Problem 4.15, compute the output response resulting from the input shown in Figure P4.16b. Assume that there is no initial energy.

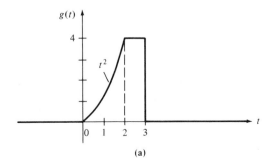

(a)

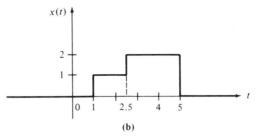

(b)

Figure P4.16

4.17 For the system shown in Figure P4.17:
(a) Compute the impulse response $h(t)$.
(b) Compute the output response $y(t)$ resulting from input $x(t) = e^{-2t}u(t)$ with no initial energy in the system at time $t = 0$.

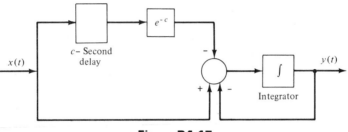

Figure P4.17

4.18 Consider the series *RLC* circuit shown in Figure P4.18a. The circuit is equivalent to the cascade connection shown in Figure P4.18b; that is, the system in Figure P4.18b has the same input/output differential equation as the *RLC* circuit.
(a) Find the impulse responses of each of the subsystems in Figure P4.18b.
(b) Using your results in part (a), compute the impulse response of the *RLC* circuit.

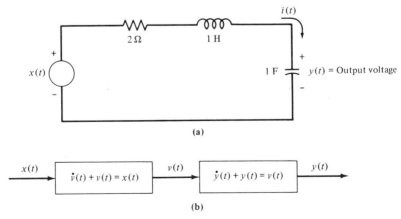

Figure P4.18

(c) Suppose that $x(t) = (\sin t)u(t)$. Using the numerical-convolution program in Figure A.5, compute the values $y(kT)$ of the resulting output response for $k = 0, 1, 2, \ldots, 50$ and $T = 0.2$.

(d) Repeat part (c) with $T = 0.1$.

(e) Again suppose that $x(t) = (\sin t)u(t)$. Use the Euler approximation and the program in Figure A.3 to compute $y(kT)$ for $k = 0, 1, 2, \ldots, 50$ with $T = 0.2$. Take $y(0) = 0$, $\dot{y}(0) = 0$. Compare your results with those obtained in part (c). Which approximation scheme gives the better results?

(f) Repeat part (e) with $T = 0.1$.

4.19 Consider the automobile on a level surface with the time-varying mass $M(t)$ and with velocity model

$$\frac{dv(t)}{dt} + \frac{k_f}{M(t)} v(t) = \frac{1}{M(t)} x(t).$$

(a) Suppose that $M(t) = 1 - 0.01t[u(t) - u(t - 10)]$. Compute the impulse responses $h(t, \lambda)$ for all λ and all $t \geq \lambda$.

(b) Now suppose that the mass is piecewise constant; that is, $M(t) = M_i$ for $iT \leq t < iT + T$. Discretize the system using the integral approximation studied in Section 3.4. Take the discretization interval to be T.

(c) For the discretization computed in part (b), compute the unit-pulse responses $h(kT, iT)$ for all integers i and integers $k \geq i$.

The Laplace Transform and the Transfer Function Representation

In this chapter we first introduce the Laplace transform, and then we apply this construct to generate the transfer function representation of a linear time-invariant continuous-time system. We shall see that the Laplace transform yields an algebraic relationship between the inputs and outputs of a system. This will give us a purely algebraic procedure for computing the output response resulting from a given input.

The Laplace transform is named after Pierre Simon Laplace (1749–1827), a French mathematician and astronomer. We define the Laplace transform of a continuous-time signal $x(t)$ in Section 5.1. Then in Section 5.2, we study the basic properties of the transform. Using these properties, we show that one can generate many new transforms from a small set of transforms. In Sections 5.3 and 5.4 we utilize partial-fraction expansions to compute the inverse Laplace transform of a given transform.

The transfer function representation of a single-input single-output linear time-invariant continuous-time system is developed in Sections 5.5 and 5.6. In Section 5.5 we show that the transfer function model can be constructed by taking the Laplace transform of the input/output convolution relationship. When the given system is finite dimensional, we show in Section 5.6 that the transfer function representation can be generated by taking the Laplace transform of the input/output differential equation.

In contrast to the time-domain models studied in the preceding chapters, the transfer function model is specified in terms of functions of a complex variable s. In a later chapter we will see that the complex variable s can be interpreted as a complex frequency variable. Since the Laplace transform and the transfer function model are given in terms of functions of a complex variable, the reader needs to be familiar with the fundamentals of complex variables. For those who need a review, a brief introductory treatment is given in Appendix B.

5.1

LAPLACE TRANSFORM

Given a function $x(t)$ of the continuous-time variable t, the *two-sided* (or *bilateral*) *Laplace transform* of $x(t)$, denoted by $X(s)$, is a function of the complex variable $s = \sigma + j\omega$ defined by

$$X(s) = \int_{-\infty}^{\infty} x(t)e^{-st}\, dt. \tag{5.1}$$

The Laplace transform $X(s)$ is a complex-valued function of the complex variable s. In other words, given a complex number s, the value $X(s)$ of the transform at the point s is in general a complex number.

The *one-sided* (or *unilateral*) *Laplace transform* of $x(t)$, also denoted by $X(s)$, is defined by

$$X(s) = \int_{0}^{\infty} x(t)e^{-st}\, dt. \tag{5.2}$$

By (5.2), we see that the one-sided transform depends only on the values of the signal $x(t)$ for $t \geq 0$. This is the reason the definition (5.2) of the transform is called the one-sided Laplace transform. We can apply the one-sided transform to signals $x(t)$ that are nonzero for $t < 0$; however, any nonzero values of $x(t)$ for $t < 0$ will not be recomputable from the one-sided transform.

If $x(t)$ is zero for all $t < 0$, the expression (5.1) reduces to (5.2), and thus in this case the one-sided and two-sided Laplace transforms are the same. In this chapter we restrict our attention to the one-sided transform, which we refer to as the Laplace transform. The two-sided Laplace transform will be encountered again in Chapter 9 when we study the Fourier transform.

In this book we shall always denote time signals by lowercase letters and the Laplace transform of time signals by uppercase letters.

Let Λ denote the set of all positive or negative real numbers σ such that

$$\int_{0}^{\infty} |x(t)|e^{-\sigma t}\, dt < \infty. \tag{5.3}$$

If the set Λ is empty, that is, there is no real number σ such that (5.3) is satisfied, the function $x(t)$ does not have a Laplace transform (which converges "absolutely"). Most functions arising in engineering do have a Laplace transform, and thus the set Λ is not empty in many cases of interest.

If Λ is not empty, let σ_{min} denote the minimal element of the set Λ; that is, σ_{min} is the smallest number such that

$$\sigma \in \Lambda \qquad \text{for all } \sigma > \sigma_{min}.$$

The set of all complex numbers s such that

$$\text{Re } s > \sigma_{min}, \tag{5.4}$$

where Re s = real part of s, is called the *region of absolute convergence* of the Laplace transform of $x(t)$. For any complex number s such that (5.4) is satisfied, the integral in (5.2) exists, and thus the Laplace transform $X(s)$ exists for this value of s. Hence the Laplace transform $X(s)$ of $x(t)$ is well defined (i.e., exists) for all values of s belonging to the region of absolute convergence. It should be stressed that the region of absolute convergence depends on the given function $x(t)$.

EXAMPLE 5.1. Suppose that $x(t)$ is the unit-step function $u(t)$. Then the Laplace transform $U(s)$ of $u(t)$ is given by

$$U(s) = \int_0^\infty u(t)e^{-st}\, dt$$

$$= \int_0^\infty e^{-st}\, dt$$

$$= \left. \frac{-1}{s} e^{-st} \right]_{t=0}^{t=\infty} \tag{5.5}$$

Now exp $(-st)$ evaluated at $t = \infty$ is defined by

$$e^{-s\infty} = \lim_{T \to \infty} e^{-sT}. \tag{5.6}$$

Setting $s = \sigma + j\omega$ in the right side of (5.6), we have

$$e^{-s\infty} = \lim_{T \to \infty} \exp\left[-(\sigma + j\omega)T\right]$$

$$= \lim_{T \to \infty} e^{-\sigma T} e^{-j\omega T}. \tag{5.7}$$

The limit in (5.7) exists if and only if $\sigma > 0$, which is equivalent to Re $s >$ 0. If Re $s > 0$, the limit in (5.7) is zero, and the expression (5.5) for $U(s)$ reduces to

$$U(s) = -\left(\frac{-1}{s}\right)e^0$$

$$= \frac{1}{s}.$$

We also have

$$\int_0^\infty u(t)e^{-\sigma t}\, dt < \infty$$

for all real numbers σ such that $\sigma > 0$. Thus the region of absolute convergence of $U(s)$ is the set of all complex numbers s such that Re $s > 0$.

EXAMPLE 5.2.   Let $x(t)$ be the unit impulse $\delta(t)$. The Laplace transform $X(s)$ is given by

$$X(s) = \int_{0^-}^\infty \delta(t)e^{-st}\, dt. \tag{5.8}$$

The lower limit of the integral in (5.8) must be taken to be 0^- since the impulse $\delta(t)$ is not defined at $t = 0$. Since $\delta(t) = 0$ for all $t \neq 0$, we have that

$$\delta(t)e^{-st} = \delta(t)e^0 = \delta(t).$$

Thus

$$X(s) = \int_{0^-}^\infty \delta(t)\, dt = 1. \tag{5.9}$$

Since

$$\int_{0^-}^\infty \delta(t)e^{-\sigma t}\, dt = e^0 = 1 \qquad \text{for all real numbers } \sigma,$$

the region of absolute convergence of the Laplace transform of $\delta(t)$ is the entire complex plane. Hence $X(s)$ exists and is equal to 1 for all complex numbers s.

EXAMPLE 5.3. Now let $x(t) = e^{-bt}u(t)$, where b is an arbitrary real number. The Laplace transform $X(s)$ is given by

$$
\begin{aligned}
X(s) &= \int_0^\infty e^{-bt}e^{-st}\, dt \\[2mm]
&= \int_0^\infty \exp\left[-(s+b)t\right]\, dt \\[2mm]
&= \frac{-1}{s+b}\exp\left[-(s+b)t\right]\Big]_{t=0}^{t=\infty}.
\end{aligned}
$$

(5.10)

To evaluate the right side of (5.10) at $t = \infty$, we have

$$
\exp\left[-(s+b)\infty\right] = \lim_{T\to\infty}\exp\left[-(s+b)T\right].
$$

(5.11)

Setting $s = \sigma + j\omega$ in the right side of (5.11), we have

$$
\exp\left[-(s+b)\infty\right] = \lim_{T\to\infty}\exp\left[-(\sigma+b)T\right]e^{-j\omega T}.
$$

(5.12)

The limit in (5.12) exists if and only if $\sigma + b > 0$, in which case the limit is zero, and the Laplace transform $X(s)$ is

$$
X(s) = \frac{1}{s+b}.
$$

(5.13)

The region of absolute convergence of the transform $X(s)$ given by (5.13) is the set of all complex numbers s such that Re $s > -b$.

The Laplace transforms of many functions of interest can be determined by table lookup. Hence it is often not necessary to have to evaluate the integral in (5.2) in order to compute the transform of a given function. Using the properties of the Laplace transform given later, we will derive the transforms for a collection of common functions. The results will be displayed in a table of transforms.

INVERSE LAPLACE TRANSFORM Given a function $x(t)$ with Laplace transform $X(s)$, we can compute $x(t)$ from $X(s)$ by taking the inverse Laplace transform of $X(s)$. The inverse transform operation is given by

$$
x(t) = \frac{1}{2\pi j}\int_{c-j\infty}^{c+j\infty} X(s)e^{st}\, ds.
$$

(5.14)

The integral in (5.14) is evaluated along the path $s = c + j\omega$ in the complex plane from $c - j\infty$ to $c + j\infty$, where c is any fixed real number for which

TABLE 5.1. Three Basic Transform Pairs.

$$u(t) \leftrightarrow \frac{1}{s}$$

$$\delta(t) \leftrightarrow 1$$

$$e^{-bt} \leftrightarrow \frac{1}{s + b}, \quad b \text{ real or complex}$$

$s = c$ is a point in the region of absolute convergence of $X(s)$. For a detailed treatment of complex integration, see Churchill et. al. [1976].

The integral in (5.14) is usually difficult to evaluate, and thus it is desirable to avoid having to use (5.14) to compute the inverse transform. A common technique for computing the inverse transform when $X(s)$ is a ratio of polynomials in s is based on a partial-fraction expansion of $X(s)$. We consider this in Section 5.3.

From here on, we denote the fact that $X(s)$ is the Laplace transform of $x(t)$ by writing the *transform pair*

$$x(t) \leftrightarrow X(s). \tag{5.15}$$

The double arrow in (5.15) means that $X(s)$ is the Laplace transform of $x(t)$, and conversely, that $x(t)$ is the inverse Laplace transform of $X(s)$. When we give a transform pair, we do not specify the region of absolute convergence of the transform. The reason for this is that in most applications of the transform, it is not necessary to consider the region of absolute convergence (as long as the transforms do have a region of convergence).

The transform pairs for the signals considered in the above examples are given in Table 5.1.

Note that in the third transform pair in Table 5.1, the scalar b may be real or complex. The verification of this transform pair in the case that b is complex is an easy modification of the derivation given in Example 5.3. We omit the details.

5.2

PROPERTIES OF THE LAPLACE TRANSFORM

The Laplace transform satisfies a number of properties that are useful in a wide range of applications. In particular, by using these properties, it is possible to derive many new transform pairs from a basic set of pairs. In this section we state and prove several fundamental properties of the transform. Examples are given to illustrate the process of generating new transform pairs from a given pair.

LINEARITY   If $x(t) \leftrightarrow X(s)$ and $v(t) \leftrightarrow V(s)$, then for any real or complex numbers $a, b,$

$$ax(t) + bv(t) \leftrightarrow aX(s) + bV(s). \tag{5.16}$$

This property says that the Laplace transform is a linear operation. To prove linearity, we shall compute the Laplace transform of $ax(t) + bv(t)$. By definition of the transform, we have

$$ax(t) + bv(t) \leftrightarrow \int_0^\infty [ax(t) + bv(t)]e^{-st}\,dt.$$

By linearity of integration, we have

$$\int_0^\infty [ax(t) + bv(t)]e^{-st}\,dt = a\int_0^\infty x(t)e^{-st}\,dt + b\int_0^\infty v(t)e^{-st}\,dt.$$

Hence

$$ax(t) + bv(t) \leftrightarrow aX(s) + bV(s).$$

EXAMPLE 5.4. Consider the signal $u(t) + e^{-t}u(t)$. From Table 5.1 we have the transform pairs

$$u(t) \leftrightarrow \frac{1}{s} \quad \text{and} \quad e^{-t}u(t) \leftrightarrow \frac{1}{s+1}.$$

Thus, using linearity, we obtain the transform pair

$$u(t) + e^{-t}u(t) \leftrightarrow \frac{1}{s} + \frac{1}{s+1} = \frac{2s+1}{s(s+1)}.$$

RIGHT SHIFT IN TIME If $x(t) \leftrightarrow X(s)$, then for any positive real number c,

$$x(t - c)u(t - c) \leftrightarrow e^{-cs} X(s). \tag{5.17}$$

The function $x(t - c)u(t - c)$ is a c-second right shift of $x(t)u(t)$. From (5.17) we see that a c-second right shift in the time domain corresponds to multiplication by e^{-cs} in the Laplace transform domain.

We shall prove the right-shift property. By definition of the Laplace transform, we have

$$x(t - c)u(t - c) \leftrightarrow \int_0^\infty x(t - c)u(t - c)e^{-st}\,dt$$

$$x(t - c)u(t - c) \leftrightarrow \int_c^\infty x(t - c)e^{-st}\,dt, \quad \text{since } u(t - c) = 0 \text{ for } t < c.$$

$$\tag{5.18}$$

Let us consider a change of variable in the integral in (5.18). Defining $\bar{t} = t - c$, we have that $t = \bar{t} + c$, $dt = d\bar{t}$, $\bar{t} = 0$ when $t = c$, and $\bar{t} = \infty$ when $t = \infty$. Thus

$$\int_c^\infty x(t - c)e^{-st}\, dt = \int_0^\infty x(\bar{t}) \exp\,[-s(\bar{t} + c)]\, d\bar{t}$$

$$= e^{-cs}\left[\int_0^\infty x(\bar{t})e^{-s\bar{t}}\, d\bar{t}\right]$$

$$= e^{-cs}X(s). \tag{5.19}$$

Therefore, combining (5.19) and (5.18), we get

$$x(t - c)u(t - c) \leftrightarrow e^{-cs}X(s).$$

EXAMPLE 5.5.   Consider the a-second pulse $p(t)$ defined by

$$p(t) = \begin{cases} 1, & 0 \le t < a \\ 0, & \text{all other } t, \end{cases}$$

where a is an arbitrary positive real number. Expressing $p(t)$ in terms of the unit-step function $u(t)$, we have

$$p(t) = u(t) - u(t - a).$$

By linearity, the Laplace transform $P(s)$ of $p(t)$ is the sum of the transform of $u(t)$ and the transform of $u(t - a)$. Now $u(t - a)$ is the a-second right shift of $u(t)$, and thus by the right-shift property (5.17), the Laplace transform of $u(t - a)$ is equal to $e^{-as}(1/s)$. Hence

$$u(t) - u(t - a) \leftrightarrow \frac{1}{s} - e^{-as}\frac{1}{s} = \frac{1 - e^{-as}}{s}.$$

It should be noted that there is no comparable result for a left shift in time. To see this, let c be an arbitrary positive real number and consider the function $x(t + c)$, where $x(t)$ is some given function. The function $x(t + c)$ is a c-second left shift of the function $x(t)$. The Laplace transform of $x(t + c)$ is equal to

$$\int_0^\infty x(t + c)e^{-st}\, dt. \tag{5.20}$$

However, (5.20) cannot be expressed in terms of the Laplace transform $X(s)$ of $x(t)$. In particular, (5.20) is not equal to $e^{cs}X(s)$.

TIME SCALING   If $x(t) \leftrightarrow X(s)$, for any positive real number a,

$$x(at) \leftrightarrow \frac{1}{a}X\left(\frac{s}{a}\right). \tag{5.21}$$

The function $x(at)$ is a time-scaled version of the given function $x(t)$. By (5.21) we see that time scaling corresponds to scaling by the factor $1/a$ in the Laplace transform domain (plus multiplication of the transform by $1/a$).

We can verify the time scaling property (5.21) by taking the Laplace transform of $x(at)$. The transform is equal to

$$\int_0^\infty x(at)e^{-st}\, dt. \tag{5.22}$$

Consider the change of variable in (5.22) given by $\bar{t} = at$. We have that $t = (1/a)\bar{t}$, $d\bar{t} = a(dt)$, $\bar{t} = 0$ when $t = 0$, and $\bar{t} = \infty$ when $t = \infty$. Then

$$\int_0^\infty x(at)e^{-st}\, dt = \int_0^\infty x(\bar{t}) \exp\left[-\left(\frac{s}{a}\right)\bar{t}\right] \frac{1}{a}\, d\bar{t}$$

$$= \frac{1}{a} X\left(\frac{s}{a}\right).$$

Hence (5.21) is verified.

EXAMPLE 5.6. Consider the time-scaled unit-step function $u(at)$, where a is an arbitrary positive real number. By (5.21) we have

$$u(at) \leftrightarrow \frac{1}{a}\frac{1}{s/a} = \frac{1}{s}.$$

This result is not unexpected since $u(at) = u(t)$ for any real number $a > 0$.

MULTIPLICATION BY A POWER OF t If $x(t) \leftrightarrow X(s)$, then for any positive integer n,

$$t^n x(t) \leftrightarrow (-1)^n \frac{d^n}{ds^n} X(s). \tag{5.23}$$

We shall prove (5.23) when $n = 1$. We have that

$$X(s) = \int_0^\infty x(t)e^{-st}\, dt. \tag{5.24}$$

Taking the derivative with respect to s of both sides of (5.24), we obtain

$$\frac{d}{ds} X(s) = \int_0^\infty x(t)\left(\frac{d}{ds} e^{-st}\right) dt$$

$$= \int_0^\infty (-t)x(t)e^{-st}\, dt. \tag{5.25}$$

The term on the right side of (5.25) is equal to the Laplace transform of $-tx(t)$. Thus we have verified (5.23) when $n = 1$. The proof for $n \geq 2$ follows by taking the second-order and higher-order derivatives of $X(s)$ with respect to s. We omit the details.

> **EXAMPLE 5.7.** Consider the unit-ramp function $r(t) = tu(t)$. From (5.23) we have
>
> $$R(s) = -\frac{d}{ds} U(s) = -\frac{d}{ds} \frac{1}{s} = \frac{1}{s^2}.$$
>
> Generalizing to the case $t^n u(t)$, $n = 1, 2, \ldots$, we have
>
> $$t^n u(t) \leftrightarrow \frac{n!}{s^{n+1}}. \tag{5.26}$$

> **EXAMPLE 5.8.** Let $v(t) = te^{-bt}$, where b is any real number. From Table 5.1 we have
>
> $$e^{-bt} \leftrightarrow \frac{1}{s + b}.$$
>
> Then from (5.23),
>
> $$V(s) = -\frac{d}{ds} \frac{1}{s + b} = \frac{1}{(s + b)^2}.$$
>
> Generalizing to the case $t^n e^{-bt}$, we have
>
> $$t^n e^{-bt} \leftrightarrow \frac{n!}{(s + b)^{n+1}}. \tag{5.27}$$

MULTIPLICATION BY AN EXPONENTIAL   If $x(t) \leftrightarrow X(s)$, then for any real or complex number a,

$$e^{at} x(t) \leftrightarrow X(s - a). \tag{5.28}$$

By the property (5.28), multiplication by an exponential function in the time domain corresponds to a shift of the s variable in the Laplace transform domain. To prove (5.28), we have that the Laplace transform of $e^{at}x(t)$ is given by

$$\int_0^\infty e^{at} x(t)e^{-st} \, dt = \int_0^\infty x(t) \exp[-(s - a)] \, dt$$

$$= X(s - a).$$

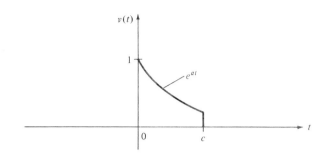

Figure 5.1. Sketch of the function $v(t)$ in Example 5.9.

EXAMPLE 5.9. Let $v(t) = [u(t) - u(t - c)]e^{at}$, where c is a positive real number and a is any real number. The function $v(t)$ is the product of the c-second pulse $u(t) - u(t - c)$ and the exponential function e^{at}. The function $v(t)$ with $a < 0$ is sketched in Figure 5.1. From Example 5.5 we have the transform pair

$$u(t) - u(t - c) \leftrightarrow \frac{1 - e^{-cs}}{s}.$$

Then, by (5.28),

$$V(s) = \frac{1 - \exp[-c(s - a)]}{s - a}.$$

MULTIPLICATION BY $\sin \omega t$ OR $\cos \omega t$ If $x(t) \leftrightarrow X(s)$, then for any real number ω,

$$x(t) \sin \omega t \leftrightarrow \frac{j}{2}[X(s + j\omega) - X(s - j\omega)], \tag{5.29}$$

$$x(t) \cos \omega t \leftrightarrow \frac{1}{2}[X(s + j\omega) + X(s - j\omega)]. \tag{5.30}$$

We shall prove (5.29). By Euler's formula,

$$\sin \omega t = \frac{j}{2}[e^{-j\omega t} - e^{j\omega t}]. \tag{5.31}$$

By (5.28),

$$e^{\mp j\omega t} x(t) \leftrightarrow X(s \pm j\omega). \tag{5.32}$$

Combining (5.31) and (5.32), we obtain (5.29). The proof of (5.30) is similar and is thus omitted.

EXAMPLE 5.10. Let $v(t) = \cos \omega t$. Since the transform depends only on the values of $v(t)$ for $t \geq 0$, we can write $v(t)$ in the form

$$v(t) = (\cos \omega t)u(t).$$

Now $u(t) \leftrightarrow 1/s$, and using (5.30) with $x(t) = u(t)$, we have

$$V(s) = \frac{1}{2}\left(\frac{1}{s + j\omega} + \frac{1}{s - j\omega}\right)$$

$$= \frac{1}{2}\frac{s - j\omega + s + j\omega}{s^2 + \omega^2}$$

$$= \frac{s}{s^2 + \omega^2}.$$

Hence we have the transform pair

$$\cos \omega t \leftrightarrow \frac{s}{s^2 + \omega^2}. \tag{5.33}$$

Similarly, we can verify the transform pair

$$\sin \omega t \leftrightarrow \frac{\omega}{s^2 + \omega^2}. \tag{5.34}$$

EXAMPLE 5.11. Now let $v(t) = e^{-bt}\cos \omega t$. To compute the Laplace transform of $v(t)$, we could set $x(t) = e^{-bt}$ and then use the multiplication by $\cos \omega t$ property. We could also set $x(t) = \cos \omega t$ and use the multiplication by an exponential property. The latter is simpler to carry out, so we shall do it that way. Replacing s by $s + b$ on the right side of (5.33), we have the transform pair

$$e^{-bt}\cos \omega t \leftrightarrow \frac{s + b}{(s + b)^2 + \omega^2}. \tag{5.35}$$

Similarly, we have the transform pair

$$e^{-bt}\sin \omega t \leftrightarrow \frac{\omega}{(s + b)^2 + \omega^2}. \tag{5.36}$$

EXAMPLE 5.12. Let $v(t) = \sin^2 \omega t$. Setting $x(t) = \sin \omega t$ and using the multiplication by $\sin \omega t$ property, we have

$$
\begin{aligned}
V(s) &= \frac{j}{2}\left[\frac{\omega}{(s + j\omega)^2 + \omega^2} - \frac{\omega}{(s - j\omega)^2 + \omega^2}\right] \\
&= \frac{j}{2}\frac{\omega(s - j\omega)^2 + \omega^3 - \omega(s + j\omega)^2 - \omega^3}{(s + j\omega)^2(s - j\omega)^2 + \omega^2(s - j\omega)^2 + \omega^2(s + j\omega)^2 + \omega^4} \\
&= \frac{j}{2}\frac{-j4\omega^2 s}{s^4 + 4\omega^2 s^2} \\
&= \frac{2\omega^2}{s(s^2 + 4\omega^2)}.
\end{aligned}
\tag{5.37}
$$

CONVOLUTION Given two functions $x(t)$ and $v(t)$ with $x(t)$ and $v(t)$ equal to zero for all $t < 0$, in Chapter 4 we defined the convolution of $x(t)$ and $v(t)$ by

$$
(x * v)(t) = \int_0^t x(\lambda)v(t - \lambda)\,d\lambda, \qquad t \geq 0.
$$

Note that since $v(t) = 0$ for all $t < 0$, we can take the upper limit in the convolution integral to be ∞ so that

$$
(x * v)(t) = \int_0^\infty x(\lambda)v(t - \lambda)\,d\lambda, \qquad t \geq 0.
\tag{5.38}
$$

Now letting $X(s)$ denote the Laplace transform of $x(t)$ and $V(s)$ denote the Laplace transform of $v(t)$, we have the transform pair

$$
(x * v)(t) \leftrightarrow X(s)V(s).
\tag{5.39}
$$

By the property (5.39), convolution in the time domain corresponds to a product in the Laplace transform domain. We shall see that as a result of the property (5.39), we shall be able to establish an algebraic relationship between the transforms of the input and output of a linear time-invariant continuous-time system.

To prove (5.39), we begin by taking the Laplace transform of the right side of (5.38). This gives the transform pair

$$
(x * v)(t) \leftrightarrow \int_0^\infty \left[\int_0^\infty x(\lambda)v(t - \lambda)\,d\lambda\right]e^{-st}\,dt.
\tag{5.40}
$$

Interchanging the integrals in (5.40), we have

$$
(x * v)(t) \leftrightarrow \int_0^\infty x(\lambda)\left[\int_0^\infty v(t - \lambda)e^{-st}\,dt\right]d\lambda.
\tag{5.41}
$$

Now with the change of variable $\bar{t} = t - \lambda$ in the second integral in (5.41), we have

$$\int_0^\infty x(\lambda) \left[\int_0^\infty v(t - \lambda) e^{-st}\, dt \right] d\lambda$$

$$= \int_0^\infty x(\lambda) \left[\int_{-\lambda}^\infty v(\bar{t}) \exp\left(-s(\bar{t} + \lambda)\right) d\bar{t} \right] d\lambda$$

$$= \int_0^\infty x(\lambda) \left[\int_0^\infty v(\bar{t}) \exp\left(-s(\bar{t} + \lambda)\right) d\bar{t} \right] d\lambda, \qquad \text{since } v(\bar{t}) = 0 \text{ for all } \bar{t} < 0$$

$$= \left[\int_0^\infty x(\lambda) e^{-s\lambda}\, d\lambda \right] \left[\int_0^\infty v(\bar{t}) e^{-s\bar{t}}\, d\bar{t} \right]$$

$$= X(s)V(s).$$

Using this result in (5.41), we obtain (5.39).

The convolution property (5.39) yields a procedure for computing the convolution $x * v$ of two signals $x(t)$ and $v(t)$. First compute the Laplace transforms $X(s)$, $V(s)$ of $x(t)$, $v(t)$, and then compute the inverse Laplace transform of the product $X(s)V(s)$. The result is the convolution $x * v$.

EXAMPLE 5.13. Let $x(t)$ denote the 1-second pulse given by $x(t) = u(t) - u(t - 1)$. We want to determine the convolution $x * x$ of this signal with itself. From Example 5.5, the transform $X(s)$ of $x(t)$ is

$$X(s) = \frac{1 - e^{-s}}{s}.$$

Thus, by (5.39), the transform of the convolution $(x * x)(t)$ is

$$X^2(s) = \left(\frac{1 - e^{-s}}{s} \right)^2 = \frac{1 - 2e^{-s} + e^{-2s}}{s^2}.$$

Setting $n = 1$ in the transform pair (5.26), we have

$$tu(t) \leftrightarrow \frac{1}{s^2}.$$

Then using linearity and the right-shift property, the inverse Laplace transform of $X^2(s)$ is

$$(x * x)(t) = tu(t) - 2(t - 1)u(t - 1) + (t - 2)u(t - 2).$$

The convolution $(x * x)(t)$ is displayed in Figure 5.2. We see that the convolution of the 1-second pulse $x(t)$ with itself is a triangular function with a

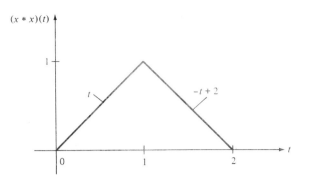

Figure 5.2. Plot of $(x * x)(t)$.

2-second duration. Of course, we could have derived this result by direct evaluation of the convolution integral given in Chapter 4.

INTEGRATION Given a function $x(t)$ with $x(t) = 0$ for all $t < 0$, as in Chapter 4 we define the integral of $x(t)$ to be the function

$$x^{(-1)}(t) = \int_0^t x(\lambda) \, d\lambda.$$

If $x(t) \leftrightarrow X(s)$, we claim that

$$x^{(-1)}(t) \leftrightarrow \frac{1}{s} X(s). \tag{5.42}$$

By (5.42), the Laplace transform of the integral of $x(t)$ is equal to $X(s)$ divided by s. The property (5.42) is actually a special case of the convolution property (5.39). To see this, consider the convolution

$$u(t) * (x(t)u(t)) = \int_0^t u(t - \lambda)x(\lambda)u(\lambda) \, d\lambda$$

$$= \int_0^t x(\lambda) \, d\lambda. \tag{5.43}$$

From (5.43) we see that the convolution of the unit-step function $u(t)$ and $x(t)u(t)$ is equal to the integral of $x(t)$. By the convolution property (5.39), the Laplace transform of $u(t) * (x(t)u(t))$ is equal to $(1/s)X(s)$. Thus we have verified (5.42).

EXAMPLE 5.14. Let $x(t) = u(t)$. Then the integral of $x(t)$ is the unit-ramp function $r(t) = tu(t)$. By (5.42), the Laplace transform of $r(t)$ is equal to $1/s$ times the transform of $u(t)$. The result is the transform pair

$$r(t) \leftrightarrow \frac{1}{s^2}.$$

Recall that we derived this transform pair previously by using the multiplication-by-t property.

DIFFERENTIATION IN THE TIME DOMAIN   If $x(t) \leftrightarrow X(s)$, then

$$\dot{x}(t) \leftrightarrow sX(s) - x(0). \tag{5.44}$$

To prove (5.44), we shall compute the transform of the derivative of $x(t)$. The transform of $\dot{x}(t)$ is

$$\int_0^\infty \dot{x}(t)e^{-st}\, dt. \tag{5.45}$$

We shall integrate (5.45) by parts. Let $v = e^{-st}$ so that $dv = -se^{-st}\, dt$. Let $w = x(t)$ so that $dw = \dot{x}(t)\, dt$. Then

$$\int_0^\infty \dot{x}(t)e^{-st}\, dt = vw \Big|_{t=0}^{t=\infty} - \int_0^\infty w\, dv$$

$$= e^{-st}x(t)\Big|_{t=0}^{t=\infty} - \int_0^\infty x(t)(-s)e^{-st}\, dt$$

$$= \lim_{t\to\infty} [e^{-st}x(t)] - x(0) + sX(s).$$

When $|x(t)| < ce^{at}$, $t > 0$, for some constants a, c, we have for any s such that Re $s > a$,

$$\lim_{t\to\infty} e^{-st}x(t) = 0.$$

Thus

$$\int_0^\infty \dot{x}(t)e^{-st}\, dt = -x(0) + sX(s).$$

Hence we have proved (5.44).

We should point out that if $x(t)$ is discontinuous at $t = 0$ or if $x(t)$ contains an impulse or a derivative of an impulse located at $t = 0$, it is necessary to take the initial time in (5.44) to be at 0^-. In other words, we have

$$\dot{x}(t) \leftrightarrow sX(s) - x(0^-). \tag{5.46}$$

If $x(t) = 0$ for $t < 0$, then $x(0^-) = 0$ and

$$\dot{x}(t) \leftrightarrow sX(s).$$

EXAMPLE 5.15. Let $x(t) = u(t)$. Then $\dot{x}(t) = \delta(t)$. Since $\dot{x}(t)$ is the unit impulse located at $t = 0$, we must use (5.46) to compute the Laplace transform of $\dot{x}(t)$. This gives

$$\dot{x}(t) \leftrightarrow s\,\frac{1}{s} - u(0^-) = 1 - 0 = 1.$$

We already knew that $\delta(t)$ has transform equal to 1, and thus the above result is consistent.

The Laplace transform of the second-order and higher-order derivatives of $x(t)$ can also be expressed in terms of $X(s)$ and initial conditions. For example, in the second-order case, we have the transform pair

$$\frac{d^2x(t)}{dt^2} \leftrightarrow s^2X(s) - sx(0) - \dot{x}(0). \tag{5.47}$$

The transform pair (5.47) can be proved by using integration by parts twice on the integral expression for the transform of the second derivative of $x(t)$. We omit the details.

If the second derivative of $x(t)$ is discontinuous or contains an impulse or a derivative of an impulse located at $t = 0$, it is necessary to take the initial conditions in (5.47) at time $t = 0^-$.

Now let n be an arbitrary positive integer and let $x^{(n)}(t)$ denote the nth derivative of a given signal $x(t)$. Then we have the transform pair

$$x^{(n)}(t) \leftrightarrow s^nX(s) - s^{n-1}x(0) - s^{n-2}\dot{x}(0) - \cdots - sx^{(n-2)}(0) - x^{(n-1)}(0). \tag{5.48}$$

INITIAL-VALUE THEOREM Given a signal $x(t)$ with transform $X(s)$, we have

$$x(0) = \lim_{s \to \infty} sX(s) \tag{5.49}$$

$$\dot{x}(0) = \lim_{s \to \infty} [s^2X(s) - sx(0)]. \tag{5.50}$$

In the general case, for an arbitrary positive integer n, we have

$$x^{(n)}(0) = \lim_{s \to \infty}[s^{n+1}X(s) - s^nx(0) - s^{n-1}\dot{x}(0) - \cdots - sx^{(n-1)}(0)]. \tag{5.51}$$

It should be noted that the relationship (5.51) is not valid if the nth derivative $x^{(n)}(t)$ contains an impulse or a derivative of an impulse at time $t = 0$.

The relationship (5.51) for $n = 0, 1, 2, \ldots$ is called the *initial-value theo-*

rem. We shall prove (5.49) [which is the same as (5.51) with $n = 0$]. Given a signal $x(t)$, by the derivative property (5.44), we have

$$\dot{x}(t) \leftrightarrow sX(s) - x(0). \tag{5.52}$$

Applying the definition of the Laplace transform to $\dot{x}(t)$, we also have

$$\dot{x}(t) \leftrightarrow \int_0^\infty \dot{x}(t)e^{-st}\, dt. \tag{5.53}$$

Combining (5.52) and (5.53), we obtain

$$sX(s) = \int_0^\infty \dot{x}(t)e^{-st}\, dt + x(0). \tag{5.54}$$

Since

$$\lim_{s\to\infty} e^{-st} = 0,$$

taking the limit as $s \to \infty$ of both sides of (5.54), we get

$$\lim_{s\to\infty} sX(s) = x(0),$$

which proves (5.49).

The initial-value theorem is useful since it allows for computation of the initial values of a function $x(t)$ and its derivatives directly from the Laplace transform $X(s)$ of $x(t)$. Hence, if we know $X(s)$ but not $x(t)$, it is possible to compute these initial values without having to compute the inverse Laplace transform of $x(t)$.

EXAMPLE 5.16.   Suppose that the function $x(t)$ has the Laplace transform

$$X(s) = \frac{-3s^2 + 2}{s^3 + s^2 + 3s + 2}.$$

Then

$$\lim_{s\to\infty} sX(s) = \lim_{s\to\infty} \frac{-3s^3 + 2s}{s^3 + s^2 + 3s + 2} = \frac{-3}{1}.$$

Thus $x(0) = -3$.

FINAL-VALUE THEOREM   Given the signal $x(t)$ with transform $X(s)$, suppose that $x(t)$ has a limit as $t \to \infty$. Then the final-value theorem states that

$$\lim_{t\to\infty} x(t) = \lim_{s\to 0} sX(s). \tag{5.55}$$

To prove (5.55), first note that by the derivative property,

$$\int_0^\infty \dot{x}(t)e^{-st}\, dt = sX(s) - x(0). \tag{5.56}$$

Taking the limit as $s \to 0$ of both sides of (5.56), we obtain

$$\lim_{s\to 0} \int_0^\infty \dot{x}(t)e^{-st}\, dt = \int_0^\infty \dot{x}(t)\, dt = \lim_{s\to 0} [sX(s) = x(0)]. \tag{5.57}$$

Now if $x(t)$ has a limit as $t \to \infty$, using integration by parts, we have

$$\int_0^\infty \dot{x}(t)\, dt = x(\infty) - x(0). \tag{5.58}$$

Combining (5.57) and (5.58), we obtain (5.55).

The final-value theorem is a very useful property since it allows us to determine the limit as $t \to \infty$ of a time signal $x(t)$ directly from the Laplace transform $X(s)$. However, one must be careful in using the final-value theorem since the limit of $sX(s)$ as $s \to 0$ may exist even though $x(t)$ does not have a limit as $t \to \infty$. For example, suppose that

$$x(s) = \frac{1}{s^2 + 1}.$$

Then

$$\lim_{s\to 0} sX(s) = \lim_{s\to 0} \frac{s}{s^2 + 1} = 0.$$

But $x(t) = \sin t$, and $\sin t$ does not have a limit as $t \to \infty$.

In many cases of interest, it is possible to determine if $x(t)$ does have a limit by checking the transform $X(s)$. In particular, suppose that $x(t)$ has the Laplace transform

$$X(s) = \frac{N(s)}{D(s)},$$

where $N(s)$ and $D(s)$ are polynomials in s with real coefficients given by

$$N(s) = b_m s^m + b_{m-1} s^{m-1} + \cdots + b_1 s + b_0$$

$$D(s) = a_n s^n + a_{n-1} s^{n-1} + \cdots + a_1 s + a_0, \qquad a_n \neq 0.$$

In the following result, we assume that the polynomials $N(s)$ and $D(s)$ do not have any common factors. If there are common factors, they should be canceled.

Factoring $D(s)$, we can write $X(s)$ in the form

$$X(s) = \frac{N(s)}{a_n(s - p_1)(s - p_2) \cdots (s - p_n)}, \tag{5.59}$$

where $p_1, p_2, \ldots, p_n$ are fixed real or complex numbers. Then

$$\lim_{t \to \infty} x(t)$$

exists if and only if the real parts of all the p_i's are strictly negative (< 0), except that one of the p_i's may be equal to zero. Restating this condition, we require that

TABLE 5.2. Properties of the Laplace Transform

Property	Transform Pair/Property
Linearity	$ax(t) + bv(t) \leftrightarrow aX(s) + bV(s)$
Right shift in time	$x(t - c)u(t - c) \leftrightarrow e^{-cs}X(s), \quad c > 0$
Time scaling	$x(at) \leftrightarrow \dfrac{1}{a} X\left(\dfrac{s}{a}\right), \quad a > 0$
Multiplication by a power of t	$t^n x(t) \leftrightarrow (-1)^n \dfrac{d^n}{ds^n} X(s), \quad n = 1, 2, \ldots$
Multiplication by an exponential	$e^{at}x(t) \leftrightarrow X(s - a), \quad a$ real or complex
Multiplication by $\sin \omega t$	$x(t) \sin \omega t \leftrightarrow \dfrac{j}{2} [X(s + j\omega) - X(s - j\omega)]$
Multiplication by $\cos \omega t$	$x(t) \cos \omega t \leftrightarrow \frac{1}{2}[X(s + j\omega) + X(s - j\omega)]$
Convolution	$(x * v)(t) \leftrightarrow X(s)V(s)$
Integration	$x^{(-1)}(t) \leftrightarrow \dfrac{1}{s} X(s)$
Differentiation in the time domain	$\dot{x}(t) \leftrightarrow sX(s) - x(0)$
Second derivative	$\ddot{x}(t) \leftrightarrow s^2 X(s) - sx(0) - \dot{x}(0)$
nth derivative	$x^{(n)}(t) \leftrightarrow s^n X(s) - s^{n-1}x(0)$
	$\qquad - \cdots - sx^{(n-2)}(0) - x^{(n-1)}(0)$
Initial-value theorem	$x(0) = \lim_{s \to \infty} sX(s)$
	$\dot{x}(0) = \lim_{s \to \infty} [s^2 X(s) - sx(0)]$
	$x^{(n)}(0) = \lim_{s \to \infty} [s^{n+1}X(s)$
	$\qquad - s^n x(0) - s^{n-1}\dot{x}(0) - \cdots - sx^{(n-1)}(0)]$
Final-value theorem	If $\lim_{t \to \infty} x(t)$ exists, then
	$\lim_{t \to \infty} x(t) = \lim_{s \to 0} sX(s)$

Re $p_i < 0$ for all i,   except p_i may be zero for one value of i.   (5.60)

If condition (5.60) is satisfied, the limit of $x(t)$ is given by (5.55). The proof that (5.60) is necessary and sufficient for the limit to exist follows from the stability analysis given in Chapter 8.

EXAMPLE 5.17. Suppose that

$$X(s) = \frac{2s^2 - 3s + 4}{s^3 + 3s^2 + 2s}.$$

TABLE 5.3. Common Transform Pairs

$u(t) \leftrightarrow \dfrac{1}{s}$

$u(t) - u(t - a) \leftrightarrow \dfrac{1 - e^{-as}}{s}$, $a > 0$

$t^n \leftrightarrow \dfrac{n!}{s^{n+1}}$, $n = 1, 2, 3, \ldots$

$\delta(t) \leftrightarrow 1$

$\delta(t - c) \leftrightarrow e^{-cs}$, $c > 0$

$e^{-bt} \leftrightarrow \dfrac{1}{s + b}$, b real or complex

$t^n e^{-bt} \leftrightarrow \dfrac{n!}{(s + b)^{n+1}}$, $n = 1, 2, 3, \ldots$

$\cos \omega t \leftrightarrow \dfrac{s}{s^2 + \omega^2}$

$\sin \omega t \leftrightarrow \dfrac{\omega}{s^2 + \omega^2}$

$\cos^2 \omega t \leftrightarrow \dfrac{s^2 + 2\omega^2}{s(s^2 + 4\omega^2)}$

$\sin^2 \omega t \leftrightarrow \dfrac{2\omega^2}{s(s^2 + 4\omega^2)}$

$e^{-bt} \cos \omega t \leftrightarrow \dfrac{s + b}{(s + b)^2 + \omega^2}$

$e^{-bt} \sin \omega t \leftrightarrow \dfrac{\omega}{(s + b)^2 + \omega^2}$

$t \cos \omega t \leftrightarrow \dfrac{s^2 - \omega^2}{(s^2 + \omega^2)^2}$

$t \sin \omega t \leftrightarrow \dfrac{2\omega s}{(s^2 + \omega^2)^2}$

Writing $X(s)$ in the form (5.59), we have

$$X(s) = \frac{2s^2 - 3s + 4}{s(s + 1)(s + 2)}.$$

Then

$$p_1 = 0, \qquad p_2 = -1, \qquad p_3 = -2,$$

and thus condition (5.60) is satisfied. Therefore,

$$\lim_{t \to \infty} x(t) = \lim_{s \to 0} sX(s) = \lim_{s \to 0} \frac{2s^2 - 3s + 4}{(s + 1)(s + 2)} = \frac{4}{1(2)} = 2.$$

For the convenience of the reader, the properties of the Laplace transform are summarized in Table 5.2. In Table 5.3 we list a collection of common transform pairs, which includes the transform pairs in Table 5.1 and the transform pairs that were derived in this section using the properties of the Laplace transform.

5.3

INVERSE LAPLACE TRANSFORM OF A RATIONAL FUNCTION

In many cases of interest, a time function $x(t)$ has a Laplace transform $X(s)$ that can be expressed in the form

$$X(s) = \frac{N(s)}{D(s)}, \tag{5.61}$$

where $N(s)$ and $D(s)$ are polynomials in s given by

$$N(s) = b_m s^m + b_{m-1} s^{m-1} + \cdots + b_1 s + b_0$$

$$D(s) = a_n s^n + a_{n-1} s^{n-1} + \cdots + a_1 s + a_0, \qquad a_n \neq 0.$$

The coefficients $b_0, b_1, \ldots, b_m$ of the polynomial $N(s)$ and the coefficients a_0, $a_1, \ldots, a_n$ of the polynomial $D(s)$ are real numbers. The function $X(s)$ given by (5.61) is said to be a *rational function* of s since it is a ratio of two polynomials in s. The degree n of $D(s)$ is called the *order of the rational function* $X(s)$.

Let $p_1, p_2, \ldots, p_n$ denote the roots of the equation

$$D(s) = 0,$$

and let $z_1, z_2, \ldots, z_m$ denote the roots of the equation

$$N(s) = 0.$$

Some or all of the roots $p_1, p_2, \ldots, p_n$ may be complex and some or all of the roots $z_1, z_2, \ldots, z_m$ may be complex. If a root p_i or z_i is complex, there must exist another root which is the complex conjugate of p_i or z_i. In other words, complex roots always appear in complex-conjugate pairs. A program for computing the roots of a polynomial is given in Figure A.6.

If $p_1, p_2, \ldots, p_n$ are the roots of $D(s) = 0$, we can express $D(s)$ in the factored form

$$D(s) = a_n(s - p_1)(s - p_2) \cdots (s - p_n).$$

The factored form of $N(s)$ is

$$N(s) = b_m(s - z_1)(s - z_2) \cdots (s - z_m).$$

We can then write $X(s)$ in factored form:

$$X(s) = \frac{b_m(s - z_1)(s - z_2) \cdots (s - z_m)}{a_n(s - p_1)(s - p_2) \cdots (s - p_n)}. \tag{5.62}$$

From (5.62) we see that if we set $s = z_i$ for any value of $i \in \{1, 2, \ldots, m\}$, and if $z_i \neq p_j$ for any j, then

$$X(s)|_{s=z_i} = X(z_i) = 0.$$

Since $X(s)$ is zero when s is set equal to z_i, the numbers $z_1, z_2, \ldots, z_m$ are called the *zeros* of the rational function $X(s)$. From (5.62) we also see that if $p_i \neq z_j$ for any j,

$$|X(s)| \to \infty \qquad \text{as } s \to p_i.$$

In other words, $|X(s)|$ grows without bound as s approaches any one of the p_i's. As a result of this property, the numbers $p_1, p_2, \ldots, p_n$ are called the *poles* of $X(s)$.

Given the rational function

$$X(s) = \frac{N(s)}{D(s)},$$

we can compute the inverse Laplace transform of $X(s)$ by first computing a partial-fraction expansion of $X(s)$. This procedure is described below. In the following development, we assume that $m < n$; that is, the degree of $N(s)$ is strictly less than the degree of $D(s)$. Such a rational function of s is said to be *strictly proper* in s. The case when $X(s)$ is not strictly proper is considered later.

DISTINCT POLES   Let us assume that the poles $p_1, p_2, \ldots, p_n$ of $X(s)$ are distinct (or nonrepeated); in other words, $p_i \neq p_j$ when $i \neq j$. Then $X(s)$ has the partial-fraction expansion

$$X(s) = \frac{c_1}{s - p_1} + \frac{c_2}{s - p_2} + \cdots + \frac{c_n}{s - p_n}, \tag{5.63}$$

where

$$c_i = [(s - p_i)X(s)]_{s=p_i}, \qquad i = 1, 2, \ldots, n. \tag{5.64}$$

We can verify (5.64) by first multiplying both sides of (5.63) by $s - p_i$. This gives

$$(s - p_i)X(s) = c_i + \sum_{\substack{r=1 \\ r \neq i}}^{n} c_r \frac{s - p_i}{s - p_r}. \tag{5.65}$$

Evaluating both sides of (5.65) at $s = p_i$, we get (5.64).

The c_i defined by (5.64) are constants that may be real or complex. The constant c_i is real if the corresponding pole p_i is real. In addition, since the poles $p_1, p_2, \ldots, p_n$ appear in complex-conjugate pairs, the c_i must also appear in complex-conjugate pairs. Hence, if c_i is complex, one of the other constants must be equal to the complex conjugate of c_i.

It is also worth noting that to compute the partial-fraction expansion (5.63), it is not necessary to factor the numerator polynomial $N(s)$; in other words, it is not necessary to compute the zeros of $X(s)$. However, we must compute the poles of $X(s)$ since the expansion is given directly in terms of the poles.

Taking the inverse Laplace transform of (5.63) term by term and using linearity, we have that the inverse Laplace transform $x(t)$ of $X(s)$ is given by

$$x(t) = c_1 e^{p_1 t} + c_2 e^{p_2 t} + \cdots + c_n e^{p_n t}, \qquad t \geq 0. \tag{5.66}$$

If the p_i are all real, the function $x(t)$ defined by (5.66) is the inverse Laplace transform of $X(s)$. If two or more of the p_i are complex, we must write the expression (5.66) for $x(t)$ in real form. We shall consider this after the following example.

EXAMPLE 5.18.   Suppose that

$$X(s) = \frac{s + 2}{s^3 + 4s^2 + 3s}.$$

Here

$$D(s) = s^3 + 4s^2 + 3s = s(s + 1)(s + 3).$$

The roots of $D(s) = 0$ are 0, -1, -3, and thus the poles of $X(s)$ are $p_1 = 0$, $p_2 = -1$, $p_3 = -3$. Therefore,

$$X(s) = \frac{c_1}{s} + \frac{c_2}{s + 1} + \frac{c_3}{s + 3},$$

where

$$c_1 = [sX(s)]_{s=0} \qquad = \left. \frac{s + 2}{(s + 1)(s + 3)} \right|_{s=0} = \frac{2}{3}$$

$$c_2 = [(s + 1)X(s)]_{s=-1} = \left. \frac{s + 2}{s(s + 3)} \right|_{s=-1} \qquad = \frac{1}{-2}$$

$$c_3 = [(s + 3)X(s)]_{s=-3} = \left. \frac{s + 2}{s(s + 1)} \right|_{s=-3} \qquad = \frac{-1}{6}.$$

Hence the inverse Laplace transform $x(t)$ of $X(s)$ is given by

$$x(t) = \tfrac{2}{3} - \tfrac{1}{2} e^{-t} - \tfrac{1}{6} e^{-3t}, \qquad t \geq 0.$$

DISTINCT POLES WITH TWO OR MORE POLES COMPLEX We continue to assume that the poles of $X(s)$ are distinct. We also assume that two or more of the poles are complex so that some of the exponentials in (5.66) are complex. As we now show, it is possible to combine the complex terms in order to write $x(t)$ in real form.

Suppose that $p_1 = \alpha + j\beta$ is complex ($\beta \neq 0$). Then the complex conjugate $\bar{p}_1 = \alpha - j\beta$ is another root of $D(s) = 0$. We set this root equal to p_2. Then $c_2 = \bar{c}_1$ and $X(s)$ has the partial-fraction expansion

$$X(s) = \frac{c_1}{s - p_1} + \frac{\bar{c}_1}{s - \bar{p}_1} + \frac{c_3}{s - p_3} + \cdots + \frac{c_n}{s - p_n}.$$

Hence

$$x(t) = c_1 e^{p_1 t} + \bar{c}_1 e^{\bar{p}_1 t} + c_3 e^{p_3 t} + \cdots + c_n e^{p_n t}.$$

We shall show that

$$c_1 e^{p_1 t} + \bar{c}_1 e^{\bar{p}_1 t}$$

can be written in real form. First, since $p_1 = \alpha + j\beta$, we have

$$c_1 e^{p_1 t} + \bar{c}_1 e^{\bar{p}_1 t} = c_1 \exp(\alpha t + j\beta t) + \bar{c}_1 \exp(\alpha t - j\beta t)$$

$$= e^{\alpha t}(c_1 e^{j\beta t} + \bar{c}_1 e^{-j\beta t}).$$

Using Euler's formula,

$$e^{\pm j\beta t} = \cos \beta t \pm j \sin \beta t,$$

we have

$$c_1 e^{p_1 t} + \bar{c}_1 e^{\bar{p}_1 t} = e^{\alpha t}(c_1 \cos \beta t + jc_1 \sin \beta t + \bar{c}_1 \cos \beta t - j\bar{c}_1 \sin \beta t)$$

$$= e^{\alpha t}[(c_1 + \bar{c}_1) \cos \beta t + j(c_1 - \bar{c}_1) \sin \beta t].$$

Now writing c_1 in rectangular form $c_1 = a + jb$, we have

$$c_1 + \bar{c}_1 = (a + jb) + (a - jb) = 2a = 2 \operatorname{Re} c_1$$

and

$$j(c_1 - \bar{c}_1) = j(a + jb) - j(a - jb) = j(j2b) = -2b = -2 \operatorname{Im} c_1,$$

where $\operatorname{Im} c_1$ is the imaginary part of c_1. Therefore,

$$c_1 e^{p_1 t} + \bar{c}_1 e^{\bar{p}_1 t} = 2e^{\alpha t}[(\operatorname{Re} c_1) \cos \beta t - (\operatorname{Im} c_1) \sin \beta t]. \qquad (5.67)$$

We can write (5.67) in a more compact form by using the trigonometric identity

$$A \cos \beta t - B \sin \beta t = \sqrt{A^2 + B^2} \cos (\beta t + \theta), \qquad (5.68)$$

where

$$\theta = \begin{cases} \tan^{-1}\left(\dfrac{B}{A}\right) & \text{when } A > 0 \\[3mm] 180° + \tan^{-1}\left(\dfrac{B}{A}\right) & \text{when } A < 0. \end{cases}$$

From (5.68) we have

$$(\operatorname{Re} c_1) \cos \beta t - (\operatorname{Im} c_1) \sin \beta t = |c_1| \cos (\beta t + \theta), \qquad (5.69)$$

where

$$\theta = \begin{cases} \tan^{-1}\left(\dfrac{\operatorname{Im} c_1}{\operatorname{Re} c_1}\right) & \text{when } \operatorname{Re} c_1 > 0 \\[3mm] 180° + \tan^{-1}\left(\dfrac{\operatorname{Im} c_1}{\operatorname{Re} c_1}\right) & \text{when } \operatorname{Re} c_1 < 0. \end{cases}$$

From the polar representation of a complex number (see Appendix B), we have

$$
\angle c_1 = \begin{cases} \tan^{-1}\left(\dfrac{\text{Im } c_1}{\text{Re } c_1}\right) & \text{when Re } c_1 > 0 \\[3mm] 180° + \tan^{-1}\left(\dfrac{\text{Im } c_1}{\text{Re } c_1}\right) & \text{when Re } c_1 < 0, \end{cases}
$$

and thus the angle θ in (5.69) is equal to $\angle c_1$.

Inserting (5.69) into (5.67), we obtain

$$c_1 e^{\bar{p}_1 t} + \bar{c}_1 e^{\bar{p}_1 t} = 2|c_1| e^{\alpha t} \cos(\beta t + \angle c_1). \tag{5.70}$$

Equation (5.70) is very useful in computing the inverse Laplace transform of any complex pair appearing in a partial-fraction expansion. The process is illustrated by the following example.

EXAMPLE 5.19. Suppose that

$$X(s) = \frac{s^2 - 2s + 1}{s^3 + 3s^2 + 4s + 2}.$$

Here

$$D(s) = s^3 + 3s^2 + 4s + 2 = (s + 1 - j)(s + 1 + j)(s + 1).$$

The roots of $D(s) = 0$ are

$$p_1 = -1 + j, \qquad p_2 = -1 - j, \qquad p_3 = -1.$$

Thus $\alpha = -1$ and $\beta = 1$, and

$$X(s) = \frac{c_1}{s + 1 - j} + \frac{\bar{c}_1}{s + 1 + j} + \frac{c_3}{s + 1},$$

where

$$c_1 = [(s + 1 - j)X(s)]_{s = -1+j} = \left.\frac{s^2 - 2s + 1}{(s + 1 + j)(s + 1)}\right|_{s = -1+j}$$

$$= \frac{-3}{2} + j2$$

$$c_3 = [(s + 1)X(s)]_{s = -1} = \left.\frac{s^2 - 2s + 1}{s^2 + 2s + 2}\right|_{s = -1} = 4.$$

Now

$$|c_1| = \sqrt{\frac{9}{4} + 4} = \frac{5}{2}$$

and

$$\angle c_1 = 180° + \tan^{-1}\left(\frac{-4}{3}\right) = 126.87°.$$

Then using (5.70), we get

$$x(t) = 5e^{-t} \cos(t + 126.87°) + 4e^{-t}, \qquad t \ge 0.$$

In the case when the poles of $X(s)$ are distinct, a program for computing the inverse Laplace transform of $X(s)$ is given in Figure A.7.

Second-Order Case. If $X(s)$ is a second-order rational function with complex poles, we can derive an expression for the inverse transform of $X(s)$ without having to carry out the partial-fraction expansion described above. The details are as follows.

Suppose that

$$X(s) = \frac{cs + d}{s^2 + es + h},$$

where c, d, e, and h are arbitrary real numbers. Here

$$D(s) = s^2 + es + h.$$

We can complete the square in $D(s)$ by writing

$$D(s) = \left(s + \frac{e}{2}\right)^2 + h - \frac{e^2}{4}.$$

From the quadratic formula, we have that the two roots of $D(s) = 0$ are complex if and only if

$$h - \frac{e^2}{4} > 0.$$

Let us suppose that this is the case. Then $h - (e^2/4)$ has a real square root which we will denote by ω; that is, let

$$\omega = \sqrt{h - \frac{e^2}{4}}.$$

Then

$$X(s) = \frac{cs + d}{(s + e/2)^2 + \omega^2}.$$

Rewriting the numerator of $X(s)$, we have

$$X(s) = \frac{c(s + e/2) + \left(\dfrac{d - (e/2)c}{\omega}\right)\omega}{(s + e/2)^2 + \omega^2}$$

$$= \frac{c(s + e/2)}{(s + e/2)^2 + \omega^2} + \frac{\left(\dfrac{d - (e/2)c}{\omega}\right)\omega}{(s + e/2)^2 + \omega^2}.$$

From the transform pairs in Table 5.3, we can compute the inverse transform of each of the two components comprising $X(s)$. Using linearity, we have that the inverse transform $x(t)$ of $X(s)$ is given by

$$x(t) = \exp\left(-\frac{e}{2}t\right)\left[c\,\cos\,\omega t - \left(\frac{(e/2)c - d}{\omega}\right)\sin\,\omega t\right], \qquad t \geq 0.$$

$$(5.71)$$

Using the trigonometric identity (5.68), we can write (5.71) in the form

$$x(t) = \sqrt{c^2 + \frac{[(e/2)c - d]^2}{\omega^2}}\,\exp\left(-\frac{e}{2}t\right)\cos\,(\omega t + \theta), \quad t \geq 0$$

$$= \sqrt{c^2 + \frac{[(e/2)c - d]^2}{h - e^2/4}}\,\exp\left(-\frac{e}{2}t\right)\cos\,(\sqrt{h - e^2/4}\,t + \theta), \quad t \geq 0,$$

$$(5.72)$$

where

$$\theta = \begin{cases} \tan^{-1}\left(\dfrac{(e/2)c - d}{c\omega}\right) & \text{when } c > 0 \\[4mm] 180° + \tan^{-1}\left(\dfrac{(e/2)c - d}{c\omega}\right) & \text{when } c < 0. \end{cases}$$

Equation (5.71) or (5.72) is the inverse Laplace transform of a general second-order rational function with complex poles.

Third-Order Case.   Now suppose that

$$X(s) = \frac{s^2 + b_1 s + b_0}{(s^2 + es + h)(s + k)},$$

where b_0, b_1, e, h, k are arbitrary real numbers. We again let $\omega^2 = h - (e^2/4)$ and assume that $\omega^2 > 0$. Then two of the poles of $X(s)$ are complex and the third pole is at $s = -k$. Now we can expand $X(s)$ into the form

$$X(s) = \frac{cs + d}{s^2 + es + h} + \frac{g}{s + k}, \tag{5.73}$$

where

$$g = [(s + k)X(s)]_{s = -k} = \left. \frac{s^2 + b_1 s + b_0}{s^2 + es + h} \right|_{s = -k} = \frac{k^2 - b_1 k + b_0}{k^2 - ek + h}.$$

To compute the coefficients c, d, in (5.73), we can combine the two terms comprising the expansion of $X(s)$ and set the numerator equal to the numerator of $X(s)$. This gives

$$(cs + d)(s + k) + g(s^2 + es + h) = s^2 + b_1 s + b_0. \tag{5.74}$$

Equating coefficients of s in (5.74), we get

$$c = 1 - g \quad \text{and} \quad d = b_1 - kc - ge.$$

Using (5.72), we have that the inverse Laplace transform $x(t)$ of $X(s)$ is given by

$$x(t) = \sqrt{c^2 + \frac{\left(d - \dfrac{ec}{2}\right)^2}{\omega^2}} \; \exp\left(-\frac{e}{2}t\right) \cos(\omega t + \theta) + g\, e^{-kt}, \qquad t \geq 0$$

$$= \sqrt{c^2 + \frac{\left(d - \dfrac{ec}{2}\right)^2}{h - \dfrac{e^2}{4}}} \; \exp\left(-\frac{e}{2}t\right) \cos\left(\sqrt{h - \frac{e^2}{4}}\, t + \theta\right)$$

$$+ g\, e^{-kt}, \qquad t \geq 0 \tag{5.75}$$

where

$$
\theta = \begin{cases}
\tan^{-1}\left(\dfrac{(e/2)c - d}{c\omega}\right) & \text{when } c > 0 \\[3ex]
180° + \tan^{-1}\left(\dfrac{(e/2)c - d}{c\omega}\right) & \text{when } c < 0.
\end{cases}
$$

Equation (5.75) is the inverse Laplace transform of a general third-order rational function with two complex poles and one real pole.

REPEATED POLES Let us return to the general case where

$$
X(s) = \frac{N(s)}{D(s)}.
$$

We continue to assume that $X(s)$ is strictly proper; that is, the degree m of $N(s)$ is strictly less than the degree n of $D(s)$. Now suppose that one of the poles of $X(s)$ is repeated r times and that the other $n - r$ poles are distinct. Let us assume that pole p_1 is repeated r times and the other poles $p_{r+1}, p_{r+2}, \cdots, p_n$ are distinct. Then $X(s)$ has the partial-fraction expansion

$$
X(s) = \frac{c_1}{s - p_1} + \frac{c_2}{(s - p_1)^2} + \cdots + \frac{c_r}{(s - p_1)^r}
$$

$$
+ \frac{c_{r+1}}{s - p_{r+1}} + \cdots + \frac{c_n}{s - p_n}. \tag{5.76}
$$

In (5.76), the constants, $c_{r+1}, c_{r+2}, \cdots, c_n$ are calculated as in the distinct-pole case; that is,

$$
c_i = [(s - p_i)X(s)]_{s=p_i}, \qquad i = r + 1, r + 2, \ldots, r.
$$

The constants $c_1, c_2, \ldots, c_r$ in (5.76) are given by

$$
c_{r-i} = \frac{1}{i!}\left[\frac{d^i}{ds^i}\left[(s - p_1)^r X(s)\right]\right]_{s=p_1}, \qquad i = 0, 1, 2, \ldots, r - 1 \tag{5.77}
$$

In particular, setting the index i equal to 0, 1, 2 in (5.77), we have

$$
c_r = [(s - p_1)^r X(s)]_{s=p_1}
$$

$$
c_{r-1} = \left[\frac{d}{ds}\left[(s - p_1)^r X(s)\right]\right]_{s=p_1}
$$

$$
c_{r-2} = \frac{1}{2}\left[\frac{d^2}{ds^2}\left[(s - p_1)^r X(s)\right]\right]_{s=p_1}.
$$

To compute the inverse Laplace transform of $X(s)$ given by the expansion (5.76), we can use linearity and the transform pairs

$$\frac{t^{n-1}}{(n-1)!}e^{-at} \leftrightarrow \frac{1}{(s+a)^n}, \qquad n = 1, 2, 3, \ldots.$$

EXAMPLE 5.20. Consider the rational function

$$X(s) = \frac{5s - 1}{s^3 - 3s - 2}.$$

The roots of $D(s) = 0$ are $-1, -1, 2$, so $r = 2$ and we have the expansion

$$X(s) = \frac{c_1}{s+1} + \frac{c_2}{(s+1)^2} + \frac{c_3}{s-2},$$

where

$$c_1 = \left[\frac{d}{ds}[(s+1)^2 X(s)]\right]_{s=-1} = \left[\frac{d}{ds}\frac{5s-1}{s-2}\right]_{s=-1}$$

$$= \frac{-9}{(s-2)^2}\bigg|_{s=-1} = -1.$$

$$c_2 = [(s+1)^2 X(s)]_{s=-1} = \frac{5s-1}{s-2}\bigg|_{s=-1} = 2$$

$$c_3 = [(s-2)X(s)]_{s=2} = \frac{5s-1}{(s+1)^2}\bigg|_{s=2} = 1.$$

Hence

$$x(t) = -e^{-t} + 2te^{-t} + e^{2t}, \qquad t \geq 0.$$

Case When $m \geq n$. Again consider $X(s) = N(s)/D(s)$ with the degree of $N(s)$ equal to m and the degree of $D(s)$ equal to n. If $m \geq n$, by long division we can write $X(s)$ in the form

$$X(s) = P(s) + \frac{R(s)}{D(s)},$$

where the quotient $P(s)$ is a polynomial in s with degree $m - n$ and the remainder $R(s)$ is a polynomial in s with degree strictly less than n. The inverse Laplace transform of $X(s)$ can then be computed by determining the inverse Laplace transform of $P(s)$ and the inverse Laplace transform of $R(s)/D(s)$. Since $R(s)/D(s)$ is strictly proper [i.e., degree $R(s) < n$], the inverse transform of $R(s)/D(s)$ can be computed by first expanding into partial fractions as given above. The inverse Laplace transform of $P(s)$ can be computed by using the transform pair

$$\frac{d^n}{dt^n} \delta(t) \leftrightarrow s^n, \qquad n = 1, 2, 3, \ldots .$$

EXAMPLE 5.21. Suppose that

$$X(s) = \frac{s^3 + 2s - 4}{s^2 + 4s - 2}.$$

Carrying out the long division

$$
\begin{array}{r}
s \quad - \ 4 \\
s^2 + 4s - 2 \overline{) s^3 + 0s^2 + \ 2s \ - \ 4} \\
\underline{s^3 + 4s^2 - \ 2s} \\
-4s^2 + \ 4s \ - \ 4 \\
\underline{-4s^2 - 16s \ + \ 8} \\
20s \ - \ 12
\end{array}
$$

we have

$$X(s) = s - 4 + \frac{20s - 12}{s^2 + 4s - 2}.$$

Thus

$$x(t) = \frac{d}{dt} \delta(t) - 4\delta(t) + v(t),$$

where $v(t)$ is the inverse Laplace transform of

$$V(s) = \frac{20s - 12}{s^2 + 4s - 2}.$$

The roots of $s^2 + 4s - 2 = 0$ are $s = -2 \pm \sqrt{6} = 0.449, -4.449$. Expanding $V(s)$ by partial fractions, we have

$$V(s) = \frac{-0.614}{s - 0.449} + \frac{20.614}{s + 4.449}.$$

Thus

$$v(t) = -0.614 \exp (0.449t) + 20.614 \exp (-4.449t), \qquad t \geq 0.$$

TRANSFORMS CONTAINING EXPONENTIALS

In many cases of interest, a function $x(t)$ will have a transform $X(s)$ of the form

$$X(s) = \frac{N_0(s)}{D_0(s)} + \frac{N_1(s)}{D_1(s)} \exp(-h_1 s) + \cdots + \frac{N_q(s)}{D_q(s)} \exp(-h_q s). \qquad (5.78)$$

In (5.78) the h_i are distinct positive real numbers, the $D_i(s)$ are polynomials in s with real coefficients, and the $N_i(s)$ are polynomials in s with real coefficients. Here we are assuming that $N_i(s) \neq 0$ for at least one value of $i \geq 1$.

The function $X(s)$ given by (5.78) is not rational in s. In other words, it is not possible to express $X(s)$ as a ratio of polynomials in s with real coefficients. This is a result of the presence of the exponential terms $\exp(-h_i s)$, which cannot be written as ratios of polynomials in s. Functions of the form (5.78) are examples of *irrational functions* of s. They are also called *transcendental functions* of s.

Functions $X(s)$ of the form (5.78) arise when we take the Laplace transform of a piecewise-continuous function $x(t)$. For instance, as shown in Example 5.5, the transform of the a-second pulse $u(t) - u(t - a)$ is equal to

$$\frac{1}{s} - \frac{1}{s} e^{-as}.$$

Clearly, this transform is in the form (5.78). Take

$$N_0(s) = 1, \qquad N_1(s) = -1, \qquad D_0(s) = D_1(s) = s, \qquad h_1 = a.$$

Since $X(s)$ given by (5.78) is not rational in s, it is not possible to apply the partial-fraction expansion directly to (5.78). However, we can still use partial-fraction expansions to compute the inverse transform of $X(s)$. The procedure is as follows.

First, we can write $X(s)$ in the form

$$X(s) = \frac{N_0(s)}{D_0(s)} + \sum_{i=1}^{q} \frac{N_i(s)}{D_i(s)} \exp(-h_i s). \qquad (5.79)$$

Now each $N_i(s)/D_i(s)$ in (5.79) is a rational function of s. If deg $N_i(s) <$ deg $D_i(s)$ for $i = 0, 1, 2, \ldots q$, each rational function $N_i(s)/D_i(s)$ can be expanded by partial fractions. In this way, we can compute the inverse Laplace transform of $N_i(s)/D_i(s)$ for $i = 0, 1, 2, \ldots, q$. Let $x_i(t)$ denote the inverse transform of $N_i(s)/D_i(s)$. Then by linearity and the right-shift property, the inverse Laplace transform $x(t)$ of $X(s)$ is given by

$$x(t) = x_0(t) + \sum_{i=1}^{q} x_i(t - h_i) u(t - h_i), \qquad t \geq 0. \qquad (5.80)$$

EXAMPLE 5.22. Suppose that

$$X(s) = \frac{s + 1}{s^2 + 1} - \frac{1}{s + 1} e^{-s} + \frac{s + 2}{s^2 + 1} \exp(-1.5s).$$

From linearity and the transform pairs in Table 5.3, we have

$$\cos t + \sin t \leftrightarrow \frac{s + 1}{s^2 + 1}$$

$$\cos t + 2 \sin t \leftrightarrow \frac{s + 2}{s^2 + 1}$$

Thus

$$x(t) = \cos t + \sin t - \exp[-(t - 1)]u(t - 1)$$
$$+ [\cos(t - 1.5) + 2 \sin(t - 1.5)]u(t - 1.5), \qquad t \geq 0.$$

5.5

TRANSFER FUNCTION REPRESENTATION

In this section we begin the application of the Laplace transform to the study of continuous-time systems. Throughout this section we assume that the system under study is a single-input single-output causal linear time-invariant continuous-time system. The system has the input/output relationship

$$y(t) = (Fx)(t), \tag{5.81}$$

where $y(t)$ is the output response resulting from input $x(t)$ with the system at rest prior to the application of $x(t)$.

Since the system is assumed to be causal, linear, and time invariant, by the results in Chapter 4 we know that the input/output operator F is a convolution operator; that is,

$$y(t) = (h * x)(t) = \int_0^{\infty} h(\lambda)x(t - \lambda) \, d\lambda,$$

where $h(t)$ is the impulse response of the system. If $x(t) = 0$ for all $t < 0$, we have

$$y(t) = \int_0^t h(\lambda)x(t - \lambda) \, d\lambda, \qquad t > 0. \tag{5.82}$$

Now we can take the Laplace transform of both sides of (5.82). Using the property that the transform of the convolution $h * x$ is the product of the transforms of h and x, we obtain

$$Y(s) = H(s)X(s). \qquad (5.83)$$

In (5.83), $Y(s)$ is the transform of the output response $y(t)$, $H(s)$ is the transform of the impulse response $h(t)$, and $X(s)$ is the transform of the input $x(t)$.

Equation (5.83) is the transfer function representation (or s-domain representation) of the given system and $H(s)$ is the transfer function (or system function) of the system. As seen from (5.83), the transfer function $H(s)$ describes the "transfer" from input to output in the s-domain, assuming no initial energy in the system at time $t = 0$ (or $t = 0^-$).

Since $H(s)$ is the Laplace transform of the impulse response $h(t)$, we have the transform pair

$$h(t) \leftrightarrow H(s). \qquad (5.84)$$

The transform pair (5.84) is of fundamental importance. In particular, it provides a bridge between the time-domain representation given by the convolution relationship and the s-domain representation given in terms of the transfer function.

It is important to note that if the input $x(t)$ is not the zero function, so that $X(s)$ is not zero, we can divide both sides of (5.83) by $X(s)$, which gives

$$H(s) = \frac{Y(s)}{X(s)}. \qquad (5.85)$$

From (5.85) we see that the transfer function $H(s)$ is equal to the ratio of the transform $Y(s)$ of the output and the transform $X(s)$ of the input.

Since $H(s)$ is the transform of the impulse response $h(t)$ and each single-input single-output system has only one $h(t)$, each system has a unique transfer function. Therefore, although $Y(s)$ will change as the input $x(t)$ ranges over some collection of input signals, by (5.85) the ratio $Y(s)/X(s)$ cannot change [assuming that there is no initial energy in the system before $x(t)$ is applied].

From (5.85) we have the interesting result that the transfer function $H(s)$ can be determined from knowledge of the response $y(t)$ to any nonzero input signal $x(t)$. It should be stressed that this result is valid only if it is known that the given system is both linear and time invariant. If the system is time varying or is nonlinear, there is no transfer function, and thus (5.85) has no meaning.

EXAMPLE 5.23.   Suppose that the input $x(t) = e^{-t}u(t)$ is applied to a linear time-invariant continuous-time system, and that the resulting output response with the system at rest at time $t = 0$ is

$$y(t) = 2 - 3e^{-t} + e^{-2t} \cos 2t, \qquad t \geq 0.$$

Then

$$Y(s) = \frac{2}{s} - \frac{3}{s+1} + \frac{s+2}{(s+2)^2 + 4}$$

and

$$X(s) = \frac{1}{s+1}.$$

Inserting these expressions for $Y(s)$ and $X(s)$ into (5.85), we have that the transfer function $H(s)$ of the system is given by

$$H(s) = \frac{2(s+1)}{s} - 3 + \frac{(s+1)(s+2)}{(s+2)^2 + 4}$$

$$= \frac{[2(s+1) - 3s][(s+2)^2 + 4] + s(s+1)(s+2)}{s[(s+2)^2 + 4]}$$

$$= \frac{s^2 + 2s + 16}{s^3 + 4s^2 + 8s}.$$

EXAMPLE 5.24. Consider the finite-time integrator defined in Example 4.10. Recall that this system can be realized using one integrator and an ideal time delay (see Figure 4.12). As shown in Example 4.10 the impulse response of the finite-time integrator is

$$h(t) = u(t) - u(t - c),$$

where c is a strictly positive real number. Taking the Laplace transform of $h(t)$, we find that the transfer function of the system is

$$H(s) = \frac{1}{s} - \frac{1}{s}e^{-cs} = \frac{1 - e^{-cs}}{s}.$$

This transfer function cannot be written as a ratio of two polynomials in s, and thus $H(s)$ is an irrational function of s. Systems with time delays such as the finite-time integrator always have irrational transfer functions.

COMPUTATION OF OUTPUT RESPONSE Again consider the transfer function representation (5.83). Taking the inverse Laplace transform of both sides of (5.83), we obtain

$$y(t) = \text{inverse transform of } H(s)X(s). \tag{5.86}$$

Using (5.86), we can compute the output response $y(t)$ resulting from an input $x(t)$ with no initial energy in the system prior to the application of $x(t)$. First, compute the transforms $X(s)$ and $H(s)$ of $x(t)$ and $h(t)$, and then form the product $H(s)X(s)$. Taking the inverse Laplace transform of $H(s)X(s)$, we get $y(t)$.

If both $H(s)$ and $X(s)$ are rational functions of s, the product $H(s)X(s)$ is also a rational function of s. In this case, the output $y(t)$ can be computed by first expanding $H(s)X(s)$ by partial fractions. The process is illustrated by the following example.

EXAMPLE 5.25.   Consider the system in Example 5.23 with transfer function

$$H(s) = \frac{s^2 + 2s + 16}{s^3 + 4s^2 + 8s}.$$

Let us compute the output response $y(t)$ resulting from input $x(t) = e^{-2t}u(t)$ with no initial energy in the system at time $t = 0$. We have

$$X(s) = \frac{1}{s + 2},$$

and thus

$$Y(s) = H(s)X(s) = \frac{s^2 + 2s + 16}{(s^3 + 4s^2 + 8s)(s + 2)} = \frac{s^2 + 2s + 16}{s(s + 2)[(s + 2)^2 + 4]}.$$

Expanding by partial fractions, we have

$$Y(s) = \frac{c_0}{s} + \frac{c_1}{s + 2} + \frac{c_2 s + c_3}{(s + 2)^2 + 4}, \qquad (5.87)$$

where

$$c_0 = [sY(s)]_{s=0} = \frac{16}{2(8)} = 1$$

$$c_1 = [(s + 2)Y(s)]_{s=-2} = \frac{(-2)^2 - (2)(2) + 16}{(-2)(4)} = -2.$$

To compute the constants c_2 and c_3 in (5.87), we can combine the terms on the right side of (5.87). This gives

$$(s + 2)(s^2 + 4s + 8) - 2s(s^2 + 4s + 8)$$
$$+ (c_2 s + c_3)s(s + 2) = s^2 + 2s + 16.$$

Collecting terms with like powers of s, we have

$$s^3 - 2s^3 + c_2 s^3 = 0. \qquad (5.88)$$

$$6s^2 - 8s^2 + (c_3 + 2c_2)s^2 = s^2. \qquad (5.89)$$

From (5.88) we get that $c_2 = 1$, and from (5.89) we have that $c_3 = 1$. Then using (5.72), we have

$$y(t) = 1 - 2e^{-2t} + \frac{\sqrt{5}}{2} e^{-2t} \cos (2t + 26.565°), \qquad t \geq 0.$$

5.6

TRANSFORM OF THE INPUT/OUTPUT DIFFERENTIAL EQUATION

In this section we consider linear time-invariant continuous-time systems that are finite dimensional. Recall from Chapter 1 that the assumption of finite dimensionality implies that the system can be modeled by an input/output differential equation. We can generate the s-domain representation of the system by taking the Laplace transform of the input/output differential equation. In particular, we will see that the system's transfer function $H(s)$ can be computed directly from the coefficients of the input/output differential equation. We begin with the first-order case.

FIRST-ORDER CASE Consider the linear time-invariant finite-dimensional continuous-time system given by the first-order input/output differential equation

$$\frac{dy(t)}{dt} + ay(t) = b_1 \frac{dx(t)}{dt} + b_0 x(t). \tag{5.90}$$

Taking the Laplace transform of both sides of (5.90), we get

$$sY(s) - y(0) + aY(s) = b_1[sX(s) - x(0)] + b_0 X(s), \tag{5.91}$$

where $Y(s)$ is the Laplace transform of the output response $y(t)$ and $X(s)$ is the Laplace transform of the input $x(t)$ applied for $t \geq 0$. If the input $x(t)$ contains an impulse at $t = 0$ or is discontinuous at $t = 0$ [i.e., $x(0^+) \neq x(0^-)$], the initial conditions in (5.91) must be taken at time $t = 0^-$ so that we have

$$sY(s) - y(0^-) + aY(s) = b_1[sX(s) - x(0^-)] + b_0 X(s). \tag{5.92}$$

If $x(0^-) = 0$, (5.92) reduces to

$$sY(s) - y(0^-) + aY(s) = b_1 sX(s) + b_0 X(s). \tag{5.93}$$

In the following development, we work with the transformed equation (5.93). Rearranging terms in (5.93), we have

$$(s + a)Y(s) = y(0^-) + (b_1 s + b_0)X(s). \tag{5.94}$$

Solving (5.94) for $Y(s)$, we obtain

$$Y(s) = \frac{y(0^-)}{s + a} + \frac{b_1 s + b_0}{s + a} X(s). \tag{5.95}$$

Equation (5.95) is the s-domain representation of the system given by the input/output differential equation (5.90). The first term on the right side of (5.95) is the Laplace transform of the part of the output response due to the initial condition $y(0^-)$, and the second term on the right side of (5.95) is the Laplace transform of the part of the output response resulting from the input $x(t)$ applied for $t \geq 0$.

The given system has no initial energy at time $t = 0^-$ if and only if $y(0^-) = 0$ [assuming that $x(0^-) = 0$]. Hence, if the system has no initial energy at time $t = 0^-$, we have

$$Y(s) = \frac{b_1 s + b_0}{s + a} X(s). \tag{5.96}$$

From the results in Section 5.5, we know that when there is no initial energy in the system, the transform $Y(s)$ is related to the transform $X(s)$ by

$$Y(s) = H(s)X(s), \tag{5.97}$$

where $H(s)$ is the transfer function of the system. Comparing (5.96) and (5.97), we have

$$H(s) = \frac{b_1 s + b_0}{s + a}.$$

Note that the transfer function is a first-order rational function of s.

Writing the transfer function $H(s)$ in the form

$$H(s) = b_1 + \frac{b_0 - ab_1}{s + a},$$

and taking the inverse Laplace transform of $H(s)$, we have

$$h(t) = b_1 \delta(t) + (b_0 - ab_1)e^{-at}, \qquad t \geq 0. \tag{5.98}$$

As discussed in Section 5.5, the function $h(t)$ is the impulse response of the system. Hence the expression (5.98) for $h(t)$ is the general form of the impulse response of a first-order system. Note that $h(t)$ is an exponential function for $t > 0$. In particular, the impulse response of a first-order system cannot be sinusoidal.

EXAMPLE 5.26. We shall compute the transfer function of the integrator. From Chapter 2 (see Section 2.2) we know that the input/output differential equation of the integrator is

$$\frac{dy(t)}{dt} = x(t). \tag{5.99}$$

Comparing (5.90) and (5.99), we see that $a = 0$, $b_1 = 0$, and $b_0 = 1$ in this example. Thus the transfer function of the integrator is

$$H(s) = \frac{0s + 1}{s + 0} = \frac{1}{s}.$$

We could have obtained $H(s)$ by taking the transform of the integrator's impulse response. Let us do it this way and check the results. First, since the integral of the unit impulse $\delta(t)$ is the unit-step function $u(t)$, the impulse response of the integrator is equal to $u(t)$. The Laplace transform of $u(t)$ is equal to $1/s$, so the transfer function $H(s)$ is equal to $1/s$, which checks with the above result.

SECOND-ORDER CASE Now consider the linear time-invariant finite-dimensional continuous-time system given by the second-order input/output differential equation

$$\frac{d^2y(t)}{dt^2} + a_1 \frac{dy(t)}{dt} + a_0 y(t) = b_2 \frac{d^2x(t)}{dt^2} + b_1 \frac{dx(t)}{dt} + b_0 x(t). \tag{5.100}$$

We assume that $x(0^-) = 0$ and $\dot{x}(0^-) = 0$. Then taking the Laplace transform of both sides of (5.100) with initial conditions at time $t = 0^-$, we have

$$s^2Y(s) - y(0^-)s - \dot{y}(0^-) + a_1[sY(s) - y(0^-)] \tag{5.101}$$
$$+ a_0Y(s) = b_2s^2X(s) + b_1sX(s) + b_0X(s).$$

Solving (5.101) for $Y(s)$, we get

$$Y(s) = \frac{y(0^-)s + \dot{y}(0^-) + a_1y(0^-)}{s^2 + a_1s + a_0} + \frac{b_2s^2 + b_1s + b_0}{s^2 + a_1s + a_0}X(s). \tag{5.102}$$

Equation (5.102) is the s-domain representation of the system with the input/output differential equation (5.100). The first term on the right side of (5.102) is the Laplace transform of the part of the output response resulting from initial conditions and the second term is the transform of the part of the response resulting from the application of the input $x(t)$ for $t \geq 0$. When $x(0^-)$

$= \dot{x}(0^-) = 0$, the system is at rest at time $t = 0^-$ if and only if $y(0^-) = 0$ and $\dot{y}(0^-) = 0$, in which case the transform of the output response is given by

$$Y(s) = \frac{b_2 s^2 + b_1 s + b_0}{s^2 + a_1 s + a_0} X(s). \qquad (5.103)$$

From (5.103) we see that the transfer function $H(s)$ of the system is

$$H(s) = \frac{b_2 s^2 + b_1 s + b_0}{s^2 + a_1 s + a_0}. \qquad (5.104)$$

In this case the transfer function is a second-order rational function of s.

EXAMPLE 5.27.   Consider the system given by the interconnection in Figure 5.3. We shall first determine the input/output differential equation of the system. From Figure 5.3,

$$\dot{y}(t) = -6y(t) + q(t) \qquad (5.105)$$

$$\dot{q}(t) = -8y(t) + 2x(t). \qquad (5.106)$$

Differentiating both sides of (5.105) and using (5.106), we obtain

$$\ddot{y}(t) = -6\dot{y}(t) + [-8y(t) + 2x(t)]. \qquad (5.107)$$

Writing (5.107) in the form (5.100) and using (5.104), we have that the transfer function $H(s)$ is given by

$$H(s) = \frac{2}{s^2 + 6s + 8}.$$

The impulse response $h(t)$ of the system is equal to the inverse Laplace transform of $H(s)$. Expanding $H(s)$, we have

$$H(s) = \frac{1}{s + 2} - \frac{1}{s + 4}.$$

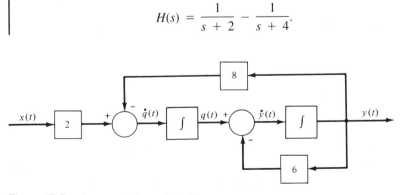

Figure 5.3.   System in Example 5.27.

Hence

$$h(t) = e^{-2t} - e^{-4t}, \qquad t \geq 0.$$

Now let us compute the output response $y(t)$ resulting from the step input $x(t) = u(t)$ with the system at rest at time $t = 0^-$. Since there is zero initial energy, $Y(s) = H(s)X(s)$. Here $X(s) = 1/s$, so we have

$$Y(s) = \frac{2}{s^2 + 6s + 8} \frac{1}{s}.$$

Expanding $Y(s)$ by partial fractions, we get

$$Y(s) = \frac{0.25}{s} - \frac{0.5}{s + 2} + \frac{0.25}{s + 4}.$$

Then the output response $y(t)$ is given by

$$y(t) = 0.25 - 0.5e^{-2t} + 0.25e^{-4t}, \qquad t \geq 0.$$

Suppose now that we want to find the output response $y(t)$ resulting from the step input $x(t) = u(t)$ with the initial conditions $y(0^-) = 1$, $\dot{y}(0^-) = 2$. The system does have initial energy at time $t = 0^-$, and thus it is *not* true that $Y(s) = H(s)X(s)$. To compute $Y(s)$, we must use (5.102). This gives

$$Y(s) = \frac{s + 8}{s^2 + 6s + 8} + \frac{2}{s^2 + 6s + 8} \frac{1}{s}$$

$$= \frac{s^2 + 8s + 2}{s(s^2 + 6s + 8)}.$$

Expanding, we have

$$Y(s) = \frac{0.25}{s} + \frac{2.5}{s + 2} - \frac{1.75}{s + 4}.$$

Thus

$$y(t) = 0.25 + 2.5e^{-2t} - 1.75e^{-4t}, \qquad t \geq 0.$$

nth-ORDER CASE   Let us now consider the general case where the system is given by the nth-order input/output differential equation

$$\frac{d^n y(t)}{dt^n} + \sum_{i=0}^{n-1} a_i \frac{d^i y(t)}{dt^i} = \sum_{i=0}^{m} b_i \frac{d^i x(t)}{dt^i}, \tag{5.108}$$

where $m \leq n$. We assume that $x^{(i)}(0^-) = 0$ for $i = 0, 1, 2, \ldots, m - 1$. By taking the Laplace transform of both sides of (5.108) with initial conditions at time $t = 0^-$, we can express the Laplace transform $Y(s)$ of the output in the form

$$Y(s) = \frac{C(s)}{D(s)} + \frac{N(s)}{D(s)} X(s), \tag{5.109}$$

where $N(s)$ and $D(s)$ are polynomials in s given by

$$N(s) = b_m s^m + b_{m-1} s^{m-1} + \cdots + b_1 s + b_0$$

$$D(s) = s^n + a_{n-1} s^{n-1} + \cdots + a_1 s + a_0.$$

The numerator $C(s)$ of the first term on the right side of (5.109) is also a polynomial in s whose coefficients are determined by the initial conditions $y(0^-)$, $y^{(1)}(0^-), \ldots, y^{(n-1)}(0^-)$. For example, if $n = 2$, then by the results above,

$$C(s) = y(0^-)s + \dot{y}(0^-) + a_1 y(0^-).$$

Equation (5.109) is the s-domain representation of the system with the nth-order input/output differential equation (5.108).

Since we are assuming that $x^{(i)}(0^-) = 0$ for $i = 0, 1, 2, \ldots, m - 1$, the system is at rest at time $t = 0^-$ if and only if the initial conditions $y(0^-)$, $y^{(1)}(0^-), \ldots, y^{(n-1)}(0^-)$ are zero, which is equivalent to the condition that $C(s) = 0$. If the system is at rest at time $t = 0$, the transform $Y(s)$ of the output response is given by

$$Y(s) = \frac{N(s)}{D(s)} X(s) = \frac{b_m s^m + \cdots + b_1 s + b_0}{s^n + a_{n-1} s^{n-1} + \cdots + a_1 s + a_0} X(s). \tag{5.110}$$

From (5.110) we have that the transfer function $H(s)$ is an nth-order rational function in s given by

$$H(s) = \frac{b_m s^m + \cdots + b_1 s + b_0}{s^n + a_{n-1} s^{n-1} + \cdots + a_1 s + a_0}. \tag{5.111}$$

RATIONAL TRANSFER FUNCTIONS AND FINITE-DIMENSIONAL SYSTEMS   From the above results we see that the transfer function $H(s)$ is rational if the system is finite dimensional. Conversely, if the transfer function is rational, the system must be finite dimensional. To see this, suppose that we are given a linear time-invariant continuous-time system with transfer function $H(s)$. Let us also suppose that $H(s)$ is rational, so that it can be written in the form (5.111). Multiplying both sides of (5.110) by $D(s)$, we obtain

$$(s^n + a_{n-1} s^{n-1} + \cdots + a_1 s + a_0)Y(s) = (b_m s^m + \cdots + b_1 s + b_0)X(s). \tag{5.112}$$

Inverse transforming both sides of (5.112), we get

$$\frac{d^n y(t)}{dt^n} + \sum_{i=0}^{n-1} a_i \frac{d^i y(t)}{dt^i} = \sum_{i=0}^{m} b_i \frac{d^i x(t)}{dt^i}.$$

Thus the system can be described by an input/output differential equation, which proves that the system is finite dimensional.

From the above observations we have the fundamental result that a linear time-invariant continuous-time system is finite dimensional if and only if the transfer function $H(s)$ is rational in s.

POLES AND ZEROS OF A FINITE-DIMENSIONAL SYSTEM If $H(s)$ is rational in s, the zeros of $H(s)$ are called the *zeros of the system* and the poles of $H(s)$ called the *poles of the system*. In other words, expressing $H(s)$ in the factored form

$$H(s) = \frac{K(s - z_1)(s - z_2) \cdots (s - z_m)}{(s - p_1)(s - p_2) \cdots (s - p_n)}, \tag{5.113}$$

we have that $z_1, z_2, \ldots, z_m$ are the zeros of the system and $p_1, p_2, \ldots, p_n$ are the poles of the system. Since

$$H(s)\big|_{s=z_i} = H(z_i) = 0, \qquad i = 1, 2, \ldots, m,$$

the transfer function is zero when evaluated at a zero of the system. In addition, since

$$|H(s)| \to \infty \text{ as } s \to p_i, \qquad i = 1, 2, \ldots, n,$$

the magnitude of the transfer function approaches infinity as s approaches a pole of the system.

From (5.113) we see that except for the constant K, the transfer function is completely determined by the values of the poles and zeros of the system. The poles and zeros of a given system are often displayed in a *pole–zero diagram*, which is a plot in the complex plane showing the location of all the poles (marked by $\times$) and all the zeros (marked by $\bigcirc$). It turns out that the location of the poles and zeros is of fundamental importance in determining the behavior of the system.

EXAMPLE 5.28. Consider the system with the transfer function

$$H(s) = \frac{2s^2 + 12s + 20}{s^3 + 6s^2 + 10s + 8}.$$

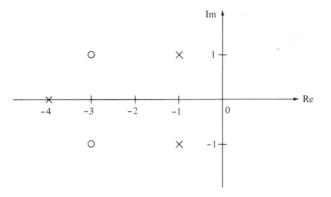

Figure 5.4. Pole–zero diagram of system in Example 5.28.

Factoring $H(s)$, we have

$$H(s) = \frac{2(s + 3 - j)(s + 3 + j)}{(s + 4)(s + 1 - j)(s + 1 + j)}.$$

Thus the zeros of the system are

$$z_1 = -3 + j \quad \text{and} \quad z_2 = -3 - j,$$

and the poles of the system are

$$p_1 = -4, \quad p_2 = -1 + j, \quad p_3 = -1 - j.$$

The pole–zero diagram is shown in Figure 5.4.

BRIDGE BETWEEN THE TIME DOMAIN AND s-DOMAIN   It is important to emphasize that the transfer function $H(s)$ given by (5.111) can be determined directly from the coefficients of the system's input/output differential equation (5.108). Hence we can easily go from the time-domain representation (5.108) to the transfer function representation (5.110). Conversely, if a system has transfer function $H(s)$ given by (5.111), the input/output differential equation of the system is given by (5.108). So we can also go from the transfer function representation to the input/output differential equation. Thus we have another link between the time domain and the s-domain.

PROBLEMS

Chapter 5

5.1   A continuous-time signal $x(t)$ has the Laplace transform

$$X(s) = \frac{s + 1}{s^2 + 5s + 7}.$$

Determine the Laplace transform $V(s)$ of the following signals.

(a) $v(t) = x(3t - 4)u(3t - 4)$.

(b) $v(t) = tx(t)$.

(c) $v(t) = \dfrac{d^2x(t)}{dt^2}$.

(d) $v(t) = x^{(-1)}(t)$.

(e) $v(t) = x(t) \sin 2t$.

(f) $v(t) = e^{-3t}x(t)$.

(g) $v(t) = (x * x)(t)$.

5.2 Compute the Laplace transform of each of the signals displayed in Figure P5.2.

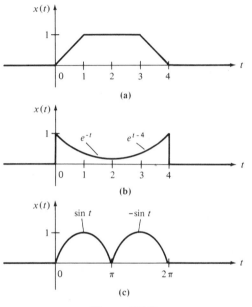

Figure P5.2

5.3 Determine the final values $[\lim_{t \to \infty} x(t)]$ of each of the signals whose Laplace transforms are given below. If there is no final value, state why not. Do not attempt to compute the inverse Laplace transforms.

(a) $X(s) = \dfrac{4}{s^2 + s}$.

(b) $X(s) = \dfrac{3s + 4}{s^2 + s}$.

(c) $X(s) = \dfrac{4}{s^2 - s}$.

(d) $X(s) = \dfrac{3s^2 + 4s + 1}{s^3 + 2s^2 + s + 2}$.

(e) $X(s) = \dfrac{3s^2 + 4s + 1}{s^3 + 3s^2 + 3s + 2}$.

(f) $X(s) = \dfrac{3s^2 + 4s + 1}{s^4 + 3s^3 + 3s^2 + 2s}$.

5.4 Determine the initial values $x(0)$ for each of the signals whose transforms are given in Problem 5.3.

5.5 Determine the inverse Laplace transform of each of the following functions.

(a) $X(s) = \dfrac{s + 2}{s^2 + 7s + 12}$.

(b) $X(s) = \dfrac{s + 1}{s^3 + 5s^2 + 7s}$.

(c) $X(s) = \dfrac{2s^2 - 9s - 35}{s^2 + 4s + 2}$.

(d) $X(s) = \dfrac{3s^2 + 2s + 1}{s^3 + 5s^2 + 8s + 4}$.

(e) $X(s) = \dfrac{s^2 + 1}{s^5 + 18s^3 + 81s}$.

(f) $X(s) = \dfrac{s + e^{-s}}{s^2 + s + 1}$.

(g) $X(s) = \dfrac{s}{s + 1} + \dfrac{se^{-s} + e^{-2s}}{s^2 + 2s + 1}$.

5.6 By using the Laplace transform, compute the convolution $(x * v)(t)$ where:
(a) $x(t) = e^{-t}u(t)$, $v(t) = (\sin t)u(t)$.
(b) $x(t) = (\cos t)u(t)$, $v(t) = (\sin t)u(t)$.
(c) $x(t) = (\sin t)u(t)$, $v(t) = (t \sin t)u(t)$.
(d) $x(t) = (\sin^2 t)u(t)$, $v(t) = tu(t)$.

5.7 A linear time-invariant continuous-time system has the impulse response $h(t) = \cos 2t + 4 \sin 2t$.
(a) Determine the transfer function $H(s)$ of the system.
(b) By using the Laplace transform, compute the output response $y(t)$ when the input $x(t)$ is equal to $\frac{5}{7}e^{-t} - \frac{12}{7}e^{-8t}$ for $t \geq 0$ with no initial energy in the system at time $t = 0$.

5.8 A linear time-invariant continuous-time system has impulse response $h(t)$ given by

$$h(t) = \begin{cases} e^{-t}, & 0 \leq t \leq 2 \\ e^{t-4}, & 2 \leq t \leq 4 \\ 0, & \text{all other } t. \end{cases}$$

By using the Laplace transform, compute the output response $y(t)$ resulting from the input $x(t) = (\sin t)u(t)$ with no initial energy.

5.9 The input

$$x(t) = \begin{cases} \sin t, & 0 \leq t \leq \pi \\ -\sin t, & \pi \leq t \leq 2\pi \\ 0, & \text{all other } t \end{cases}$$

is applied to a linear time-invariant continuous-time system with no initial energy in the system at time $t = 0$. The resulting response is displayed in Figure P5.9. Determine the transfer fuunction $H(s)$ of the system.

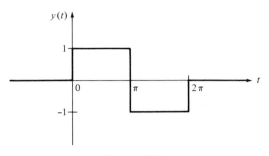

Figure P5.9

5.10 Consider the simple pendulum given by the small-signal model

$$I \frac{d^2\theta(t)}{dt^2} + MgL\,\theta(t) = Lx(t).$$

Recall that the output $y(t)$ of the system is equal to the angle $\theta(t)$ of the pendulum from the vertical reference.

(a) Determine the transfer function $H(s)$ of the system.

(b) Suppose that $\theta(0) = 0$, $\dot\theta(0) = 1$, and the input force $x(t) = A\delta(t)$, where A is a constant and $\delta(t)$ is the unit impulse. Determine the value of A so that the resulting response $\theta(t)$ is zero for *all* $t > 0$. In other words, we want the impulsive input force $A\delta(t)$ to "cancel" the nonzero initial velocity $\dot\theta(0)$.

(c) Now suppose that $\theta(0) = 30°$ ($\pi/6$ radians) and $\dot\theta(0) = 1$. Find an input force $x(t)$ so that $\theta(t) = 30°$ for all $t > 0$; in other words, the input force holds the pendulum at its initial position $\theta(0)$.

5.11 Consider the automobile on a level surface given by the input/output differential equation

$$\frac{d^2y(t)}{dt^2} + \frac{k_f}{M}\frac{dy(t)}{dt} = \frac{1}{M}x(t).$$

(a) Determine the transfer function of the system.

(b) Suppose that $y(0) = 0$ and $\dot y(0) = 0$. By using the Laplace transform, compute the position $y(t)$ of the car for *all* $t \geq 0$ when the input force is $x(t) = t$, $0 \leq t \leq 10$, $x(t) = 20 - t$, $10 \leq t \leq 20$, $x(t) = 0$ for all other t.

(c) Suppose that $k_f = 0$. Compute a force of the form $x(t) = au(t) - bu(t - 10) + (b - a)u(t - 20)$ so that when $y(0) = 0$, $\dot y(0) = 5$, the car stops at time $t = 20$ at position $y(20) = 50$.

5.12 Consider the field-controlled dc motor first defined in Example 2.3. The input/output differential equation of the motor is

$$JL_f \frac{d^3y(t)}{dt^3} + (fL_f + R_f J)\frac{d^2y(t)}{dt^2} + R_f f \frac{dy(t)}{dt} = kx(t),$$

where $x(t)$ is the voltage applied to the field winding and $y(t)$ is the angle of the motor shaft and load.

(a) Determine the transfer function of the system.

(b) Find the impulse response $h(t)$ of the system.

5.13 For each of the continuous-time systems defined below, determine the system's transfer function $H(s)$ if the system has a transfer function. If there is no transfer function, state why not.

(a) $\dfrac{dy(t)}{dt} + e^{-t}y(t) = x(t)$.

(b) $\dfrac{dy(t)}{dt} + (v * y)(t) = x(t)$, where $v(t) = (\sin t)u(t)$.

(c) $\dfrac{d^2y(t)}{dt} + \displaystyle\int_0^t y(\lambda)\, d\lambda = \dfrac{dx(t)}{dt} - x(t)$.

(d) $\dfrac{dy(t)}{dt} = (y * x)(t)$.

(e) $\dfrac{dy(t)}{dt} - 2y(t) = tx(t)$.

5.14 A linear time-invariant continuous-time system has transfer function $H(s) = (s + 7)/(s^2 + 4)$. Derive an expression for the output response $y(t)$ in terms of $y(0^-)$, $\dot{y}(0^-)$ and the input $x(t)$. Assume that $x(0^-) = 0$.

5.15 The transfer function of a linear time-invariant continuous-time system is given by

$$H(s) = \frac{2s - 1}{s^2 + 3s + 2}.$$

The following parts are independent.

(a) An input $x(t)$ with $x(0^-) = 1$ produces the output response

$$y(t) = -\tfrac{1}{2} + 3e^{-t} - \tfrac{5}{2}e^{-2t}, \qquad t \geq 0^-.$$

Was the system at rest at time $t = 0^-$? Justify your answer.

(b) An input $x(t)$ with $x(0^-) = 1$ produces the output response

$$y(t) = -\tfrac{1}{2} + 5e^{-t} - \tfrac{9}{2}e^{-2t}, \qquad t \geq 0^-.$$

Was the system at rest at time $t = 0^-$? Justify your answer.

5.16 The input $x_1(t) = e^{-t}u(t)$ is applied to a linear time-invariant continuous-time system with nonzero initial conditions $y(0)$, $\dot{y}(0)$. The resulting response is $y_1(t) = 3t + 2 - e^{-t}$, $t \geq 0$. A second input $x_2(t) = e^{-2t}u(t)$ is applied to the system with the *same* initial conditions $y(0)$, $\dot{y}(0)$. The resulting response is $y_2(t) = 2t + 2 - e^{-2t}$, $t \geq 0$. Compute $y(0)$, $\dot{y}(0)$, and the impulse response $h(t)$ of the system.

5.17   A continuous-time system is given by the input/output differential equation

$$\frac{d^2y(t)}{dt^2} + 2\frac{dy(t-1)}{dt} - y(t) + 3y(t-2) = \frac{dx(t)}{dt} + x(t-2).$$

(a) Compute the transfer function $H(s)$ of the system.
(b) Compute the impulse response $h(t)$.

5.18   A continuous-time system is given by the input/output differential equation

$$\frac{d^2y(t)}{dt^2} + 4\frac{dy(t)}{dt} + 3y(t) = 2\frac{d^2x(t)}{dt^2} - 4\frac{dx(t)}{dt} - x(t).$$

In each of the following parts, compute the response $y(t)$ for all $t \geq 0$.
(a) $y(0^-) = -2$, $\dot{y}(0^-) = 1$, $x(t) = 0$ for all $t \geq 0^-$.
(b) $y(0^-) = 0$, $\dot{y}(0^-) = 0$, $x(t) = \delta(t)$, $\delta(t) =$ unit impulse.
(c) $y(0^-) = 0$, $\dot{y}(0^-) = 0$, $x(t) = u(t)$.
(d) $y(0^-) = -2$, $\dot{y}(0^-) = 1$, $x(t) = u(t)$.
(e) $y(0^-) = -2$, $\dot{y}(0^-) = 1$, $x(t) = u(t+1)$.

5.19   In this problem the objective is to design the oscillator illustrated in Figure P5.19. Using two integrators and subtracters, adders, and scalar multipliers, design the oscillator so that when the switch is closed at time $t = 0$, the output voltage $v(t)$ is sin $200t$, $t \geq 0$. We are assuming that the oscillator is at rest at time $t = 0$.

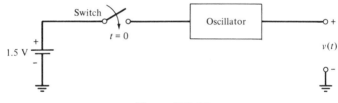

Figure P5.19

5.20   A linear time-invariant continuous-time system has impulse response $h(t) = [e^{-t} \cos(2t - 45°)]u(t) - tu(t)$. Determine the input/output differential equation of the system.

Applications of the Laplace Transform

In addition to giving an algebraic procedure for computing the output response resulting from an input, the Laplace transform can also be used to solve a number of problems in system analysis and design. Some of these applications are considered in this chapter.

We begin in Section 6.1 by showing that the transfer function of an *RLC* circuit or an interconnection of integrators can be determined directly from the circuit diagram. In Sections 6.2 and 6.3 we consider the computation of the transfer function of a system given by a block diagram. In the Sections 6.4 to 6.6, we utilize the transfer function model to study cascade synthesis, feedback control, and the inverse problem. In Section 6.7 we derive the transfer function representation for multi-input multi-output linear time-invariant continuous-time systems.

6.1

TRANSFER FUNCTION OF RLC CIRCUITS AND INTERCONNECTIONS OF INTEGRATORS

The transfer function of a linear time-invariant finite-dimensional system is often computed directly from the signal-flow diagram of the system, so it is not always

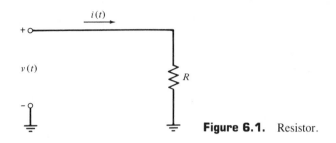

Figure 6.1. Resistor.

necessary that we first determine the input/output differential equation of the system. This can be done for *RLC* circuits and systems consisting of interconnections of integrators, adders, subtracters, and scalar multipliers. We begin with *RLC* circuits.

Consider the resistor shown in Figure 6.1. Taking the input to be the current $i(t)$ into the resistor and the output to be the voltage $v(t)$ across the resistor, we have the input/output relationship

$$v(t) = Ri(t). \tag{6.1}$$

Taking the Laplace transform of both sides of (6.1), we have

$$V(s) = RI(s). \tag{6.2}$$

From (6.2) we see that the transfer function of the resistor is equal to R. If we had taken the input to be $v(t)$ and the output to be $i(t)$, the transfer function would have been the conductance $1/R$.

Consider the capacitor shown in Figure 6.2. The input/output differential equation with the input equal to the current $i(t)$ and output equal to the voltage $v(t)$ is

$$\frac{dv(t)}{dt} = \frac{1}{C} i(t). \tag{6.3}$$

Taking the Laplace transform of both sides of (6.3) with $v(0^-) = 0$, we get

$$V(s) = \frac{1}{Cs} I(s). \tag{6.4}$$

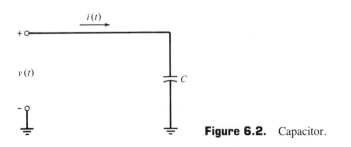

Figure 6.2. Capacitor.

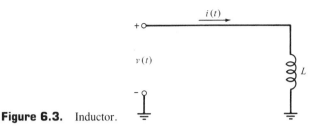

Figure 6.3. Inductor.

From (6.4) we see that the transfer function of the capacitor is equal to the impedance $1/Cs$. If we had taken the input to be $v(t)$ and the output to be $i(t)$, the transfer function would have been equal to the admittance Cs.

Now consider the inductor shown in Figure 6.3. With the current $i(t)$ equal to the input and with the voltage $v(t)$ equal to the output, the input/output differential equation of the inductor is

$$v(t) = L \frac{di(t)}{dt}. \tag{6.5}$$

Note that (6.5) cannot be viewed as a special case of the input/output differential equation (5.108) with $m \leq n$, since here $n = 0$, $m = 1$, and thus $m > n$. Nevertheless, we can still generate an s-domain representation for the inductor by taking the Laplace transform of both sides of (6.5). The result is the transformed equation

$$V(s) = LsI(s). \tag{6.6}$$

Thus we can define the transfer function of the inductor to be Ls. The quantity Ls is the impedance of the inductor.

INTERCONNECTION OF RLCs To determine the transfer function of an interconnection of R, L, and Cs, we can first write the circuit in the s-domain; that is, we can redraw the circuit in terms of the Laplace transforms of the voltages and currents in the circuit and in terms of the impedances of the circuit components. We can then write loop or node equations in the s-domain. These equations can be combined algebraically to yield the transfer function relationship of the circuit. The process is illustrated by the following example.

EXAMPLE 6.1. Consider the series RLC shown in Figure 6.4. As shown, the input is the voltage $x(t)$ applied to the series connection and the output is the voltage $v_c(t)$ across the capacitor. The s-domain representation of the circuit is shown in Figure 6.5. Working with the s-domain representation, by voltage division we have

$$V_c(s) = \frac{1/Cs}{Ls + R + (1/Cs)} X(s) = \frac{1/LC}{s^2 + (R/L)s + (1/LC)} X(s). \tag{6.7}$$

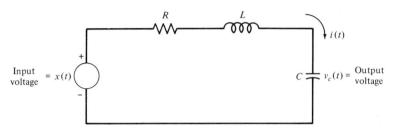

Figure 6.4.   Series *RLC* circuit.

From (6.7) we have that the transfer function $H(s)$ of the circuit is

$$H(s) = \frac{1/LC}{s^2 + (R/L)s + (1/LC)}.$$ (6.8)

Multiplying both sides of (6.7) by the denominator of $H(s)$, we have

$$\left(s^2 + \frac{R}{L}s + \frac{1}{LC}\right)V_c(s) = \frac{1}{LC}X(s).$$ (6.9)

Inverse transforming both sides of (6.9), we have that the circuit is given by the input/output differential equation

$$\frac{d^2v_c(t)}{dt^2} + \frac{R}{L}\frac{dv_c(t)}{dt} + \frac{1}{LC}v_c(t) = \frac{1}{LC}x(t).$$ (6.10)

We could have computed the input/output differential equation (6.10) by writing the integrodifferential equations for the circuit using Kirchhoff's laws (see Example 2.2). The *s*-domain approach discussed above is usually much faster than working in the time domain with integrodifferential equations. It is also interesting to note that if we had taken a different choice for the input and output of the circuit, the transfer function would not equal the result given in (6.8). For instance, if we keep the definition of the input the same,

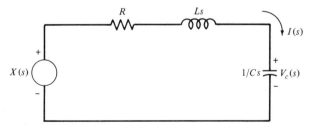

Figure 6.5.   Representation of series *RLC* circuit in *s*-domain.

but we take the output to be the voltage $v_R(t)$ across the resistor, by voltage division we have

$$V_R(s) = \frac{R}{Ls + R + (1/Cs)} X(s) = \frac{(R/L)s}{s^2 + (R/L)s + (1/LC)} X(s).$$

The resulting transfer function is

$$H(s) = \frac{(R/L)s}{s^2 + (R/L)s + (1/LC)},$$

which differs from (6.8).

INTERCONNECTIONS OF INTEGRATORS Let us now consider an interconnection of integrators, adders, subtracters, and scalar multipliers. From Example 5.26 we know that an integrator has transfer function $1/s$, so in the s-domain we can describe an integrator by its transfer function $1/s$. To compute the transfer function of an interconnection of integrators, we can first redraw the interconnection in the s-domain by taking transforms of all signals in the interconnection and by representing integrators by $1/s$. We can then write an equation for the Laplace transform of the output of each integrator in the interconnection. We can also write an equation for the transform of the output in terms of the transforms of the outputs of the integrators. These equations can then be combined algebraically to derive the transfer function relationship. The procedure is illustrated by the following example.

EXAMPLE 6.2. Consider the system shown in Figure 6.6. We have labeled the output of the first integrator $q_1(t)$ and the output of the second integrator $q_2(t)$. The s-domain representation of the system is shown in Figure 6.7. We have

$$sQ_1(s) = -4Q_1(s) + X(s) \tag{6.11}$$

$$sQ_2(s) = Q_1(s) - 3Q_2(s) + X(s) \tag{6.12}$$

$$Y(s) = Q_2(s) + X(s). \tag{6.13}$$

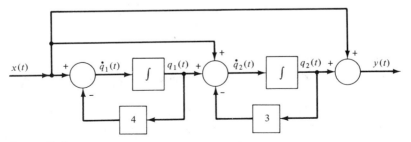

Figure 6.6. System with two integrators.

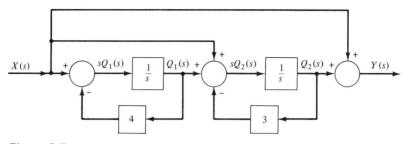

Figure 6.7. Representation of system in s-domain.

Solving (6.11) for $Q_1(s)$, we have

$$Q_1(s) = \frac{1}{s + 4} X(s). \tag{6.14}$$

Solving (6.12) for $Q_2(s)$ and using (6.14), we get

$$Q_2(s) = \frac{1}{s + 3} \left[Q_1(s) + X(s) \right] = \frac{1}{s + 3} \left(\frac{1}{s + 4} + 1 \right) X(s)$$

$$= \frac{s + 5}{(s + 3)(s + 4)} X(s). \tag{6.15}$$

Inserting the expression (6.15) for $Q_2(s)$ into (6.13) gives

$$Y(s) = \left[\frac{s + 5}{(s + 3)(s + 4)} + 1 \right] X(s) = \frac{s^2 + 8s + 17}{(s + 3)(s + 4)} X(s). \tag{6.16}$$

Thus the transfer function $H(s)$ is

$$H(s) = \frac{s^2 + 8s + 17}{(s + 3)(s + 4)} = \frac{s^2 + 8s + 17}{s^2 + 7s + 12}.$$

From (6.16) we have that the input/output differential equation of the system is

$$\frac{d^2y(t)}{dt^2} + 7 \frac{dy(t)}{dt} + 12y(t) = \frac{d^2x(t)}{dt^2} + 8 \frac{dx(t)}{dt} + 17x(t).$$

6.2

TRANSFER FUNCTIONS OF BASIC INTERCONNECTIONS

In the first part of this section we determine the transfer functions for three basic types of interconnections. Then we consider the effect of initial conditions in computing the transform of the output response of a given interconnection.

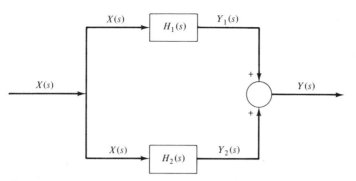

Figure 6.8. Parallel interconnection of two systems.

PARALLEL INTERCONNECTION   Consider a parallel interconnection of two linear time-invariant continuous-time systems with transfer functions $H_1(s)$ and $H_2(s)$. The interconnection is shown in Figure 6.8. The Laplace transform $Y(s)$ of the output of the parallel connection is given by

$$Y(s) = Y_1(s) + Y_2(s). \tag{6.17}$$

If each system in the connection has no initial energy, then

$$Y_1(s) = H_1(s)X(s) \quad \text{and} \quad Y_2(s) = H_2(s)X(s).$$

Inserting these expressions into (6.17), we have

$$Y(s) = H_1(s)X(s) + H_2(s)X(s)$$

$$= (H_1(s) + H_2(s))X(s). \tag{6.18}$$

From (6.18) we see that the transfer function $H(s)$ of the parallel interconnection is equal to the sum of the transfer functions of the systems in the connection; that is,

$$H(s) = H_1(s) + H_2(s). \tag{6.19}$$

CASCADE CONNECTION   Now consider the cascade connection shown in Figure 6.9. We assume that each system in the cascade connection has no initial energy and that the second system does not load the first system. No loading means that

$$Y_1(s) = H_1(s)X(s). \tag{6.20}$$

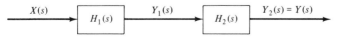

Figure 6.9. Cascade connection.

If $y_1(t)$ is a voltage waveform, one may assume that there is no loading if the output impedance of the first system is much less than the input impedance of the second system.

Now since

$$Y_2(s) = H_2(s)Y_1(s),$$

using (6.20) we obtain

$$Y(s) = Y_2(s) = H_2(s)H_1(s)X(s). \tag{6.21}$$

From (6.21) we have that the transfer function $H(s)$ of the cascade connection is equal to the product of the transfer functions of the systems in the connection; that is,

$$H(s) = H_2(s)H_1(s).$$

Since $H_1(s)$ and $H_2(s)$ are scalar-valued functions of s,

$$H_2(s)H_1(s) = H_1(s)H_2(s),$$

and thus $H(s)$ can also be expressed in the form

$$H(s) = H_1(s)H_2(s). \tag{6.22}$$

FEEDBACK CONNECTION   Now consider the interconnection shown in Figure 6.10. In this connection the output of the first system is fed back to the input through the second system, and thus the connection is called a *feedback connection*. Note that if the feedback loop is disconnected, the transfer function from $X(s)$ to $Y(s)$ is $H_1(s)$. The system with transfer function $H_1(s)$ is called the *open-loop system* since the transfer function from $X(s)$ to $Y(s)$ is equal to $H_1(s)$ if the feedback is disconnected. [Some authors refer to $H_1(s)H_2(s)$ as the *open-loop transfer function*.] The system with transfer function $H_2(s)$ is called the *feedback system*, and the feedback connection is called the *closed-loop system*. We shall compute the transfer function of the closed-loop system.

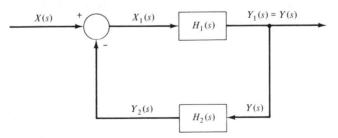

Figure 6.10.   Feedback connection.

We assume that there is no initial energy in either system and that the feedback system does not load the open-loop system. Then from Figure 6.10, we have

$$Y(s) = H_1(s)X_1(s) \tag{6.23}$$

$$X_1(s) = X(s) - Y_2(s) = X(s) - H_2(s)Y(s). \tag{6.24}$$

Inserting the expression (6.24) for $X_1(s)$ into (6.23), we get

$$Y(s) = H_1(s)[X(s) - H_2(s)Y(s)]. \tag{6.25}$$

Rearranging terms in (6.25), we have

$$[1 + H_1(s)H_2(s)]Y(s) = H_1(s)X(s). \tag{6.26}$$

Solving (6.26) for $Y(s)$ gives

$$Y(s) = \frac{H_1(s)}{1 + H_1(s)H_2(s)} X(s). \tag{6.27}$$

From (6.27) we have that the transfer function $H(s)$ of the feedback connection is given by

$$H(s) = \frac{H_1(s)}{1 + H_1(s)H_2(s)}. \tag{6.28}$$

From (6.28) we see that the closed-loop transfer function $H(s)$ is equal to the open-loop transfer function $H_1(s)$ divided by 1 plus the product $H_1(s)H_2(s)$ of the transfer functions of the open-loop system and feedback system. Note that if the subtracter in Figure 6.10 were changed to an adder, the transfer function $H(s)$ of the closed-loop system would change to

$$H(s) = \frac{H_1(s)}{1 - H_1(s)H_2(s)}. \tag{6.29}$$

INTERCONNECTIONS WITH INITIAL ENERGY   If the systems in the interconnections above contain initial energy, the transform $Y(s)$ of the output response will have the form

$$Y(s) = \frac{C(s)}{D(s)} + \frac{N(s)}{D(s)} X(s), \tag{6.30}$$

where $N(s)/D(s)$ is equal to the transfer function $H(s)$ of the interconnection and where the coefficients of the polynomial $C(s)$ are determined by the initial conditions. Here we are assuming that the input $x(t)$ applied to the interconnection satisfies the condition $x^{(i)}(0^-) = 0$, $i \geq 0$. As a result, any initial energy in the interconnection is characterized by the values of $y(t)$ and its derivatives at time $t = 0^-$.

If the transfer function $H(s) = N(s)/D(s)$ of the interconnection is in reduced form (all common factors have been canceled), and if the degree of $D(s)$ is equal to q, the coefficients of $C(s)$ are determined by the initial conditions $y(0^-)$, $y^{(1)}(0^-), \ldots, y^{(q-1)}(0^-)$. If these values are known, the complete output response can be computed by taking the inverse Laplace transform of (6.30). However, the initial conditions may be given in terms of the outputs of the systems in the interconnection. If this is the case, the given information must be used to determine the initial values of the output $y(t)$ and the derivatives of $y(t)$. This can be carried out by working with the input/output differential equations describing the systems in the interconnection. The process is illustrated by the following two examples.

EXAMPLE 6.3.   Consider the cascade connection shown in Figure 6.9 with the transfer functions of the two systems given by

$$H_1(s) = \frac{b_1}{s + a_1} \quad \text{and} \quad H_2(s) = \frac{b_2}{s + a_2}.$$

The transfer function $H(s)$ of the cascade connection is given by

$$H(s) = H_1(s)H_2(s) = \frac{b_1 b_2}{(s + a_1)(s + a_2)} = \frac{b_1 b_2}{s^2 + (a_1 + a_2)s + a_1 a_2}.$$

In this example $H(s) = N(s)/D(s)$, where

$$D(s) = s^2 + (a_1 + a_2)s + a_1 a_2.$$

The degree of $D(s)$ is 2 and thus any initial energy in the cascade connection is characterized by the initial values $y(0^-)$ and $\dot{y}(0^-)$, where $y(t)$ is the output of the connection. Suppose that these initial values are known. Then using the expression (5.102), we have that the transform $Y(s)$ of the output of the cascade connection is given by

$$Y(s) = \frac{y(0^-)s + \dot{y}(0^-) + (a_1 + a_2)y(0^-)}{s^2 + (a_1 + a_2)s + a_1 a_2}$$

$$+ \frac{b_1 b_2}{s^2 + (a_1 + a_2)s + a_1 a_2} X(s). \tag{6.31}$$

The output response $y(t)$ can then be computed by taking the inverse Laplace transform of (6.31). Now suppose that we know the initial values $y_1(0^-)$ and $y_2(0^-)$ of the outputs of the systems comprising the cascade connection. Since $y(t) = y_2(t)$, we have

$$y(0^-) = y_2(0^-). \tag{6.32}$$

But to use (6.31), we must determine the initial condition $\dot{y}(0^-)$. This can be computed as follows. First, since $Y(s) = H_2(s)Y_1(s)$, the second system is described by the input/output differential equation

$$\frac{dy(t)}{dt} + a_2 y(t) = b_2 y_1(t). \tag{6.33}$$

Setting $t = 0^-$ in (6.33) gives

$$\dot{y}(0^-) = -a_2 y(0^-) + b_2 y_1(0^-). \tag{6.34}$$

Thus the initial conditions $y(0^-)$ and $\dot{y}(0^-)$ can be computed from (6.32) and (6.34).

EXAMPLE 6.4. Now consider the feedback connection shown in Figure 6.10 with $H_1(s)$ and $H_2(s)$ given by

$$H_1(s) = \frac{b_1}{s + a_1} \quad \text{and} \quad H_2(s) = \frac{b_2}{s + a_2}.$$

The closed-loop transfer function $H(s)$ is

$$H(s) = \frac{\dfrac{b_1}{s + a_1}}{1 + \dfrac{b_1 b_2}{(s + a_1)(s + a_2)}} = \frac{b_1(s + a_2)}{s^2 + (a_1 + a_2)s + a_1 a_2 + b_1 b_2}.$$

Here $H(s) = N(s)/D(s)$ with

$$D(s) = s^2 + (a_1 + a_2)s + a_1 a_2 + b_1 b_2.$$

Since the degree of $D(s)$ is again 2, any initial energy in the closed-loop system is characterized by the initial values $y(0^-)$ and $\dot{y}(0^-)$, where $y(t)$ is the output of the feedback connection. If these initial values are known, the transform $Y(s)$ of $y(t)$ is given by

$$Y(s) = \frac{y(0^-)s + \dot{y}(0^-) + (a_1 + a_2)y(0^-)}{s^2 + (a_1 + a_2)s + a_1 a_2 + b_1 b_2} \tag{6.35}$$

$$+ \frac{b_1 s + b_1 a_2}{s^2 + (a_1 + a_2)s + a_1 a_2 + b_1 b_2} X(s).$$

Now suppose that we are given the initial values $y_1(0^-)$ and $y_2(0^-)$. Since $y(t) = y_1(t)$, we have

$$y(0^-) = y_1(0^-). \tag{6.36}$$

To be able to use (6.35), we also need to calculate $\dot{y}(0^-)$. Since $Y(s) = H_1(s)X_1(s)$, the input/output differential equation of the open-loop system is

$$\frac{dy(t)}{dt} + a_1 y(t) = b_1 x_1(t). \tag{6.37}$$

Inserting

$$x_1(t) = x(t) - y_2(t)$$

into (6.37), we get

$$\frac{dy(t)}{dt} + a_1 y(t) = b_1 x(t) - b_1 y_2(t). \tag{6.38}$$

Setting $t = 0^-$ in (6.38), we have [assuming that $x(0^-) = 0$]

$$\dot{y}(0^-) = -a_1 y(0^-) - b_1 y_2(0^-). \tag{6.39}$$

Therefore, the initial conditions $y(0^-)$ and $\dot{y}(0^-)$ can be determined from (6.36) and (6.39).

6.3

BLOCK DIAGRAMS

A linear time-invariant continuous-time system is sometimes specified by a block diagram consisting of an interconnection of "blocks," with each block represented by a transfer function. The blocks can be thought of as subsystems comprising the given system. The transfer function of a system given by a block diagram can be determined by combining blocks in the diagram. The process is called *block-diagram reduction*. Block-diagram reduction is greatly facilitated by using the following rules for moving pick-off points and adders or subtracters.

MOVING A PICK-OFF POINT As illustrated in Figure 6.11, a pick-off point may be moved from one side of a system to another. The symbol "$\equiv$" in the figure means that the two diagrams are equivalent. Let us verify that this is the case for the diagrams in Figure 6.11b. From the diagram on the left side of Figure 6.11b, we have

$$Y_1(s) = H(s)X(s)$$

and

$$Y_2(s) = X(s).$$

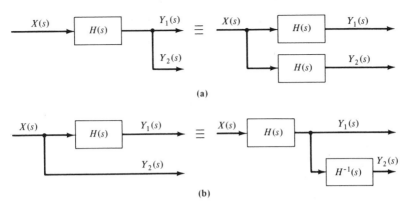

Figure 6.11. Moving a pick-off point.

From the diagram on the right side of Figure 6.11b, we have

$$Y_1(s) = H(s)X(s)$$

and

$$Y_2(s) = H^{-1}(s)H(s)X(s) = X(s).$$

The equations for the two diagrams in Figure 6.11b are the same, and thus the two systems in Figure 6.11b are equivalent. It is obvious that the two systems in Figure 6.11a are equivalent.

MOVING AN ADDER OR SUBTRACTER An adder or subtracter may be moved from one side of a system to another as illustrated in Figure 6.12. The verification that the diagrams in Figure 6.12a and b are equivalent is an easy exercise that is left to the reader.

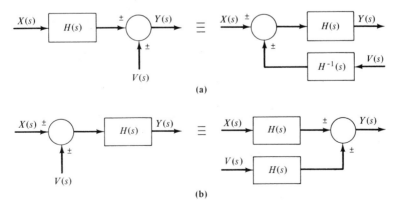

Figure 6.12. Moving an adder or subtracter.

BLOCK-DIAGRAM REDUCTION The transfer function of a system specified by a block diagram can be determined by using the equivalences in Figures
6.11 and 6.12 and the expressions for the transfer functions of parallel, cascade,
and feedback connections. The process is illustrated by the following example.

> **EXAMPLE 6.5.** We shall compute the transfer function of the system
> given by the block diagram shown in Figure 6.13. We can begin the block-
> diagram reduction by moving the pick-off point before the integrator to the
> other side of the integrator. The result is displayed in Figure 6.14. In this
> diagram the blocks with transfer functions $2/(s + 2)$ and s are in cascade,
> and thus they can be combined. Also, there is a feedback connection to the
> left of the last adder in the diagram. Using the expression for the transfer
> function of a feedback connection, the feedback connection has transfer
> function
>
> $$\frac{1/s}{1 + 3/s} = \frac{1}{s + 3}.$$
>
> The resulting block diagram is shown in Figure 6.15. In this diagram we can
> move the pick-off point before the $1/(s + 3)$ block to the other side of the
> block. This results in the diagram shown in Figure 6.16. Combining the
> blocks $1/(s + 1)$ and $1/(s + 3)$, which are in cascade, and computing the
> transfer function of the parallel connection after the $1/(s + 3)$ block, we get
> the diagram in Figure 6.17. Now the transfer function of the feedback con-
> nection is
>
> $$\frac{\dfrac{1}{(s + 1)(s + 3)}}{1 + \dfrac{2s}{(s + 1)(s + 2)(s + 3)}} = \frac{s + 2}{(s + 1)(s + 2)(s + 3) + 2s}.$$
>
> Thus the diagram reduces to the result shown in Figure 6.18. We see that
> the transfer function of the system is
>
> $$H(s) = \frac{(s + 2)(s + 4)}{(s + 1)(s + 2)(s + 3) + 2s} = \frac{s^2 + 6s + 8}{s^3 + 6s^2 + 13s + 6}.$$

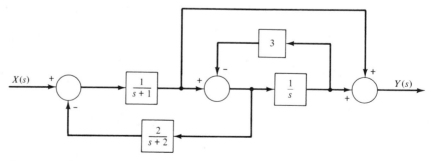

Figure 6.13. System in Example 6.5.

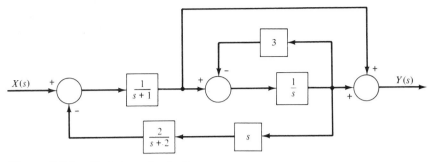

Figure 6.14. First stage of reduction.

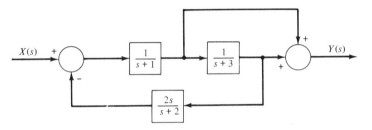

Figure 6.15. Next stage of reduction.

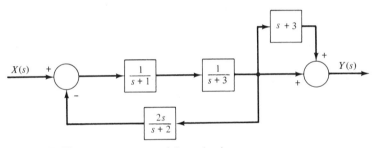

Figure 6.16. Another stage of the reduction.

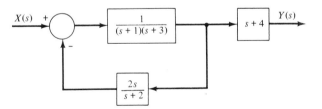

Figure 6.17. Next-to-last step in reduction.

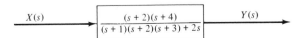

Figure 6.18. Reduced diagram.

MASON'S THEOREM   Another technique for determining the transfer function of a block diagram is based on Mason's theorem. This result is specified in terms of the transfer functions of the paths from the input to the output and the transfer functions of the feedback loops comprising the block diagram. In particular, suppose that there are N paths from the input to the output, and that there are Q feedback loops. Let $P_i(s)$ denote the transfer function of the ith path from the input to the output, and let $L_i(s)$ denote the transfer function of the ith feedback loop. Let $\Delta(s)$ denote the system determinant defined by

$$\Delta(s) = 1 - \sum_{i=1}^{Q} L_i(s) + \sum L_q(s)L_r(s) - \sum L_m(s)L_n(s)L_p(s) + \cdots ,$$

where the second sum consists of the products $L_q(s)L_r(s)$ of the transfer functions of all nontouching feedback loops. (Two loops are nontouching if the signals in one loop do not appear in the other loop, and conversely.) The third sum consists of all the products $L_m(s)L_n(s)L_p(s)$, where each loop in a product does not touch the other two loops in the product.

Mason's theorem states that the transfer function $H(s)$ of the system is given by

$$H(s) = \frac{1}{\Delta(s)} \sum_{i=1}^{N} P_i(s)\, \Delta_i(s),$$

where $\Delta_i(s)$ is the system determinant after excluding all feedback loops that intersect the ith path from the input to the output.

We shall apply Mason's formula to the system with the block diagram shown in Figure 6.13. This system has two paths from the input to the output with the transfer functions of the paths given by

$$P_1(s) = \frac{1}{s + 1}, \qquad P_2(s) = \frac{1}{s(s + 1)}.$$

There are two feedback loops in the block diagram with transfer functions

$$L_1(s) = \frac{-2}{(s + 1)(s + 2)}, \qquad L_2(s) = \frac{-3}{s}.$$

Note the minus signs that result from the subtracters in the feedback loops. In this example, the two feedback loops are touching, and thus the system determinant is

$$\Delta(s) = 1 - L_1(s) - L_2(s).$$

Inserting the expressions for $L_1(s)$ and $L_2(s)$, we obtain

$$\Delta(s) = 1 + \frac{2}{(s + 1)(s + 2)} + \frac{3}{s}.$$

Now the path with transfer function $1/(s + 1)$ intersects the feedback loop with transfer function $-2/[(s + 1)(s + 2)]$, but does not intersect the feedback loop with transfer function $-3/s$. Thus $\Delta_1(s)$ is the system determinant with the first feedback loop excluded. This gives

$$\Delta_1(s) = 1 - L_2(s) = 1 + \frac{3}{s}.$$

The other path from the input to the output intersects both of the feedback loops, so $\Delta_2(s)$ is the system determinant with both loops excluded. Hence $\Delta_2(s) = 1$. Finally, from Mason's formula we have

$$H(s) = \frac{\dfrac{1}{s + 1}\left(1 + \dfrac{3}{s}\right) + \dfrac{1}{s(s + 1)}}{1 + \dfrac{2}{(s + 1)(s + 2)} + \dfrac{3}{s}} \quad (1)$$

$$= \frac{s(s + 2)\left(1 + \dfrac{3}{s}\right) + s + 2}{s(s + 1)(s + 2) + 2s + 3(s + 1)(s + 2)}$$

$$= \frac{s^2 + 6s + 8}{s^3 + 6s^2 + 13s + 6}$$

This result agrees with the result derived above using block-diagram reduction.

SIGNAL-FLOW GRAPHS   Linear time-invariant systems consisting of an interconnection of blocks or subsystems are sometimes represented by a signal-flow graph. A *signal-flow graph* consists of a collection of nodes interconnected by branches. The nodes represent signal points in the system, while the branches represent the transfer functions between the nodes. Each branch has an arrow showing the "flow" of signals from node to node. As an example the signal-flow graph of the system in Example 6.5 is displayed in Figure 6.19.

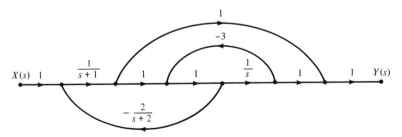

Figure 6.19.   Signal-flow graph of system in Example 6.5.

CASCADE SYNTHESIS

Consider the linear time-invariant finite-dimensional continuous-time system given by the input/output differential equation

$$\frac{d^n y(t)}{dt^n} + \sum_{i=0}^{n-1} a_i \frac{d^i y(t)}{dt^i} = \sum_{i=0}^{m} b_i \frac{d^i x(t)}{dt^i}, \tag{6.40}$$

where $m \leq n$. As discussed in Section 2.3, the system defined by (6.40) can be realized using n integrators and combinations of adders, subtracters, and scalar multipliers. Two different realizations are given in Figures 2.15 and 2.16.

The realizations constructed in Chapter 2 have feedback loops from one end of the realization to the other. As a result of these loops, it is difficult to analyze the effect of using nonideal components in the actual (op-amp) implementation of a realization. We would like to keep the effect of using nonideal components as "localized" as possible. This means that in realizing a system, we should avoid using feedback loops that "span" a large portion of the system diagram. This can be accomplished by employing a cascade realization that is based on the transfer function representation of the system. The cascade realization is constructed as follows.

From the results in Section 5.6, the transfer function of the system is

$$H(s) = \frac{N(s)}{D(s)} = \frac{b_m s^m + b_{m-1} s^{m-1} + \cdots + b_1 s + b_0}{s^n + a_{n-1} s^{n-1} + \cdots + a_1 s + a_0}. \tag{6.41}$$

The rational function $H(s)$ given by (6.41) is said to be a *proper* rational function of s since the degree m of $N(s)$ is less than or equal to the degree n of $D(s)$.

We can always factor $H(s)$ in the form

$$H(s) = H_1(s)H_2(s) \cdots H_r(s)H_{r+1}(s) \cdots H_q(s), \tag{6.42}$$

where for $1 \leq i \leq r$, each $H_i(s)$ is a first-order rational function of the form

$$H_i(s) = \frac{b_i s + c_i}{s + a_i}, \tag{6.43}$$

and for $r + 1 \leq i \leq q$, each $H_i(s)$ is a second-order rational function with complex poles. For $r + 1 \leq i \leq q$, the $H_i(s)$ are of the form

$$H_i(s) = \frac{f_i s^2 + k_i s + p_i}{s^2 + d_i s + e_i}. \tag{6.44}$$

The constants a_i, b_i, c_i in (6.43) are real numbers and the constants d_i, e_i, f_i, k_i, p_i in (6.44) are real numbers.

X(s) $\to$ $H_q(s)$ $\to$ $H_{q-1}(s)$ $\to$ ••• $\to$ $H_r(s)$ $\to$ ••• $\to$ $H_1(s)$ $\to$ Y(s)

Figure 6.20. Cascade realization of $H(s)$.

It should be noted that the factorization (6.42) is not unique, in general, since the zeros of $H(s)$ can be placed in the factors $H_i(s)$ in different ways [as long as each $H_i(s)$ is proper in s].

After determining the factorization (6.42), we can realize each factor $H_i(s)$ using the realizations for first-order and second-order systems given in Section 2.3. By cascading the realizations of the $H_i(s)$, we obtain a realization of $H(s)$. Such a realization of $H(s)$ is referred to as a *cascade realization*. The cascade realization corresponding to the factorization (6.42) is illustrated in Figure 6.20. Owing to the form of a cascade realization, there are no feedback loops from one factor to another in the realization. Hence the effect of using nonideal components is localized.

In building a cascade realization, one must be concerned with the problem of one cascaded factor loading another. If the input and output signals of the subsystems in a cascade realization are voltage signals, the effect of loading can be ignored if the output impedances of the cascaded subsystems are much smaller than the input impedances of the cascaded subsystems. Subsystems built by interconnecting op amps have this characteristic.

The process of determining a cascade realization is illustrated in the following example.

EXAMPLE 6.6. Consider the system with transfer function

$$H(s) = \frac{2s^2 + 12s + 20}{s^3 + 6s^2 + 10s + 8} = \frac{2(s^2 + 6s + 10)}{(s + 4)(s^2 + 2s + 2)}.$$

This system has two complex zeros, two complex poles, and one real pole. Writing $H(s)$ in the form (6.42), we obtain

$$H(s) = H_1(s)H_2(s),$$

where

$$H_1(s) = \frac{2}{s + 4} \quad \text{and} \quad H_2(s) = \frac{s^2 + 6s + 10}{s^2 + 2s + 2}.$$

Note that the quadratic term $s^2 + 6s + 10$ cannot be put into the numerator of the $H_1(s)$ factor, since this would result in an improper rational function. Realizing $H_1(s)$ and $H_2(s)$ using the results in Chapter 2, we get the cascade realization shown in Figure 6.21.

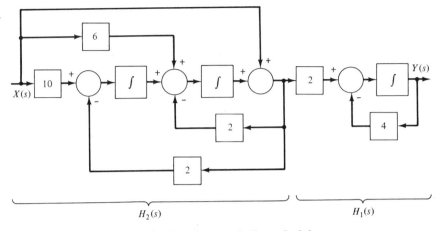

Figure 6.21. Cascade realization of system in Example 6.6.

★6.5

APPLICATION TO CONTROL

One of the major applications of the transfer function framework is in the study of control. In this section we consider a particular type of control problem that can be solved using the transfer function model. For an in-depth treatment of control, we suggest that the reader consult a textbook on control systems; for example, see Ogata [1970], Kuo [1982], or Dorf [1980].

Suppose that we have a linear time-invariant finite-dimensional continuous-time system with transfer function

$$H(s) = \frac{N(s)}{D(s)}.$$

The given system is sometimes called the *plant*.

We assume that the output $y(t)$ of the given system can be measured by means of some type of sensor. The measured output $y(t)$ can then be fed back as illustrated in Figure 6.22. Here the signal $r(t)$ is a reference signal and the signal $d(t)$ is a *disturbance signal*. The signal $e(t)$ is an *error signal* given by

$$e(t) = r(t) - y(t).$$

As shown in Figure 6.22, the error signal is applied to another linear time-invariant continuous-time system with transfer function $H_c(s)$. This second system is called the *controller*.

Now the objective is to design the controller transfer function $H_c(s)$ so that the error $e(t)$ goes to zero as $t \to \infty$; that is, we want

$$e(t) \to 0 \qquad \text{as } t \to \infty. \tag{6.45}$$

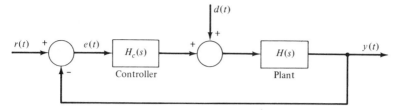

Figure 6.22. System with output feedback.

The condition (6.45) implies that the system output $y(t)$ is approximately equal to the reference signal $r(t)$ for suitably large values of t. This type of system performance is called *tracking* in the presence of disturbances. The problem of achieving tracking can be approached in the following manner.

We first compute the transfer function model of the closed-loop system in Figure 6.22. If we assume that $d(t) = 0$, the block diagram in Figure 6.22 reduces to the diagram in Figure 6.23. From our results in Section 6.2, the transfer function model of this system is

$$Y(s) = \frac{H_c(s)H(s)}{1 + H_c(s)H(s)} R(s),\qquad(6.46)$$

where $R(s)$ is the Laplace transform of the reference signal $r(t)$.

Now suppose that $r(t) = 0$ and $d(t) \neq 0$. Then the diagram in Figure 6.22 reduces to the diagram in Figure 6.24. Again using the results in Section 6.2, we have that the transfer function representation of the system is

$$Y(s) = \frac{H(s)}{1 + H_c(s)H(s)} D(s),\qquad(6.47)$$

where $D(s)$ is the transform of the disturbance signal $d(t)$.

By linearity, we can combine (6.46) and (6.47). This gives

$$Y(s) = \frac{H_c(s)H(s)}{1 + H_c(s)H(s)} R(s) + \frac{H(s)}{1 + H_c(s)H(s)} D(s).\qquad(6.48)$$

Equation (6.48) is the transform of the output response resulting from the application of the reference signal $r(t)$ and the disturbance signal $d(t)$ with no initial energy in the system at time $t = 0$.

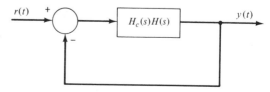

Figure 6.23. Closed-loop system when $d(t) = 0$.

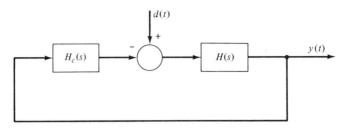

Figure 6.24.   Closed-loop system with $r(t) = 0$.

Now suppose that both $r(t)$ and $d(t)$ are step functions given by

$$r(t) = r_0, \quad t \geq 0 \quad \text{and} \quad d(t) = d_0, \quad t \geq 0.$$

Then $R(s) = r_0/s$ and $D(s) = d_0/s$. Let us write $H_c(s)$ in the form

$$H_c(s) = \frac{N_c(s)}{D_c(s)},$$

where $N_c(s)$ and $D_c(s)$ are polynomials in s with deg $N_c(s) \leq$ deg $D_c(s)$. Then the expression (6.48) for $Y(s)$ becomes

$$Y(s) = \frac{H_c(s)H(s)r_0}{[1 + H_c(s)H(s)]s} + \frac{H(s)d_0}{[1 + H_c(s)H(s)]s} = \frac{H(s)(H_c(s)r_0 + d_0)}{[1 + H_c(s)H(s)]s}$$

$$= \frac{\dfrac{N(s)}{D(s)}\left[\dfrac{N_c(s)}{D_c(s)}r_0 + d_0\right]}{\left[1 + \dfrac{N(s)N_c(s)}{D(s)D_c(s)}\right]s} = \frac{N(s)(N_c(s)r_0 + D_c(s)d_0)}{[D(s)D_c(s) + N(s)N_c(s)]s}. \tag{6.49}$$

With the above choice for $r(t)$, we have tracking if

$$y(t) \rightarrow r_0 \quad \text{as } t \rightarrow \infty. \tag{6.50}$$

The condition (6.50) can be satisfied by choosing $H_c(s) = N_c(s)/D_c(s)$ so that

1. $D_c(s) = s\overline{D}_c(s)$; that is, $D(s)$ has a zero at zero.
2. The real parts of all the zeros of $D(s)D_c(s) + N(s)N_c(s)$ are strictly negative.

It follows from the final-value theorem (see Section 5.2) that the tracking condition (6.50) is satisfied if conditions 1 and 2 are satisfied. To see this, suppose that condition 2 is satisfied. Then $y(t)$ has a limit as $t \rightarrow \infty$ and by the final-value theorem,

$$\lim_{t \to \infty} y(t) = \lim_{s \to 0} sY(s). \tag{6.51}$$

Hence, using (6.49), we have

$$\lim_{t \to \infty} y(t) = \frac{N(0)(N_c(0)r_0 + D_c(0)d_0)}{D(0)D_c(0) + N(0)N_c(0)}. \qquad (6.52)$$

Now if condition 1 is also satisfied, $D_c(0) = 0$, and from (6.52), we have

$$\lim_{t \to \infty} y(t) = \frac{N(0)N_c(0)r_0}{N(0)N_c(0)} = r_0.$$

Thus we do have tracking.

There are systematic methods for computing polynomials $N_c(s)$ and $D_c(s)$ which satisfy conditions 1 and 2, and for which deg $N_c(s) \leq$ deg $D_c(s)$. For the details, we refer the reader to Chen [1984].

EXAMPLE 6.7 (Speed Control). Again consider an automobile given by the velocity model

$$\dot{v}(t) + \frac{k_f}{M} v(t) = \frac{1}{M} x(t), \qquad (6.53)$$

where $v(t)$ is the velocity of the car at time t and $x(t)$ is the drive or braking force applied to the car at time t. The transfer function $H(s)$ of the car is

$$H(s) = \frac{1/M}{s + k_f/M}.$$

Now we want to design a speed controller that maintains the car at a fixed velocity v_0, where v_0 is arbitrary. The closed-loop system with the controller $H_c(s)$ is shown in Figure 6.25.

In this example the output $v(t)$ is measured using a tachometer (or speedometer). The disturbance $d(t)$ in Figure 6.25 is an external force acting on the car. For example, $d(t)$ could be a force resulting from wind gusts or it could be a gravitational force resulting from the car moving up and down hills. We shall take $d(t)$ to be the step function $d(t) = d_0, t \geq 0$, where the

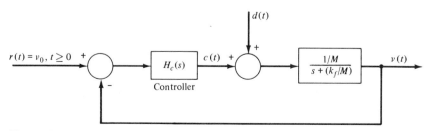

Figure 6.25. Car with speed controller.

magnitude d_0 is arbitrary. In this example the expression (6.49) for the transform of the output is

$$V(s) = \frac{(1/M)[N_c(s)v_0 + D_c(s)d_0]}{[(s + k_f/M)D_c(s) + (1/M)N_c(s)]s},$$

where

$$H_c(s) = \frac{N_c(s)}{D_c(s)}$$

and $V(s)$ is the transform of the velocity $v(t)$. If we choose

$$D_c(s) = s \quad \text{and} \quad N_c(s) = s + 1,$$

then

$$D(s)D_c(s) + N(s)N_c(s) = \left(s + \frac{k_f}{M}\right)D_c(s) + \frac{1}{M}N_c(s)$$

$$= s^2 + \frac{k_f + 1}{M}s + \frac{1}{M}.$$

The zeros of $D(s)D_c(s) + N(s)N_c(s)$ are

$$s = \frac{-(k_f + 1)}{2M} \pm \frac{\sqrt{(k_f + 1)^2 - 4M}}{2M}.$$

If $4M > (k_f + 1)^2$, these zeros are complex with real part equal to $-(k_f + 1)/2M$. Then since the mass M is strictly positive ($M > 0$) and the coefficient of friction k_f is nonnegative, the real parts of the zeros are both strictly negative, and thus condition 2 given above is satisfied. Further, since $D_c(s) = s$, condition 1 is also satisfied. Therefore,

$$v(t) \to v_0 \quad \text{as } t \to \infty,$$

and we have speed control. From the block diagram in Figure 6.25, the transform $C(s)$ of the control force $c(t)$ applied to the car is

$$C(s) = \frac{s + 1}{s}\left[\frac{v_0}{s} - V(s)\right] = \frac{v_0}{s} - V(s) + \frac{1}{s}\left[\frac{v_0}{s} - V(s)\right].$$

Taking the inverse Laplace transform of $C(s)$, we have that the control force $c(t)$ acting on the car is given by

$$c(t) = v_0 - v(t) + \int_0^t (v_0 - v(\lambda))\, d\lambda, \quad t \geq 0.$$

Defining the error

$$e(t) = v_0 - v(t),$$

we have

$$c(t) = e(t) + \int_0^t e(\lambda) \, d\lambda, \qquad t \geq 0.$$

Thus the control force $c(t)$ is equal to the error $e(t)$ plus the integral of the error. This type of control is called *proportional-plus-integral control*.

Now suppose that the (normalized) mass is $M = 1$, $k_f = 0.1$, and $v_0 = d_0 = 1$. Then from the above expression for the transform $V(s)$ of the velocity, we have

$$V(s) = \frac{2s + 1}{(s^2 + 1.1s + 1)s}.$$

Since we are assuming that the system is at rest at time $t = 0$, the initial velocity $v(0)$ and initial acceleration $\dot{v}(0)$ of the car are both zero. Taking the inverse Laplace transform of $V(s)$, we obtain

$$v(t) = 1 + 2.00358 \exp{(-0.55t)} \cos{(0.83516t - 119.94°)}, \qquad t \geq 0.$$

The velocity $v(t)$ is plotted in Figure 6.26. We see that the car does reach the desired velocity ($v_0 = 1$), but there is a sizable amount of overshoot in doing so. The amount of overshoot and other aspects of the response can be adjusted by selecting different values of gains in the controller transfer function $H_c(s)$. We refer the reader to Ogata [1970] for a detailed treatment of control system design.

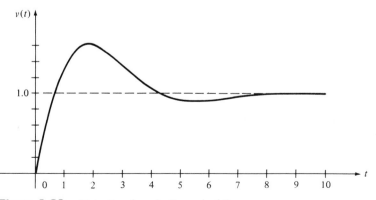

Figure 6.26.   Velocity of car in Example 6.7.

INVERSE PROBLEM

In some applications we would like to be able to compute an input $x(t)$ that produces a desired output $y(t)$. This problem is called the *inverse problem*. It is very difficult to solve in the time domain (e.g., using the input/output differential equation). However, the inverse problem can be solved very easily in the s-domain as follows.

Suppose that we have a linear time-invariant continuous-time system with transfer function $H(s)$. Our objective is to compute an input $x(t)$ that produces a given output $y(t)$ for $t > 0$ with the system at rest at time $t = 0^-$. Since there is no initial energy at time $t = 0^-$, we have that

$$Y(s) = H(s)X(s).$$

Solving for $X(s)$, we have

$$X(s) = H^{-1}(s)Y(s). \tag{6.54}$$

The input $x(t)$ that produces the desired output $y(t)$ can then be computed by taking the inverse Laplace transform of (6.54).

EXAMPLE 6.8.   Consider the car with the input/output differential equation

$$\frac{d^2y(t)}{dt^2} + \frac{k_f}{M}\frac{dy(t)}{dt} = \frac{1}{M}x(t), \tag{6.55}$$

where now the output $y(t)$ is the position of the car at time t. We want to determine a drive force $x(t)$ that will move the car from initial position $y(0) = 0$ to position $y(10) = 100$ at time $t = 10$. We assume that the initial velocity $\dot{y}(0)$ is zero and we want the velocity $\dot{y}(10)$ at $t = 10$ to be zero also. So we want to move the car from a stopped position to a stopped position. Since the initial conditions are zero, we have that

$$Y(s) = H(s)X(s).$$

From (6.55)

$$H(s) = \frac{1/M}{s^2 + (k_f/M)s}.$$

Then

$$X(s) = H(s)^{-1}Y(s) = \frac{s^2 + (k_f/M)s}{1/M}Y(s). \tag{6.56}$$

We do not want the input $x(t)$ to contain any impulses, and thus the transform $X(s)$ must be strictly proper in s; that is,

$$X(s) = \frac{N_x(s)}{D_x(s)},$$

where $N_x(s)$ and $D_x(s)$ are polynomials in s with degree $N_x(s) <$ degree $D_x(s)$. From (6.56) we see that $X(s)$ will be strictly proper in s if and only if $Y(s)$ is of the form

$$Y(s) = \frac{N_y(s)}{D_y(s)},$$

with

$$\deg N_y(s) < \deg D_y(s) - 2. \tag{6.57}$$

The degree constraint (6.57) is satisfied if we take

$$y(t) = at^3 + bt^2 \tag{6.58}$$

for some constants a, b. Taking the Laplace transform of (6.58), we obtain

$$Y(s) = \frac{6a}{s^4} + \frac{2b}{s^3} = \frac{6a + 2bs}{s^4}.$$

Thus (6.57) is satisfied with $y(t)$ given by (6.58). Now

$$y(0) = 0$$

and

$$\dot{y}(t) = 3at^2 + 2bt = 0 \qquad \text{when } t = 0,$$

and thus we have the desired initial values. We also require that

$$y(10) = 100 \qquad \text{and} \qquad \dot{y}(10) = 0.$$

Inserting these values into the expressions for $y(t)$ and $\dot{y}(t)$, we have that

$$a(10^3) + b(10^2) = 100$$
$$3a(10^2) + 2b(10) = 0.$$

Solving for a and b, we obtain

$$a = -0.2 \quad \text{and} \quad b = 3.$$

Then

$$Y(s) = \frac{6(-0.2) + 2(3)s}{s^4} = \frac{-1.2 + 6s}{s^4},$$

and

$$X(s) = H^{-1}(s)Y(s) = \frac{s^2 + (k_f/M)s}{1/M} \frac{-1.2 + 6s}{s^4}$$

$$= \frac{M[s + (k_f/M)](-1.2 + 6s)}{s^3}$$

$$= \frac{6Ms^2 + (6k_f - 1.2M)s - 1.2k_f}{s^3}$$

$$= \frac{6M}{s} + \frac{6k_f - 1.2M}{s^2} - \frac{1.2k_f}{s^3}.$$

Thus

$$x(t) = 6M + (6k_f - 1.2M)t - 0.6k_f t^2, \qquad 0 \le t \le 10.$$

The drive force is plotted in Figure 6.27 for the case $M = 1$, $k_f = 0.1$. Note that the drive force is positive for $0 < t < 6.18$ and is negative for $6.18 < t < 10$. The negative force means that it is necessary to apply the brakes in order to stop the car at the position $y(10) = 100$ at time $t = 10$.

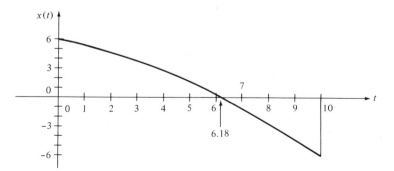

Figure 6.27. Force applied to car in Example 6.8.

We should note that the solution for the drive force $x(t)$ in Example 6.8 is not unique. In fact, there are an infinite number of solutions to this problem. This raises the question as to which solutions are better than other solutions in some sense (such as the amount of fuel consumed in moving the car). Determining solutions which are "optimal" in some sense is studied in control theory. The reader may consult Chen [1984] for results on this topic.

INVERSE SYSTEM Given the transfer function model $Y(s) = H(s)X(s)$, let us again solve for $X(s)$, which results in the relationship

$$X(s) = H^{-1}(s)Y(s). \tag{6.59}$$

Now (6.59) can be viewed as the transfer function model of a linear time-invariant continuous-time system with transfer function $H^{-1}(s)$. This system is called the *inverse* of the given system with transfer function $H(s)$.

For every linear time-invariant continuous-time system with transfer function $H(s)$, there is a unique inverse system with transfer function $H^{-1}(s)$. However, the inverse system may contain differentiators, and thus may be difficult to implement. More precisely, suppose that $H(s)$ is a proper rational function in s; that is,

$$H(s) = \frac{N(s)}{D(s)},$$

where $N(s)$ and $D(s)$ are polynomials in s with deg $N(s) \le$ deg $D(s)$. Then the transfer function of the inverse system is

$$H^{-1}(s) = \frac{D(s)}{N(s)}.$$

If the degree of $N(s)$ is less than the degree of $D(s)$, $H^{-1}(s)$ is not a proper function of s, and therefore any realization of the inverse system must contain one or more differentiators. The transfer function $H^{-1}(s)$ of the inverse system is proper in s if deg $N(s) =$ deg $D(s)$. In this case the inverse system can be realized using only integrators, adders, subtracters, and scalar multipliers.

When the inverse system has a proper transfer function, we can put the inverse system in cascade with the given system as illustrated in Figure 6.28. Since $H^{-1}(s)H(s) = 1$, the output response of the cascade connection in Figure 6.28 is equal to the input $x(t)$ to the connection. In other words, the inverse system "undoes" what the given system does to an input $x(t)$. So the effect of the given system can be eliminated by cascading with the inverse system. This property is very useful in many applications.

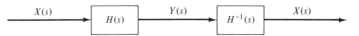

Figure 6.28. Cascade of given system with inverse system.

MULTI-INPUT MULTI-OUTPUT SYSTEMS

In Section 1.7 we considered the input/output operator representation of multi-input multi-output systems. In this section we show that the transfer function representation can be generalized to linear time-invariant continuous-time systems with p inputs and q outputs, where p and q are arbitrary positive integers.

Given a p-input q-output linear time-invariant continuous-time system, we shall let $x_i(t)$ denote the input signal applied to the ith input terminal and $y_i(t)$ denote the response at the ith output terminal. As we now show, we can generate the transfer function representation of the system by first computing the responses at the output terminals resulting from the application of the unit impulse $\delta(t)$ to the input terminals.

Given a fixed positive integer i, where $1 \leq i \leq p$, let $x_i(t) = \delta(t)$ with all the other input signals equal to zero. For each fixed positive integer r, where $1 \leq r \leq q$, let $h_{ri}(t)$ denote the resulting response at the rth output terminal with the system at rest at time $t = 0^-$. The response $h_{ri}(t)$ is the impulse response from the ith input to the rth output. Since there are p inputs and q outputs, there are a total of pq impulse responses in the collection $\{h_{ri}(t)\}$.

Now suppose that an arbitrary input $x_i(t)$ is applied to the ith input terminal with no input applied to the other input terminals. Since the system is linear and time invariant, by the convolution representation the resulting response at the rth output terminal is given by

$$y_r(t) = (h_{ri} * x_i)(t), \qquad r = 1, 2, \ldots, q. \tag{6.60}$$

Here we are assuming that the system is at rest prior to the application of $x_i(t)$.

Taking the Laplace transform of both sides of (6.60), we have that

$$Y_r(s) = H_{ri}(s)X_i(s), \qquad r = 1, 2, \ldots, q, \tag{6.61}$$

where $Y_r(s)$ is the transform of $y_r(t)$, $H_{ri}(s)$ is the transform of $h_{ri}(t)$ and $X_i(s)$ is the transform of $x_i(t)$. The function $H_{ri}(s)$ is the transfer function from the ith input terminal to the rth output terminal.

If we now apply inputs to all the input terminals with the system initially at rest, by linearity the resulting response at the rth output terminal is

$$y_r(t) = \sum_{i=1}^{p} (h_{ri} * x_i)(t), \qquad r = 1, 2, \ldots, q. \tag{6.62}$$

Taking the transform of both sides of (6.62), we obtain

$$Y_r(s) = \sum_{i=1}^{p} H_{ri}(s)X_i(s), \qquad r = 1, 2, \ldots, q. \tag{6.63}$$

The expression (6.63) for $Y_r(s)$ is the transform of the response at the rth output terminal resulting from the application of the inputs $x_1(t)$, $x_2(t)$, . . . , $x_p(t)$ to the p input terminals with the system at rest prior to the application of the inputs.

The collection of q equations given by (6.63) defines the transfer function representation of the given p-input q-output system. Usually, (6.63) is rewritten in matrix form as follows. (A brief review of matrix algebra is given in Appendix C.) Let $Y(s)$ denote the q-element column vector defined by

$$Y(s) = \begin{bmatrix} Y_1(s) \\ Y_2(s) \\ \cdot \\ \cdot \\ \cdot \\ Y_q(s) \end{bmatrix},$$

and let $X(s)$ denote the p-element column vector defined by

$$X(s) = \begin{bmatrix} X_1(s) \\ X_2(s) \\ \cdot \\ \cdot \\ \cdot \\ X_p(s) \end{bmatrix}.$$

The vector $Y(s)$ is the vector of transformed output signals and $X(s)$ is the vector of transformed input signals. Now let $H(s)$ denote the q by p matrix defined by

$$H(s) = \begin{bmatrix} H_{11}(s) & H_{12}(s) & \cdots & H_{1p}(s) \\ H_{21}(s) & H_{22}(s) & \cdots & H_{2p}(s) \\ \cdot & \cdot & & \cdot \\ \cdot & \cdot & & \cdot \\ \cdot & \cdot & & \cdot \\ H_{q1}(s) & H_{q2}(s) & \cdots & H_{qp}(s) \end{bmatrix}.$$

Then we can rewrite (6.63) in the matrix-equation form

$$Y(s) = H(s)X(s) = \begin{bmatrix} H_{11}(s) & H_{12}(s) & \cdots & H_{1p}(s) \\ H_{21}(s) & H_{22}(s) & \cdots & H_{2p}(s) \\ \cdot & \cdot & & \cdot \\ \cdot & \cdot & & \cdot \\ \cdot & \cdot & & \cdot \\ H_{q1}(s) & H_{q2}(s) & \cdots & H_{qp}(s) \end{bmatrix} \begin{bmatrix} X_1(s) \\ X_2(s) \\ \cdot \\ \cdot \\ \cdot \\ X_p(s) \end{bmatrix}. \qquad (6.64)$$

The matrix multiplication on the far right side of (6.64) is the usual multiplication of a matrix times a column vector. In particular, writing out the matrix equation (6.64) component by component, we have

$$Y_1(s) = H_{11}(s)X_1(s) + H_{12}(s)X_2(s) + \cdots + H_{1p}(s)X_p(s)$$

$$Y_2(s) = H_{12}(s)X_1(s) + H_{22}(s)X_2(s) + \cdots + H_{2p}(s)X_p(s)$$

$$\cdot$$
$$\cdot$$
$$\cdot$$

$$Y_q(s) = H_{q1}(s)X_1(s) + H_{q2}(s)X_2(s) + \cdots + H_{qp}(s)X_p(s).$$

The q by p matrix $H(s)$ in (6.64) is called the *transfer function matrix* of the system. The matrix equation (6.64) is the *transfer function representation* of the system.

If the multi-input multi-output system under study is given by a collection of coupled input/output differential equations, the transfer function representation (6.64) can be generated by taking the Laplace transform of the input/output differential equations. The process is illustrated by the following example.

EXAMPLE 6.9. Consider the two-car system that was studied in Examples 1.20 and 2.9. The system is described by the input/output differential equations

$$\dot{v}_1(t) + \frac{k_f}{M} v_1(t) = \frac{1}{M} x_1(t)$$

$$\dot{v}_2(t) + \frac{k_f}{M} v_2(t) = \frac{1}{M} x_2(t)$$

$$\dot{w}(t) = v_2(t) - v_1(t).$$

Here $v_1(t)$ is the velocity of the first car, $v_2(t)$ is the velocity of the second car, $w(t)$ is the distance between the cars, $x_1(t)$ is the force applied to the first car, and $x_2(t)$ is the force applied to the second car. We shall take the inputs of the system to be $x_1(t)$ and $x_2(t)$, and the outputs of the system to be $v_1(t)$, $v_2(t)$, and $w(t)$. Therefore, in this example the number p of inputs is equal to 2 and the number q of outputs is equal to 3. Taking the Laplace transform of the foregoing input/output differential equations with all initial conditions equal to zero, we obtain

$$\left(s + \frac{k_f}{M}\right) V_1(s) = \frac{1}{M} X_1(s)$$

$$\left(s + \frac{k_f}{M}\right) V_2(s) = \frac{1}{M} X_2(s)$$

$$sW(s) = V_2(s) - V_1(s).$$

Writing these equations in the form (6.64), we have

$$Y(s) = \begin{bmatrix} V_1(s) \\ V_2(s) \\ W(s) \end{bmatrix} = \begin{bmatrix} \dfrac{1/M}{s + k_f/M} & 0 \\ 0 & \dfrac{1/M}{s + k_f/M} \\ \dfrac{-1/M}{s(s + k_f/M)} & \dfrac{1/M}{s(s + k_f/M)} \end{bmatrix} \begin{bmatrix} X_1(s) \\ X_2(s) \end{bmatrix} \qquad (6.65)$$

Equation (6.65) is the transfer function representation of the two-car system. From (6.65) we see that the transfer function matrix $H(s)$ of the system is

$$H(s) = \begin{bmatrix} \dfrac{1/M}{s + k_f/M} & 0 \\ 0 & \dfrac{1/M}{s + k_f/M} \\ \dfrac{-1/M}{s(s + k_f/M)} & \dfrac{1/M}{s(s + k_f/M)} \end{bmatrix}.$$

The transfer function representation of a p-input q-output linear time-invariant system can be determined directly from a signal-flow diagram of the system. The procedure is a straightforward generalization of the s-domain approach we used in the single-input single-output case (see Section 6.1).

EXAMPLE 6.10. Consider the two-input two-output system given by the interconnection in Figure 6.29. The s-domain representation of the system is displayed in Figure 6.30. We have

$$sQ_1(s) = -3Y_1(s) + X_1(s) \qquad (6.66)$$

$$sQ_2(s) = X_2(s) \qquad (6.67)$$

$$Y_1(s) = Q_1(s) + Q_2(s) \qquad (6.68)$$

$$Y_2(s) = Q_2(s) = \frac{1}{s} X_2(s). \qquad (6.69)$$

Solving (6.66) and (6.67) for $Q_1(s)$ and $Q_2(s)$ and inserting the results into (6.68), we obtain

$$Y_1(s) = \frac{-3}{s} Y_1(s) + \frac{1}{s} [X_1(s) + X_2(s)].$$

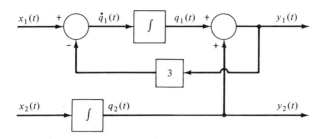

Figure 6.29. System in Example 6.10.

Thus

$$Y_1(s) = \frac{1}{s + 3}[X_1(s) + X_2(s)]. \qquad (6.70)$$

Writing (6.69) and (6.70) in matrix form, we get

$$Y(s) = \begin{bmatrix} Y_1(s) \\ Y_2(s) \end{bmatrix} = \begin{bmatrix} \dfrac{1}{s+3} & \dfrac{1}{s+3} \\[2mm] 0 & \dfrac{1}{s} \end{bmatrix} \begin{bmatrix} X_1(s) \\ X_2(s) \end{bmatrix}.$$

Hence the transfer function matrix of the system is

$$H(s) = \begin{bmatrix} \dfrac{1}{s+3} & \dfrac{1}{s+3} \\[2mm] 0 & \dfrac{1}{s} \end{bmatrix}.$$

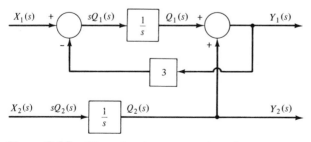

Figure 6.30. The s-domain representation of system.

PROBLEMS

Chapter 6

6.1 Using the s-domain representation, compute the transfer functions of the circuits shown in Figure P6.1.

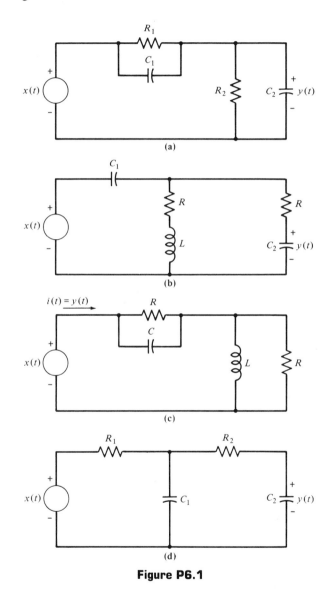

(a)

(b)

(c)

(d)

Figure P6.1

6.2 For the circuit in Figure P6.1c, determine all values of R, L, and C such that $H(s) = K$, where K is a constant.

6.3 Using the s-domain representation, compute the transfer function for the systems displayed in Figure P6.3.

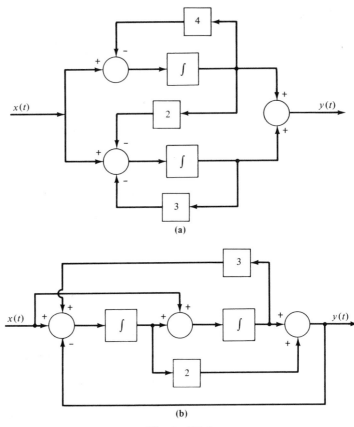

(a)

(b)

Figure P6.3

6.4 Consider the system shown in Figure P6.4.
(a) Using block-diagram reduction, determine the transfer function of the system.
(b) Using the Laplace transform, compute $y(t)$ for $t \geq 0$ when $x(t) = u(t)$ with $q_1(0^-) = 1$ and $q_2(0^-) = -3$.

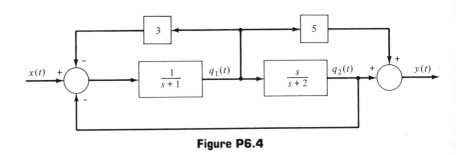

Figure P6.4

6.5 Repeat Problem 6.4 for the system shown in Figure P6.5.

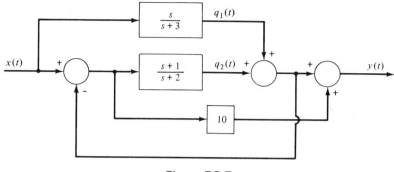

Figure P6.5

6.6 Using block-diagram reduction, reduce the system in Figure P6.6 to a single-block transfer function.

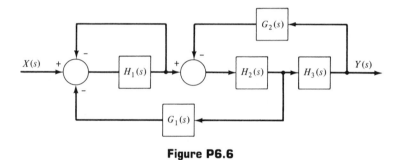

Figure P6.6

6.7 Using Mason's theorem, determine the transfer function of the system shown in Figure P6.7.

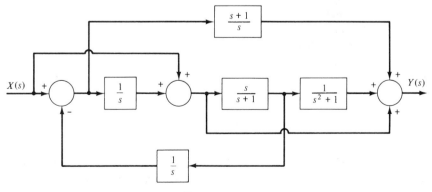

Figure P6.7

6.8 For each of the following systems, construct a cascade realization with the components of the realization having the smallest possible order.

(a) $H(s) = \dfrac{s^3}{s^3 + 4s^2 + 5s + 2}$.

(b) $H(s) = \dfrac{s^2 + 7s + 12}{s^3 + 4s^2 + 5s + 2}$.

(c) $H(s) = \dfrac{s^2 + 2s + 2}{s^3 + 4s^2 + 5s + 2}$.

6.9 Consider the field-controlled dc motor with the input/output differential equation

$$JL_f \frac{d^3y(t)}{dt^3} + (fL_f + R_fJ) \frac{d^2y(t)}{dt^2} + R_ff \frac{dy(t)}{dt} = kx(t).$$

Construct a cascade realization with each component of the realization having the smallest possible order.

6.10 Consider the feedback control system shown in Figure P6.10. Assume that there is no initial energy in the system at time $t = 0$.

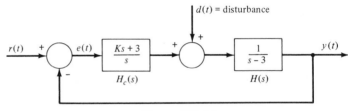

Figure P6.10

(a) Derive an expression for $E(s)$ in terms of $D(s)$ and $Y(s)$, where $E(s)$ is the Laplace transform of the error signal $e(t) = r(t) - y(t)$.

(b) Suppose that $r(t) = u(t)$ and $d(t) = 0$ for all t. Determine all (real) values of K so that $e(t) \rightarrow 0$ as $t \rightarrow \infty$.

(c) Suppose that $r(t) = u(t)$ and $d(t) = u(t)$. Determine all (real) values of K so that $e(t) \rightarrow 0$ as $t \rightarrow \infty$.

(d) Suppose that $r(t) = u(t)$ and $d(t) = (\sin t)u(t)$. Determine all (real) values of K so that $e(t) \rightarrow 0$ as $t \rightarrow \infty$.

(e) Repeat part (d) with the controller transfer function given by

$$H_c(s) = \frac{K}{s(s^2 + 1)}.$$

(f) With $H_c(s) = K/s(s^2 + 1)$, compute the output response $y(t)$ for $t \geq 0$ when $r(t) = u(t)$ and $d(t) = (\sin t)u(t)$.

6.11 The transfer function of the small-signal model of the simple pendulum is given by (with $M = L = I = 1$ and $g = 9.8$)

$$H(s) = \frac{1}{s^2 + 9.8}.$$

The pendulum is placed in the closed-loop system illustrated in Figure P6.11.

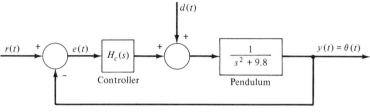

Figure P6.11

Design the controller transfer function $H_c(s)$ so that when $r(t) = Au(t)$, where A is a constant, and $d(t) = 0$ for all t, we have that

$$e(t) \rightarrow 0 \qquad \text{as } t \rightarrow \infty.$$

6.12 A linear time-invariant continuous-time system has transfer function $H(s) = 1/s^2$. An input $x(t)$, with $x(t) = 0$ for all $t < 0$, is applied to the system with no initial energy at time $t = 0$. The resulting output response is shown in Figure P6.12. Compute $x(t)$ for all $t \geq 0$.

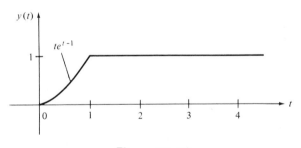

Figure P6.12

6.13 A continuous-time system has the input/output differential equation

$$\frac{d^2 y(t)}{dt^2} + y(t) = \frac{dx(t)}{dt} + 2x(t).$$

An input $x(t)$, with $x(t) = 0$ for all $t < 0$, is applied to the system with initial conditions $y(0^-) = 0$ and $\dot{y}(0^-) = 2$. The resulting output response is

$$y(t) = \begin{cases} -2e^{-2t} + \sin t, & 0 \leq t \leq 2 \\ -2e^{-2t}, & t > 2. \end{cases}$$

Compute $x(t)$ for all $t \geq 0$.

6.14   Consider the automobile on a level surface given by the transfer function

$$H(s) = \frac{1/M}{s^2 + (k_f/M)s}.$$

(a) Suppose that the car has initial position $y(0) = 0$ and velocity $\dot{y}(0) = 0$ at $t = 0$. Find an input force $x(t)$ that produces the output

$$y(t) = \begin{cases} \dfrac{t^2}{2}, & 0 \le t \le 1 \\[2mm] -\dfrac{t^2}{2} + 2t - 1, & 1 \le t \le 2 \\[2mm] 1, & t \ge 2. \end{cases}$$

Note that this input force $x(t)$ moves the car from position $y(0) = 0$ to position $y(2) = 1$ with $\dot{y}(2) = 0$.

(b) Suppose that the car has initial velocity $\dot{y}(0) = 55$ with $y(0) = 0$. Find an input force $x(t)$ (which does not contain any impulses) that stops the car at time $t = 10$ at position $y(10) = 100$.

6.15   A signal $x(t)$ is applied to a continuous-time system which distorts the signal. The output $y(t)$ of the system is given by

$$y(t) = Ax(t) + B \int_{t-h}^{t} x(\lambda)\, d\lambda,$$

where A, B, h are constants with $h > 0$. We are assuming that there is no initial energy in the system prior to the application of $x(t)$.

(a) Determine the transfer function of the inverse system which reproduces $x(t)$ from $y(t)$.

(b) Construct a wiring diagram of the inverse system using integrators, time delays, adders, and subtracters.

6.16   A linear time-invariant continuous-time system has the transfer function

$$H(s) = \frac{s(s + 1)}{s^2 + 2s + 4}.$$

With $h(t) = $ impulse response of the system, compute a time function $v(t)$ such that

$$(h * v)(t) = \delta(t),$$

where $\delta(t)$ is the unit impulse.

6.17   Consider the two-car system given by the input/output differential equations

$$\ddot{y}_1(t) + \frac{k_f}{M}\dot{y}_1(t) = \frac{1}{M}x_1(t)$$

$$\ddot{y}_2(t) + \frac{k_f}{M}\dot{y}_2(t) = \frac{1}{M}x_2(t)$$

$$w(t) = y_2(t) - y_1(t),$$

where $x_1(t)$ [$x_2(t)$] is the force applied to the first (second) car, and $y_1(t)$ [$y_2(t)$] is the position of the first (second) car. With the inputs equal to $x_1(t)$, $x_2(t)$, and the outputs equal to $y_1(t)$, $\dot{y}_1(t)$, $y_2(t)$, $\dot{y}_2(t)$, $w(t)$, $\dot{w}(t)$, compute the transfer function matrix of the system.

6.18 Consider the two-input three-output continuous-time system shown in Figure P6.18.

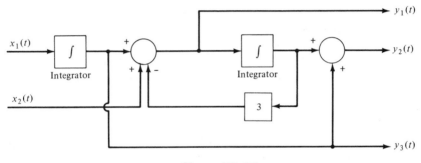

Figure P6.18

(a) Determine the transfer function matrix of the system.
(b) The inputs $x_1(t)$ and $x_2(t)$ produce the output responses $y_1(t) = 2 + e^{-3t}$ for $t \geq 0$ and $y_2(t) = 3e^{-2t} - 3e^{-3t}$ for $t \geq 0$ with no initial energy in the system. Determine $x_1(t)$, $x_2(t)$, and $y_3(t)$.
(c) Determine the transfer function matrix of the inverse system.

6.19 Consider an inverted pendulum on a motor-driven cart as illustrated in Figure P6.19. Here $\theta(t)$ is the angle of the pendulum from the vertical position, $d(t)$ is the position of the cart at time t, $x(t)$ is the drive or braking force applied to the cart, and M is the mass of the cart. The mass of the pendulum is m. From the laws of mechanics, the process is described by the following differential equations

$$(J + mL^2)\ddot{\theta}(t) - mgL \sin\theta(t) + mL\ddot{d}(t)\cos\theta(t) = 0$$
$$(M + m)\ddot{d}(t) + mL\ddot{\theta}(t) = x(t),$$

where J is the moment of inertia of the inverted pendulum about the center of mass, g is the gravity constant, and L is one-half the length of the pendulum. We assume that the angle $\theta(t)$ is small, so that $\cos\theta(t) \approx 1$ and $\sin\theta(t) \approx \theta(t)$.

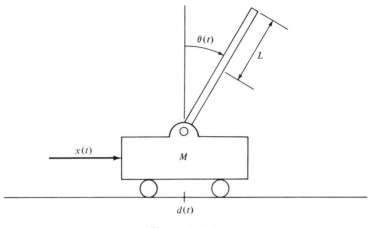

Figure P6.19

(a) Taking the input to be $x(t)$ and the outputs to be $\theta(t)$ and $d(t)$, determine the transfer function matrix of the system.

(b) Now suppose that $J = 1$, $L = 1$, $g = 9.8$, $M = 1$, and $m = 0.1$. Using the transfer function representation, compute $\theta(t)$ and $d(t)$ for all $t > 0$ when $x(t) = \delta(t)$ with $\theta(0) = \dot{\theta}(0) = d(0) = \dot{d}(0) = 0$. This corresponds to someone hitting the cart with an object (e.g., a hammer) at time $t = 0$.

(c) Repeat part (b) with $x(t) = 0$ for $t \geq 0$, $\theta(0) = 10°$, and $\dot{\theta}(0) = d(0) = \dot{d}(0) = 0$. This corresponds to someone moving the pendulum to the position $\theta(0) = 10°$ and then releasing the pendulum without imparting any initial velocity.

(d) Repeat part (b) with $x(t) = 0$ for $t \geq 0$, $\theta(0) = 0$, $\dot{\theta}(0) = 1$, and $d(0) = \dot{d}(0) = 0$. This corresponds to someone tapping the pendulum at time $t = 0$.

The *z*-Transform and Discrete-Time Systems

In this chapter we study the *z*-transform, which is the discrete-time counterpart of the Laplace transform. The *z*-transform operates on a discrete-time signal $x(kT)$, in contrast to the Laplace transform, which operates on a continuous-time or analog signal $x(t)$. As pointed out in Section 1.3, many discrete-time signals arise in practice by sampling a continuous-time signal $x(t)$ every T seconds, where T is a fixed positive number called the sampling interval.

In Sections 7.1 and 7.2 we define the *z*-transform of the discrete-time signal $x(kT)$, and then we study the basic properties of the *z*-transform. In Section 7.3 we consider the problem of computing the inverse *z*-transform. A computer program is given for calculating the first Q values of the inverse transform of a rational function in z.

In Sections 7.4 and 7.5 we use the *z*-transform to generate the *z*-domain representation of a linear time-invariant discrete-time system. As in the Laplace-transform approach to linear time-invariant continuous-time systems, the *z*-transform yields an algebraic relationship between the inputs and outputs of a discrete-time system. The transfer function of interconnections of unit delayers and interconnections of subsystems is studied in Section 7.6.

The theory of the *z*-transform and its application to linear time-invariant discrete-time systems closely resembles the theory of the Laplace transform and its application to linear time-invariant continuous-time systems. In particular,

265

the results and techniques in this chapter closely parallel the results and techniques given in Chapters 5 and 6 on the Laplace transform. However, there are some differences between the transform theory in the continuous-time case and the transform theory in the discrete-time case, although for the most part, these differences are minor. In reading this chapter, the reader should look for the similarities and differences in the two cases.

7.1

z-TRANSFORM

Consider the discrete-time signal $x(kT)$, where T is a fixed positive real number and k is an integer variable; that is, $k = \ldots, -2, -1, 0, 1, 2, \ldots$. The one-sided z-transform of $x(kT)$, denoted by $X(z)$, is a function of a complex variable z defined by

$$X(z) = x(0) + x(T)z^{-1} + x(2T)z^{-2} + \cdots. \qquad (7.1)$$

Using the summation symbol, we can write (7.1) in the form

$$X(z) = \sum_{k=0}^{\infty} x(kT)z^{-k}. \qquad (7.2)$$

From (7.2) we see that the z-transform $X(z)$ is a power series in z^{-1} whose coefficients are the values of the signal $x(kT)$. If we insert a particular complex number z into the power series (7.2), the resulting value of $X(z)$ will be a complex number in general. Thus the z-transform $X(z)$ is a complex-valued function of the complex variable z. In working with the z-transform, we shall always use uppercase letters to denote the transform of signals denoted by lowercase letters.

The transform $X(z)$ defined above is called the *one-sided z-transform* since $X(z)$ depends only on the values of the signal $x(kT)$ for $k = 0, 1, 2, \ldots$. The one-sided z-transform can be applied to signals $x(kT)$ that are nonzero for $k = -1, -2, \ldots$, but any nonzero values of $x(kT)$ for $k < 0$ cannot be recovered from the one-sided z-transform.

The *two-sided z-transform* of $x(kT)$, also denoted by $X(z)$, is defined by

$$X(z) = \sum_{k=-\infty}^{\infty} x(kT)z^{-k}.$$

The two-sided z-transform consists of a power series in z and a power series in z^{-1}. Note that if $x(kT) = 0$ for $k = -1, -2, \ldots$, the one-sided and two-sided z-transforms of $x(kT)$ are identical. In this chapter we concentrate on the one-sided z-transform, which we shall refer to as the z-transform. The two-sided z-transform will reappear in Chapter 11 when we study the discrete-time Fourier transform.

To be precise from a mathematical standpoint, we should consider the question of convergence of the power series in (7.2). Convergence of the power series is defined in the following manner.

Given the discrete-time signal $x(kT)$, let P denote the set of all positive real numbers ρ such that

$$\lim_{N \to \infty} \sum_{k=0}^{N} |x(kT)| \rho^{-k} = c < \infty, \tag{7.3}$$

where c is a positive constant that may depend on ρ. The condition (7.3) means that the power series

$$X(\rho) = X(z)|_{z=\rho} = \sum_{k=0}^{\infty} x(kT) \rho^{-k}$$

converges absolutely. If there is no positive number ρ such that (7.3) is satisfied, the signal $x(kT)$ does not have a one-sided z-transform (which converges absolutely). The great majority of discrete-time signals arising in engineering do have a z-transform. For such signals, the set P is not empty.

Now let $\rho_{\min}$ denote the minimal element for the set P; that is, $\rho_{\min}$ is the smallest positive number such that

$$\rho \in P \qquad \text{for all } \rho > \rho_{\min}.$$

The set of all complex numbers z such that

$$|z| > \rho_{\min} \tag{7.4}$$

is called the *region of absolute convergence* of the z-transform $X(z)$. For any complex number z satisfying (7.4), the power series $X(z)$ converges, and thus the z-transform $X(z)$ exists for this value of z. In other words, the z-transform $X(z)$ is well defined for all values of z belonging to the region of absolute convergence.

INVERSE z-TRANSFORM If $X(z)$ is the z-transform of the discrete-time signal $x(kT)$, we can compute $x(kT)$ from $X(z)$ by taking the *inverse z-transform* of $X(z)$, given by

$$x(kT) = \frac{1}{2\pi j} \int X(z) z^{k-1} \, dz. \tag{7.5}$$

The integral in (7.5) is evaluated by integrating along a counterclockwise closed circular contour in the complex plane centered at the origin and with radius $c > \rho_{\min}$.

When the transform $X(z)$ is a rational function of z, the inverse z-transform can be computed without having to evaluate the integral in (7.5). We consider this in Section 7.3.

TRANSFORM PAIRS　From here on, we shall usually use the transform pair notation

$$x(kT) \leftrightarrow X(z) \tag{7.6}$$

to denote the fact that $X(z)$ is the z-transform of $x(kT)$ and that $x(kT)$ is the inverse z-transform of $X(z)$. Many transform pairs can be determined from a few basic transform pairs. We generate a table of common transform pairs in Section 7.2.

In the remainder of this section we illustrate the computation of the z-transform using the definition (7.2) of $X(z)$.

EXAMPLE 7.1.　Let $\Delta(kT)$ denote the unit pulse concentrated at $kT = 0$ given by

$$\Delta(kT) = \begin{cases} 1, & k = 0 \\ 0, & k \neq 0. \end{cases}$$

Since $\Delta(kT)$ is zero for all k except $k = 0$, the z-transform is

$$\sum_{k=0}^{\infty} \Delta(kT)z^{-k} = \Delta(0)z^{-0} = 1.$$

So we have the transform pair

$$\Delta(kT) \leftrightarrow 1. \tag{7.7}$$

In addition, since

$$\sum_{k=0}^{\infty} |\Delta(kT)|\rho^{-k} = 1 < \infty \qquad \text{for all } \rho > 0,$$

the region of absolute convergence is the set of all complex numbers z such that $|z| > 0$.

From the transform pair (7.7), we see that the unit pulse $\Delta(kT)$ is the discrete-time counterpart to the unit impulse $\delta(t)$ in the sense that the z-transform of $\Delta(kT)$ is 1 and the Laplace transform of $\delta(t)$ is also 1. However, it should be stressed that $\Delta(kT)$ is not a sampled version of $\delta(t)$; in particular, $\Delta(0) = 1$, whereas $\delta(0)$ is not defined.

EXAMPLE 7.2.　Given a fixed positive integer N, consider the unit pulse $\Delta(kT - NT)$ located at $k = NT$. For example, when $N = 2$, $\Delta(kT - 2T)$ is the pulse shown in Figure 7.1. The z-transform of $\Delta(kT - NT)$ is

$$\sum_{k=0}^{\infty} \Delta(kT - NT)z^{-k} = \Delta(0)z^{-N} = z^{-N} = \frac{1}{z^N}.$$

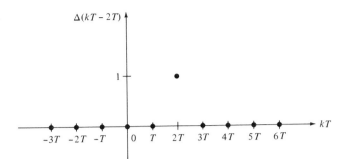

Figure 7.1. Unit pulse $\Delta(kT - 2T)$ located at $kT = 2T$.

Thus we have the transform pair

$$\Delta(kT - NT) \leftrightarrow \frac{1}{z^N}, \qquad N = 0, 1, 2, \ldots \quad (7.8)$$

As in Example 7.1, the region of absolute convergence is the set of all complex numbers z such that $|z| > 0$.

EXAMPLE 7.3. Consider the discrete-time unit-step function $u(kT)$ given by

$$u(kT) = \begin{cases} 1, & k = 0, 1, 2, \ldots \\ 0, & k = -1, -2, \ldots \end{cases}$$

The z-transform $U(z)$ is

$$\begin{aligned} U(z) &= \sum_{k=0}^{\infty} u(kT)z^{-k} \\ &= \sum_{k=0}^{\infty} z^{-k} \\ &= 1 + z^{-1} + z^{-2} + z^{-3} + \cdots. \end{aligned} \quad (7.9)$$

Multiplying both sides of (7.9) by $(z - 1)$, we obtain

$$\begin{aligned} (z - 1)U(z) &= (z + 1 + z^{-1} + z^{-2} + \cdots) \\ &\quad - (1 + z^{-1} + z^{-2} + \cdots) \\ &= z. \end{aligned} \quad (7.10)$$

From (7.10) we have

$$U(z) = \frac{z}{z - 1}.$$

Thus we have the transform pair

$$u(kT) \leftrightarrow \frac{z}{z - 1}. \tag{7.11}$$

Note that the form of the z-transform of the discrete-time unit-step function $u(kT)$ is different from the form of the Laplace transform $U(s) = 1/s$ of the unit-step function $u(t)$. The region of absolute convergence for the z-transform $U(z)$ is the set of all complex numbers z such that $|z| > 1$. This follows from the result that

$$\sum_{k=0}^{\infty} (1)\rho^{-k} < \infty \qquad \text{if and only if } \rho > 1. \tag{7.12}$$

To prove (7.12), first note that for any positive integer N,

$$\sum_{k=0}^{N} \rho^{-k} = \frac{(1/\rho)^{N+1} - 1}{(1/\rho) - 1}, \tag{7.13}$$

where we are assuming that $\rho \neq 1$. The relationship (7.13) can be verified by multiplying both sides of (7.13) by $(1/\rho) - 1$. Now the right side of (7.13) has a finite limit as $N \to \infty$ if and only if $\rho > 1$. Thus we have proved (7.12).

EXAMPLE 7.4. Given a fixed real or complex number a, let $x(kT) = a^k$. When $a = e^{-bT}$ where b is a real or complex number, the discrete-time signal $x(kT) = a^k$ can be viewed as a sampled version of the continuous-time signal $x(t) = e^{-bt}$. Now the z-transform $X(z)$ of $x(kT)$ is given by

$$X(z) = \sum_{k=0}^{\infty} a^k z^{-k}$$
$$= 1 + az^{-1} + a^2 z^{-2} + \cdots. \tag{7.14}$$

Multiplying both sides of (7.14) by $(z - a)$, we get

$$(z - a)X(z) = (z + a + a^2 z^{-1} + a^3 z^{-2} + \cdots)$$
$$- (a + a^2 z^{-1} + a^3 z^{-2} + \cdots)$$
$$= z.$$

Hence we have the transform pair

$$a^k \leftrightarrow \frac{z}{z - a}. \tag{7.15}$$

Note that if we take $a = 1$, we get the transform pair (7.11). The region of absolute convergence for the transform $X(z) = z/(z - a)$ is the set of all complex numbers z such that $|z| > |a|$. This follows by taking the limit as $N \to \infty$ of both sides of the relationship

$$\sum_{k=0}^{N} a^k \rho^{-k} = \frac{(a/\rho)^{N+1} - 1}{(a/\rho) - 1},$$

where $a/\rho \neq 1$.

Although a transform is defined only for values of z belonging to the region of absolute convergence, from here on we shall not consider the region of absolute convergence. As in the Laplace transform theory, it is usually not necessary to compute the region of absolute convergence in order to apply the transform. Thus we can proceed without having to consider the region of convergence.

The transform pairs derived in the examples of this section are shown in Table 7.1.

7.2

PROPERTIES OF THE z-TRANSFORM

The z-transform possesses a number of properties that are useful in deriving transform pairs and in the application of the transform to the study of linear time-invariant discrete-time systems. These properties are very similar to the properties of the Laplace transform that were given in Section 5.2.

In this section we state and prove the properties of the z-transform. As an illustration of the use of the properties, we generate a collection of common transform pairs from the basic set of pairs given in Table 7.1.

LINEARITY If $x(kT) \leftrightarrow X(z)$ and $v(kT) \leftrightarrow V(z)$, then for any real or complex numbers a, b,

$$ax(kT) + bv(kT) \leftrightarrow aX(z) + bV(z). \tag{7.16}$$

TABLE 7.1. Basic Transform Pairs.

$$\Delta(kT) \leftrightarrow 1$$

$$\Delta(kT - NT) \leftrightarrow \frac{1}{z^N}, \quad N = 1, 2, \ldots$$

$$u(kT) \leftrightarrow \frac{z}{z - 1}$$

$$a^k \leftrightarrow \frac{z}{z - a}, \quad a \text{ real or complex}$$

The transform pair (7.16) implies that the z-transform is a linear operation. To prove linearity, we shall compute the z-transform of $ax(kT) + bv(kT)$. From the definition (7.2) of the transform, we have that

$$ax(kT) + bv(kT) \leftrightarrow \sum_{k=0}^{\infty} [ax(kT) + bx(kT)]z^{-k}.$$

Now

$$\sum_{k=0}^{\infty} [ax(kT) + bv(kT)]z^{-k} = \sum_{k=0}^{\infty} [ax(kT)z^{-k} + bv(kT)z^{-k}]$$

$$= a \sum_{k=0}^{\infty} x(kT)z^{-k} + b \sum_{k=0}^{\infty} v(kT)z^{-k}.$$

Thus

$$ax(kT) + bv(kT) \leftrightarrow aX(z) + bV(z).$$

EXAMPLE 7.5. Let $x(kT) = u(kT)$ and $v(kT) = a^k$, $a \neq 1$. From Table 7.1 we have

$$u(kT) \leftrightarrow \frac{z}{z-1} \quad \text{and} \quad a^k \leftrightarrow \frac{z}{z-a}.$$

Hence, by linearity,

$$u(kT) + a^k \leftrightarrow \frac{z}{z-1} + \frac{z}{z-a} = \frac{2z^2 - (1+a)z}{(z-1)(z-a)}.$$

RIGHT SHIFT IN TIME   Suppose that $x(kT) \leftrightarrow X(z)$. Given a positive integer q, consider the discrete-time signal $x(kT - qT)u(kT - qT)$, which is a qT-second right shift of $x(kT)u(kT)$. Then we claim that

$$x(kT - qT)u(kT - qT) \leftrightarrow z^{-q}X(z). \tag{7.17}$$

By (7.17) we have that a qT-second right shift of $x(kT)u(kT)$ corresponds to multiplication by z^{-q} in the z-transform domain. To prove this property, first note that by definition of the z-transform,

$$x(kT - qT)u(kT - qT) \leftrightarrow \sum_{k=0}^{\infty} x(kT - qT)u(kT - qT)z^{-k}$$

$$\leftrightarrow \sum_{k=q}^{\infty} x(kT - qT)z^{-k} \tag{7.18}$$

since $u(kT - qT) = 0$ for $k < q$

We shall consider a change of index in the summation in (7.18). Defining $\bar{k} = k - q$, so that $k = \bar{k} + q$, we have that $\bar{k} = 0$ when $k = q$, and $\bar{k} = \infty$ when $k = \infty$. Hence

$$\sum_{k=q}^{\infty} x(kT - qT)z^{-k} = \sum_{\bar{k}=0}^{\infty} x(\bar{k}T)z^{-(\bar{k}+q)}$$

$$= z^{-q} \sum_{\bar{k}=0}^{\infty} x(\bar{k}T)z^{-\bar{k}}$$

$$= z^{-q}X(z).$$

Thus

$$x(kT - qT)u(kT - qT) \leftrightarrow z^{-q}X(z).$$

EXAMPLE 7.6. Let N be a fixed positive integer. We want to determine the z-transform of the pulse $p(kT)$ defined by

$$p(kT) = \begin{cases} 1, & k = 0, 1, 2, \ldots, N - 1 \\ 0, & \text{all other } k. \end{cases}$$

Writing $p(kT)$ in terms of the discrete-time unit-step function $u(kT)$, we have

$$p(kT) = u(kT) - u(kT - NT).$$

From Table 7.1, the z-transform of $u(kT)$ is equal to $z/(z - 1)$. By the right-shift property (7.17), the z-transform of $u(kT - NT)$ is equal to

$$z^{-N} \frac{z}{z - 1} = \frac{z^{-N+1}}{z - 1}.$$

Therefore, by linearity the z-transform of $p(kT)$ is

$$\frac{z}{z - 1} - \frac{z^{-N+1}}{z - 1} = \frac{z(1 - z^{-N})}{z - 1} = \frac{z^N - 1}{z^{N-1}(z - 1)}.$$

SECOND VERSION OF THE RIGHT-SHIFT PROPERTY There is a second version of the right-shift property that we will need when we apply the z-transform to the input/output difference equation of a finite-dimensional linear time-invariant discrete-time system. The property is as follows. Suppose that $x(kT) \leftrightarrow X(z)$. Then we have the transform pairs

$$x(kT - T) \leftrightarrow z^{-1}X(z) + x(-T)$$

$$x(kT - 2T) \leftrightarrow z^{-2}X(z) + x(-2T) + z^{-1}x(-T)$$

.
.
.

$$x(kT - qT) \leftrightarrow z^{-q}X(z) + x(-qT) + z^{-1}x(-qT + T)$$
$$+ \cdots + z^{-q+1} x(-T). \tag{7.19}$$

Note that if $x(kT) = 0$ for $k = -1, -2, \ldots, -q$, the transform pair (7.19) reduces to

$$x(kT - qT) \leftrightarrow z^{-q}X(z), \tag{7.20}$$

which is identical to the transform pair (7.17).

We shall prove (7.19) in the case $q = 1$. By definition of the z-transform, we have

$$x(kT - T) \leftrightarrow \sum_{k=0}^{\infty} x(kT - T)z^{-k}. \tag{7.21}$$

Defining the change of index $\bar{k} = k - 1$ in the summation in (7.21), we obtain

$$x(kT - T) \leftrightarrow \sum_{\bar{k}=-1}^{\infty} x(\bar{k}T)z^{-(k+1)} = \sum_{\bar{k}=0}^{\infty} x(\bar{k}T)z^{-(k+1)} + x(-T)$$

$$\leftrightarrow z^{-1} \sum_{\bar{k}=0}^{\infty} x(\bar{k}T)z^{-k} + x(-T)$$

$$\leftrightarrow z^{-1}X(z) + x(-T).$$

Thus we have verified the transform pair (7.19) in the case $q = 1$. The verification of (7.19) for $q > 1$ can be demonstrated in a similar manner. We leave the details to the interested reader.

LEFT SHIFT IN TIME   In contrast to the Laplace transform, the z-transform does have a left-shift property as follows. Given the discrete-time signal $x(kT)$, the qT-second left shift of $x(kT)$ is the signal $x(kT + qT)$. Now suppose that $x(kT) \leftrightarrow X(z)$. Then we have the transform pairs

$$x(kT + T) \leftrightarrow zX(z) - x(0)z$$

$$x(kT + 2T) \leftrightarrow z^2X(z) - x(0)z^2 - x(T)z$$

.
.
.

$$x(kT + qT) \leftrightarrow z^q X(z) - x(0)z^q - x(T)z^{q-1}$$
$$- \cdots - x(qT - T)z. \tag{7.22}$$

We shall prove (7.22) when $q = 1$. We have

$$x(kT + T) \leftrightarrow \sum_{k=0}^{\infty} x(kT + T)z^{-k}. \tag{7.23}$$

Defining the change of index $\bar{k} = k + 1$ in the summation in (7.23), we obtain

$$x(kT + T) \leftrightarrow \sum_{\bar{k}=1}^{\infty} x(\bar{k}T)z^{-(\bar{k}-1)}$$
$$\leftrightarrow z \sum_{\bar{k}=1}^{\infty} x(\bar{k}T)z^{-\bar{k}} = z\left[\sum_{\bar{k}=0}^{\infty} x(\bar{k}T)z^{-\bar{k}} - x(0) \right]$$
$$\leftrightarrow z[X(z) - x(0)].$$

EXAMPLE 7.7. Consider the T-second left shift $u(kT + T)$ of the discrete-time unit-step function $u(kT)$. By the left-shift property, the z-transform of $u(kT + T)$ is equal to

$$zU(z) - u(0)z = \frac{z^2}{z - 1} - z = \frac{z^2 - z(z - 1)}{z - 1} = \frac{z}{z - 1}.$$

Hence the z-transform of $u(kT + T)$ is equal to the z-transform of $u(kT)$. This result is not unexpected since $u(kT + T) = u(kT)$ for $k = 0, 1, 2, \ldots$.

MULTIPLICATION BY k AND k^2 If $x(kT) \leftrightarrow X(z)$, we claim that

$$kx(kT) \leftrightarrow -z \frac{d}{dz} X(z) \tag{7.24}$$

and

$$k^2 x(kT) \leftrightarrow z \frac{d}{dz} X(z) + z^2 \frac{d^2}{dz^2} X(z). \tag{7.25}$$

We shall first prove (7.24). By definition of the z-transform, we have

$$X(z) = \sum_{k=0}^{\infty} x(kT)z^{-k}. \tag{7.26}$$

Taking the derivative with respect to z of both sides of (7.26), we obtain

$$\frac{d}{dz} X(z) = \sum_{k=0}^{\infty} (-k)x(kT)z^{-k-1}$$

$$= -z^{-1} \sum_{k=0}^{\infty} kx(kT)z^{-k}. \tag{7.27}$$

Thus

$$-z \frac{d}{dz} X(z) = \sum_{k=0}^{\infty} kx(kT)z^{-k} = z\text{-transform of } kx(kT), \tag{7.28}$$

and we have proved (7.24). Now taking the derivative with respect to z of both sides of (7.27), we have

$$\frac{d^2}{dz^2} X(z) = \sum_{k=0}^{\infty} k(k+1)x(kT)z^{-k-2}$$

$$= \sum_{k=0}^{\infty} k^2 x(kT)z^{-k-2} + \sum_{k=0}^{\infty} kx(kT)z^{-k-2}.$$

Thus

$$z^2 \frac{d^2}{dz^2} X(z) = \sum_{k=0}^{\infty} k^2 x(kT)z^{-k} + \sum_{k=0}^{\infty} kx(kT)z^{-k}$$

$$= z\text{-transform of } k^2 x(kT) + z\text{-transform of } kx(kT). \tag{7.29}$$

Combining (7.28) and (7.29), we have that the z-transform of $k^2x(kT)$ is equal to

$$z \frac{d}{dz} X(z) + z^2 \frac{d^2}{dz^2} X(z),$$

which proves (7.25).

EXAMPLE 7.8.   Let $x(kT) = a^k$, where a is any nonzero real or complex number. From Table 7.1

$$X(z) = \frac{z}{z-a}.$$

Then

$$z \frac{d}{dz} X(z) = z \left[\frac{-z}{(z-a)^2} + \frac{1}{z-a} \right] = \frac{-az}{(z-a)^2}.$$

Thus we have the transform pair

$$ka^k \leftrightarrow \frac{az}{(z - a)^2}. \tag{7.30}$$

Note that if $a = 1$, we get the transform pair

$$ku(kT) \leftrightarrow \frac{z}{(z - 1)^2}. \tag{7.31}$$

EXAMPLE 7.9. Now suppose that we want to compute the z-transform of k^2a^k, where again a is any nonzero real or complex number. Setting $x(kT) = a^k$, so that

$$X(z) = \frac{z}{z - a},$$

we have

$$\frac{d^2}{dz^2} X(z) = \frac{2a}{(z - a)^3}.$$

Then using (7.25), we have the transform pair

$$k^2a^k \leftrightarrow \frac{-az}{(z - a)^2} + \frac{2az^2}{(z - a)^3} = \frac{az(z + a)}{(z - a)^3}. \tag{7.32}$$

Setting $a = 1$ in (7.32), we get the transform pair

$$k^2u(kT) \leftrightarrow \frac{z(z + 1)}{(z - 1)^3}. \tag{7.33}$$

MULTIPLICATION BY a^k If $x(kT) \leftrightarrow X(z)$, then for any nonzero real or complex number a,

$$a^kx(kT) \leftrightarrow X\left(\frac{z}{a}\right). \tag{7.34}$$

By (7.34), multiplication by a^k in the time domain corresponds to scaling of the z-variable in the transform domain. To prove (7.34), observe that

$$a^kx(kT) \leftrightarrow \sum_{k=0}^{\infty} a^kx(kT)z^{-k}$$

$$\leftrightarrow \sum_{k=0}^{\infty} x(kT) \left(\frac{z}{a}\right)^{-k} = X\left(\frac{z}{a}\right).$$

EXAMPLE 7.10. Let $p(kT)$ denote the pulse defined by $p(kT) = u(kT) - u(kT - NT)$. From Example 7.6 we have

$$p(kT) \leftrightarrow \frac{z(1 - z^{-N})}{z - 1}.$$

Then for any nonzero number a, by (7.34) we have the transform pair

$$a^k p(kT) \leftrightarrow \frac{(z/a)[1 - (z/a)^{-N}]}{(z/a) - 1} = \frac{z(1 - a^N z^{-N})}{z - a}.$$

MULTIPLICATION BY cos (ωkT) AND sin (ωkT) If $x(kT) \leftrightarrow X(z)$, then for any positive real numbers ω, T,

$$\cos(\omega kT)x(kT) \leftrightarrow \tfrac{1}{2}[X(e^{j\omega T}z) + X(e^{-j\omega T}z)] \qquad (7.35)$$

and

$$\sin(\omega kT)x(kT) \leftrightarrow \frac{j}{2}[X(e^{j\omega T}z) - X(e^{-j\omega T}z)]. \qquad (7.36)$$

To prove (7.35), we use Euler's formula,

$$\cos(\omega kT) = \tfrac{1}{2}(e^{-j\omega kT} + e^{j\omega kT}).$$

Then

$$\cos(\omega kT)x(kT) = \tfrac{1}{2}[e^{-j\omega kT}x(kT) + e^{j\omega kT}x(kT)].$$

By (7.34)

$$e^{\pm j\omega kT}x(kT) = (e^{\pm j\omega T})^k \, x(kT) \leftrightarrow X(e^{\mp j\omega T}z).$$

Hence

$$\cos(\omega kT)x(kT) \leftrightarrow \tfrac{1}{2}[X(e^{j\omega T}z) + X(e^{-j\omega T}z)].$$

The proof of (7.36) is similar and is thus omitted.

EXAMPLE 7.11. Let $v(kT) = \cos(\omega kT)$. Since the z-transform of $v(kT)$ depends only on the values of $v(kT)$ for $k = 0, 1, 2, \ldots$, we can write $v(kT)$ in the form

$$v(kT) = \cos(\omega kT)u(kT).$$

Taking $x(kT) = u(kT)$, we have that

$$X(z) = \frac{z}{z - 1}.$$

Using (7.35), the z-transform of $v(kT)$ is

$$V(z) = \frac{1}{2}\left(\frac{e^{j\omega T}z}{e^{j\omega T}z - 1} + \frac{e^{-j\omega T}z}{e^{-j\omega T}z - 1}\right)$$

$$= \frac{1}{2}\left[\frac{e^{j\omega T}z(e^{-j\omega T}z - 1) + e^{-j\omega T}z(e^{j\omega T}z - 1)}{(e^{j\omega T}z - 1)(e^{-j\omega T}z - 1)}\right]$$

$$= \frac{1}{2}\left[\frac{z^2 - e^{j\omega T}z + z^2 - e^{-j\omega T}z}{z^2 - (e^{j\omega T} + e^{-j\omega T})z + 1}\right]$$

$$= \frac{z^2 - (\cos \omega T)z}{z^2 - (2\cos \omega T)z + 1}.$$

Hence we have the transform pair

$$\cos (\omega kT) \leftrightarrow \frac{z^2 - (\cos \omega T)z}{z^2 - (2\cos \omega T)z + 1}. \tag{7.37}$$

Similarly, we can verify the transform pair

$$\sin (\omega kT) \leftrightarrow \frac{(\sin \omega T)z}{z^2 - (2\cos \omega T)z + 1}. \tag{7.38}$$

EXAMPLE 7.12. Now let $v(kT) = a^k \cos (\omega kT)$. To compute the z-transform of $v(kT)$, we could set $x(kT) = a^k$ and use the multiplication by $\cos (\omega kT)$ property. However, it is easier to set $x(kT)$ equal to $\cos (\omega kT)$ and use the multiplication-by-a^k property. From (7.37) we have

$$V(z) = \frac{(z/a)^2 - (\cos \omega T)(z/a)}{(z/a)^2 - (2\cos \omega T)(z/a) + 1}$$

$$= \frac{z^2 - (a\cos \omega T)z}{z^2 - (2a\cos \omega T)z + a^2}.$$

Thus we have the transform pair

$$a^k \cos (\omega kT) \leftrightarrow \frac{z^2 - (a\cos \omega T)z}{z^2 - (2a\cos \omega T)z + a^2}.$$

In a similar fashion, we can derive the transform pair

$$a^k \sin (\omega kT) \leftrightarrow \frac{(a\sin \omega T)z}{z^2 - (2a\cos \omega T)z + a^2}.$$

CONVOLUTION   Given two discrete-time functions $x(kT)$ and $v(kT)$ with $x(kT)$ and $v(kT)$ equal to zero for $k = -1, -2, \ldots$, in Chapter 4 we defined the convolution of $x(kT)$ and $v(kT)$ by

$$(x * v)(kT) = \sum_{i=0}^{k} x(iT)v(kT - iT), \qquad k \geq 0.$$

Note that since $v(kT) = 0$ for $k = -1, -2, \ldots$, we can take the convolution sum from $k = 0$ to $k = \infty$; that is, we can write

$$(x * v)(kT) = \sum_{i=0}^{\infty} x(iT)v(kT - iT), \qquad k \geq 0.$$

Taking the z-transform of both sides, we get the transform pair

$$(x * v)(kT) \leftrightarrow \sum_{k=0}^{\infty} \left[\sum_{i=0}^{\infty} x(iT)v(kT - iT) \right] z^{-k}$$

$$\leftrightarrow \sum_{i=0}^{\infty} x(iT) \left[\sum_{k=0}^{\infty} v(kT - iT)z^{-k} \right].$$

With the change of index $\bar{k} = k - i$ in the second summation, we have

$$(x * v)(kT) \leftrightarrow \sum_{i=0}^{\infty} x(iT) \left[\sum_{\bar{k}=-i}^{\infty} v(\bar{k})z^{-\bar{k}-i} \right]$$

$$\leftrightarrow \sum_{i=0}^{\infty} x(iT) \left[\sum_{\bar{k}=0}^{\infty} v(\bar{k})z^{-\bar{k}-i} \right], \qquad \text{since } v(\bar{k}) = 0, \quad \bar{k} < 0$$

$$\leftrightarrow \left[\sum_{i=0}^{\infty} x(iT)z^{-i} \right] \left[\sum_{k=0}^{\infty} v(\bar{k})z^{-k} \right]$$

$$\leftrightarrow X(z)V(z). \tag{7.39}$$

By (7.39) we see that the z-transform of the convolution $x * v$ is equal to the product $X(z)V(z)$, where $X(z)$ and $V(z)$ are the z-transforms of $x(kT)$ and $v(kT)$, respectively. Hence convolution in the discrete-time domain corresponds to a product in the z-transform domain.

SUMMATION   Given the discrete-time signal $x(kT)$ with $x(kT) = 0$ for $k = -1, -2, \ldots$, let $v(kT)$ denote the discrete-time signal defined by

$$v(kT) = \sum_{i=0}^{k} x(iT).$$

The signal $v(kT)$ is called the *sum* of $x(kT)$. Now if $x(kT) \leftrightarrow X(z)$, then

$$v(kT) \leftrightarrow \frac{z}{z-1} X(z). \tag{7.40}$$

The proof of (7.40) follows by observing that $v(kT)$ can be expressed in terms of the convolution

$$v(kT) = u(kT) * [x(kT)u(kT)]. \tag{7.41}$$

To verify (7.41), first note that by definition of the convolution operation, we have

$$u(kT) * [x(kT)u(kT)] = \sum_{i=-\infty}^{\infty} u(kT - iT)x(iT)u(iT). \tag{7.42}$$

Since $u(kT - iT) = 0$ for $k < i$, $u(kT - iT) = 1$ for $k \geq i$, $u(iT) = 0$ for $i < 0$, and $u(iT) = 1$ for $i \geq 0$, the summation in (7.42) reduces to

$$\sum_{i=0}^{k} x(iT).$$

Thus we have verified (7.41). Now taking the z-transform of both sides of (7.41) and using the convolution property (7.39), we obtain

$$V(z) = \frac{z}{z-1} X(z),$$

which proves (7.40).

EXAMPLE 7.13. Let $x(kT) = u(kT)$. Then

$$v(kT) = \sum_{i=0}^{k} u(iT) = (k + 1)u(kT).$$

By (7.40)

$$V(z) = \frac{z}{z-1} X(z) = \frac{z}{z-1} \frac{z}{z-1} = \frac{z^2}{(z-1)^2}.$$

Hence we have the transform pair

$$(k + 1)u(kT) \leftrightarrow \frac{z^2}{(z-1)^2}. \tag{7.43}$$

INITIAL-VALUE THEOREM   If $x(kT) \leftrightarrow X(z)$, then

$$x(0) = \lim_{z \to \infty} X(z)$$

$$x(T) = \lim_{z \to \infty} [zX(z) - zx(0)]$$

.

.

.

$$x(qT) = \lim_{z \to \infty} [z^q X(z) - z^q x(0) - z^{q-1} x(T) - \cdots - zx(qT - T)]. \quad (7.44)$$

We shall prove (7.44) when $q = 1$. We have

$$zX(z) = z \sum_{k=0}^{\infty} x(kT)z^{-k}$$

$$= zx(0) + x(T) + \sum_{k=2}^{\infty} x(kT)z^{-k+1}.$$

Thus

$$zX(z) - zx(0) = x(T) + \sum_{k=2}^{\infty} x(kT)z^{-k+1}. \quad (7.45)$$

Now since

$$z^{-k+1} \to 0 \qquad \text{as } z \to \infty \text{ for all } k \geq 2,$$

it follows that

$$x(kT)z^{-k+1} \to 0 \qquad \text{as } z \to \infty \text{ for all } k \geq 2.$$

Thus, taking the limit as $z \to \infty$ of both sides of (7.45), we get

$$\lim_{z \to \infty} [zX(z) - zx(0)] = x(T),$$

which completes the proof in the $q = 1$ case.

The initial-value theorem allows us to compute the initial values of a discrete-time function $x(kT)$ directly from the z-transform of $x(kT)$. In Section 7.3 we will show that if the transform $X(z)$ of $x(kT)$ is a rational function of z, the initial values of $x(kT)$ can be calculated by a long-division operation.

FINAL-VALUE THEOREM   Given the discrete-time signal $x(kT)$ with z-transform $X(z)$, suppose that $x(kT)$ has a limit as $k \to \infty$. Then the final-value theorem states that

$$\lim_{k\to\infty} x(kT) = \lim_{z\to 1} \left[\frac{z-1}{z} X(z) \right]. \qquad (7.46)$$

The proof of (7.46) is analogous to the proof we gave of the final-value theorem in the continuous-time case. We shall not consider the details.

As in the continuous-time case, we must be careful in using the final-value theorem since the limit on the right side of (7.46) may exist even though $x(kT)$ does not have a limit as $k \to \infty$.

If the transform $X(z)$ is rational, the existence of a limit of $x(kT)$) as $k \to \infty$ can be determined as follows. First write $X(z)$ in the form

$$X(z) = \frac{N(z)}{a_n(z - p_1)(z - p_2) \cdots (z - p_n)}.$$

Then $x(kT)$ has a limit as $k \to \infty$ if and only if the magnitudes $|p_1|, |p_2|, \ldots, |p_n|$ are all strictly less than 1 (i.e., $|p_i| < 1$ for all i), except that one of the p_is may be equal to 1. The proof that this condition is necessary and sufficient for the existence of the limit follows from the stability analysis given in Chapter 8. If this condition on the poles of $X(z)$ is satisfied, the limit of $x(kT)$ as $k \to \infty$ is given by (7.46).

As in the continuous-time case, the final-value theorem makes it possible to determine the limiting value of a time signal directly from the transform of the signal.

EXAMPLE 7.14. Suppose that

$$X(z) = \frac{3z^2 - 2z + 4}{z^3 - 2z^2 + 1.5z - 0.5}.$$

Factoring the denominator of $X(z)$, we have

$$X(z) = \frac{3z^2 - 2z + 4}{(z - 1)(z - 0.5 + j0.5)(z - 0.5 - j0.5)}.$$

In this example, $p_1 = 1$, $p_2 = 0.5 - j0.5$, and $p_3 = 0.5 + j0.5$. The mangitude of p_2 and p_3 is 0.707, and thus the condition given above is satisfied. Hence

$$\lim_{k\to\infty} x(kT) = \lim_{z\to 1} \left[\frac{z-1}{z} X(z) \right] = \lim_{z\to 1} \frac{3z^2 - 2z + 4}{z(z^2 - z + 0.5)} = \frac{5}{0.5} = 10.$$

The foregoing properties of the z-transform are summarized in Table 7.2. In Table 7.3, we give a collection of common z-transform pairs which includes the transform pairs that were derived above using the properties of the z-transform.

TABLE 7.2. Properties of the z-Transform

Property	Transform Pair/Property
Linearity	$ax(kT) + bv(kT) \leftrightarrow aX(z) + bV(z)$
Right shift in time	$x(kT - qT)u(kT - qT) \leftrightarrow z^{-q}X(z)$
Second version of right shift	$x(kT - T) \leftrightarrow z^{-1}X(z) + x(-T)$ $x(kT - 2T) \leftrightarrow z^{-2}X(z) + x(-2T) + z^{-1}x(-T)$
	. . .
	$x(kT - qT) \leftrightarrow z^{-q}X(z) + x(-qT) + z^{-1}x(-qT + T)$ $\qquad\qquad + \cdots + z^{-q+1}x(-T)$
Left shift in time	$x(kT + T) \leftrightarrow zX(z) - x(0)z$ $x(kT + 2T) \leftrightarrow z^2X(z) - x(0)z^2 - x(T)z$
	. . .
	$x(kT + qT) \leftrightarrow z^qX(z) - x(0)z^q - x(T)z^{q-1} - \cdots$ $\qquad\qquad - x(qT - T)z$
Multiplication by k	$kx(kT) \leftrightarrow -z\dfrac{d}{dz}X(z)$
Multiplication by k^2	$k^2x(kT) \leftrightarrow z\dfrac{d}{dz}X(z) + z^2\dfrac{d^2}{dz^2}X(z)$
Multiplication by a^k	$a^kx(kT) \leftrightarrow X\left(\dfrac{z}{a}\right)$
Multiplication by $\cos \omega kT$	$\cos(\omega kT)x(kT) \leftrightarrow \frac{1}{2}[X(e^{j\omega T}z) + X(e^{-j\omega T}z)]$
Multiplication by $\sin \omega kT$	$\sin(\omega kT)x(kT) \leftrightarrow \dfrac{j}{2}[X(e^{j\omega T}z) - X(e^{-j\omega T}z)]$
Convolution	$(x * v)(kT) \leftrightarrow X(z)V(z)$
Summation	$v(kT) \leftrightarrow \dfrac{z}{z-1}X(z)$, where $v(kT) = \displaystyle\sum_{i=0}^{k} x(iT)$
Initial-value theorem	$x(0) = \lim_{z \to \infty} X(z)$ $x(T) = \lim_{z \to \infty} [zX(z) - zx(0)]$
	. . .
	$x(qT) = \lim_{z \to \infty} [z^qX(z) - z^qx(0) - z^{q-1}x(T)$ $\qquad\qquad - \cdots - zx(qT - T)]$
Final-value theorem	If $\lim_{k \to \infty} x(kT)$ exists, then $\lim_{k \to \infty} x(kT) = \lim_{z \to 1}\left[\dfrac{z-1}{z}X(z)\right]$

TABLE 7.3. Common z-Transform Pairs

$$\Delta(kT) \leftrightarrow 1$$

$$\Delta(kT - qT) \leftrightarrow \frac{1}{z^q}, \quad q = 1, 2, \ldots$$

$$u(kT) \leftrightarrow \frac{z}{z - 1}$$

$$u(kT) - u(kT - NT) \leftrightarrow \frac{z^N - 1}{z^{N-1}(z - 1)}, \quad N = 1, 2, 3, \ldots$$

$$a^k \leftrightarrow \frac{z}{z - a}, \quad a \text{ real or complex}$$

$$k \leftrightarrow \frac{z}{(z - 1)^2}$$

$$k + 1 \leftrightarrow \frac{z^2}{(z - 1)^2}$$

$$k^2 \leftrightarrow \frac{z(z + 1)}{(z - 1)^3}$$

$$ka^k \leftrightarrow \frac{az}{(z - a)^2}$$

$$k^2 a^k \leftrightarrow \frac{az(z + a)}{(z - a)^3}$$

$$k(k + 1)a^k \leftrightarrow \frac{2az^2}{(z - a)^3}$$

$$\cos(\omega kT) \leftrightarrow \frac{z^2 - (\cos \omega T)z}{z^2 - (2 \cos \omega T)z + 1}$$

$$\sin(\omega kT) \leftrightarrow \frac{(\sin \omega T)z}{z^2 - (2 \cos \omega T)z + 1}$$

$$a^k \cos(\omega kT) \leftrightarrow \frac{z^2 - (a \cos \omega T)z}{z^2 - (2a \cos \omega T)z + a^2}$$

$$a^k \sin(\omega kT) \leftrightarrow \frac{(a \sin \omega T)z}{z^2 - (2a \cos \omega T)z + a^2}$$

7.3

INVERSE z-TRANSFORM OF A RATIONAL FUNCTION

Suppose that the z-transform $X(z)$ of the discrete-time signal $x(kT)$ is a rational function of z; that is, $X(z)$ can be written in the form

$$X(z) = \frac{N(z)}{D(z)}, \tag{7.47}$$

where $N(z)$ and $D(z)$ are polynomials in z with real coefficients, and where the degree of $N(z)$ is less than or equal to the degree of $D(z)$. As was the case for

Laplace transforms, the z-transform is often a rational function of z, so the rational form (7.47) of $X(z)$ is very common.

As in the continuous-time case, the roots of $D(z) = 0$ are called the poles of the rational function $X(z)$, and the roots of $N(z) = 0$ are the zeros of $X(z)$. The degree of $D(z)$ is called the *order* of the rational function $X(z)$.

Given $X(z)$ in the rational form (7.47), in this section we consider two methods for computing the inverse z-transform $x(kT)$ of $X(z)$. The first method is based on the expansion of $X(z)$ into a power series in z^{-1}, while the second method is based on a partial-fraction expansion.

EXPANSION BY LONG DIVISION   Let $X(z)$ be given in the rational form (7.47) with both $N(z)$ and $D(z)$ written in descending powers of z. To compute the inverse z-transform $x(kT)$ for a finite range of values of k, we can expand $X(z)$ into a power series in z^{-1} by dividing $D(z)$ into $N(z)$ using long division. The values of the function $x(kT)$ are then "read off" from the coefficients of the power-series expansion. The process is illustrated by the following example.

EXAMPLE 7.15.   Suppose that

$$X(z) = \frac{z^2 - 1}{z^3 + 2z + 4}.$$

Dividing $D(z)$ into $N(z)$, we obtain

$$
\begin{array}{r}
z^{-1} + 0z^{-2} - 3z^{-3} - 4z^{-4} + \cdots \\
\hline
z^3 + 2z + 4 \,\overline{)\, z^2 \quad\quad - 1} \\
z^2 \quad + 2 \quad + 4z^{-1} \\
\hline
- 3 \quad - 4z^{-1} \\
- 3 \quad\quad\quad - 6z^{-2} - 12z^{-3} \\
\hline
- 4z^{-1} + 6z^{-2} + 12z^{-3} \\
- 4z^{-1} \quad\quad\quad - 8z^{-3} - 16z^{-4} \\
\hline
6z^{-2} + 20z^{-3} + 16z^{-4} \\
\vdots
\end{array}
$$

Thus

$$X(z) = z^{-1} - 3z^{-3} - 4z^{-4} + \cdots . \qquad (7.48)$$

By definition of the z-transform,

$$X(z) = x(0) + x(T)z^{-1} + x(2T)z^{-2} + \cdots . \qquad (7.49)$$

Equating (7.48) and (7.49), we have

$$x(0) = 0, \quad x(T) = 1, \quad x(2T) = 0, \quad x(3T) = -3, \quad x(4T) = -4, \quad \cdots.$$

A program for computing the inverse z-transform by long division is given in Figure A.8. Given a positive integer Q, the program computes $x(kT)$ for $k = 0, 1, 2, \ldots, Q$. The user inputs the coefficients $b(i)$ of $N(z)$ and the coefficients $a(i)$ of $D(z)$, where

$$X(z) = \frac{N(z)}{D(z)} = \frac{b(n)z^n + b(n-1)z^{n-1} + \cdots + b(0)}{a(n)z^n + a(n-1)z^{n-1} + \cdots + a(0)}.$$

For example, running the program with $X(z)$ equal to the rational function in Example 7.15 and with $Q = 20$, we obtain the results displayed in Table 7.4. Note that the values of $|x(kT)|$ appear to be growing without bound as k increases.

TABLE 7.4. Inverse z-transform of the Rational Function in Example 7.15

k	$x(kT)$
0	0
1	1
2	0
3	-3
4	-4
5	6
6	20
7	4
8	-64
9	-88
10	112
11	432
12	128
13	$-1,312$
14	$-1,984$
15	2,112
16	9,216
17	3,712
18	$-26,880$
19	$-44,288$
20	38,912

Initial Values: From the results in Example 7.15, we see that by carrying out the first few steps of the expansion of $X(z)$ using long division, we obtain the initial values $x(0), x(T), \ldots$ of the signal $x(kT)$. In particular, note that the initial value $x(0)$ is nonzero if and only if the degree of $N(z)$ is equal to the degree of $D(z)$. If the degree of $N(z)$ is strictly less than the degree of $D(z)$ minus 1, both $x(0)$ and $x(T)$ are zero, and so on.

INVERSION VIA PARTIAL-FRACTION EXPANSION   Using long division as described above, we can calculate $x(kT)$ for any finite range of integer values of k. However, if we want to determine an analytical expression for $x(kT)$ that is valid for all integers $k \geq 0$, it is necessary to use partial-fraction expansions as we did in the case of rational Laplace transforms. The steps are as follows.

Again suppose that $X(z)$ is given in the rational form

$$X(z) = \frac{N(z)}{D(z)}.$$

If the degree of $N(z)$ is equal to the degree of $D(z)$, the partial-fraction expansion described in Section 5.3 cannot be directly applied to $X(z)$. However, by dividing $D(z)$ into $N(z)$, one can write $X(z)$ in the form

$$X(z) = x(0) + \frac{R(z)}{D(z)},$$

where $x(0)$ is the initial value of the signal $x(kT)$ at time $kT = 0$, and where the degree of $R(z)$ is strictly less than the degree of $D(z)$. The rational function $R(z)/D(z)$ can then be expanded by partial fractions.

There is another approach that avoids having to divide $D(z)$ into $N(z)$; namely, we can first consider expanding

$$\frac{X(z)}{z} = \frac{N(z)}{zD(z)}.$$

The rational function $X(z)/z$ can be expanded into partial fractions since the degree of $N(z)$ is strictly less than the degree of $zD(z)$ in the case when $N(z)$ and $D(z)$ have the same degrees. After $X(z)/z$ has been expanded, we can multiply by z and then compute the inverse z-transform term by term. There are two cases to consider.

Distinct Poles.   Let $p_1, p_2, \ldots, p_n$ denote the poles of $X(z)$. The poles can be computed by using the root-finding program in Figure A.6. We assume that the poles are distinct and are all nonzero. Then $X(z)/z$ has the partial-fraction expansion

$$\frac{X(z)}{z} = \frac{c_0}{z} + \frac{c_1}{z - p_1} + \frac{c_2}{z - p_2} + \cdots + \frac{c_n}{z - p_n}, \qquad (7.50)$$

where

$$c_0 = \left[z \frac{X(z)}{z} \right]_{z=0} = X(0)$$

and

$$c_i = \left[(z - p_i) \frac{X(z)}{z} \right]_{z=p_i}, \qquad i = 1, 2, \ldots, n.$$

Multiplying both sides of (7.50) by z, we obtain

$$X(z) = c_0 + \frac{c_1 z}{z - p_1} + \frac{c_2 z}{z - p_2} + \cdots + \frac{c_n z}{z - p_n}. \qquad (7.51)$$

Now using the transform pairs in Table 7.3, we have that the inverse z-transform of c_0 is equal to $c_0 \Delta(kT)$, where $\Delta(kT)$ is the unit pulse located at $kT = 0$, and the inverse z-transform of

$$\frac{c_i z}{z - p_i}$$

is equal to $c_i p_i^k$, $k \geq 0$. Taking the inverse z-transform term by term of the right side of (7.51), we get

$$x(kT) = c_0 \Delta(kT) + c_1 p_1^k + c_2 p_2^k + \cdots \qquad (7.52)$$
$$+ c_n p_n^k, \qquad k = 0, 1, 2, \ldots.$$

If one of the poles $p_1, p_2, \ldots, p_n$ of $X(z)$ is complex, it is necessary to rewrite (7.52) in real form. This can be accomplished as follows.

Suppose that p_1 is complex; that is, $p_1 = a + jb$, $b \neq 0$. Then one of the other poles of $X(z)$ must be equal to the complex conjugate $\bar{p}_1$ of p_1. Let us suppose that $p_2 = \bar{p}_1$. Then in the expansion (7.52), it must be true that $c_2 = \bar{c}_1$. Hence the second and third terms on the right side of (7.52) are equal to

$$c_1 p_1^k + \bar{c}_1 \bar{p}_1^k, \qquad k = 0, 1, 2, \ldots. \qquad (7.53)$$

We write (7.53) in real form. Let $p_1 = \rho e^{j\theta}$ denote the polar form of the complex number p_1. Then $\bar{p}_1 = \rho e^{-j\theta}$, and

$$p_1^k = \rho^k e^{jk\theta}$$

$$\bar{p}_1^k = \rho^k e^{-jk\theta}.$$

Therefore,

$$c_1 p_1^k + \bar{c}_1 \bar{p}_1^k = \rho^k (c_1 e^{jk\theta} + \bar{c}_1 e^{-jk\theta}). \qquad (7.54)$$

Using Euler's formula, we have

$$c_1 e^{jk\theta} + \bar{c}_1 e^{-jk\theta} = c_1 \cos k\theta + jc_1 \sin k\theta + \bar{c}_1 \cos k\theta - j\bar{c}_1 \sin k\theta$$

$$= (c_1 + \bar{c}_1) \cos k\theta + j(c_1 - \bar{c}_1) \sin k\theta$$

$$= 2(\text{Re } c_1) \cos k\theta - 2 (\text{Im } c_1) \sin k\theta. \tag{7.55}$$

Inserting (7.55) into (7.54) gives

$$c_1 p_1^k + \bar{c}_1 \bar{p}_1^k = 2\rho^k[(\text{Re } c_1) \cos k\theta - (\text{Im } c_1) \sin k\theta]. \tag{7.56}$$

Finally, using the trigonometric identity (5.68), we can rewrite (7.56) in the form

$$c_1 p_1^k + \bar{c}_1 \bar{p}_1^k = 2|c_1|\rho^k \cos (k\theta + \sphericalangle c_1), \qquad k = 0, 1, 2, \ldots . \tag{7.57}$$

EXAMPLE 7.16. Suppose that

$$X(z) = \frac{z^3 + 1}{z^3 - z^2 - z - 2}.$$

Here

$$D(z) = z^3 - z^2 - z - 2$$

$$= \left(z + \frac{1}{2} + j\frac{\sqrt{3}}{2}\right)\left(z + \frac{1}{2} - j\frac{\sqrt{3}}{2}\right)(z - 2).$$

The roots of $D(z) = 0$ are

$$p_1 = -\frac{1}{2} - j\frac{\sqrt{3}}{2}$$

$$p_2 = \bar{p}_1 = -\frac{1}{2} + j\frac{\sqrt{3}}{2}$$

$$p_3 = 2.$$

Expanding $X(z)/z$, we have

$$\frac{X(z)}{z} = \frac{c_0}{z} + \frac{c_1}{z + \dfrac{1}{2} + j\dfrac{\sqrt{3}}{2}} + \frac{\bar{c}_1}{z + \dfrac{1}{2} - j\dfrac{\sqrt{3}}{2}} + \frac{c_3}{z - 2},$$

where

$$c_0 = X(0) = \frac{-1}{2}$$

$$c_1 = \left[\left(z + \frac{1}{2} + j\frac{\sqrt{3}}{2} \right) \frac{X(z)}{z} \right]_{z=-\frac{1}{2}-j\frac{\sqrt{3}}{2}} = \frac{3}{7} + j\frac{\sqrt{3}}{21}$$

$$c_3 = \left[(z - 2) \frac{X(z)}{z} \right]_{z=2} = \frac{9}{14}.$$

From (7.52) we have that the inverse z-transform $x(kT)$ of $X(z)$ is

$$x(kT) = -\frac{1}{2} \Delta(kT) + c_1 \left(-\frac{1}{2} - j\frac{\sqrt{3}}{2} \right)^k$$

$$+ \bar{c}_1 \left(-\frac{1}{2} + j\frac{\sqrt{3}}{2} \right)^k + \frac{9}{14} 2^k, \qquad k \geq 0 \tag{7.58}$$

To write the second and third terms of $x(kT)$ in real form, we can use (7.57). Here the polar form of p_1 is

$$p_1 = (1) \exp\left(\frac{j4\pi}{3} \right),$$

so $\rho = 1$ and $\theta = 4\pi/3$ rad. Now

$$|c_1| = \sqrt{\frac{9}{49} + \frac{3}{441}} = \sqrt{\frac{84}{441}} = \frac{2}{\sqrt{21}} = 0.436$$

and

$$\angle c_1 = \tan^{-1} \frac{7}{21\sqrt{3}} = 10.89°.$$

Then rewriting the second and third terms of (7.58) using (7.57), we obtain

$$x(kT) = -\frac{1}{2} \Delta(kT) + 0.873 \cos\left(\frac{4\pi}{3} k + 10.89° \right)$$

$$+ \frac{9}{14} (2)^k, \qquad k \geq 0.$$

Repeated Poles. Again let $p_1, p_2, \ldots, p_n$ denote the n poles of $X(z) = N(z)/D(z)$. Then the $n + 1$ poles of $X(z)/z$ are $0, p_1, p_2, \ldots, p_n$. Suppose that one of the poles of $X(z)/z$ is repeated r times and that the other $n + 1 -$

r poles of $X(z)/z$ are distinct. Let us assume that the pole p_1 of $X(z)/z$ is the repeated pole. Then $X(z)/z$ has the partial-fraction expansion

$$\frac{X(z)}{z} = \frac{c_0}{z} + \frac{c_1}{z - p_1} + \frac{c_2}{(z - p_1)^2} + \cdots + \frac{c_r}{(z - p_1)^r} + \frac{c_{r+1}}{z - p_{r+1}}$$
$$+ \cdots + \frac{c_n}{z - p_n}. \tag{7.59}$$

In (7.59), the constants $c_0, c_1, \ldots, c_n$ are given by

$$c_0 = X(0)$$

$$c_i = \left[(z - p_i) \frac{X(z)}{z} \right]_{z=p_i}, \qquad i = r + 1, r + 2, \ldots, n$$

$$c_r = \left[(z - p_1)^r \frac{X(z)}{z} \right]_{z=p_1}$$

$$c_{r-1} = \left[\frac{d}{dz} (z - p_1)^r \frac{X(z)}{z} \right]_{z=p_1}$$

$$c_{r-2} = \frac{1}{2!} \left[\frac{d^2}{dz^2} (z - p_1)^r \frac{X(z)}{z} \right]_{z=p_1}$$

$$\cdot$$
$$\cdot$$
$$\cdot$$

$$c_{r-i} = \frac{1}{i!} \left[\frac{d^i}{dz^i} (z - p_1)^r \frac{X(z)}{z} \right]_{z=p_1}.$$

Multiplying both sides of (7.59) by z, we have

$$X(z) = c_0 + \frac{c_1 z}{z - p_1} + \frac{c_2 z}{(z - p_1)^2} + \cdots + \frac{c_r z}{(z - p_1)^r}$$

$$+ \frac{c_{r+1} z}{z - p_{r+1}} + \cdots + \frac{c_n z}{z - p_n}.$$

The inverse z-transform of the terms

$$\frac{c_i z}{(z - p_1)^i} \tag{7.60}$$

can be computed for $i = 2, 3$ using the transform pairs in Table 7.3. This results in the transform pairs

$$c_2 k(p_1^{k-1}) \leftrightarrow \frac{c_2 z}{(z - p_1)^2} \tag{7.61}$$

$$\frac{1}{2} c_3 k(k - 1)(p_1^{k-2})u(kT - T) \leftrightarrow \frac{c_3 z}{(z - p_1)^3}. \tag{7.62}$$

The inverse transform of (7.60) for $i > 3$ can be computed by repeatedly using the multiplication by k property of the z-transform. This results in the transform pair

$$\frac{c_i}{(i - 1)!} k(k - 1) \cdots (k - i + 2)(p_1^{k-i+1})u(kT - iT + 2T)$$

$$\leftrightarrow \frac{c_i z}{(z - p_1)^i}, \qquad i > 3.$$

EXAMPLE 7.17. Suppose that

$$X(z) = \frac{6z^3 + 2z^2 - z}{z^3 - z^2 - z + 1}.$$

Then

$$\frac{X(z)}{z} = \frac{6z^2 + 2z - 1}{z^3 - z^2 - z + 1} = \frac{6z^2 + 2z - 1}{(z - 1)^2(z + 1)}.$$

Note that we have canceled the common factor z in the numerator and denominator of $X(z)/z$. To eliminate unnecessary computations, any common factors should be canceled before performing a partial-fraction expansion. Now the poles of $X(z)/z$ are $p_1 = 1$, $p_2 = 1$, $p_3 = -1$. We then have the partial-fraction expansion

$$\frac{X(z)}{z} = \frac{c_1}{z - 1} + \frac{c_2}{(z - 1)^2} + \frac{c_3}{z + 1},$$

where

$$c_2 = \left[(z - 1)^2 \frac{X(z)}{z} \right]_{z=1} = \frac{6 + 2 - 1}{2} = 3.5$$

$$c_1 = \left[\frac{d}{dz} (z - 1)^2 \frac{X(z)}{z} \right]_{z=1} = \left[\frac{d}{dz} \frac{6z^2 + 2z - 1}{z + 1} \right]_{z=1}$$

$$= \frac{(z + 1)(12z + 2) - (6z^2 + 2z - 1)(1)}{(z + 1)^2} \Big|_{z=1}$$

$$= \frac{2(14) - 7}{4} = 5.25$$

$$c_3 = \left[(z + 1) \frac{X(z)}{z} \right]_{z=-1} = \frac{6 - 2 - 1}{(-2)^2} = 0.75.$$

Hence

$$X(z) = \frac{5.25z}{z - 1} + \frac{3.5z}{(z - 1)^2} + \frac{0.75z}{z + 1}.$$

Using the transform pair (7.61), we obtain the inverse z-transform

$$x(kT) = 5.25(1)^k + 3.5k(1)^{k-1} + 0.75(-1)^k, \qquad k = 0, 1, 2, \ldots$$
$$= 5.25 + 3.5k + 0.75(-1)^k, \qquad k = 0, 1, 2, \ldots.$$

SECOND-ORDER CASE WITH COMPLEX POLES As in the Laplace transform theory, if $X(z)$ is a second-order rational function in z with complex poles, we can determine the inverse z-transform without having to perform a partial-fraction expansion. We shall derive a general expression for the inverse transform in the second-order case.

Let us first consider the case when

$$X(z) = \frac{cz^2 + dz}{z^2 + ez + h},$$

where c, d, e, and h are arbitrary real numbers. From the quadratic formula, the poles of $X(z)$ are complex if and only if

$$e^2 - 4h < 0$$

or

$$e^2 < 4h. \tag{7.63}$$

Since e^2 is always nonnegative, the condition (7.63) implies that $h > 0$. In addition, if $h > 0$, (7.63) implies that $e^2/4h < 1$. Thus (7.63) implies that

$$h > 0 \qquad \text{and} \qquad \frac{e^2}{4h} < 1. \tag{7.64}$$

Conversely, it is easy to see that (7.64) implies (7.63), and hence (7.63) and (7.64) are equivalent. Therefore, the poles of $X(z)$ are complex if and only if (7.64) is satisfied. Let us assume that this is the case. Then h has a real square root $\sqrt{h}$ and we can write $X(z)$ in the form

$$X(z) = \frac{c(z/\sqrt{h})^2 + (d/\sqrt{h})(z/\sqrt{h})}{(z/\sqrt{h})^2 + (e/\sqrt{h})(z/\sqrt{h}) + 1}. \tag{7.65}$$

Now define the rational function

$$W(z) = \frac{cz^2 + (d/\sqrt{h})z}{z^2 + (e/\sqrt{h})z + 1}.$$

Then, by (7.65),

$$X(z) = W\left(\frac{z}{\sqrt{h}}\right). \tag{7.66}$$

We shall derive an expression for the inverse z-transform of $W(z)$, and then using (7.66) we will be able to compute the inverse transform of $X(z)$.

Since $e^2/4h < 1$, it must be true that $e/2\sqrt{h} < 1$, and we can define the angle θ by

$$\theta = \cos^{-1}\frac{-e}{2\sqrt{h}}. \tag{7.67}$$

Then rewriting $W(z)$ in terms of $\cos\theta$, we have

$$
\begin{aligned}
W(z) &= \frac{cz^2 + (d/\sqrt{h})z}{z^2 - (2\cos\theta)z + 1} \\
&= \frac{c(z^2 - (\cos\theta)z) + \left[\dfrac{(d/\sqrt{h}) + c\cos\theta}{\sin\theta}\right](\sin\theta)z}{z^2 - (2\cos\theta)z + 1}
\end{aligned}
$$

From the transform pairs in Table 7.3, the inverse z-transform $w(kT)$ of $W(z)$ is given by

$$w(kT) = c\cos k\theta + \left[\frac{(d/\sqrt{h}) + c\cos\theta}{\sin\theta}\right]\sin k\theta, \qquad k = 0, 1, 2, \ldots. \tag{7.68}$$

Using (7.66) and the multiplication by a^k property, we have

$$x(kT) = (\sqrt{h})^k w(kT). \tag{7.69}$$

Inserting the expression (7.68) for $w(kT)$ into (7.69), we obtain

$$x(kT) = (\sqrt{h})^k \left[c\cos k\theta + \left[\frac{(d/\sqrt{h}) + c\cos\theta}{\sin\theta}\right]\sin k\theta\right], \tag{7.70}$$
$$k = 0, 1, 2, \ldots.$$

Now suppose that

$$X(z) = \frac{az^2 + bz + f}{z^2 + ez + h},$$

where a, b, f, e, and h are arbitrary constants. We can reduce this case to the above case by dividing $h + ez + z^2$ into $f + bz + az^2$. This gives

$$X(z) = \frac{f}{h} + \frac{cz^2 + dz}{z^2 + ez + h},$$

where

$$c = a - \frac{f}{h} \quad \text{and} \quad d = b - \frac{ef}{h}. \tag{7.71}$$

We again assume that (7.64) is satisfied. Then using (7.70), we have the inverse transform

$$x(kT) = \frac{f}{h} \Delta(kT)$$

$$+ (\sqrt{h})^k \left\{ c \cos k\theta + \left[\frac{(d/\sqrt{h}) + c \cos \theta}{\sin \theta} \right] \sin k\theta \right\}, \qquad k \geq 0, \tag{7.72}$$

where θ is given by (7.67) and c, d are given by (7.71). Equation (7.72) is the inverse z-transform of a general second-order rational function with complex poles.

EXAMPLE 7.18. Let

$$X(z) = \frac{z^2 + 2z + 2}{z^2 + 2z + 4}.$$

Here $e = 2$, $h = 4$, and thus $e^2 = 4 < 16 = 4h$. Therefore, the poles of $X(z)$ are complex. Dividing $4 + 2z + z^2$ into $2 + 2z + z^2$, we get

$$X(z) = \frac{2}{4} + \frac{(1 - \frac{2}{4})z^2 + (2 - (2)(2)/4)z}{z^2 + 2z + 4}$$

$$= \frac{1}{2} + \frac{0.5z^2 + z}{z^2 + 2z + 4}.$$

Now

$$\theta = \cos^{-1}\left(\frac{-2}{4}\right) = \frac{2\pi}{3} \text{ rad.}$$

Then, from (7.72),

$$x(kT) = 0.5\Delta(kT)$$

$$+ 2^k\left[0.5\cos\left(\frac{2\pi k}{3}\right) + \frac{(\frac{1}{2}) + 0.5(-0.5)}{0.866}\sin\left(\frac{2\pi k}{3}\right)\right] \quad k \geq 0$$

$$= 0.5\Delta(kT)$$

$$+ 2^k\left[0.5\cos\left(\frac{2\pi k}{3}\right) + 0.289\sin\left(\frac{2\pi k}{3}\right)\right], \quad k \geq 0.$$

7.4

TRANSFER FUNCTION REPRESENTATION

Recall from Chapter 5 that we were able to generate the transfer function representation of a linear time-invariant continuous-time system by taking the Laplace transform of the input/output convolution integral representation. In this section we consider the discrete-time counterpart to the transfer function model, which can be generated by taking the z-transform of the input/output convolution sum representation. As will be seen, the transfer function approach to linear time-invariant discrete-time systems closely resembles the transfer function approach to linear time-invariant continuous-time systems.

Suppose that we have a single-input single-output causal linear time-invariant discrete-time system given by the input/output relationship

$$y(kT) = (Fx)(kT), \tag{7.73}$$

where F is the input/output operator of the system. The output response $y(kT)$ given by (7.73) is the response resulting from input $x(kT)$ with no initial energy in the system prior to the application of $x(kT)$.

As shown in Chapter 4, the input/output operator F is a convolution operator given by

$$y(kT) = (Fx)(kT) = (h * x)(kT) = \sum_{i=0}^{\infty} h(iT)x(kT - iT), \tag{7.74}$$

where $h(kT)$ is the unit-pulse response of the system (see Section 4.2). If the input $x(kT)$ is zero for all integers $k < 0$, the sum in (7.74) may be taken from $i = 0$ to $i = k$, so that

$$y(kT) = (h * x)(kT) = \sum_{i=0}^{k} h(iT)x(kT - iT). \tag{7.75}$$

Taking the z-transform of both sides of (7.75), we have that

$$Y(z) = H(z)X(z), \tag{7.76}$$

where $Y(z)$ is the z-transform of the output $y(kT)$, $H(z)$ is the transform of the unit-pulse response $h(kT)$, and $X(z)$ is the transform of the input $x(kT)$. In going from (7.75) to (7.76), we used the property that the z-transform of the convolution of two discrete-time signals is equal to the product of the transforms of the two signals.

Equation (7.76) is the transfer function representation of the given discrete-time system and $H(z)$ is the transfer function of the system. From (7.76) we see that the transfer function $H(z)$ represents the "transfer" from the input to the output in the z-domain, assuming that there is no initial energy in the system at time $kT = 0$.

Since the transfer function $H(z)$ is the z-transform of the unit-pulse response $h(kT)$, we have the transform pair

$$h(kT) \leftrightarrow H(z).$$

This transform pair is a major link between the time domain and the z-domain in the theory of linear time-invariant discrete-time systems.

If the input $x(kT)$ is not zero for at least one positive value of k, so that $X(z) \neq 0$, we can divide both sides of (7.76) by $X(z)$, which gives

$$H(z) = \frac{Y(z)}{X(z)}. \tag{7.77}$$

Thus the transfer function $H(z)$ is equal to the ratio of the transforms of the output and input. Note that since $H(z)$ is unique, the ratio $Y(z)/X(z)$ cannot change as the input $x(kT)$ ranges over some collection of input signals. From (7.77) we also see that $H(z)$ can be determined from the output response to any input that is not identically zero for $k \geq 0$.

COMPUTATION OF OUTPUT RESPONSE   Taking the inverse z-transform of both sides of (7.76), we have that

$$y(kT) = \text{inverse } z\text{-transform of } H(z)X(z). \tag{7.78}$$

Equation (7.78) can be used to compute the output response $y(kT)$ resulting from input $x(kT)$ with no initial energy in the system at time $kT = 0$. In particular, if $H(z)$ and $X(z)$ are rational functions in z, so that $H(z)X(z)$ is a rational function in z, we can compute a closed-form expression for $y(kT)$ by first expanding $H(z)X(z)$ [or $H(z)X(z)/z$] by partial fractions. The computations are illustrated by the following example.

EXAMPLE 7.19.   A linear time-invariant discrete-time system has unit-pulse response

$$h(k) = 3(2^{-k}) \cos\left(\frac{\pi k}{6} + \frac{\pi}{12}\right), \qquad k = 0, 1, 2, \ldots \tag{7.79}$$

Here $T = 1$ and the argument of the cosine in (7.79) is in radians. Let us first determine the transfer function $H(z)$ of the system, which is equal to the z-transform of $h(kT)$. We can compute $H(z)$ from the transform pairs in Table 7.3. First, we need to expand the cosine in (7.79). This gives

$$
h(k) = 3(2^{-k}) \left[\cos\left(\frac{\pi k}{6}\right) \cos\left(\frac{\pi}{12}\right) - \sin\left(\frac{\pi k}{6}\right) \sin\left(\frac{\pi}{12}\right) \right], \quad k \geq 0
$$

$$
= 2.898 \left(\frac{1}{2}\right)^k \cos\left(\frac{\pi k}{6}\right) - 0.776 \left(\frac{1}{2}\right)^k \sin\left(\frac{\pi k}{6}\right), \quad k \geq 0.
$$

$$(7.80)$$

Taking the z-transform of (7.80), we obtain

$$
H(z) = 2.898 \frac{z^2 - [0.5 \cos(\pi/6)]z}{z^2 - [\cos(\pi/6)]z + 0.25}
$$

$$
- 0.776 \frac{[0.5 \sin(\pi/6)]z}{z^2 - [\cos(\pi/6)]z + 0.25}
$$

$$
= \frac{2.898z^2 - 1.449z}{z^2 - 0.866z + 0.25}.
$$

$$(7.81)$$

Now let us compute the step response $g(k)$ of the system. Recall that $g(k)$ is the output response when the input $x(k)$ is the discrete-time step function $u(k)$ with no initial energy in the system at time $k = 0$. The z-transform $G(z)$ of the step response is given by

$$
G(z) = H(z)U(z) = H(z) \frac{z}{z - 1}.
$$

$$(7.82)$$

Inserting the expression (7.81) for the transfer function into (7.82), we have

$$
G(z) = \frac{2.898z^3 - 1.449z^2}{(z - 1)(z^2 - 0.866z + 0.25)}.
$$

We can expand $G(z)/z$ in the form

$$
\frac{G(z)}{z} = \frac{\dfrac{2.898 - 1.449}{1 - 0.866 + 0.25}}{z - 1} + \frac{cz + d}{z^2 - 0.866z + 0.25}
$$

$$
= \frac{3.773}{z - 1} + \frac{cz + d}{z^2 - 0.866z + 0.25}
$$

$$(7.83)$$

Putting the right side of (7.83) over a common denominator and equating coefficients of polynomials gives

$$c = -0.875, \qquad d = 0.943.$$

Then multiplying both sides of (7.83) by z, we have

$$G(z) = \frac{3.773z}{z - 1} + \frac{-0.875z^2 + 0.943z}{z^2 - 0.866z + 0.25}. \qquad (7.84)$$

The inverse z-transform of the second term on the right side of (7.84) can be computed by using (7.70) with

$$e = -0.866, \qquad h = 0.25.$$

Taking the inverse z-transform of (7.84), we have

$$
\begin{aligned}
g(k) &= 3.773 + \left(\frac{1}{2}\right)^k \left\{ -0.875 \cos\left(\frac{\pi k}{6}\right) \right. \\
&\quad \left. + \left[\frac{(0.943/0.5) - 0.757}{.5} \right] \sin\left(\frac{\pi k}{6}\right) \right\}, \qquad k \geq 0 \\
&= 3.773 + \left(\frac{1}{2}\right)^k \left[-0.875 \cos\left(\frac{\pi k}{6}\right) \right. \\
&\quad \left. + 2.258 \sin\left(\frac{\pi k}{6}\right) \right], \qquad k \geq 0.
\end{aligned}
$$

7.5

TRANSFORM OF INPUT/OUTPUT DIFFERENCE EQUATION

In this section we consider finite-dimensional discrete-time systems specified by a linear constant-coefficient input/output difference equation. We will show that the transfer function model of a finite-dimensional linear time-invariant discrete-time system can be constructed directly from the input/output difference equation. The results are analogous to those given in Section 5.6 on the transformation of input/output differential equations. We begin with the first-order case.

FIRST-ORDER CASE   Consider the linear time-invariant finite-dimensional discrete-time system given by the first-order input/output difference equation

$$y(kT + T) + ay(kT) = b_1 x(kT + T) + b_0 x(kT). \qquad (7.85)$$

If we take the z-transform of (7.85) and use the left-shift property of the transform, the initial conditions will be at time $kT = 0$. We prefer to have the initial conditions at negative values of time, so before transforming (7.85), we need to shift both sides of the equation to the right by T seconds. This is equivalent to replacing kT by $kT - T$ in (7.85), which gives

$$y(kT) + ay(kT - T) = b_1 x(kT) + b_0 x(kT - T). \tag{7.86}$$

Taking the z-transform of both sides of (7.86) and using the right-shift property of the transform, we obtain

$$Y(z) + a[z^{-1}Y(z) + y(-T)] = b_1 X(z) + b_0[z^{-1}X(z) + x(-T)], \tag{7.87}$$

where $Y(z)$ is the z-transform of the output response $y(kT)$ and $X(z)$ is the z-transform of the input $x(kT)$. Solving (7.87) for $Y(z)$, we have

$$Y(z) = \frac{-ay(-T) + b_0 x(-T)}{1 + az^{-1}} + \frac{b_1 + b_0 z^{-1}}{1 + az^{-1}} X(z). \tag{7.88}$$

Multiplying the terms on the right side of (7.88) by z/z, we obtain

$$Y(z) = \frac{-ay(-T)z + b_0 x(-T)z}{z + a} + \frac{b_1 z + b_0}{z + a} X(z). \tag{7.89}$$

Equation (7.89) is the z-domain representation of the discrete-time system defined by the input/output difference equation (7.85). The first term on the right side of (7.89) is the z-transform of the part of the output response resulting from the initial conditions $y(-T)$, $x(-T)$, and the second term on the right side of (7.89) is the z-transform of the part of the output response resulting from the input $x(kT)$ applied for $k = 0, 1, 2, \ldots$.

If $x(-T) = 0$, (7.89) reduces to

$$Y(z) = \frac{-ay(-T)z}{z + a} + \frac{b_1 z + b_0}{z + a} X(z). \tag{7.90}$$

If $x(-T) = 0$, the system has no initial energy at time $kT = 0$ if and only if $ay(-T) = 0$, in which case (7.90) reduces to

$$Y(z) = \frac{b_1 z + b_0}{z + a} X(z). \tag{7.91}$$

From the results in Section 7.4, we know that when there is no initial energy in the system at time $kT = 0$, the transform $Y(z)$ is given by

$$Y(z) = H(z)X(z), \tag{7.92}$$

where $H(z)$ is the transfer function of the system. Comparing (7.91) and (7.92), we see that the transfer function is

$$H(z) = \frac{b_1 z + b_0}{z + a}. \tag{7.93}$$

Dividing the transfer function $H(z)$ by z and expanding by partial fractions, we have that

$$\frac{H(z)}{z} = \frac{b_0/a}{z} + \frac{b_1 - (b_0/a)}{z + a}.$$

Then

$$H(z) = \frac{b_0}{a} + \frac{[b_1 - (b_0/a)]z}{z + a},$$

and taking the inverse z-transform, we have that the unit pulse $h(kT)$ is

$$h(kT) = \frac{b_0}{a} \Delta(kT) + \left(b_1 - \frac{b_0}{a}\right)(-a)^k, \qquad k = 0, 1, 2, \ldots . \quad (7.94)$$

Equation (7.94) is the general form of the unit-pulse response of a first-order discrete-time system.

EXAMPLE 7.20. Consider the unit delayer which was defined in Section 3.2. The input/output difference equation of the unit delayer is

$$y(kT + T) = x(kT). \quad (7.95)$$

The difference equation (7.95) is a special case of (7.85) with $a = 0$, $b_1 = 0$, and $b_0 = 1$. Thus from (7.93), the transfer function of the unit delayer is

$$H(z) = \frac{0z + 1}{z + 0} = \frac{1}{z}.$$

Note that the unit delayer is the discrete-time counterpart to the integrator in the sense that the unit delayer has transfer function $1/z$ and the integrator has transfer function $1/s$.

SECOND-ORDER CASE Now consider the discrete-time system given by the second-order input/output difference equation

$$y(kT + 2T) + a_1 y(kT + T) + a_0 y(kT) \quad (7.96)$$
$$= b_2 x(kT + 2T) + b_1 x(kT + T) + b_0 x(kT).$$

Shifting (7.96) to the right by $2T$ seconds (i.e., replacing kT by $kT - 2T$), we get

$$y(kT) + a_1 y(kT - T) + a_0 y(kT - 2T) \quad (7.97)$$
$$= b_2 x(kT) + b_1 x(kT - T) + b_0 x(kT - 2T).$$

We assume that $x(-2T) = 0$ and $x(-T) = 0$. Then applying the z-transform to both sides of (7.97), we obtain

$$Y(z) + a_1[z^{-1}Y(z) + y(-T)] + a_0[z^{-2}Y(z) + z^{-1}y(-T) + y(-2T)]$$
$$= b_2X(z) + b_1z^{-1}X(z) + b_0z^{-2}X(z). \tag{7.98}$$

Solving (7.98) for $Y(z)$, we have

$$Y(z) = \frac{[-a_1y(-T) - a_0y(-2T)] - a_0y(-T)z^{-1}}{1 + a_1z^{-1} + a_0z^{-2}} \tag{7.99}$$
$$+ \frac{b_2 + b_1z^{-1} + b_0z^{-2}}{1 + a_1z^{-1} + a_0z^{-2}} X(z).$$

Multiplying the terms on the right side of (7.99) by z^2/z^2, we have

$$Y(z) = \frac{[-a_1y(-T) - a_0y(-2T)]z^2 - a_0y(-T)z}{z^2 + a_1z + a_0} + \frac{b_2z^2 + b_1z + b_0}{z^2 + a_1z + a_0} X(z). \tag{7.100}$$

Equation (7.100) is the z-domain representation of the discrete-time system given by the second-order input/output difference equation (7.96). The first term on the right side of (7.100) is the z-transform of the part of the output response resulting from the initial conditions $y(-T)$, $y(-2T)$, and the second term on the right side of (7.100) is the z-transform of the part of the output response resulting from the input $x(kT)$ applied for $k = 0, 1, 2, \ldots$.

If $x(-T) = 0$ and $x(-2T) = 0$, there is no initial energy in the system at time $kT = 0$ if and only if the first term on the right side of (7.100) is zero, in which case (7.100) reduces to

$$Y(z) = \frac{b_2z^2 + b_1z + b_0}{z^2 + a_1z + a_0} X(z). \tag{7.101}$$

From (7.101) we see that the transfer function $H(z)$ of the system is

$$H(z) = \frac{b_2z^2 + b_1z + b_0}{z^2 + a_1z + a_0}. \tag{7.102}$$

Note that $H(z)$ is a second-order rational function of z.

EXAMPLE 7.21. Consider the discrete-time system given by the input/output difference equation

$$y(k + 2) + y(k + 1) = x(k + 1) - x(k).$$

For this system, $T = 1$. By (7.102) the transfer function of the system is

$$H(z) = \frac{z - 1}{z^2 + z}.$$

Let us compute the step response $g(k)$ of the system. Since there is no initial energy in the system at time $k = 0$,

$$G(z) = H(z)U(z) = \frac{z - 1}{z^2 + z} \frac{z}{z - 1} = \frac{1}{z + 1}.$$

We can compute the inverse z-transform of $G(z)$ using table lookup as follows. From Table 7.3, we have the transform pair

$$(-a)^k \leftrightarrow \frac{z}{z + a}.$$

Now

$$\frac{a}{z + a} = 1 - \frac{z}{z + a},$$

which results in the transform pair

$$\Delta(k) - (-a)^k \leftrightarrow \frac{a}{z + a}. \tag{7.103}$$

Using (7.103) with $a = 1$, we have that the inverse z-transform of $G(z)$ is

$$g(k) = \Delta(k) - (-1)^k, \qquad k = 0, 1, 2, \ldots .$$

Hence we have the step response $g(k)$ of the system. Let us now find the response $y(k)$ resulting from the input

$$x(k) = \left(\frac{1}{2}\right)^k, \qquad k = 0, 1, 2, \ldots$$

$$x(-2) = x(-1) = 0,$$

with the initial conditions $y(-2) = 1$, $y(-1) = 2$. In this case there is initial energy in the system, so $Y(z) \neq H(z)X(z)$. To compute $Y(z)$, we can use the expression (7.100) for $Y(z)$. Inserting the values for a_1, a_0, b_2, b_1, b_0 into (7.100), we obtain

$$Y(z) = \frac{[-(1)(2) - (0)(1)]z^2 - (0)(2)z}{z^2 + z} + \frac{z - 1}{z^2 + z} X(z)$$

$$Y(z) = \frac{-2z}{z+1} + \frac{z-1}{z^2 + z} \frac{z}{z - 1/2}$$

$$= \frac{-2z}{z+1} + \frac{4/3}{z+1} - \frac{1/3}{z - 1/2}.$$

Using the transform pair (7.103), we have

$$y(k) = -2(-1)^k + \tfrac{4}{3}(\Delta(k) - (-1)^k) + 2(\tfrac{1}{3})[\Delta(k) - (\tfrac{1}{2})^k]$$

$$y(k) = -\tfrac{10}{3}(-1)^k + 2\Delta(k) - \tfrac{2}{3}(\tfrac{1}{2})^k,$$

nth-ORDER CASE   Suppose that the discrete-time system under study is specified by the nth-order input/output difference equation

$$y(kT + nT) + \sum_{i=0}^{n-1} a_i y(kT + iT) = \sum_{i=0}^{m} b_i x(kT + iT), \qquad (7.104)$$

where $m \le n$. In the following development, we assume that $x(kT) = 0$ for $k = -1, -2, \ldots, -n$.

Shifting (7.104) to the right by nT seconds and taking the z-transform, we can express the transform $Y(z)$ of the response in the form

$$Y(z) = \frac{C(z)}{D(z)} + \frac{N(z)}{D(z)} X(z), \qquad (7.105)$$

where

$$N(z) = b_m z^m + b_{m-1} z^{m-1} + \cdots + b_1 z + b_0$$

and

$$D(z) = z^n + a_{n-1} z^{n-1} + \cdots + a_1 z + a_0,$$

and where $C(z)$ is a polynomial in z whose coefficients are determined by the initial conditions $y(-T), y(-2T), \ldots, y(-nT)$. As shown above, in the $n = 2$ case, $C(z)$ is given by

$$C(z) = [-a_1 y(-T) - a_0 y(-2T)]z^2 - a_0 y(-T)z.$$

With the assumption that $x(kT) = 0$ for $k = -1, -2, \ldots, -n$, the system defined by (7.104) has no initial energy at time $kT = 0$ if and only if $C(z) = 0$, in which case the expression (7.105) reduces to

$$Y(z) = \frac{N(z)}{D(z)} X(z). \qquad (7.106)$$

From (7.106) we see that the transfer function $H(z)$ of the system is

$$H(z) = \frac{N(z)}{D(z)} = \frac{b_m z^m + b_{m-1} z^{m-1} + \cdots + b_1 z + b_0}{z^n + a_{n-1} z^{n-1} + \cdots + a_1 z + a_0}. \qquad (7.107)$$

By (7.107) the transfer function of a linear time-invariant finite-dimensional discrete-time system is rational in z. Conversely, if a linear time-invariant discrete-time system has the rational transfer function (7.107), the system can be modeled by the input/output difference equation (7.104), and thus the system is finite dimensional. Therefore, a linear time-invariant discrete-time system is finite dimensional if and only if its transfer function is rational in z.

If the transfer function $H(z)$ is rational, we can define the notion of the poles and zeros of the given discrete-time system. In particular, the zeros are the values of z for which $H(z)$ is zero and the poles are the values of z for which $H(z)$ is infinite in magnitude. We can also define the pole–zero diagram of a discrete-time system as we did in the continuous-time case (see Section 5.6).

7.6

TRANSFER FUNCTION OF INTERCONNECTIONS

In the first part of this section we consider the computation of the transfer functions of interconnections of unit delayers and interconnections of subsystems. The results are directly analogous to those derived in Chapter 6 for linear time-invariant continuous-time systems. In the last part of the section we generalize the transfer function framework to multi-input multi-output linear time-invariant discrete-time systems.

INTERCONNECTIONS OF UNIT DELAYERS Suppose that we have a discrete-time system given by an interconnection of unit delayers, adders, subtracters, and scalar multipliers. The transfer function for any such interconnection can be computed by working in the z-domain with the unit delayers represented by their transfer function $1/z$. The procedure is very similar to that considered in Section 6.1 for continuous-time systems consisting of interconnections of integrators.

EXAMPLE 7.22. Consider the discrete-time system given by the interconnection in Figure 7.2. The z-domain representation of the system is shown in Figure 7.3. We have

$$zQ_1(z) = Q_2(z) + X(z) \qquad (7.108)$$

$$zQ_2(z) = Q_1(z) - 3Y(z) \qquad (7.109)$$

$$Y(z) = 2Q_1(z) + Q_2(z). \qquad (7.110)$$

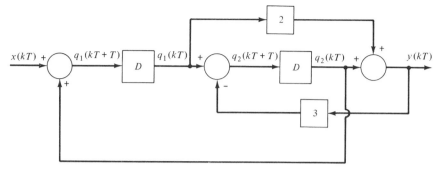

Figure 7.2. System in Example 7.22.

Solving (7.108) for $Q_1(z)$ and inserting the result into (7.109) and (7.110), we have

$$zQ_2(z) = z^{-1}Q_2(z) + z^{-1}X(z) - 3Y(z) \qquad (7.111)$$

$$Y(z) = 2[z^{-1}Q_2(z) + z^{-1}X(z)] + Q_2(z). \qquad (7.112)$$

Solving (7.111) for $Q_2(z)$ and inserting the result into (7.112), we obtain

$$Y(z) = \frac{2z^{-1} + 1}{z - z^{-1}}[z^{-1}X(z) - 3Y(z)] + 2z^{-1}X(z).$$

Then

$$\left[1 + \frac{3(2z^{-1} + 1)}{z - z^{-1}}\right]Y(z) = \frac{z^{-1}(2z^{-1} + 1)}{z - z^{-1}}X(z) + 2z^{-1}X(z)$$

$$\frac{z + 5z^{-1} + 3}{z - z^{-1}}Y(z) = \left[\frac{z^{-1}(2z^{-1} + 1)}{z - z^{-1}} + 2z^{-1}\right]X(z)$$

$$= \frac{z^{-1} + 2}{z - z^{-1}}X(z).$$

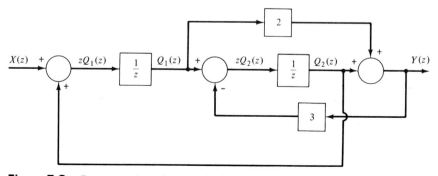

Figure 7.3. Representation of system in the z-domain.

Thus

$$Y(z) = \frac{z^{-1} + 2}{z + 5z^{-1} + 3} X(z) = \frac{2z + 1}{z^2 + 3z + 5} X(z),$$

so the transfer function is

$$H(z) = \frac{2z + 1}{z^2 + 3z + 5}.$$

TRANSFER FUNCTION OF BASIC INTERCONNECTIONS The transfer function of the cascade, parallel, and feedback connections have exactly the same form as in the continuous-time case. The results are displayed in Figure 7.4.

The transfer function equivalences for moving pick-off points and adders or subtracters also have exactly the same form as in the continuous-time case. In particular, simply replace s by z in Figures 6.11 and 6.12 to obtain the equivalences in the discrete-time case.

BLOCK-DIAGRAM REDUCTION Using the equivalences in Figure 7.4 and the equivalences for moving pick-off points and adders/subtracters, we can reduce a given block diagram in order to compute the transfer function of the

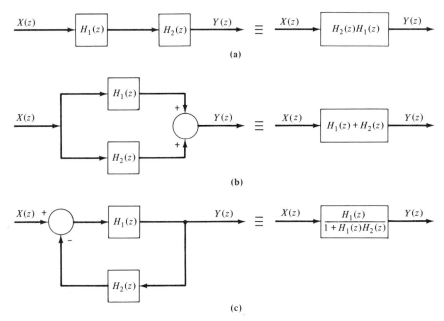

Figure 7.4. Transfer functions of basic interconnections: (a) cascade connection; (b) parallel connection; (c) feedback connection.

system. Again, the process is directly analogous to block-diagram reduction in the continuous-time case (see Section 6.3). The transfer function of a block diagram can also be determined by using Mason's theorem. The procedure is analogous to the continuous-time case discussed in Section 6.3.

MULTI-INPUT MULTI-OUTPUT DISCRETE-TIME SYSTEMS Now suppose that the given discrete-time system has p inputs and q outputs. We assume that the system is linear and time invariant. Then by a generalization of the results in Chapter 4, the response $y_r(kT)$ at the rth output terminal is given by

$$y_r(kT) = \sum_{i=1}^{p} (h_{ri} * x_i)(kT), \qquad r = 1, 2, \ldots, q, \qquad (7.113)$$

where $x_i(kT)$ is the signal applied to the ith input terminal and $h_{ri}(kT)$ is the unit-pulse response from the ith input terminal to the rth output terminal. Here we are assuming that the system has no initial energy prior to the application of the inputs.

Taking the z-transform of both sides of (7.113), we obtain

$$Y_r(z) = \sum_{i=1}^{p} H_{ri}(z)X_i(z), \qquad r = 1, 2, \ldots, q, \qquad (7.114)$$

where $H_{ri}(z)$ is the transfer function from the ith input terminal to the rth output terminal.

The collection of q equations given by (7.114) can be written in matrix form as follows. Let $X(z)$ denote the p-element column vector whose ith entry is $X_i(z)$, and let $Y(z)$ denote the q-element column vector whose rth entry is $Y_r(z)$. Finally, let $H(z)$ denote the q by p matrix whose r, i entry is $H_{ri}(z)$; that is,

$$H(z) = \begin{bmatrix} H_{11}(z) & H_{12}(z) & \cdots & H_{1p}(z) \\ H_{21}(z) & H_{22}(z) & \cdots & H_{2p}(z) \\ \vdots & & & \\ & & & \\ H_{q1}(z) & H_{q2}(z) & \cdots & H_{qp}(z) \end{bmatrix}.$$

Then we can write (7.114) in the matrix form

$$Y(z) = H(z)X(z). \qquad (7.115)$$

Equation (7.115) is the transfer function representation of the given p-input q-output discrete-time system.

APPLICATIONS OF THE TRANSFER FUNCTION MODEL

In addition to giving a method for computing the output response of a discrete-time system, the transfer function model can be applied to many problems in the analysis and design of discrete-time systems. For example, this framework can be used to solve discrete-time versions of the problems studied in Chapter 6 dealing with cascade synthesis, feedback control, and system inversion. In fact, all the results derived in Sections 6.4 to 6.6 have counterparts in the discrete-time case. The details are left to the reader.

It is worth noting that z-transform techniques play a very important role in the analysis and design of digital control systems (also called sampled-data control systems). This theory is used to design microprocessor-implemented controllers for a wide range of applications. A brief introduction to digital control is given in Section 12.5.

PROBLEMS

Chapter 7

7.1   Consider the discrete-time signal $x(kT)$ where

$$x(kT) = \begin{cases} b^k & \text{for } k = 0, 1, 2, \ldots, N - 1 \\ 0 & \text{for all other } k. \end{cases}$$

Here b is an arbitrary real number and N is a positive integer.
(a) For what real values of b does $x(kT)$ possess a z-transform?
(b) For the values of b determined in part (a), show that the z-transform of $x(kT)$ is

$$X(z) = \frac{z^N - b^N}{z^{N-1}(z - b)}.$$

7.2   A discrete-time signal $x(kT)$ has z-transform

$$X(z) = \frac{z}{8z^2 - 2z - 1}.$$

Determine the z-transform $V(z)$ of the following signals.
(a) $v(kT) = x(kT - 4T)u(kT - 4T)$.
(b) $v(kT) = x(kT + 2T)u(kT + 2T)$.
(c) $v(kT) = \cos(2kT)x(kT)$.
(d) $v(kT) = e^{3kT}x(kT)$.
(e) $v(kT) = k^2x(kT)$.
(f) $v(kT) = (x * x)(kT)$.
(g) $v(kT) = x(0) + x(T) + x(2T) + \cdots + x(kT)$.

7.3 Compute the z-transform of the following discrete-time signals. Express your answer as a ratio of polynomials in z whenever possible.
(a) $x(k) = \Delta(k) + 2\Delta(k - 1)$.
(b) $x(k) = 1$ for $k = 0, 1$ and $x(k) = 2$ for all $k \geq 2$ (i.e., $k = 2, 3, 4, \ldots$).
(c) $x(k) = e^{0.5k}u(k) + u(k - 2)$.
(d) $x(k) = e^{0.5k}$ for $k = 0, 1$, and $x(k) = 1$ for all $k \geq 2$.
(e) $x(k) = \left(\sin \dfrac{\pi k}{2} \right) u(k - 2)$.
(f) $x(k) = (0.5)^k k u(k)$
(g) $x(k) = u(k) - ku(k - 1) + (1/3)^k u(k - 2)$.
(h) $x(k) = k$ for $k = 0, 1, 2$, and $x(k) = -k$ for all $k \geq 3$.
(i) $x(k) = (k - 1)u(k) - ku(k - 3)$.

7.4 Let $x(k)$ be a discrete-time signal with $x(k) = 0$ for $k = -1, -2, \ldots$. The signal $x(k)$ is said to be summable if

$$\sum_{k=0}^{\infty} x(k) < \infty.$$

If $x(k)$ is summable, the sum x_{sum} of $x(k)$ is defined by

$$x_{\text{sum}} = \sum_{k=0}^{\infty} x(k).$$

Now suppose that the z-transform $X(z)$ of $x(k)$ can be expressed in the form

$$X(z) = \frac{N(z)}{a_n(z - p_1)(z - p_2) \cdots (z - p_n)},$$

where $N(z)$ is a polynomial in z. By using the final-value theorem, show that if $|p_i| < 1$ for $i = 1, 2, \ldots, n$, $x(k)$ is summable and

$$x_{\text{sum}} = \lim_{z \to 1} X(z).$$

7.5 Using the result in Problem 7.4, compute x_{sum} for the following signals. In each case, assume that $x(k) = 0$ for all $k < 0$.
(a) $x(k) = a^k, |a| < 1$.
(b) $x(k) = k(a^k), |a| < 1$.
(c) $x(k) = a^k \cos \pi k, |a| < 1$.
(d) $x(k) = a^k \sin \dfrac{\pi k}{2}, |a| < 1$.

7.6 Compute the inverse z-transform $x(k)$ of the following transforms. Determine $x(k)$ for all integers $k \geq 0$.
(a) $X(z) = \dfrac{z}{z^2 + 1}$.

(b) $X(z) = \dfrac{z^2}{z^2 + 1}.$

(c) $X(z) = \dfrac{1}{z^2 + 1} + \dfrac{1}{z^2 - 1}.$

(d) $X(z) = \dfrac{z^2}{z^2 + 1} + \dfrac{z}{z^2 - 1}.$

(e) $X(z) = \dfrac{z^2 - 1}{z^2 + 1}.$

(f) $X(z) = \dfrac{z + 2}{(z - 1)(z^2 + 1)}.$

(g) $X(z) = \dfrac{z^2 + 2}{(z - 1)(z^2 + 1)}.$

7.7   Find the inverse z-transform $x(k)$ of the following transforms. Determine $x(k)$ for all k.

(a) $X(z) = \dfrac{z + 0.3}{z^2 + 0.75z + 0.125}.$

(b) $X(z) = \dfrac{5z + 1}{4z^2 + 4z + 1}.$

(c) $X(z) = \dfrac{4z + 1}{z^2 - z + 0.5}.$

(d) $X(z) = \dfrac{z}{16z^2 + 1}.$

(e) $X(z) = \dfrac{2z + 1}{z(10z^2 - z - 2)}.$

7.8   By using the z-transform, compute the convolution $(x * v)(k)$ for all $k \geq 0$ where:

(a) $x(k) = u(k) + 3\Delta(k - 1),\ v(k) = u(k - 2).$

(b) $x(k) = u(k),\ v(k) = ku(k).$

(c) $x(k) = \left(\sin \dfrac{\pi k}{2} \right) u(k),\ v(k) = e^{-k}u(k - 2).$

(d) $x(k) = u(k - 1) + \Delta(k),\ v(k) = e^{-k}u(k) - 2e^{-2k}u(k - 2).$

7.9   A linear time-invariant discrete-time system has unit-pulse response

$$
h(k) = \begin{cases}
\dfrac{1}{k} & \text{for } k = 1, 2, 3 \\[2mm]
k - 2 & \text{for } k = 4, 5 \\[1mm]
0 & \text{for all other } k.
\end{cases}
$$

(a) Compute the transfer function $H(z)$.

(b) By using the z-transform, compute the output response $y(k)$ resulting from the input $x(k) = (1/k)[u(k - 1) - u(k - 3)]$ with no initial energy in the system.

7.10 The input $x(k) = u(k) - 2u(k - 2) + u(k - 4)$ is applied to a linear time-invariant discrete-time system. The resulting response with no initial energy is $y(k) = ku(k) - ku(k - 4)$. Compute the transfer function $H(z)$.

7.11 Consider a car on a level surface given by the input/output differential equation

$$\frac{d^2y(t)}{dt^2} + \frac{k_f}{M}\frac{dy(t)}{dt} = \frac{1}{M}x(t).$$

Recall that $y(t)$ is the position of the car at time t and $x(t)$ is the drive or braking force applied to the car. By using the Euler approximation of the derivatives (see Problem 3.13), we obtain the discrete-time simulation of the car given by

$$y(kT + 2T) + \left(\frac{Tk_f}{M} - 2\right)y(kT + T) + \left(1 - \frac{Tk_f}{M}\right)y(kT) = \frac{T^2}{M}x(kT).$$

Determine the transfer function $H_d(z)$ of the discrete-time simulation. Here d stands for "discretized."

7.12 Again consider the car defined in Problem 7.11. With $M = 1$, $k_f = 0.1$, and $T = 1$, consider the discrete-time simulation resulting from the procedure given in Problem 3.16.

(a) Determine the transfer function $H_d(z)$ of this discretization. Compare your answer with that obtained in Problem 7.11 with $M = 1$, $k_f = 0.1$, and $T = 1$.

(b) Using the z-transform and the transfer function $H_d(z)$ found in part (a), compute $y(k)$ when $x(t) = u(t)$ with no initial energy in the system.

7.13 Consider the field-controlled dc motor given by the input/output differential equation

$$JL_f \frac{d^3y(t)}{dt^3} + (fL_f + R_fJ)\frac{d^2y(t)}{dt^2} + R_ff \frac{dy(t)}{dt} = kx(t).$$

If we take $J = 1$, $k = 1$, $L_f = 0.01$, $f = 0.1$, and $R_f = 1$, the transfer function of the motor can be expressed in the form

$$H(s) = H_1(s)H_2(s)H_3(s),$$

where

$$H_1(s) = \frac{1}{s + 0.1}, \qquad H_2(s) = \frac{1}{0.01s + 1}, \qquad H_3(s) = \frac{1}{s}.$$

Discretize each $H_i(s)$ by using the integral-approximation procedure given in Section 3.4. Take T to be arbitrary. Compute the transfer function $H_d(z)$ of the resulting discrete-time simulation of the dc motor.

7.14   A linear time-invariant discrete-time system has transfer function

$$H(z) = \frac{z^2 - z - 2}{z^2 + 1.5z - 1}.$$

(a) Compute the unit-pulse response $h(k)$ for all $k \geq 0$.

(b) Compute the output response $y(k)$ for all $k \geq 0$ resulting from the input $x(k) = 1$ for all $k \geq 0$ with zero initial energy at time $k = 0$.

(c) Compute the output values $y(0)$, $y(1)$, $y(2)$ resulting from the input $x(k) = 2^k \sin(\pi k/4) + \tan(\pi k/3)$, $k = 0, 1, 2, \ldots$, with the system at rest at time $k = 0$.

(d) If possible, find an input $x(k)$ with $x(k) = 0$ for all $k < 0$ and such that the output response $y(k)$ resulting from $x(k)$ is given by $y(0) = 2$, $y(1) = -3$, $y(k) = 0$ for all $k \geq 2$. Assume that the system is at rest at $k = 0$.

7.15   The input $x(kT) = (0.5)^k u(kT)$ is applied to a linear time-invariant discrete-time system with the initial conditions $y(-T) = 8$ and $y(-2T) = 4$. The resulting output response is

$$y(kT) = 4(0.5)^k u(kT) - 0.5k(0.5)^{k-1} u(kT - T) - (-0.5)^k u(kT).$$

Find the transfer function $H(z)$.

7.16   Find the unit-pulse response $h(kT)$ of the linear time-invariant discrete-time system with transfer function

$$H(z) = \frac{z}{(z - 0.5)^2(z^2 + 0.25)}.$$

7.17   A linear time-invariant discrete-time system has transfer function

$$H(z) = \frac{3z}{(z + 0.5)(z - 0.5)}.$$

The output response resulting from the input $x(kT) = u(kT)$ and initial conditions $y(-T)$, $y(-2T)$ is

$$y(kT) = [(0.5)^k - 3(-0.5)^k + 4]u(kT).$$

Determine the initial conditions $y(-T)$, $y(-2T)$ and the part of the output response due to the initial conditions.

7.18   A linear time-invariant discrete-time system is given by the input/output difference equation

$$y(k + 2) + y(k) = 2x(k + 1) - x(k).$$

(a) Compute the unit-pulse response $h(k)$.

(b) Compute the output response $y(k)$ for all $k \geq 0$ when $x(k) = u(k)$ with zero initial energy.

(c) Compute $y(k)$ for all $k \geq 0$ when $x(k) = 2^k u(k)$ with $y(-1) = 3$ and $y(-2) = 2$.

(d) An input $x(k)$ with $x(-2) = x(-1) = 0$ produces the output response $y(k) = \sin(\pi k)u(k)$ with no initial energy at time $k = 0$. Determine $x(k)$.

(e) An input $x(k)$ with $x(-2) = x(-1) = 0$ produces the output response $y(k) = \Delta(k - 1)$. Compute $x(k)$.

7.19 A linear time-invariant discrete-time system has unit-pulse response $h(k)$ equal to the Fibonacci sequence; that is, $h(0) = 0$, $h(1) = 1$, $h(2) = 1$, $h(3) = 2$, $h(4) = 3$, $h(5) = 5$, $h(6) = 8$, $h(7) = 13$, and so on. Show that the system's transfer function $H(z)$ is rational in z. Express $H(z)$ as a ratio of polynomials in positive powers of z.

7.20 A linear time-invariant discrete-time system is given by the input/output difference equation

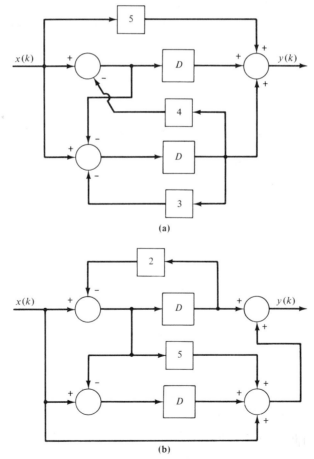

(a)

(b)

Figure P7.21

$$y(k) + y(k - 1) - 2y(k - 2) = 2x(k) - x(k - 1).$$

Find an input $x(k)$ with $x(k) = 0$ for $k < 0$ that gives the output response $y(k) = 2[u(k) - u(k - 3)]$ with initial conditions $y(-2) = 2$, $y(-1) = 0$.

7.21   By using the z-domain representation, determine the transfer functions of the discrete-time systems shown in Figure P7.21.

7.22   Determine the transfer function matrix of the three-input two-output discrete-time system shown in Figure P7.22.

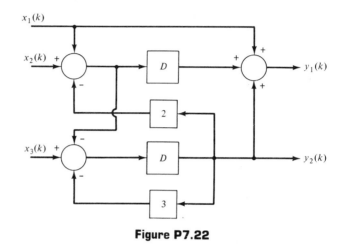

Figure P7.22

7.23   Consider the cascade connection shown in Figure P7.23. Determine the unit-pulse response $h_2(k)$ of the system with transfer function $H_2(z)$ so that when $x(k) = \Delta(k)$ with no initial energy, the response $y(k)$ is equal to $\Delta(k)$.

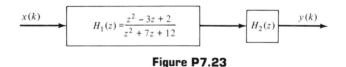

Figure P7.23

7.24   A linear time-invariant discrete-time system is given by the feedback connection shown in Figure P7.24. In Figure P7.24, $X(z)$ is the z-transform of the system's input $x(k)$, $Y(z)$ is the z-transform of the system's output $y(k)$, and $H_1(z)$, $H_2(z)$ are the transfer functions of the subsystems given by

$$H_1(z) = \frac{z}{z + 1}, \qquad H_2(z) = \frac{9}{z - 8}.$$

(a) Determine the unit-pulse response of the overall system.

(b) Compute the output response when $x(k) = u(k)$ with zero initial energy at time $k = 0$.

(c) Compute $y(k)$ when $x(k) = (0.5)^k u(k)$ with $y(-1) = -3$, $y(-2) = 4$.

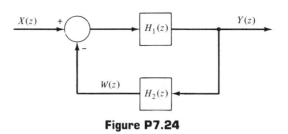

Figure P7.24

(d) Compute $y(k)$ when $x(k) = (0.5)^k u(k)$ with $y(-2) = 1$, $w(-1) = 2$, where $w(k)$ is the output of the feedback system in Figure P7.24.

7.25 A linear time-invariant discrete-time system is given by the cascade connection shown in Figure P7.25.
(a) Compute the unit-pulse response of the overall system.
(b) Compute the input/output difference equation of the overall system.
(c) Compute $y(k)$ when $x(k) = u(k)$ with no initial energy.
(d) Compute $y(k)$ when $x(k) = u(k)$ and $y(-1) = 3$, $q(-1) = 2$.
(e) Compute $y(k)$ when $x(k) = (0.5)^k u(k)$ and $y(-2) = 2$, $q(-2) = 3$.

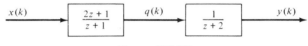

Figure P7.25

Stability and the Response to Sinusoidal Inputs

One of the most fundamental and important concepts in the study of systems is the notion of stability. Actually, there is more than one type of stability. In the first part of this chapter we consider three types of stability: asymptotic stability, marginal stability, and bounded-input bounded-output (BIBO) stability. Asymptotic stability and marginal stability involve the behavior of the output response resulting from initial energy (or initial conditions), while BIBO stability involves the behavior of the output response resulting from the application of inputs. Stability is a time-domain property of a system since it involves the behavior of system responses.

Asymptotic stability is often defined in terms of a state representation of the system. The definition of asymptotic stability given in this chapter is equivalent to the definition based on a state model.

Both asymptotic stability and BIBO stability are of importance in a wide range of applications. It turns out that asymptotic stability always implies BIBO stability for linear time-invariant finite-dimensional systems. This result is proved in Section 8.2.

The stability of linear time-invariant finite-dimensional continuous-time systems is studied in Sections 8.1 to 8.3. The discrete-time version of this theory is given in Section 8.8. We show that stability or instability of a given system can be characterized in terms of the location of the poles of the system. Since the poles of a system are a feature of the transfer function representation, the characterization of stability in terms of pole locations is an excellent illustration of how the transfer function model can be utilized to study time-domain properties.

If the system under study is asymptotically stable, in Sections 8.4 and 8.9 we show that the steady-state part of the output response resulting from a sinusoidal input is also a sinusoid with the same frequency as the input. The steady-state output response is a magnitude-scaled and phase-shifted version of the sinusoidal input. The amount of the magnitude scaling and the amount of the phase shift depend in general on the frequency of the sinusoidal input. This leads to the fundamental notion of the frequency response of a linear time-invariant system. We can characterize systems in terms of the properties of the frequency response. For example, a filter is a system with a frequency response that discriminates against certain frequencies. Examples of lowpass and bandpass continuous-time filters are considered in Sections 8.5 to 8.7.

8.1

STABILITY OF CONTINUOUS-TIME SYSTEMS

Consider a single-input single-output linear time-invariant continuous-time system with input $x(t)$ and output $y(t)$. Throughout this section we assume that the system is finite dimensional, and thus the system's transfer function $H(s)$ is rational in s; that is,

$$H(s) = \frac{N(s)}{D(s)},$$

where $N(s)$ and $D(s)$ are polynomials in s given by

$$N(s) = b_m s^m + b_{m-1} s^{m-1} + \cdots + b_1 s + b_0$$

$$D(s) = a_n s^n + a_{n-1} s^{n-1} + \cdots + a_1 s + a_0, \qquad a_n \neq 0.$$

In the following development, we assume that $m \leq n$ and $x(0^-)$, $x^{(1)}(0^-)$, $\ldots$, $x^{(m-1)}(0^-)$ are zero. Then as discussed in Section 5.6, the Laplace transform $Y(s)$ of the output response resulting from initial conditions $y(0^-)$, $y^{(1)}(0^-)$, $\ldots$, $y^{(n-1)}(0^-)$ and the input $x(t)$ applied for $t \geq 0$ can be expressed in the form

$$Y(s) = \frac{C(s)}{D(s)} + \frac{N(s)}{D(s)} X(s). \tag{8.1}$$

In (8.1), $C(s)$ is a polynomial in s whose coefficients depend on the initial conditions $y(0^-)$, $y^{(1)}(0^-)$, $\ldots$, $y^{(n-1)}(0^-)$. If $y(0^-) \neq 0$, the degree of $C(s)$ is equal to $n - 1$. If $y(0^-) = 0$, the degree of $C(s)$ is less than $n - 1$.

If the input $x(t)$ is zero for all $t \geq 0$, then $X(s) = 0$ and (8.1) reduces to

$$Y(s) = \frac{C(s)}{D(s)}. \tag{8.2}$$

The inverse Laplace transform of $C(s)/D(s)$ is called the *zero-input response* since it is the output response when the input $x(t)$ is zero for all $t \geq 0$. We shall denote the zero-input response by $y_{zi}(t)$. By definition, $y_{zi}(t)$ is that part of the output response resulting from initial conditions only.

DEFINITIONS OF STABILITY Consider the system with the s-domain representation (8.1). The system is said to be *asymptotically stable* if

$$y_{zi}(t) \to 0 \qquad \text{as } t \to \infty \tag{8.3}$$

for *all* initial conditions $y(0^-)$, $y^{(1)}(0^-)$, $\ldots$, $y^{(n-1)}(0^-)$. Here the term "asymptotic" means that the graph of $y_{zi}(t)$ versus t is asymptotic to the line $y = 0$ as t becomes large. Asymptotic stability simply means that the output response resulting from any initial energy in the system decays to zero as t becomes large.

The system is said to be *marginally stable* if for each set of initial conditions $y(0^-)$, $y^{(1)}(0^-)$, $\ldots$, $y^{(n-1)}(0^-)$, there is a finite positive constant M (depending on the initial conditions in general) such that

$$|y_{zi}(t)| \leq M \qquad \text{for all } t \geq 0. \tag{8.4}$$

By condition (8.4), marginal stability means that for any set of initial conditions, the zero-input response $y_{zi}(t)$ is bounded in magnitude.

The system is unstable if there are values of the initial conditions $y(0^-)$, $y^{(1)}(0^-)$, $\ldots$, $y^{(n-1)}(0^-)$ for which

$$|y_{zi}(t)| \to \infty \qquad \text{as } t \to \infty. \tag{8.5}$$

So the system is *unstable* if the magnitude of the zero-input response grows without bound for some values of the initial conditions.

Instability is definitely not a desirable property since the possibility of unbounded responses implies that an actual system can "self-destruct." A good example is an aircraft or spacecraft control system where instability can result in a crash.

POLE CRITERIA FOR STABILITY Let $p_1, p_2, \ldots, p_n$ denote the poles of $H(s)$. Then we have the following pole criteria for stability.

The system is asymptotically stable if and only if

$$\text{Re } p_i < 0 \qquad \text{for } i = 1, 2, \ldots, n. \tag{8.6}$$

In other words, the system is asymptotically stable if and only if all the poles are in the *open left-half plane,* where the open left-half plane is the region of the complex plane consisting of all points to the left of the $j\omega$-axis, but not including the $j\omega$-axis.

The system is marginally stable if and only if Re $p_i \leq 0$ for all nonrepeated poles and Re $p_i < 0$ for all repeated poles. Thus the system is marginally stable if and only if all the poles are in the open left-half plane, except that there can be nonrepeated poles on the $j\omega$-axis.

Finally, the system is unstable if there is at least one pole p_i with Re $p_i > 0$ or if there is at least one repeated pole p_i with Re $p_i = 0$. So the system is unstable if there are one or more poles in the *open right-half plane* (the region to the right of the $j\omega$-axis) or if there are repeated poles on the $j\omega$-axis.

The pole criteria for stability are not difficult to prove. In particular, suppose that the poles $p_1, p_2, \ldots, p_n$ are distinct so that $C(s)/D(s)$ has the partial-fraction expansion

$$\frac{C(s)}{D(s)} = \sum_{i=1}^{n} \frac{d_i}{s - p_i},$$ (8.7)

where the d_i are real or complex constants. The particular values of the constants $d_1, d_2, \ldots, d_n$ depend on the values of the initial conditions $y(0^-), y^{(1)}(0^-),$ $\ldots, y^{(n-1)}(0^-)$.

Taking the inverse Laplace transform of (8.7), we have that the zero-input response is

$$y_{zi}(t) = \sum_{i=1}^{n} d_i e^{p_i t}, \qquad t \geq 0.$$ (8.8)

It follows from (8.8) that $y_{zi}(t) \to 0$ as $t \to \infty$ for all initial conditions if and only if (8.6) is satisfied. So we have verified the pole criteria for asymptotic stability in the case of distinct poles.

We continue to assume that the p_i are distinct, but now we allow nonrepeated poles on the $j\omega$-axis. In particular, suppose that $p_1 = j\omega_1, p_2 = -j\omega_1$, and Re $p_i < 0$ for $i = 3, 4, \ldots, n$. Then in (8.8), $\bar{d}_2 = d_1$, and using the trigonometric identity (5.70), we can rewrite (8.8) in the form

$$y_{zi}(t) = 2|d_1| \cos(\omega_1 t + \sphericalangle d_1) + \sum_{i=3}^{n} d_i e^{p_i t}, \qquad t \geq 0.$$ (8.9)

From (8.9), if Re $p_i < 0$ for $i = 3, 4, \ldots, n$, we have

$$|y_{zi}(t)| \leq 2|d_1| + \sum_{i=3}^{n} |d_i|, \qquad t \geq 0.$$

and thus the system is marginally stable. So if all the poles are in the open left-half plane except for nonrepeated poles on the $j\omega$-axis, the system is marginally stable.

Finally, suppose that there are one or more poles in the open right-half plane. Then from (8.8), we have that $|y_{zi}(t)| \to \infty$ as $t \to \infty$, and the system is unstable. We have therefore verified the pole criteria for stability and instability in the case of distinct poles.

Now if there are a pair of repeated poles at $s = \sigma + j\omega$ and $s = \sigma - j\omega$, $y_{zi}(t)$ will contain a term of the form

$$At e^{\sigma t} \cos (\omega t + \theta),$$

where A and θ are constants. If all the other poles are distinct and are in the open left-half plane, $y_{zi}(t)$ will converge to zero if and only if $\sigma < 0$, and thus the system is asymptotically stable if and only if the repeated pole is in the open left-half plane. If the repeated pole is on the $j\omega$-axis or is in the open right-half plane (i.e., $\sigma = 0$ or $\sigma > 0$), then $|y_{zi}(t)| \to \infty$ as $t \to \infty$ and the system is unstable. This proves the pole criteria for stability and instability in the repeated-pole case.

EXAMPLE 8.1. Consider the automobile on a level surface given by the input/output differential equation

$$\frac{d^2 y(t)}{dt^2} + \frac{k_f}{M} \frac{dy(t)}{dt} = \frac{1}{M} x(t).$$

Here $y(t)$ is the position of the car at time t. The transfer function of the system is

$$H(s) = \frac{1/M}{s^2 + (k_f/M)s} = \frac{1/M}{s(s + k_f/M)}.$$

Thus the poles of the system are $p_1 = 0$ and $p_2 = -k_f/M$. Since $k_f > 0$ and $M > 0$, the pole p_2 is in the open left-half plane. But due to the pole $p_1 = 0$, the system is not asymptotically stable, although it is marginally stable. Since the system is not asymptotically stable, the zero-input response $y_{zi}(t)$ will not decay to zero for all initial conditions $y(0^-)$, $\dot{y}(0^-)$. Let us compute $y_{zi}(t)$ for arbitrary values of the initial position $y(0^-)$ and initial velocity $\dot{y}(0^-)$. From equation (5.102) we have

$$C(s) = y(0^-)s + \dot{y}(0^-) + \frac{k_f}{M} y(0^-)$$

and

$$Y_{zi}(s) = \frac{C(s)}{D(s)} = \frac{y(0^-)s + \dot{y}(0^-) + (k_f/M)y(0^-)}{s(s + k_f/M)}.$$

Expanding by partial fractions gives

$$Y_{zi}(s) = \frac{(M/k_f)\dot{y}(0^-) + y(0^-)}{s} - \frac{(M/k_f)\dot{y}(0^-)}{s + k_f/M}. \qquad (8.10)$$

Taking the inverse Laplace transform of (8.10), we have

$$y_{zi}(t) = \frac{M}{k_f}\dot{y}(0^-)\left[1 - \exp\left(-\frac{k_f}{M}t\right)\right] + y(0^-), \qquad t \geq 0, \quad (8.11)$$

and therefore,

$$y_{zi}(t) \rightarrow \frac{M}{k_f}\dot{y}(0^-) + y(0^-) \qquad \text{as } t \rightarrow \infty. \qquad (8.12)$$

From (8.11) we see that $|y_{zi}(t)|$ is bounded for all $t \geq 0$, but by (8.12), $y_{zi}(t)$ does not decay to zero as $t \rightarrow 0$. Note that for the car system to be asymptotically stable, the car would always have to return to the "zero position" given any initial position $y(0^-)$ and any initial velocity $\dot{y}(0^-)$ with no force acting on the car. Of course, this is not possible, so it is intuitively clear that the car system cannot be asymptotically stable.

EXAMPLE 8.2. Consider the series *RLC* circuit that was studied in Example 6.1. The circuit is redrawn in Figure 8.1. In the following analysis, we assume that $R > 0$, $L > 0$, and $C > 0$. As computed in Example 6.1, the transfer function of the circuit is

$$H(s) = \frac{1/LC}{s^2 + (R/L)s + 1/LC}.$$

From the quadratic formula, the poles of the system are

$$p_1, p_2 = -\frac{R}{2L} \pm \sqrt{b},$$

where

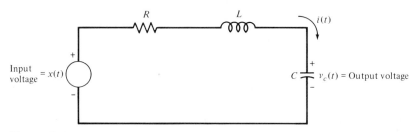

Figure 8.1. Series *RLC* circuit.

$$b = \left(\frac{R}{2L}\right)^2 - \frac{1}{LC}.$$

Now if $b < 0$, both poles are complex with real part equal to $-R/2L$, and thus in this case the circuit is asymptotically stable. If $b > 0$, both poles are real. In this case,

$$-\frac{R}{2L} - \sqrt{b} < 0.$$

In addition, $b > 0$ implies that

$$b < \left(\frac{R}{2L}\right)^2,$$

and thus

$$\sqrt{b} < \frac{R}{2L}.$$

Therefore,

$$-\frac{R}{2L} + \sqrt{b} < 0,$$

and thus the circuit is still asymptotically stable. So the circuit is asymptotically stable for any values of $R, L, C > 0$. This means that if the input voltage $x(t)$ is zero for all $t \geq 0$, the voltage $v_c(t)$ across the capacitor decays to zero as $t \to \infty$ for any initial voltage $v_c(0^-)$ and for any initial value $\dot{v}_c(0^-)$ of the derivative of the voltage. Since the current $i(t)$ in the series connection is given by

$$i(t) = C\frac{dv_c(t)}{dt},$$

the initial condition $\dot{v}_c(0^-)$ is equivalent to the initial current $(1/C)i(0^-)$. Thus for any initial values $v_c(0^-)$ and $i(0^-)$, the voltage $v_c(t)$ decays to zero as $t \to \infty$.

★8.2

BIBO STABILITY

In Section 8.1 we defined the notions of stability and instability in terms of the output response resulting from initial energy in the system. In this section we consider another notion of stability, which is characterized in terms of the output response resulting from the application of input signals with no initial energy in the system.

Suppose that we have a linear time-invariant finite-dimensional continuous-time system with impulse response $h(t)$. Let $y_{zs}(t)$ denote the output response of the system resulting from input $x(t)$ applied for $t \geq 0$ with no initial energy in the system at time $t = 0^-$. The output $y_{zs}(t)$ is called the *zero-state response* since it is the response to the input $x(t)$ with the system in the zero state at time $t = 0^-$ (i.e., no initial energy in the system at time $t = 0^-$). From the results in Chapter 4, we know that

$$y_{zs}(t) = (h * x)(t) = \int_0^t h(t - \lambda)x(\lambda) \, d\lambda.$$

Given a signal $v(t)$, we say that $v(t)$ is bounded if there exists a positive constant c_v (which may depend on v) such that

$$|v(t)| \leq c_v \qquad \text{for all } t.$$

Now the given system is said to be *bounded-input bounded-output* (BIBO) *stable* if whenever the input $x(t)$ is bounded, the resulting zero-state response $y_{zs}(t)$ is also bounded. In other words, bounded input signals produce bounded output signals in a BIBO stable system.

We claim that the system is BIBO stable if and only if

$$\int_0^\infty |h(t)| \, dt < \infty. \tag{8.13}$$

The condition (8.13) says that the area under the magnitude of the impulse response function of the system is finite. We shall prove that (8.13) implies that the system is BIBO stable.

Let $y_{zs}(t)$ denote the zero-state response resulting from the input $x(t)$ applied for $t \geq 0$. Then taking the absolute value of both sides of the convolution relationship $y_{zs}(t) = (h * x)(t)$, we have

$$|y_{zs}(t)| \leq \int_0^t |h(t - \lambda)x(\lambda)| \, d\lambda = \int_0^t |h(t - \lambda)||x(\lambda)| \, d\lambda. \tag{8.14}$$

Now if $x(t)$ is bounded,

$$|x(t)| \leq c_x \qquad \text{all } t \geq 0,$$

for some positive constant c_x. Using this in (8.14), we obtain

$$|y_{zs}(t)| \leq c_x \int_0^t |h(t - \lambda)| \, d\lambda. \tag{8.15}$$

Setting $\bar{\lambda} = t - \lambda$ in the integral in (8.15), we have

$$\int_0^t |h(t - \lambda)| \, d\lambda = \int_t^0 |h(\bar{\lambda})| \, (-1) \, d\bar{\lambda} = \int_0^t |h(\bar{\lambda})| \, d\bar{\lambda} = \int_0^t |h(\lambda)| \, d\lambda.$$

$$(8.16)$$

Now suppose that (8.13) is satisfied. Then there exists a positive constant B such that

$$\int_0^t |h(\lambda)| \, d\lambda = B < \infty \qquad \text{for all } t \geq 0. \tag{8.17}$$

Combining (8.15)–(8.17), we have

$$|y_{zs}(t)| \leq c_x B \qquad \text{for all } t \geq 0,$$

and thus the output response is bounded. This proves that the system is BIBO stable.

The proof that BIBO stability implies (8.13) is beyond the scope of this work, so we shall not consider this.

RELATIONSHIP WITH ASYMPTOTIC STABILITY If a system is asymptotically stable as defined in Section 8.1, it must be BIBO stable. We shall prove this by showing that asymptotic stability implies condition (8.13), which in turn implies that the system is BIBO stable.

Let $H(s)$ denote the transfer function of the given system. We assume that the poles $p_1, p_2, \ldots, p_n$ of $H(s)$ are distinct. Writing $H(s)$ in the form

$$H(s) = \frac{N(s)}{(s - p_1)(s - p_2) \cdots (s - p_n)} \qquad \text{with deg } N(s) < n,$$

we can expand by partial fractions, which gives

$$H(s) = \sum_{i=1}^n \frac{c_i}{s - p_i}, \tag{8.18}$$

where the c_i are real or complex constants. Taking the inverse Laplace transform of both sides of (8.18), we have that the impulse response $h(t)$ is

$$h(t) = \sum_{i=1}^n c_i e^{p_i t}, \qquad t \geq 0. \tag{8.19}$$

Taking the magnitude of both sides of (8.19), we get

$$|h(t)| \leq \sum_{i=1}^n |c_i e^{p_i t}| = \sum_{i=1}^n |c_i| \, |e^{p_i t}|.$$

Then

$$\int_0^\infty |h(t)|\, dt \le \int_0^\infty \sum_{i=1}^n |c_i|\, |e^{p_i t}|\, dt = \sum_{i=1}^n |c_i| \int_0^\infty |e^{p_i t}|\, dt. \qquad (8.20)$$

Now if the system is asymptotically stable, Re $p_i < 0$ for all i, and thus

$$\int_0^\infty |e^{p_i t}|\, dt = \frac{1}{|\text{Re } p_i|} < \infty \qquad \text{for all } i. \qquad (8.21)$$

Combining (8.20) and (8.21) gives

$$\int_0^\infty |h(t)|\, dt \le \sum_{i=1}^n \frac{|c_i|}{|\text{Re } p_i|},$$

and thus (8.13) is satisfied.

If the transfer function $H(s)$ does not have any common poles and zeros, it turns out that BIBO stability implies asymptotic stability. Thus in this case, BIBO stability and asymptotic stability are equivalent. The proof that BIBO stability implies asymptotic stability when there are no common poles and zeros is left to a more advanced treatment of stability.

EXAMPLE 8.3.   Consider the system given by the input/output differential equation

$$\frac{d^2 y(t)}{dt^2} + \frac{dy(t)}{dt} - 2y(t) = \frac{dx(t)}{dt} - x(t).$$

The transfer function of this system is

$$H(s) = \frac{s-1}{s^2 + s - 2} = \frac{s-1}{(s-1)(s+2)}. \qquad (8.22)$$

Since there is a pole at $s = 1$, the system is not asymptotically stable. So for some values of the initial conditions $y(0^-)$, $\dot{y}(0^-)$, the magnitude of the zero-input response $y_{zi}(t)$ will grow without bound as $t \to \infty$. In particular, by equation (5.102), the Laplace transform $Y_{zi}(s)$ of $y_{zi}(t)$ is

$$\begin{aligned} Y_{zi}(s) &= \frac{y(0^-)(s+1) + \dot{y}(0^-)}{s^2 + s - 2} \\ &= \frac{(\tfrac{1}{3})[2y(0^-) + \dot{y}(0^-)]}{s-1} + \frac{(\tfrac{1}{3})[y(0^-) - \dot{y}(0^-)]}{s+2}. \end{aligned} \qquad (8.23)$$

Taking the inverse Laplace transform of (8.23), we have

$$y_{zi}(t) = \tfrac{1}{3}\{[2y(0^-) + \dot{y}(0^-)]e^t + [y(0^-) - \dot{y}(0^-)]e^{-2t}\}, \qquad t \ge 0.$$

Thus $|y_{zi}(t)| \to \infty$ as $t \to \infty$ if $\dot{y}(0^-) \neq -2y(0^-)$.

Now taking the inverse Laplace transform of (8.22), we have that the impulse response is

$$h(t) = e^{-2t}u(t).$$

Hence

$$\int_0^\infty |h(t)| \, dt = -\tfrac{1}{2} e^{-2t}\big]_{t=0}^{t=\infty} = \tfrac{1}{2} < \infty.$$

Therefore, (8.13) is satisfied and the system is BIBO stable. Note that this system has a common pole and zero at $s = 1$.

★**8.3**

ROUTH STABILITY TEST

By the results in Section 8.1, asymptotic stability of the system with transfer function $H(s) = N(s)/D(s)$ can be checked by first computing the poles of $H(s)$, which are the roots of $D(s) = 0$. The poles can be computed by using the root-finding program in Figure A.6.

It turns out that there are procedures for testing for stability that do not require the computation of the poles of the system. One such procedure is the Routh stability test, which is based on simple computations involving the coefficients of the polynomial $D(s)$. The details are as follows.

Suppose that

$$D(s) = a_n s^n + a_{n-1} s^{n-1} + \cdots + a_1 s + a_0, \qquad a_n > 0. \quad (8.24)$$

Note that the leading coefficient a_n of $D(s)$ may be any nonzero positive number. By the results in Section 8.1, the system is asymptotically stable if and only if all the zeros of $D(s)$ are in the open left-half plane (OLHP). A necessary (but in general not sufficient) condition for this to be the case is that all the coefficients of $D(s)$ must be strictly positive; that is,

$$a_i > 0 \qquad \text{for } i = 0, 1, 2, \ldots, n - 1. \quad (8.25)$$

Thus, if $D(s)$ has one or more coefficients that are zero or negative, there is at least one pole not in the OLHP and the system is not asymptotically stable. When we refer to a "pole not in the OLHP," we mean a pole located on the $j\omega$-axis or located in the open right-half plane.

It should be stressed that the condition (8.25) is not a sufficient condition for stability in general. In other words, there are unstable systems for which (8.25) is satisfied.

Now we shall state the Routh stability test, which gives the necessary and sufficient conditions for stability. Given the polynomial $D(s)$ defined by (8.24), we first construct the Routh array shown in Table 8.1

TABLE 8.1.   Routh Array

s^n	a_n	a_{n-2}	a_{n-4}	$\cdots$
s^{n-1}	a_{n-1}	a_{n-3}	a_{n-5}	$\cdots$
s^{n-2}	b_{n-2}	b_{n-4}	b_{n-6}	$\cdots$
s^{n-3}	c_{n-3}	c_{n-5}	c_{n-7}	$\cdots$
$\cdot$	$\cdot$	$\cdot$	$\cdot$	
$\cdot$	$\cdot$	$\cdot$	$\cdot$	
$\cdot$	$\cdot$	$\cdot$	$\cdot$	
s^2	d_2	d_0	0	$\cdots$
s^1	e_1	0	0	$\cdots$
s^0	f_0	0	0	$\cdots$

As seen from Table 8.1, the Routh array has $n + 1$ rows, with the rows indexed by the powers of s. The number of columns of the array is $(n/2) + 1$ if n is even or $(n + 1)/2$ if n is odd. The first two rows of the Routh array are filled by the coefficients of $D(s)$, starting with the leading coefficient a_n. The elements in the third row are given by

$$b_{n-2} = \frac{a_{n-1}a_{n-2} - a_n a_{n-3}}{a_{n-1}} = a_{n-2} - \frac{a_n a_{n-3}}{a_{n-1}}$$

$$b_{n-4} = \frac{a_{n-1}a_{n-4} - a_n a_{n-5}}{a_{n-1}} = a_{n-4} - \frac{a_n a_{n-5}}{a_{n-1}}$$

$$\cdot$$
$$\cdot$$
$$\cdot$$

The elements in the fourth row are given by

$$c_{n-3} = \frac{b_{n-2}a_{n-3} - a_{n-1}b_{n-4}}{b_{n-2}} = a_{n-3} - \frac{a_{n-1}b_{n-4}}{b_{n-2}}$$

$$c_{n-5} = \frac{b_{n-2}a_{n-5} - a_{n-1}b_{n-6}}{b_{n-2}} = a_{n-5} - \frac{a_{n-1}b_{n-6}}{b_{n-2}}$$

$$\cdot$$
$$\cdot$$
$$\cdot$$

The other rows (if there are any) are computed in a similar fashion. As a check on the computations, it should turn out that the last nonzero element in each column of the array is equal to the coefficient a_0 of $D(s)$.

In calculating the Routh array, it may happen that $b_{n-2} = 0$, in which case one cannot perform the division in computing the elements in the fourth row. If $b_{n-2} = 0$, we can set $b_{n-2} = 0.00001$ (or any very small positive number), and then continue. Similarly, if $c_{n-3} = 0$, we can set $c_{n-3} = 0.00001$, and continue. If any zero elements are set equal to small positive numbers, the last nonzero element of the columns of the Routh array will, in general, not be equal to a_0.

The Routh stability test states that the system is asymptotically stable (all poles in OLHP) if and only if all the elements in the first column of the Routh array are strictly positive (> 0). In addition, the number of poles not in the OLHP is equal to the number of sign changes in the first column. We illustrate the application of the Routh stability test in the examples given below. The proof of the Routh stability test is well beyond the scope of this book, so we shall not verify this result.

As we now show in the following examples, when the degree n of $D(s)$ is less than or equal to 4, using the Routh test we can derive simple conditions for stability given directly in terms of the coefficients of $D(s)$.

EXAMPLE 8.4. Let $n = 2$ and $a_2 = 1$, so that

$$D(s) = s^2 + a_1 s + a_0.$$

The Routh array for this case is given in Table 8.2.

TABLE 8.2. Routh Array in the $n = 2$ Case

s^2	1	a_0
s^1	a_1	0
s^0	$\dfrac{a_1 a_0 - (1)(0)}{a_1} = a_0$	0

The elements in the first column of the Routh array are 1, a_1, a_0, and thus the poles are in the OLHP if and only if the coefficients a_1 and a_0 are both positive. So in this case, the positive-coefficient condition (8.25) is necessary and sufficient for asymptotic stability. Now suppose that $a_1 > 0$ and $a_0 < 0$. Then there is one sign change in the first column of the Routh array, which means that there is one pole not in the OLHP. If $a_1 < 0$ and $a_0 < 0$, there still is one sign change and thus there still is one pole not in the OLHP. If $a_1 < 0$ and $a_0 > 0$, there are two sign changes in the first column, and therefore both poles are not in the OLHP.

EXAMPLE 8.5. Consider the third-order case

$$D(s) = s^3 + a_2 s^2 + a_1 s + a_0.$$

The Routh array is displayed in Table 8.3.

TABLE 8.3.   The $n = 3$ Case.

s^3	1	a_1
s^2	a_2	a_0
s^1	$\dfrac{a_2 a_1 - (1)a_0}{a_2} = a_1 - \dfrac{a_0}{a_2}$	0
s^0	a_0	0

Since

$$a_1 - \frac{a_0}{a_2} > 0$$

if and only if

$$a_1 > \frac{a_0}{a_2},$$

all three poles are in the OLHP if and only if

$$a_2 > 0, \qquad a_1 > \frac{a_0}{a_2}, \qquad a_0 > 0.$$

This result shows that when $n = 3$, it is not true in general that positivity of a_2, a_1, and a_0 implies that the system is asymptotically stable. Note that if $a_2 < 0$, $a_1 > a_0/a_2$, and $a_0 > 0$, there are two sign changes in the first column of the Routh array, and thus there are two poles not in the OLHP. If $a_2 < 0$, $a_1 > a_0/a_0$, and $a_0 < 0$, there are three sign changes, and therefore all three poles are not in the OLHP. If $a_2 < 0$, $a_1 < a_0/a_2$, and $a_0 < 0$, there is one sign change, which means that there is one pole not in the OLHP.

EXAMPLE 8.6.   Now let

$$D(s) = s^4 + a_3 s^3 + a_2 s^2 + a_1 s + a_0.$$

The Routh array is displayed in Table 8.4.

TABLE 8.4.   Routh Array When $n = 4$ and $a_4 = 1$

s^4	1		a_2	a_0
s^3	a_3		a_1	0
s^2	$\dfrac{a_3 a_2 - a_1}{a_3} = a_2 - \dfrac{a_1}{a_3}$		$\dfrac{a_3 a_0 - 0}{a_3} = a_0$	0
s^1	$\dfrac{[a_2 - (a_1/a_3)]a_1 - a_3 a_0}{a_2 - (a_1/a_3)} = a_1 - \dfrac{a_3^2 a_0}{a_3 a_2 - a_1}$		0	0
s^0	a_0		0	0

In this case the system is asymptotically stable if and only if

$$a_3 > 0$$

$$a_2 > \frac{a_1}{a_3}$$

$$a_1 > \frac{a_3^2 a_0}{a_3 a_2 - a_1}$$

$$a_0 > 0.$$

As n is increased above the value $n = 4$, the conditions for stability in terms of the coefficients of $D(s)$ get rather complicated. For $n > 4$ we can still apply the Routh test on a case-by-case basis. A program for this is given in Figure A.9. For any value of n, this program outputs the first column of the Routh array. It also gives the number of poles not in the OLHP. Because of the simplicity of the computations, the program requires very little time to run even when n is large (e.g., $n = 20$). This is in direct contrast to a root-finding routine, which (when $n = 20$) may require several minutes of computation on a personal computer.

It is interesting to observe that for values of n greater than or equal to 4, it is difficult to choose (in a random fashion) positive coefficients $a_0, a_1, \ldots,$ a_n such that the resulting polynomial has all roots in the OLHP. One can attempt this using the program in Figure A.9.

EXAMPLE 8.7. Suppose that

$$D(s) = 6s^5 + 5s^4 + 4s^3 + 3s^2 + 2s + 1.$$

Then $n = 5$ and

$$a_0 = 1, \qquad a_1 = 2, \qquad a_2 = 3, \qquad a_3 = 4, \qquad a_4 = 5, \qquad a_5 = 6.$$

Running the program in Figure A.9, we get the following results.

First column of Routh array is

6
5
.4000001
-6.999997
.8571429
1

There are 2 poles not in the
open left-half plane.

RESPONSE TO A SINUSOIDAL INPUT

Again consider the linear time-invariant continuous-time system with the transfer function $H(s) = N(s)/D(s)$. We assume that the transfer function $H(s)$ is proper in s; that is, the degree of $N(s)$ is less than or equal to the degree of $D(s)$. In the following development, we also assume that the system is asymptotically stable, so that the poles of the system are in the open left-half plane.

Our objective is to determine the output response of the system when the input $x(t)$ is a sinusoid; that is,

$$x(t) = A \cos \omega t, \qquad t \geq 0,$$

where the magnitude A and the frequency ω (in radians/sec) are arbitrary constants. From Table 5.3, the Laplace transform of the input is

$$X(s) = \frac{As}{s^2 + \omega^2}.$$

Then if the system has no initial energy at time $t = 0$, the transform $Y(s)$ of the resulting output response is given by

$$Y(s) = H(s) \frac{As}{s^2 + \omega^2} = \frac{AsN(s)}{D(s)(s^2 + \omega^2)}.$$

Factoring $s^2 + \omega^2$ into the product $(s - j\omega)(s + j\omega)$, we have

$$Y(s) = \frac{AsN(s)}{D(s)(s - j\omega)(s + j\omega)}. \tag{8.26}$$

We can pull out the terms $s + j\omega$ and $s - j\omega$ in (8.26) by using the partial-fraction expansion. This gives

$$Y(s) = \frac{\gamma(s)}{D(s)} + \frac{c}{s - j\omega} + \frac{\bar{c}}{s + j\omega}, \tag{8.27}$$

where $\gamma(s)$ is a polynomial in s with deg $\gamma(s) <$ deg $D(s)$, and

$$c = [(s - j\omega)Y(s)]_{s=j\omega} = \left[\frac{AsN(s)}{D(s)(s + j\omega)} \right]_{s=j\omega} = \frac{jA\omega N(j\omega)}{D(j\omega)(j2\omega)} = \frac{A}{2} H(j\omega).$$

Thus

$$Y(s) = \frac{\gamma(s)}{D(s)} + \frac{(A/2)H(j\omega)}{s - j\omega} + \frac{(A/2)\overline{H(j\omega)}}{s + j\omega}. \tag{8.28}$$

Now let $y_{tr}(t)$ denote the inverse Laplace transform of $\gamma(s)/D(s)$. Then taking the inverse Laplace transform of both sides of (8.28), we obtain

$$y(t) = y_{tr}(t) + \frac{A}{2} [H(j\omega)e^{j\omega t} + \overline{H(j\omega)}e^{-j\omega t}], \qquad t \geq 0.$$

Using equation (5.70), we can write this expression for $y(t)$ in the form

$$y(t) = y_{tr}(t) + A|H(j\omega)| \cos [\omega t + \measuredangle H(j\omega)], \qquad t \geq 0. \qquad (8.29)$$

Since the roots of $D(s) = 0$ are all in the open left-half plane, by the final-value theorem

$$\lim_{t \to \infty} y_{tr}(t) = \lim_{s \to 0} s \frac{\gamma(s)}{D(s)}.$$

But

$$\lim_{s \to 0} s \frac{\gamma(s)}{D(s)} = \frac{0}{D(0)} = 0$$

since $D(0) \neq 0$. Thus

$$y_{tr}(t) \to 0 \qquad \text{as } t \to \infty.$$

The response $y_{tr}(t)$ is called the *transient response* since it decays to zero as t approaches infinity. From (8.29), for t large we have

$$y(t) \approx A|H(j\omega)| \cos [\omega t + \measuredangle H(j\omega)]. \qquad (8.30)$$

The term on the right side of (8.30) is called the *steady-state response* since it does not decay to zero as t approaches infinity. We shall denote the steady-state response by $y_{ss}(t)$, so that

$$y_{ss}(t) = A|H(j\omega)| \cos [\omega t + \measuredangle H(j\omega)], \qquad t \geq 0. \qquad (8.31)$$

By the above result, we see that the steady-state response $y_{ss}(t)$ to the sinusoidal input $x(t) = A \cos \omega t$, $t \geq 0$, has the same frequency as the input, but it is phase shifted by the amount $\measuredangle H(j\omega)$ and it is scaled in magnitude by the amount $|H(j\omega)|$. Note that the phase shift $\measuredangle H(j\omega)$ and the magnitude scale factor $|H(j\omega)|$ depend in general on the frequency ω of the input sinusoid.

Examples illustrating the computation of the transient and steady-state responses are given in Sections 8.5 and 8.6.

It is important to point out that the foregoing characterization of the steady-state response is valid only for linear time-invariant systems. In other words, if the system is time varying or if it is nonlinear, the steady-state response cannot be written in the general form of (8.31). In particular, a distinguishing feature of a nonlinear system is that the steady-state response to the sinusoidal input $A \cos \omega t$ usually contains sinusoidal terms whose frequencies differ from ω. So new frequencies can be generated from a given frequency in a nonlinear system.

It should also be mentioned that if the given system is not asymptotically stable, in general $y_{tr}(t)$ will not decay to zero. In this case, the steady-state part of the output response $y(t)$ consists of all terms in $y_{tr}(t)$ which do not decay to zero plus the response $y_{ss}(t)$ defined by (8.31). For example, if all the poles of the system are in the open left-half plane except for a pair of poles at $s = \pm j\alpha$, $\alpha \neq \omega$, $y_{tr}(t)$ will contain a term of the form $B \cos (\alpha t + \theta)$. Hence the steady-state response resulting from the input $x(t) = A \cos \omega t$ will be of the form

$$B \cos (\alpha t + \theta) + A|H(j\omega)| \cos [\omega t + \sphericalangle H(j\omega)].$$

The frequency α is called a *natural frequency* of the system.

SYSTEMS WITH INITIAL ENERGY   If the system has initial energy before the application of the input $x(t) = A \cos \omega t$, it is still possible to define the transient and steady-state parts of the output response. In this case, the Laplace transform of the output response has the form

$$Y(s) = \frac{C(s)}{D(s)} + H(s) \frac{As}{s^2 + \omega^2},$$

where $C(s)$ is a polynomial in s whose coefficients depend on the initial conditions [see (8.1)]. From the above results,

$$H(s) \frac{As}{s^2 + \omega^2} = \frac{\gamma(s)}{D(s)} + \frac{(A/2)H(j\omega)}{s - j\omega} + \frac{(A/2)\overline{H(j\omega)}}{s + j\omega},$$

and thus

$$Y(s) = \frac{C(s) + \gamma(s)}{D(s)} + \frac{(A/2)H(j\omega)}{s - j\omega} + \frac{(A/2)\overline{H(j\omega)}}{s + j\omega}.$$

The transient response $y_{tr}(t)$ is then equal to the inverse Laplace transform of $[C(s) + \gamma(s)]/D(s)$. We again have that $y_{tr}(t) \rightarrow 0$ as $t \rightarrow \infty$ since the roots of $D(s) = 0$ are in the open left-half plane.

The steady-state response $y_{ss}(t)$ is equal to the inverse transform of

$$\frac{(A/2)H(j\omega)}{s - j\omega} + \frac{(A/2)\overline{H(j\omega)}}{s + j\omega}.$$

Thus

$$y_{ss}(t) = A|H(j\omega)| \cos [\omega t + \sphericalangle H(j\omega)],$$

and therefore the steady-state response is identical to the steady-state response in the case when there is no initial energy.

For any input $x(t)$ and any initial conditions, we can define the steady-state response as that part of the output that does not decay to zero as $t \rightarrow \infty$. The transient response is that part of the output response that does decay to zero as $t \rightarrow \infty$.

COMPLEX FREQUENCY We can generalize the above results by considering inputs of the form

$$x(t) = Ae^{\sigma t} \cos \omega t, \qquad t \geq 0,$$

where A, σ, and ω are fixed but arbitrary real constants. If $\sigma = 0$, this input is the same as the sinusoidal input considered above.

From Table 5.3 the Laplace transform of $x(t)$ is

$$X(s) = \frac{A(s - \sigma)}{(s - \sigma)^2 + \omega^2}.$$

Note that $s = \sigma + j\omega$ is a pole of the rational function $X(s)$; that is,

$$|X(s)| \rightarrow \infty \qquad \text{as } s \rightarrow \sigma + j\omega.$$

The complex number $s = \sigma + j\omega$ is called the *complex frequency* of the signal $x(t) = Ae^{\sigma t} \cos \omega t$.

Now the Laplace transform of the output response resulting from $x(t)$ with no initial energy is

$$Y(s) = H(s) \frac{A(s - \sigma)}{(s - \sigma)^2 + \omega^2}.$$

Generalizing the analysis we gave for the case $\sigma = 0$, we have that the output response is

$$y(t) = y_{tr}(t) + A|H(\sigma + j\omega)|e^{\sigma t} \cos [\omega t + \measuredangle H(\sigma + j\omega)], \qquad t \geq 0.$$

So in this case, the input signal $x(t)$ is scaled in magnitude by the amount $|H(\sigma + j\omega)|$ and is phase shifted by the amount $\measuredangle H(\sigma + j\omega)$.

It is important to note that the magnitude scale factor $|H(\sigma + j\omega)|$ and the phase shift $\measuredangle H(\sigma + j\omega)$ are given in terms of the evaluation of the transfer function $H(s)$ at the complex frequency $s = \sigma + j\omega$ of the input signal $x(t)$. As a result of this property, we can view the variable s in $H(s)$ as a complex-frequency variable.

FREQUENCY RESPONSE FUNCTION

Consider the linear time-invariant continuous-time system with transfer function

$$H(s) = \frac{b_m s^m + b_{m-1} s^{m-1} + \cdots + b_1 s + b_0}{a_n s^n + a_{n-1} s^{n-1} + \cdots + a_1 s + a_0}, \qquad a_n \neq 0. \quad (8.32)$$

We continue to assume that the poles of the system are in the open left-half plane.

Recall from Section 8.4 that the steady-state response $y_{ss}(t)$ resulting from the sinusoidal input $x(t) = A \cos \omega t$, $t \geq 0$, is given by

$$y_{ss}(t) = A|H(j\omega)| \cos [\omega t + \sphericalangle H(j\omega)], \qquad t \geq 0. \quad (8.33)$$

In (8.33), $H(j\omega)$ is the transfer function $H(s)$ evaluated at $s = j\omega$. The function $H(j\omega)$ is a complex-valued function of the real frequency variable ω; that is, given a real number ω, $H(j\omega)$ is a complex number in general. To simplify the notation, from here on we shall denote the function $H(j\omega)$ by $H(\omega)$.

Since the steady-state response to a sinusoidal input can be determined from the magnitude $|H(\omega)|$ and phase $\sphericalangle H(\omega)$ of $H(\omega)$, the function $H(\omega)$ is called the *frequency function* of the system. The magnitude $|H(\omega)|$ of $H(\omega)$ is called the *magnitude function* of the system and the angle $\sphericalangle H(\omega)$ of $H(\omega)$ is called the *phase function* of the system. The plots of $|H(\omega)|$ and $\sphericalangle H(\omega)$ versus ω are called the *frequency response curves* of the system. Note that by (8.33), the steady-state response $y_{ss}(t)$ can be computed directly from the frequency response curves of the system.

Sometimes the magnitude function $|H(\omega)|$ is given in decibels, denoted by $|H(\omega)|_{dB}$ and defined by

$$|H(\omega)|_{dB} = 20 \log_{10} |H(\omega)|.$$

Note that

$$|H(\omega)|_{dB} < 0 \text{ dB} \qquad \text{when } |H(\omega)| < 1$$

$$|H(\omega)|_{dB} = 0 \text{ dB} \qquad \text{when } |H(\omega)| = 1$$

$$|H(\omega)|_{dB} > 0 \text{ dB} \qquad \text{when } |H(\omega)| > 1.$$

Therefore, when $|H(\omega)|_{dB} < 0$ dB, the system attenuates the input $x(t) = A \cos \omega t$; when $|H(\omega)|_{dB} = 0$ dB, the system passes $x(t)$ with no attenuation; and when $|H(\omega)|_{dB} > 0$ dB, the system amplifies $x(t)$.

The plot of $|H(\omega)|_{dB}$ versus ω with ω plotted on a logarithmic scale and the plot of $\sphericalangle H(\omega)$ versus ω with ω plotted on a logarithmic scale are called the *Bode diagrams* of the system. The Bode diagrams can be easily sketched by first determining asymptotes for the curves. We shall illustrate this construction in the examples given later.

The frequency response curves of a linear time-invariant system are often computed experimentally by measuring the steady-state response resulting from the sinusoidal input $x(t) = A \cos \omega t$. By performing this experiment for various values of ω, we can extrapolate the results to obtain the magnitude function $|H(\omega)|$ and phase function $\angle H(\omega)$ for all values of $\omega (\omega \geq 0)$. This then determines the frequency function $H(\omega)$ since

$$H(\omega) = |H(\omega)| \exp [j \angle H(\omega)].$$

If the number of poles of the system is known to be n, the frequency function $H(\omega)$ can be determined by measuring the steady-state responses to the sinusoid $A \cos \omega t$ with a finite number of different values of the frequency ω. We consider this approach to system identification in the single-pole case given below.

If the transfer function $H(s)$ is given in the rational form (8.32), we can determine the frequency response curves by first setting $s = j\omega$ in (8.32). The result can be expressed in the form

$$H(\omega) = \frac{(b_0 - b_2\omega^2 + b_4\omega^4 - \cdots) + j(b_1\omega - b_3\omega^3 + b_5\omega^5 - \cdots)}{(a_0 - a_2\omega^2 + a_4\omega^4 - \cdots) + j(a_1\omega - a_3\omega^3 + a_5\omega^5 - \cdots)}.$$

$$(8.34)$$

By calculating the magnitude and angle of $H(\omega)$ given by (8.34), we can compute the frequency response curves of the system. A program for carrying out this computation is given in Figure A.10. The program computes the magnitude and angle of $H(\omega)$ for

$$\omega = WL + \frac{k(WH - WL)}{Q}, \qquad k = 0, 1, 2, \ldots, Q,$$

where WL is the minimum frequency in rad/sec, WH is the maximum frequency in rad/sec, and Q is the number of points.

We should note that the computation of $H(\omega)$ using (8.34) may not yield good numerical accuracy when the order n of the transfer function $H(s)$ is large. Better accuracy can be obtained by first factoring the polynomials in the numerator and denominator of $H(s)$. However, this requires a root-finding routine which would result in a much longer program than the one given in Figure A.10. We invite the reader to write a program that is based on a factorization of $H(s)$.

If the number of poles and zeros of the system is not large, the general "shape" of the frequency response curves can be determined from the vector representations of the factors comprising $H(\omega)$. We illustrate this in the examples given below.

SINGLE-POLE SYSTEMS   Consider the system with the transfer function

$$H(s) = \frac{B}{s + B},$$

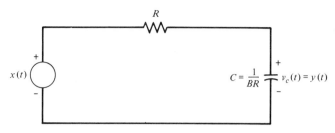

Figure 8.2. *RC* circuit realization of $H(s) = B/(s + B)$.

where $B > 0$. This system has a single pole at $s = -B$. It does not have any (finite) zeros. Since $B > 0$, the pole is in the open left-half plane, and thus the system is asymptotically stable.

The system with transfer function $H(s) = B/(s + B)$ could be the *RC* circuit shown in Figure 8.2 or it could be the integrator connection shown in Figure 8.3. In other words, both the *RC* circuit and the integrator connection are realizations of the transfer function $H(s) = B/(s + B)$.

Setting $s = j\omega$ in the transfer function $H(s) = B/(s + B)$ and taking the magnitude, we obtain

$$|H(\omega)| = \left|\frac{B}{j\omega + B}\right| = \frac{B}{|j\omega + B|} = \frac{B}{\sqrt{\omega^2 + B^2}}. \qquad (8.35)$$

The phase function is

$$\measuredangle H(\omega) = -\measuredangle (j\omega + B) = -\tan^{-1}\frac{\omega}{B}. \qquad (8.36)$$

We can determine the frequency response curves by evaluating (8.35) and (8.36) for various values of ω. Instead of doing this, we want to show that the shape of the frequency response curves can be determined from the vector representation of the factor $j\omega + B$ comprising $H(\omega)$. The analysis is given below.

The magnitude $|j\omega + B|$ and the angle $\measuredangle (j\omega + B)$ can be computed from the vector representation of $j\omega + B$ shown in Figure 8.4. Here the magnitude $|j\omega + B|$ is the length of the vector from the pole $s = -B$ to the point $s = j\omega$ on the imaginary axis, and the angle $\measuredangle (j\omega + B)$ is the angle between this vector and the real axis of the complex plane. From Figure 8.4 we see that

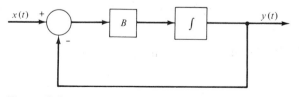

Figure 8.3. Integrator realization of $H(s) = B/(s + B)$.

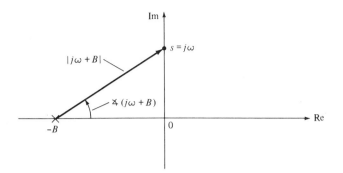

Figure 8.4. Vector representation of $j\omega + B$.

$|j\omega + B|$ becomes infinite as $\omega \rightarrow \infty$ and $\measuredangle (j\omega + B)$ approaches 90° as $\omega \rightarrow \infty$. Then from (8.35) and (8.36), we see that the magnitude function $|H(\omega)|$ starts with value 1 when $\omega = 0$ and approaches zero as $\omega \rightarrow \infty$, while the phase $\measuredangle H(\omega)$ starts with value 0° when $\omega = 0$ and approaches $-90°$ as $\omega \rightarrow \infty$.

The exact frequency response curves can be computed using the program in Figure A.10. For example, when $B = 2$, we get the curves shown in Figure 8.5. The magnitude plot in Figure 8.5a reveals that the system with transfer function $H(s) = B/(s + B)$ is a lowpass filter since it passes sinusoids whose frequency is less than B rad/sec, while it attenuates sinusoids whose frequency is above B rad/sec. The point $\omega = B$ is called the *3-dB point* since this is the value of ω for which $|H(\omega)|_{dB}$ is down by 3 dB from the peak value of 0 dB.

This lowpass filter is said to have a bandwidth of B rad/sec since the system passes sinusoids whose frequency is less than B rad/sec. The passband of the filter is the frequency range from 0 to B. The stopband of the filter is the frequency range from B to ∞. As seen from the magnitude plot, in this example the cutoff between the passband and stopband is not very sharp. We will see later that we can get a sharper cutoff by increasing the number of poles of the system.

From the phase plot in Figure 8.5b, we see that the phase of the one-pole lowpass filter is approximately linear over the passband.

BODE DIAGRAMS OF SINGLE-POLE SYSTEM   Again consider the system with transfer function $H(s) = B/(s + B)$. The frequency response curves with ω plotted on a log scale can easily be sketched using the asymptote construction in the Bode diagram procedure. In this procedure, we first approximate $|H(\omega)|$ and $\measuredangle H(\omega)$ by a low-frequency and high-frequency asymptote. The steps are as follows.

The frequency $\omega = B$ is called the *break frequency* or the *corner frequency* for the factor $j\omega + B$. For $\omega < B$, we approximate $j\omega + B$ by B, so that $H(\omega)$ is given approximately by

$$H(\omega) \approx \frac{B}{B} = 1, \qquad \omega < B.$$

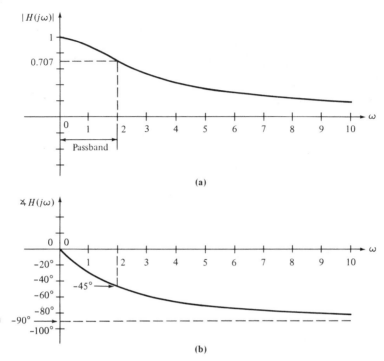

Figure 8.5. Frequency response curves for $H(s) = 2/(s + 2)$: (a) magnitude curve; (b) phase curve.

Then

$$20 \log_{10}|H(\omega)| = |H(\omega)|_{dB} \approx 20 \log_{10}(1) = 0 \text{ dB}, \qquad \omega < B \quad (8.37)$$

and

$$\angle H(\omega) \approx 0°, \qquad \omega < B. \tag{8.38}$$

As shown in Figure 8.6a, (8.37) determines a line at 0 dB, called the *low-frequency asymptote* for the magnitude curve. The approximation (8.38) determines a line at 0°, which is shown in Figure 8.6b. This is the low-frequency asymptote for the phase curve.

For $\omega > B$, we approximate $j\omega + B$ by $j\omega$ so that

$$H(\omega) \approx \frac{B}{j\omega}, \qquad \omega > B.$$

Then

$$|H(\omega)|_{dB} \approx 20 \log_{10}\left|\frac{B}{j\omega}\right| = 20 \log_{10}B - 20 \log_{10}\omega, \qquad \omega > B \quad (8.39)$$

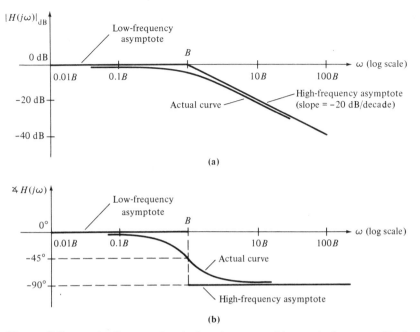

Figure 8.6. Bode diagrams for single-pole system: (a) magnitude curve; (b) phase curve.

and

$$\measuredangle \, H(\omega) \approx -90°, \qquad \omega > B. \qquad (8.40)$$

Equation (8.39) determines a line with slope -20 dB/decade, where a decade is a factor of 10 in frequency. Note that the line is located at 0 dB when $\omega = B$. This line is the high-frequency asymptote for the magnitude curve. The high-frequency asymptote is plotted in Figure 8.6a. Equation (8.40) determines a line at $-90°$, which is plotted in Figure 8.6b. This is the high-frequency asymptote for the phase curve.

Now the actual magnitude and phase curves approach the asymptotes as ω becomes much less than B and as ω becomes much greater than B. At the break frequency $\omega = B$, the actual magnitude curve is 3 dB off from the asymptotes and the actual phase curve is 45° off from the asymptotes. From this information it is possible to sketch the magnitude and phase curves with a fair degree of accuracy. The result is shown in Figure 8.6.

TRANSIENT AND STEADY-STATE RESPONSE OF SINGLE-POLE SYSTEMS For the system with transfer function $H(s) = B/(s + B)$, we shall now derive an explicit expression for the transient and steady-state responses resulting from the sinusoidal input $x(t) = A \cos \omega t$, $t \geq 0$, with no initial energy in the system at time $t = 0$. The Laplace transform of the resulting output response is given by

$$Y(s) = \frac{B}{s + B} \frac{As}{s^2 + \omega^2}.$$

We can write $Y(s)$ in the form

$$Y(s) = \frac{b}{s + B} + \frac{cs + d}{s^2 + \omega^2}, \tag{8.41}$$

where

$$b = [(s + B)Y(s)]_{s = -B} = \frac{BA(-B)}{(-B)^2 + \omega^2} = \frac{-B^2A}{B^2 + \omega^2}.$$

Comparing (8.41) and (8.27), we have that

$$\frac{\gamma(s)}{D(s)} = \frac{-B^2A/(B^2 + \omega^2)}{s + B},$$

and thus the first term on the right side of (8.41) is the Laplace transform of the transient response $y_{\text{tr}}(t)$. Taking the inverse Laplace transform gives

$$y_{\text{tr}}(t) = \frac{-B^2A}{B^2 + \omega^2} e^{-Bt}, \qquad t \geq 0.$$

Again comparing (8.41) and (8.27), we must have that the second term on the right side of (8.41) is the transform of the steady-state response $y_{\text{ss}}(t)$. So the output response $y(t)$ is

$$y(t) = \frac{-B^2A}{B^2 + \omega^2} e^{-Bt} + A|H(\omega)| \cos [\omega t + \sphericalangle H(\omega)], \qquad t \geq 0. \tag{8.42}$$

Inserting the expressions (8.35) and (8.36) for $|H(\omega)|$ and $\sphericalangle H(\omega)$ into (8.42), we obtain

$$y(t) = \frac{-B^2A}{B^2 + \omega^2} e^{-Bt} + \frac{BA}{\sqrt{B^2 + \omega^2}} \cos \left[\omega t - \tan^{-1} \left(\frac{\omega}{B} \right) \right], \qquad t \geq 0.$$

SINGLE-POLE SYSTEMS WITH A ZERO   From the above results we know that a single-pole system with no zero is a lowpass filter. We can change this frequency response characteristic by adding a zero to the system. In particular, consider the single-pole system with the transfer function

$$H(s) = \frac{s + C}{s + B}.$$

We continue to assume that $B > 0$, so that the system is still asymptotically stable. The constant C may be positive or negative, so that the zero $s = -C$ may be in the left-half plane or the right-half plane.

An integrator realization of the system with transfer function $H(s) = (s + C)/(s + B)$ is given in Figure 8.7.

Setting $s = j\omega$ in $H(s) = (s + C)/(s + B)$, we have

$$H(\omega) = \frac{j\omega + C}{j\omega + B}.$$

Then the magnitude and phase functions are given by

$$|H(\omega)| = \frac{|j\omega + C|}{|j\omega + B|} = \sqrt{\frac{\omega^2 + C^2}{\omega^2 + B^2}} \qquad (8.43)$$

$$\sphericalangle H(\omega) = \sphericalangle (j\omega + C) - \sphericalangle (j\omega + B)$$

$$= \tan^{-1}\left(\frac{\omega}{C}\right) - \tan^{-1}\left(\frac{\omega}{B}\right). \qquad (8.44)$$

We shall determine the Bode diagrams in the case when $0 < C < B$. We begin by determining the asymptotes that are based on the following approximations.

The break frequency associated with the factor $j\omega + C$ is $\omega = C$ and the break frequency associated with the term $j\omega + B$ is $\omega = B$. Then for $\omega < C$, we can approximate $j\omega + C$ and $j\omega + B$ by C and B, respectively. Thus we can approximate $H(\omega)$ by

$$H(\omega) \approx \frac{C}{B}, \qquad \omega < C. \qquad (8.45)$$

When $C < \omega < B$, we can approximate $j\omega + C$ by $j\omega$ and $j\omega + B$ by B, so

$$H(\omega) \approx \frac{j\omega}{B}, \qquad C < \omega < B. \qquad (8.46)$$

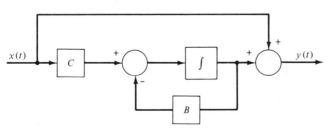

Figure 8.7. Integrator realization of single-pole system with zero.

Finally, for $\omega > B$, both $j\omega + C$ and $j\omega + B$ can be approximated by $j\omega$, and we have

$$H(\omega) \approx \frac{j\omega}{j\omega} = 1, \qquad \omega > B. \tag{8.47}$$

The approximations given by (8.45)–(8.47) determine the asymptotes for the Bode diagrams. These are plotted in Figure 8.8. The actual magnitude and phase curves are also shown in Figure 8.8.

From Figure 8.8a we see that the system is a highpass filter since it passes with little attenuation all frequencies above B rad/sec. If $C/B < 0.707$, the filter has a 3-dB point equal to the value of frequency for which the magnitude is down by 3 dB from the peak value of 0 dB. If $C = 0$, the magnitude curve is down by 3 dB at the point $\omega = B$.

Although highpass filters exist in theory, they do not exist in practice, since actual systems cannot pass sinusoids with arbitrarily large frequencies. In other words, no actual system can have an infinite bandwidth. Thus any implementation of the realization in Figure 8.7 would only be an approximation of a highpass filter (in the case $0 < C < B$).

From Figure 8.8b we see that the phase is positive. As a result of this characteristic, the system is sometimes called a *lead system* or a *lead network*.

To conclude this example, we now suppose that $C > B$ and that $H(s)$ is given by

$$H(s) = \frac{B}{C} \frac{s + C}{s + B}.$$

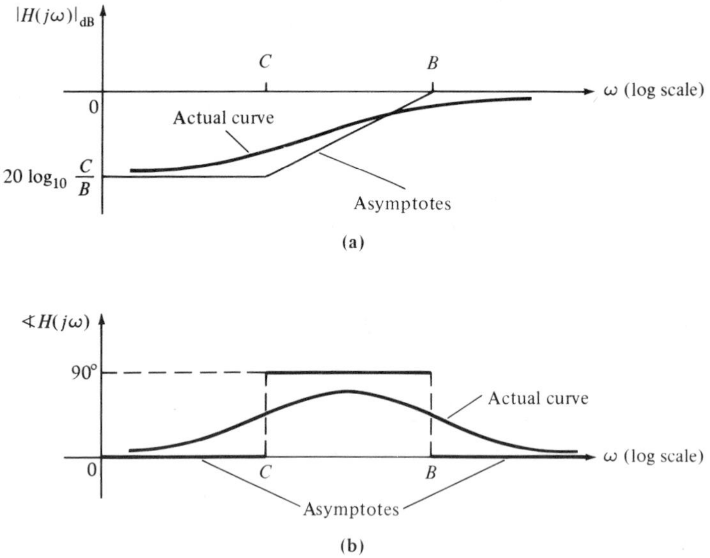

Figure 8.8. Bode diagrams for single-pole system with zero at $s = -C$, $0 < C < B$: (a) magnitude curve; (b) phase curve.

By sketching the Bode diagrams for this case, we can verify that the system is a lowpass filter with negative phase. Because of the negative-phase characteristic, the system is called a *lag system* or a *lag network*.

SYSTEM IDENTIFICATION   Frequency response data are often used to determine the system model. To illustrate this, we consider a simple case in which the system parameters are identified by measuring the responses to sinusoidal inputs.

Again consider the one-pole system with the frequency function

$$H(\omega) = \frac{j\omega + C}{j\omega + B},$$

where $B > 0$ and $C > 0$. In some applications the parameters B and C in $H(\omega)$ may not be known. We will show that B and C can be determined by measuring the magnitudes of the steady-state responses to two sinusoidal inputs $x(t) = A \cos \omega_0 t$ and $x(t) = A \cos \omega_1 t$. Here ω_0 and ω_1 are any two frequencies for which the input $A \cos (\omega t)$ can actually be applied (e.g., using a sine-wave generator) to the system under study. Let the resulting steady-state responses be given by $M_0 \cos (\omega_0 t + \theta_0)$ and $M_1 \cos (\omega_1 t + \theta_1)$. We can determine the magnitudes M_0 and M_1 by measuring the steady-state responses to our sinusoidal inputs.

Now by the above analysis,

$$M_0 = A|H(\omega_0)| = A \sqrt{\frac{\omega_0^2 + C}{\omega_0^2 + B}}$$

and

$$M_1 = A|H(\omega_1)| = A \sqrt{\frac{\omega_1^2 + C}{\omega_1^2 + B}}.$$

Note that these equations are nonlinear equations in B and C. Nevertheless, we can solve for B and C, which gives

$$B = \sqrt{\frac{A^2}{M_1^2 - M_0^2} (\omega_1^2 - \omega_0^2) + \frac{M_0^2 \omega_0^2 - M_1^2 \omega_1^2}{M_1^2 - M_0^2}}$$

and

$$C = \sqrt{\frac{M_0^2 M_1^2}{(M_1^2 - M_0^2)A^2} (\omega_0^2 - \omega_1^2) + \frac{M_0^2 \omega_1^2 - M_1^2 \omega_0^2}{M_1^2 - M_0^2}}.$$

Thus it is possible to identify the system parameters B, C by measuring the magnitudes of the responses to two sinusoidal inputs.

If it is possible to measure the phase of the steady-state response, we could determine B and C by measuring the response to a single sinusoidal input.

This method of system identification generalizes to n-pole systems, although the computations may get complicated for n large since it is necessary to solve a system of nonlinear equations. Also note that to be able to use this method, we must know the number of poles of the system. In practice, it may suffice to make a "good guess" as to the number of poles.

8.6

TWO-POLE SYSTEMS

Now consider the system with transfer function

$$H(s) = \frac{\omega_n^2}{s^2 + 2\zeta\omega_n s + \omega_n^2}, \tag{8.48}$$

where $\zeta > 0$ and $\omega_n > 0$. From the Routh stability test, this system is asymptotically stable if and only if

$$2\zeta\omega_n > 0 \quad \text{and} \quad \omega_n^2 > 0.$$

These conditions are satisfied since we are assuming that $\zeta > 0$ and $\omega_n > 0$. Thus the system is asymptotically stable.

An RLC circuit realization of the system is shown in Figure 8.9. The verification that the circuit has the transfer function (8.48) can be carried out by working with the s-domain representation of the circuit (see Section 6.1). We invite the reader to check this.

From the results in Section 2.3, we also have the integrator realization shown in Figure 8.10. This realization can be built using the op-amp circuit shown in Figure 2.13.

The poles of the system with the transfer function (8.48) are located at

$$s = \frac{-2\zeta\omega_n \pm \sqrt{4\zeta^2\omega_n^2 - 4\omega_n^2}}{2} = -\zeta\omega_n \pm \omega_n\sqrt{\zeta^2 - 1}. \tag{8.49}$$

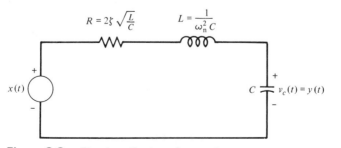

Figure 8.9. Circuit realization of two-pole system.

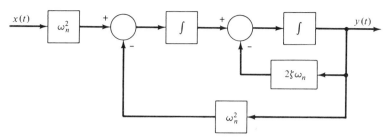

Figure 8.10. Integrator realization of two-pole system.

Letting

$$p_1 = -\zeta\omega_n - \omega_n\sqrt{\zeta^2 - 1}, \qquad p_2 = -\zeta\omega_n + \omega_n\sqrt{\zeta^2 - 1},$$

we have

$$H(s) = \frac{\omega_n^2}{(s - p_1)(s - p_2)}.$$

Then

$$|H(\omega)| = \frac{\omega_n^2}{|j\omega - p_1||j\omega - p_2|} \tag{8.50}$$

and

$$\angle H(\omega) = -\angle(j\omega - p_1) - \angle(j\omega - p_2). \tag{8.51}$$

Now, from (8.49), we see that the poles of the system are real when $\zeta \geq 1$. We can determine the shape of the frequency response curves in this case by considering the vector representations of $j\omega - p_1$ and $j\omega - p_2$ shown in Figure 8.11. Here the magnitudes $|j\omega - p_1|$ and $|j\omega - p_2|$ become infinite as $\omega \to \infty$, and the angles $\angle(j\omega - p_1)$ and $\angle(j\omega - p_2)$ approach 90° as $\omega \to \infty$. Then, from (8.50) and (8.51), we have that the magnitude $|H(\omega)|$ starts with value 1 at $\omega = 0$ and approaches zero as $\omega \to \infty$. The phase $\angle H(\omega)$ starts with value 0° when $\omega = 0$ and approaches $-180°$ as $\omega \to \infty$. Thus the system is a low-pass filter. The 3-dB bandwidth of the filter depends on ζ and ω_n. When $\zeta = 1$, the 3-dB bandwidth is equal to $\sqrt{\sqrt{2} - 1}\,\omega_n$. (This can be computed from the vector representation in Figure 8.11.)

With $\zeta = 1$ and $\omega_n = 3.1$ rad/sec, the 3-dB bandwidth is approximately equal to 2 rad/sec. The frequency response curves for this case were computed using the program in Figure A.10. The results are shown in Figure 8.12. Also displayed in Figure 8.12 are the frequency response curves of the one-pole low-pass filter with transfer function $H(s) = 2/(s + 2)$. Note that the two-pole filter has a sharper cutoff than the one-pole filter.

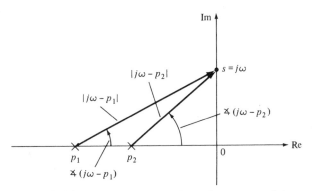

Figure 8.11. Vector representations of $j\omega - p_1$ and $j\omega - p_2$ when poles are real.

TWO-POLE SYSTEMS WITH A RESONANCE

We now assume that $0 \leq \zeta < 1$, so that the poles are complex. From (8.49) we have that the poles of the system are located at $s = -\zeta\omega_n \pm j\omega_n\sqrt{1 - \zeta^2}$. We can then write $H(\omega)$ in the form

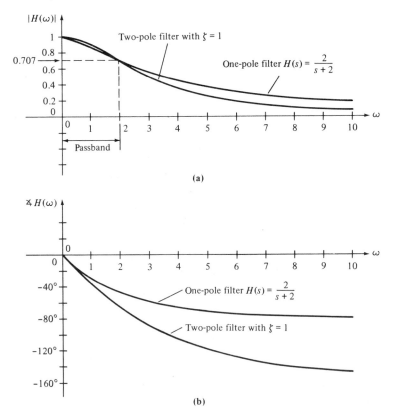

Figure 8.12. Frequency response curves of two-pole filter with $\zeta = 1$ and one-pole filter: (a) magnitude curve; (b) phase curve.

$$H(\omega) = \frac{\omega_n^2}{(j\omega + \zeta\omega_n + j\omega_n\sqrt{1 - \zeta^2})(j\omega + \zeta\omega_n - j\omega_n\sqrt{1 - \zeta^2})}. \quad (8.52)$$

The vector representations of $j\omega + \zeta\omega_n + j\omega_n\sqrt{1 - \zeta^2}$ and $j\omega + \zeta\omega_n - j\omega_n\sqrt{1 - \zeta^2}$ are shown in Figure 8.13. Note that as ω increases from $\omega = 0$, the magnitude $|j\omega + \zeta\omega_n - j\omega_n\sqrt{1 - \zeta^2}|$ decreases, while the magnitude $|j\omega + \zeta\omega_n + j\omega_n\sqrt{1 - \zeta^2}|$ increases. For $\omega > \omega_n\sqrt{1 - \zeta^2}$, both these magnitudes grow until they become infinite. Since the magnitude $|H(\omega)|$ is equal to ω_n^2 divided by $|j\omega + \zeta\omega_n + j\omega_n\sqrt{1 - \zeta^2}|$ and $|j\omega + \zeta\omega_n - j\omega_n\sqrt{1 - \zeta^2}|$, we see that $|H(\omega)|$ "rolls off" to zero for $\omega > \omega_n\sqrt{1 - \zeta^2}$. However, it is not clear if $|H(\omega)|$ first increases or decreases as ω is increased from $\omega = 0$. We claim that when

$$\zeta < \frac{1}{\sqrt{2}}, \quad (8.53)$$

the magnitude $|H(\omega)|$ increases as ω is increased from 0 and when

$$\zeta \geq \frac{1}{\sqrt{2}},$$

the magnitude $|H(\omega)|$ decreases as ω is increased from 0. The proof of this follows by taking the derivative of $|H(\omega)|$ with respect to ω. If this derivative is zero for some positive real value ω_r of ω, then $|H(\omega)|$ must have a peak at this value of ω, which implies that $|H(\omega)|$ must increase as ω is varied from 0 to ω_r. Taking the derivative of the magnitude function (8.50), we obtain

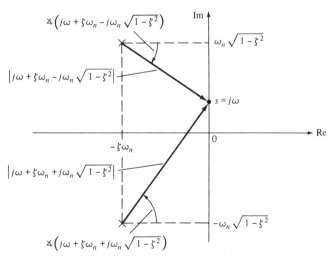

Figure 8.13. Complex-pole case.

$$\frac{d}{d\omega}|H(\omega)| = \frac{d}{d\omega}\frac{\omega_n^2}{\sqrt{(\omega_n^2 - \omega^2)^2 + 4\zeta^2\omega_n^2\omega^2}}$$

$$= \omega_n^2\,[2(\omega_n^2 - \omega^2)(-2\omega) + 8\zeta^2\omega_n^2\omega][(\omega_n^2 - \omega^2)^2 + 4\zeta^2\omega_n^2\omega^2]^{-3/2}$$

$$= 4\omega_n^2\omega(-\omega_n^2 + \omega^2 + 2\zeta^2\omega_n^2)[(\omega_n^2 - \omega^2)^2 + 4\zeta^2\omega_n^2\omega^2]^{-3/2}. \tag{8.54}$$

From (8.54) we see that the derivative is zero when $\omega = 0$ or when

$$-\omega_n^2 + \omega^2 + 2\zeta^2\omega_n^2 = 0. \tag{8.55}$$

Solving (8.55) for ω gives

$$\omega = \pm\omega_n\sqrt{1 - 2\zeta^2}.$$

Thus $d|H(\omega)|/d\omega$ is zero for some positive real value of ω if and only if

$$1 - 2\zeta^2 > 0$$

or

$$\zeta^2 < \frac{1}{2},$$

which is equivalent to (8.53). Hence $|H(\omega)|$ increases as ω is increased from 0 if and only if (8.53) is satisfied, in which case we say that the system has a *resonance* at the frequency

$$\omega_r = \omega_n\sqrt{1 - 2\zeta^2}.$$

The frequency ω_r is called a *resonant frequency* of the system. When $1/\sqrt{2} \leq \zeta < \infty$, the system does not have a resonance, and there is no resonant frequency.

Let us compute the magnitude $|H(\omega)|$ at the resonant frequency ω_r. Setting $\omega = \omega_r$ in the magnitude function $|H(\omega)|$, we obtain

$$|H(\omega_r)| = \frac{\omega_n^2}{\sqrt{[\omega_n^2 - \omega_n^2(1 - 2\zeta^2)]^2 + 4\zeta^2\omega_n^2\omega_n^2(1 - 2\zeta^2)}}$$

$$= \frac{\omega_n^2}{\omega_n^2\sqrt{1 - 2(1 - 2\zeta^2) + (1 - 2\zeta^2)^2 + 4\zeta^2(1 - 2\zeta^2)}}$$

$$= \frac{1}{\sqrt{1 - (1 - 2\zeta^2)^2}}$$

$$= \frac{1}{2\zeta\sqrt{1 - \zeta^2}}.$$

Let M_p denote the peak magnitude of $|H(\omega)|$; that is, $M_p = |H(\omega_r)|$. A plot of M_p versus ζ for $\zeta < 1/\sqrt{2}$ is given in Figure 8.14. From the plot we see that the peak magnitude M_p approaches infinity as ζ approaches zero. Thus the resonance is more pronounced the smaller ζ is. The parameter ζ can therefore be interpreted as a damping factor, since the smaller it is, the greater the resonance. The parameter ζ is called the *damping ratio*. In the *RLC* circuit realization shown in Figure 8.9, the resistance R is directly proportional to the damping ratio ζ. This makes sense since the resistance determines the amount of damping in the circuit.

From Figure 8.13 we also see that the closer ζ is to zero, the closer the poles are to the $j\omega$-axis of the complex plane. Therefore, the resonance is stronger the closer the poles are to the $j\omega$-axis.

Note that if $\zeta = 0$, so that there is no damping, the resonant frequency ω_r is equal to ω_n. As a result of this property, ω_n is called the *undamped natural frequency* of the system.

If $\zeta < 1/\sqrt{2}$, the system behaves like a bandpass filter since it amplifies input sinusoids whose frequencies are in a neighborhood of ω_r. The resonant frequency ω_r is the center frequency of the filter. We define the 3-dB bandwidth of the filter to be all those frequencies ω for which the magnitude $|H(\omega)|$ is greater than or equal to $M_p/\sqrt{2}$. In decibels, this is equivalent to requiring that

$$|H(\omega)|_{dB} \geq 20 \log M_p - 3 \text{ dB}.$$

We shall now derive an expression for the approximate 3 dB bandwidth. In the following analysis, we assume that $\omega_r \approx \omega_n\sqrt{1 - \zeta^2}$. This will be the case if $\zeta < 0.2$.

Now when $\omega = \omega_r$, the vector representations of $j\omega + \zeta\omega_n + j\omega_n\sqrt{1 - \zeta^2}$ and $j\omega + \zeta\omega_n - j\omega_n\sqrt{1 - \zeta^2}$ are shown in Figure 8.15. We see that when we change the value of s from $s = j\omega_r$ to $s = j(\omega_r - \zeta\omega_n)$, the magnitude of $j\omega + \zeta\omega_n - j\omega_n\sqrt{1 - \zeta^2}$ increases by a factor of $\sqrt{2}$, while the magnitude of $j\omega + \zeta\omega_n + j\omega_n\sqrt{1 - \zeta^2}$ does not change very much. Thus by

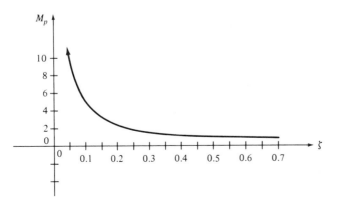

Figure 8.14. Plot of peak magnitude M_p versus ζ for $\zeta < 1/\sqrt{2}$.

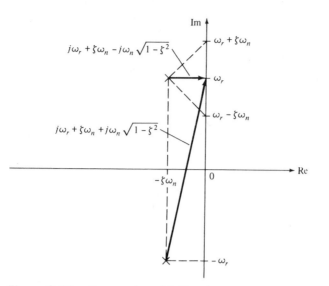

Figure 8.15. Computation of 3-dB points.

(8.52), when $\omega = \omega_r - \zeta\omega_n$ the magnitude $|H(\omega)|_{dB}$ is down by approximately 3 dB from the peak value $20 \log M_p$. So $\omega = \omega_r - \zeta\omega_n$ is a 3-dB point of the system. By a similar argument, we have that $\omega = \omega_r + \zeta\omega_n$ is another 3-dB point. The approximate 3-dB bandwidth of the system is then equal to

$$\omega_r + \zeta\omega_n - (\omega_r - \zeta\omega_n) = 2\zeta\omega_n.$$

EXAMPLE 8.8. As a specific example, let us design a bandpass filter with center frequency ω_r equal to 10 rad/sec and with 3-dB bandwidth equal to 2 rad/sec. Then

$$10 = \omega_r = \omega_n\sqrt{1 - 2\zeta^2}$$

and

$$2 = 2\zeta\omega_n.$$

Solving the second equation for ω_n and inserting the result into the first equation, we have

$$10 = \frac{\sqrt{1 - 2\zeta^2}}{\zeta},$$

and thus

$$\frac{1 - 2\zeta^2}{\zeta^2} = 100.$$

Solving for ζ gives

$$\zeta = \frac{1}{\sqrt{102}} \approx 0.1.$$

Then

$$\omega_n = \frac{1}{\zeta} = 10,$$

and the transfer function is

$$H(s) = \frac{100}{s^2 + 2s + 100}.$$

Using the program in Figure A.10, we obtain the frequency response curves shown in Figure 8.16. Note that the phase is approximately linear over the passband of the filter ($9 \leq \omega \leq 11$). We shall conclude this example by computing the transient response of the system resulting from the sinusoidal input $x(t) = A \cos \omega t$, $t \geq 0$. The transform of the complete response resulting from this input is

$$Y(s) = \frac{100As}{(s^2 + 2s + 100)(s^2 + \omega^2)}$$

$$= \frac{100As}{(s + 1 - j\sqrt{99})(s + 1 + j\sqrt{99})(s^2 + \omega^2)}.$$

We can write $Y(s)$ in the form

$$Y(s) = \frac{c}{s + 1 - j\sqrt{99}} + \frac{\bar{c}}{s + 1 + j\sqrt{99}} + \frac{ds + e}{s^2 + \omega^2}, \qquad (8.56)$$

where

$$c = [(s + 1 - j\sqrt{99})Y(s)]_{s = -1 + j\sqrt{99}}$$

$$= \frac{100A(-1 + j\sqrt{99})}{(j2\sqrt{99})[(-1 + j\sqrt{99})^2 + \omega^2]}.$$

The transient response $y_{tr}(t)$ is equal to the inverse Laplace transform of the sum of the first two terms on the right side of (8.56). Using the identity (5.70), we have

$$y_{tr}(t) = 2|c|e^{-t} \cos (\sqrt{99}t + \sphericalangle c), \qquad t \geq 0.$$

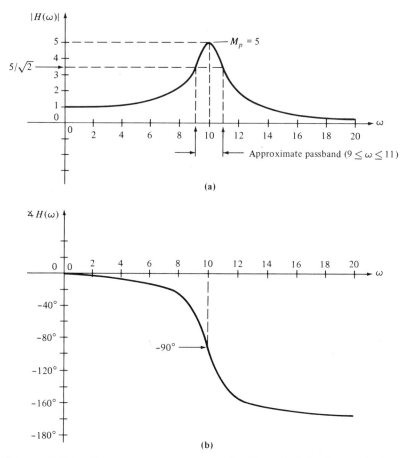

Figure 8.16. Frequency response curves for Example 8.8: (a) magnitude curve; (b) phase curve.

BUTTERWORTH FILTERS

For the two-pole system with the transfer function

$$H(s) = \frac{\omega_n^2}{s^2 + 2\zeta\omega_n s + \omega_n^2},$$

in Section 8.6 we showed that the system is a lowpass filter when $\zeta \geq 1/\sqrt{2}$. If we take $\zeta = 1/\sqrt{2}$, the resulting lowpass filter is said to be *maximally flat* since the variation in the magnitude $|H(\omega)|$ is as small as possible across the passband of the filter. This filter is called the *two-pole Butterworth filter*.

The transfer function of the two-pole Butterworth filter is

$$H(s) = \frac{\omega_n^2}{s^2 + \sqrt{2}\,\omega_n s + \omega_n^2}.$$

Factoring the denominator of $H(s)$, we see that the poles are located at

$$s = -\frac{\omega_n}{\sqrt{2}} \pm j\frac{\omega_n}{\sqrt{2}}.$$

Note that the magnitude of each of the poles is equal to ω_n.

Setting $s = j\omega$ in $H(s)$, we have that the magnitude function of the two-pole Butterworth filter is

$$\begin{aligned}
|H(\omega)| &= \frac{\omega_n^2}{\sqrt{(\omega_n^2 - \omega^2)^2 + 2\omega_n^2\omega^2}} \\[2mm]
&= \frac{\omega_n^2}{\sqrt{\omega_n^4 - 2\omega_n^2\omega^2 + \omega^4 + 2\omega_n^2\omega^2}} \\[2mm]
&= \frac{\omega_n^2}{\sqrt{\omega_n^4 + \omega^4}} \\[2mm]
&= \frac{1}{\sqrt{1 + (\omega/\omega_n)^4}}.
\end{aligned} \tag{8.57}$$

From (8.57) we see that the 3-dB bandwidth of the Butterworth filter is equal to ω_n. For the case $\omega_n = 2$ rad/sec, the frequency response curves of the Butterworth filter are plotted in Figure 8.17. Also displayed are the frequency response curves for the one-pole lowpass filter with transfer function $H(s) = 2/(s + 2)$, and the two-pole lowpass filter with $\zeta = 1$ and with 3-dB bandwidth equal to 2 rad/sec. Note that the Butterworth filter has the sharpest cutoff of all three filters.

There are two-pole lowpass filters with a sharper cutoff than the Butterworth filter, but all these other filters (such as the Chebyshev filter) have "ripple" in the passband. We shall not consider these types of filters.

***n*-POLE BUTTERWORTH FILTER** For any positive integer n, the *n-pole Butterworth filter* is the lowpass filter of order n with a maximally flat frequency response across the passband. The distinguishing characteristic of the Butterworth filter is that the poles lie on a semicircle in the open left-half plane. The radius of the semicircle is equal to ω_c, where ω_c is the 3-dB bandwidth of the filter. In the third-order case, the poles are as displayed in Figure 8.18.

The transfer function of the three-pole Butterworth filter is

$$H(s) = \frac{\omega_c^3}{(s + \omega_c)(s^2 + \omega_c s + \omega_c^2)} = \frac{\omega_c^3}{s^3 + 2\omega_c s^2 + 2\omega_c^2 s + \omega_c^3}.$$

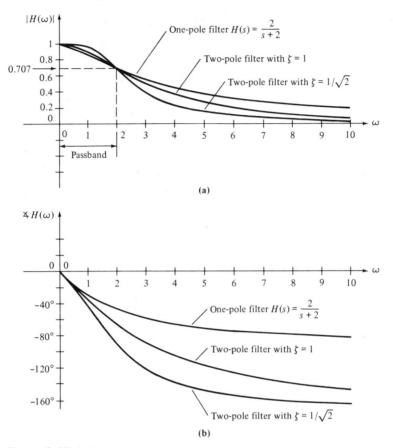

(a)

(b)

Figure 8.17. Frequency curves of one and two-pole low-pass filters: (a) magnitude curves; (b) phase curves.

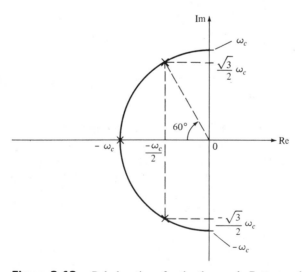

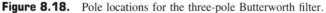

Figure 8.18. Pole locations for the three-pole Butterworth filter.

Setting $s = j\omega$ in $H(s)$ and taking the magnitude, we have that the magnitude function of the three-pole filter is

$$|H(\omega)| = \frac{1}{\sqrt{1 + (\omega/\omega_c)^6}}.$$

The magnitude function is plotted in Figure 8.19 for the case $\omega_c = 2$. Also plotted is the magnitude function of the two-pole Butterworth filter with band-width equal to 2. Clearly, the three-pole filter has a sharper cutoff than the two-pole filter.

In the general case, the magnitude function of the n-pole Butterworth filter is

$$|H(\omega)| = \frac{1}{\sqrt{1 + (\omega/\omega_c)^{2n}}}.$$

The transfer function can be determined from a table for Butterworth polynomials. For example, when $n = 4$, the transfer function is

$$H(s) = \frac{\omega_c^4}{(s^2 + 0.765\omega_c s + \omega_c^2)(s^2 + 1.85\omega_c s + \omega_c^2)}.$$

When $n = 5$, the transfer function is

$$H(s) = \frac{\omega_c^5}{(s + \omega_c)(s^2 + 0.618\omega_c s + \omega_c^2)(s^2 + 1.62\omega_c s + \omega_c^2)}.$$

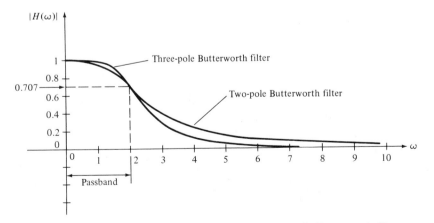

Figure 8.19. Magnitude curves of two-pole and three-pole Butterworth filters.

STABILITY OF DISCRETE-TIME SYSTEMS

In this section and Section 8.9 we develop the discrete-time version of the results derived in earlier sections. We begin with the definition of stability of linear time-invariant discrete-time systems. As in the continuous-time case, asymptotic stability is characterized in terms of the behavior of the zero-input response. We define this response below.

Consider the linear time-invariant finite-dimensional discrete-time system given by the transfer function

$$H(z) = \frac{N(z)}{D(z)}$$

$$= \frac{b_m z^m + b_{m-1} z^{m-1} + \cdots + b_1 z + b_0}{a_n z^n + a_{n-1} z^{n-1} + \cdots + a_1 z + a_0}, \qquad a_n \neq 0, \quad (8.58)$$

where $m \leq n$.

Let $Y(z)$ denote the z-transform of the output response resulting from the input $x(kT)$ and the initial values $y(-T)$, $y(-2T)$, . . . , $y(-nT)$. From the results in Section 7.5, if $x(kT) = 0$ for $k = -1, -2, \ldots, -n$, we can write $Y(z)$ in the form

$$Y(z) = \frac{C(z)}{D(z)} + \frac{N(z)}{D(z)} X(z). \qquad (8.59)$$

Recall that the first term on the right side of (8.59) is the z-transform of the part of the output response resulting from the initial values $y(-T)$, $y(-2T)$, . . . , $y(-nT)$. The numerator $C(z)$ of this term is a polynomial in z whose coefficients depend on the initial values $y(-T)$, $y(-2T)$, . . . , $y(-nT)$. The degree of $C(z)$ is less than or equal to n.

If $x(kT) = 0$ for $k = 0, 1, 2, \ldots$, then $X(z) = 0$ and the z-transform of the output response becomes

$$Y(z) = \frac{C(z)}{D(z)}.$$

The inverse z-transform of $C(z)/D(z)$ is called the *zero-input response* since it is the response resulting from just the initial values $y(-T)$, $y(-2T)$, . . . , $y(-nT)$. We shall denote the zero-input response by $y_{zi}(kT)$; that is, $y_{zi}(kT)$ is the inverse z-transform of $C(z)/D(z)$.

DEFINITIONS OF STABILITY   The discrete-time system with transfer function (8.58) is asymptotically stable if

$$y_{zi}(kT) \rightarrow 0 \qquad \text{as } k \rightarrow \infty \qquad (8.60)$$

for all initial values $y(-T), \ldots, y(-nT)$. Since any initial energy in the system at time $kT = 0$ is characterized by the initial values $y(-T), y(-2T), \ldots, y(-nT)$, asymptotic stability means that the output response resulting from initial energy always decays to zero as the discrete-time index k becomes large.

The system is marginally stable if for each set of initial values $y(-T), y(-2T), \ldots, y(-nT)$, there is a finite positive constant M such that

$$|y_{zi}(kT)| \leq M \qquad \text{for } k = 0, 1, 2, \ldots.$$

The system is unstable if for some initial values $y(-T), y(-2T), \ldots, y(-nT)$, the magnitude of the zero-input response grows without bound as $k \to \infty$; that is,

$$|y_{zi}(kT)| \to \infty \qquad \text{as } k \to \infty.$$

POLE CRITERIA FOR STABILITY Let $p_1, p_2, \ldots, p_n$ denote the poles of the system with transfer function (8.58). By definition of the poles, we have

$$H(z) = \frac{N(z)}{a_n(z - p_1)(z - p_2) \cdots (z - p_n)}.$$

Now the system is asymptotically stable if and only if

$$|p_i| < 1 \qquad \text{for } i = 1, 2, \ldots, n. \tag{8.61}$$

The system is marginally stable if and only if $|p_i| \leq 1$ for all nonrepeated poles, and $|p_i| < 1$ for all repeated poles. The system is unstable if $|p_i| > 1$ for one or more poles, or if $|p_i| = 1$ for a repeated pole.

By (8.61), the system is asymptotically stable if and only if all the poles are located in the *open unit-disc* of the complex plane. The open unit-disc is that part of the complex plane consisting of all complex numbers whose magnitude is strictly less than 1. The open unit-disc is the hatched region shown in Figure 8.20.

The system is marginally stable if all the poles lie in the open unit-disc, except that there may be nonrepeated poles on the *unit-circle,* where the unit-circle is the set of all complex numbers whose magnitude is equal to 1.

Defining the *closed unit-disc* to be the open unit-disc plus the unit-circle, we have that the system is unstable if there are one or more poles outside the closed unit-disc, or if there are repeated poles on the unit-circle.

The proof of the pole criteria for stability follows from an analysis of the inverse z-transform of $C(z)/D(z)$. The details resemble our derivation of the pole criteria for stability in the continuous-time case (see Section 8.1). We leave this to the interested reader.

From the above results, we see that the stability boundary in the discrete-time case is the unit-circle of the complex plane. In contrast, in the continuous-time case the stability boundary is the imaginary axis ($j\omega$-axis) of the complex plane.

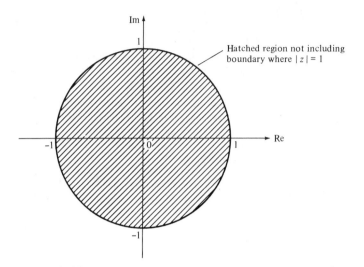

Figure 8.20. Open unit-disc.

Thus the pole criteria for stability differ from the discrete-time case to the continuous-time case. In Chapter 12 we show that there is a mapping between the stability region in the continuous-time case and the stability region in the discrete-time case.

BIBO STABILITY As in the continuous-time case, asymptotic stability of a linear time-invariant finite-dimensional discrete-time system implies BIBO (bounded-input bounded-output) stability of the system. The definition of BIBO stability for discrete-time systems is as follows.

Let $y_{zs}(kT)$ denote the output response resulting from the input $x(kT)$ with $y(kT) = 0$ for $k = -1, -2, \ldots, -n$. The response $y_{zs}(kT)$ is called the zero-state response since it is the output resulting from $x(kT)$ with the system in the zero state (no initial energy) at time $kT = 0$. From the results in Chapter 4, we have

$$y_{zs}(kT) = (h * x)(kT) = \sum_{i=0}^{k} h(kT - iT)x(iT),$$

where $h(kT)$ is the unit-pulse response of the system.

Now the system is said to be BIBO stable if whenever

$$|x(kT)| \le c_x \qquad \text{for } k = 0, 1, 2, \ldots,$$

for some finite positive constant c_x, there is a finite positive constant c_y such that

$$|y_{zs}(kT)| \le c_y \qquad \text{for } k = 0, 1, 2, \ldots.$$

This simply means that bounded inputs produce bounded outputs, which is the reason for the term "BIBO" stability.

As noted above, asymptotic stability always implies BIBO stability. In addition, if there are no common poles and zeros, asymptotic stability and BIBO stability turn out to be equivalent. The proof of this result is beyond the scope of this book and thus will not be given.

★JURY STABILITY TEST Suppose that

$$H(z) = \frac{N(z)}{D(z)},$$

where

$$D(z) = a_n z^n + a_{n-1} z^{n-1} + \cdots + a_1 z + a_0, \qquad a_n > 0.$$

By using the Routh stability test as discussed in Section 8.3, we were able to test for asymptotic stability of a continuous-time system without having to compute the poles of the system. There is a discrete-time counterpart to the Routh test. This test is based on the Jury table, which is displayed in Table 8.5. Here we are assuming that the degree n of $D(z)$ is greater than or equal to 2.

As in the Routh array, the first two rows of the Jury table are filled by the coefficients of $D(z)$. However, the manner in which the coefficients are distributed in the first two rows of the Jury table is different from that in the Routh array. The coefficients in the third and fourth rows of the Jury table are given by

$$b_i = a_0 a_i - a_{n-i} a_n, \qquad i = 0, 1, 2, \ldots, n - 1.$$

TABLE 8.5. Jury Table

Row	z^0	z^1	z^2	$\cdots$	z^{n-2}	z^{n-1}	z^n
1	a_0	a_1	a_2	$\cdots$	a_{n-2}	a_{n-1}	a_n
2	a_n	a_{n-1}	a_{n-2}	$\cdots$	a_2	a_1	a_0
3	b_0	b_1	b_2	$\cdots$	b_{n-2}	b_{n-1}	
4	b_{n-1}	b_{n-2}	b_{n-3}	$\cdots$	b_1	b_0	
5	c_0	c_1	c_2	$\cdots$	c_{n-2}		
6	c_{n-2}	c_{n-3}	c_{n-4}	$\cdots$	c_0		
.	.	.	.				
.	.	.	.				
.	.	.	.				
$2n-5$	d_0	d_1	d_2	d_3			
$2n-4$	d_3	d_2	d_1	d_0			
$2n-3$	e_0	e_1	e_2				

The coefficients in the fifth and sixth rows are given by

$$c_i = b_0 b_i - b_{n-i-1} b_{n-1}, \qquad i = 0, 1, 2, \ldots, n - 2.$$

The process continues down to the $(2n - 3)$th row, whose last element e_2 is given by

$$e_2 = d_0 d_2 - d_1 d_3.$$

Having computed the Jury table, by the Jury test we have that the system is asymptotically stable (all poles are in the open unit-disc) if and only if all the following conditions are satisfied:

$$D(1) > 0 \qquad \text{and} \qquad (-1)^n D(-1) > 0$$
$$a_n > |a_0|$$
$$|b_0| > |b_{n-1}|$$
$$|c_0| > |c_{n-2}|$$
$$.$$
$$.$$
$$.$$
$$|e_0| > |e_2|.$$

EXAMPLE 8.9. Suppose that

$$D(z) = z^2 + a_1 z + a_0.$$

The Jury table is shown in Table 8.6.

TABLE 8.6. Jury Table for Example 8.9

Row	z^0	z^1	z^2
1	a_0	a_1	1

From the conditions given above, we have that the roots of $D(z) = 0$ are in the open unit-disc if and only if

$$D(1) = 1 + a_1 + a_0 > 0$$
$$(-1)^2 D(-1) = 1 - a_1 + a_0 > 0$$
$$1 > |a_0|.$$

The first two conditions are equivalent to

$$-a_1 < 1 + a_0 \qquad \text{and} \qquad a_1 < 1 + a_0,$$

which in turn is equivalent to

$$|a_1| < 1 + a_0.$$

Hence the roots of $D(z) = 0$ are in the open unit-disc if and only if

$$|a_1| < 1 + a_0 \quad \text{and} \quad |a_0| < 1. \tag{8.62}$$

This is an interesting result since it is not obvious that (8.62) is a necessary and sufficient condition for both roots of a second-degree polynomial to be in the open unit-disc.

8.9

FREQUENCY RESPONSE OF DISCRETE-TIME SYSTEMS

Again consider the linear time-invariant finite-dimensional discrete-time system with the transfer function

$$H(z) = \frac{N(z)}{D(z)}$$

$$= \frac{b_m z^m + b_{m-1} z^{m-1} + \cdots + b_1 z + b_0}{a_n z^n + a_{n-1} z^{n-1} + \cdots + a_1 z + a_0}, \quad a_n \neq 0. \tag{8.63}$$

Throughout this section we assume that the system is asymptotically stable.

The frequency response characteristics of the system with the transfer function (8.63) are determined from the response of the system to the discrete-time sinusoidal input

$$x(kT) = A \cos \omega k, \quad k = 0, 1, 2, \ldots, \tag{8.64}$$

where A and ω are fixed (but arbitrary) real constants.

The analysis of the system response resulting from the input (8.64) is a natural discrete-time counterpart to the development we gave in Section 8.4 in the continuous-time case. The details are as follows.

From Table 7.3 the z-transform of the sinusoidal input (8.64) is

$$X(z) = \frac{A[z^2 - (\cos \omega)z]}{z^2 - (2 \cos \omega)z + 1}.$$

If there is no initial energy in the system at time $kT = 0$, the z-transform of the resulting output response is

$$Y(z) = \frac{AN(z)[z^2 - (\cos \omega)z]}{D(z)[z^2 - (2 \cos \omega)z + 1]}.$$

Now

$$z^2 - (2 \cos \omega)z + 1 = (z - \cos \omega - j \sin \omega)(z - \cos \omega + j \sin \omega)$$
$$= (z - e^{j\omega})(z - e^{-j\omega}),$$

and thus

$$Y(z) = \frac{AN(z)[z^2 - (\cos \omega)z]}{D(z)(z - e^{j\omega})(z - e^{-j\omega})}.$$

Dividing $Y(z)$ by z, we obtain

$$\frac{Y(z)}{z} = \frac{AN(z)(z - \cos \omega)}{D(z)(z - e^{j\omega})(z - e^{-j\omega})}.$$

Pulling out the terms $z - e^{j\omega}$ and $z - e^{-j\omega}$, we have

$$\frac{Y(z)}{z} = \frac{\eta(z)}{D(z)} + \frac{c}{z - e^{j\omega}} + \frac{\bar{c}}{z - e^{-j\omega}},$$

where $\eta(z)$ is a polynomial in z with the degree of $\eta(z)$ less than n. The constant c is given by

$$c = \left[(z - e^{j\omega})\frac{Y(z)}{z}\right]_{z=e^{j\omega}} = \frac{AN(z)(z - \cos \omega)}{D(z)(z - e^{-j\omega})}\Bigg]_{z=e^{j\omega}}$$

$$= \frac{AN(e^{j\omega})(e^{j\omega} - \cos \omega)}{D(e^{j\omega})(e^{j\omega} - e^{-j\omega})}$$

$$= \frac{AN(e^{j\omega})(j \sin \omega)}{D(e^{j\omega})(j2 \sin \omega)}$$

$$= \frac{AN(e^{j\omega})}{2D(e^{j\omega})} = \frac{A}{2} H(e^{j\omega}).$$

Multiplying the expression for $Y(z)/z$ by z, we get

$$Y(z) = \frac{z\eta(z)}{D(z)} + \frac{(A/2)H(e^{j\omega})z}{z - e^{j\omega}} + \frac{(A/2)\overline{H(e^{j\omega})}z}{z - e^{-j\omega}}. \qquad (8.65)$$

Let $y_{tr}(kT)$ denote the inverse z-transform of $z\eta(z)/D(z)$. Since the system is asymptotically stable, the roots of $D(z) = 0$ are within the open unit-disc of the complex plane. Then by the final-value theorem (see Table 7.2),

$$\lim_{k\to\infty} y_{tr}(kT) = \lim_{z\to1} \left[\frac{z-1}{z}\frac{z\eta(z)}{D(z)}\right]$$

$$= \lim_{z\to1} \frac{(z-1)\eta(z)}{D(z)} = 0 \qquad \text{since } D(1) \neq 0.$$

Since $y_{tr}(kT)$ decays to zero as $k \to \infty$, it is called the transient response resulting from the sinusoidal input $x(kT) = A \cos \omega k$.

Now let $y_{ss}(kT)$ denote the inverse z-transform of the second and third terms on the right side of (8.65). Using the identity (7.57), we can write $y_{ss}(kT)$ in the form

$$y_{ss}(kT) = A|H(e^{j\omega})| \cos [\omega k + \sphericalangle H(e^{j\omega})], \qquad k = 0, 1, 2, \ldots . \qquad (8.66)$$

The response $y_{ss}(kT)$ clearly does not decay to zero as $k \to \infty$, and thus it is the steady-state response resulting from the input $x(kT) = A \cos \omega k$. The complete output response resulting from this input is

$$y(kT) = y_{tr}(kT) + y_{ss}(kT).$$

Note that since $y_{tr}(kT) \to 0$ as $k \to \infty$, for k suitably large we have

$$y(kT) \approx y_{ss}(kT) = A|H(e^{j\omega})| \cos [\omega k + \sphericalangle H(e^{j\omega})], \qquad k = 0, 1, \ldots .$$

As in the continuous-time case, we see that if the given discrete-time system is asymptotically stable, the steady-state response to a sinusoidal input is also a sinusoid with the same frequency, but which is amplitude scaled and phase shifted. In the discrete-time case, the amplitude is scaled by the amount $|H(e^{j\omega})|$ and the phase shift is equal to $\sphericalangle H(e^{j\omega})$. As a result of these properties, $H(e^{j\omega})$ is called the *frequency function* of the system.

FREQUENCY RESPONSE CURVES   The plot of $|H(e^{j\omega})|$ versus ω is called the magnitude plot of the system and the plot of $\sphericalangle H(e^{j\omega})$ versus ω is called the phase plot of the system. The magnitude and phase plots are referred to as the frequency response curves of the discrete-time system.

Unlike the frequency response curves in the continuous-time case, the functions $|H(e^{j\omega})|$ and $\sphericalangle H(e^{j\omega})$ are periodic functions of ω with period 2π; that is,

$$|H(e^{j(\omega + 2\pi)})| = |H(e^{j\omega})| \qquad \text{for all } \omega \qquad (8.67)$$

and

$$\sphericalangle H(e^{j(\omega + 2\pi)}) = \sphericalangle H(e^{j\omega}) \qquad \text{for all } \omega. \qquad (8.68)$$

The properties (8.67) and (8.68) follow directly from periodicity of $e^{j\omega}$, that is,

$$e^{j(\omega + 2\pi)} = e^{j\omega} \qquad \text{for all } \omega.$$

As a result of the properties (8.67) and (8.68), the magnitude and phase curves of a discrete-time system are completely determined once they have been computed for any interval of length 2π. For example, if the plots of $|H(e^{j\omega})|$ and $\sphericalangle H(e^{j\omega})$ have been computed for $0 \leq \omega \leq 2\pi$, the plots of $|H(e^{j\omega})|$

and $\sphericalangle H(e^{j\omega})$ for $\omega > 2\pi$ are simply repetitions of the plots for the interval $0 \leq \omega \leq 2\pi$.

From the above results, we have that the frequency response curves of a discrete-time system with transfer function $H(z)$ are determined by computing the magnitude and phase of $H(z)$ with z equal to $e^{j\omega}$, where $0 \leq \omega \leq 2\pi$. The values of z for which $z = e^{j\omega}$ are the points on the unit-circle of the complex plane. To see this, for any value of ω we have

$$|e^{j\omega}| = |\cos \omega + j \sin \omega| = \sqrt{\cos^2\omega + \sin^2\omega} = 1.$$

The vector representation of the complex number $z = e^{j\omega}$ is shown in Figure 8.21. The frequency response curves of a discrete-time system are computed by evaluating the magnitude and phase of $H(z)$ as z moves around the unit-circle displayed in Figure 8.21.

As shown in Section 8.5, the frequency response curves of a continuous-time system with transfer function $H(s)$ are determined by computing the magnitude and phase of $H(s)$ at values of s equal to $j\omega$. Thus, in the continuous-time case, we compute $H(s)$ for values of s on the $j\omega$-axis of the complex plane, while in the discrete-time case we compute $H(z)$ for values of z on the unit-circle of the complex plane. This is one of the fundamental differences between the theory of continuous-time systems and discrete-time systems.

COMPUTATION OF FREQUENCY RESPONSE CURVES Let $z = e^{j\omega}$. Then

$$z = \cos \omega + j \sin \omega,$$

and for any positive integer i,

$$z^i = \cos i\omega + j \sin i\omega.$$

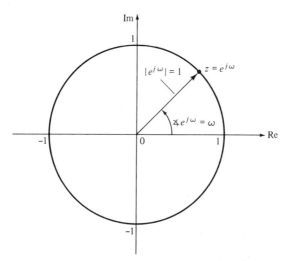

Figure 8.21. Vector representation of $z = e^{j\omega}$.

If we then set $z = e^{j\omega}$ in (8.63), we obtain

$H(e^{j\omega})$

$$= \frac{(b_0 + b_1 \cos \omega + \cdots + b_m \cos m\omega) + j(b_1 \sin \omega + \cdots + b_m \sin m\omega)}{(a_0 + a_1 \cos \omega + \cdots + a_n \cos n\omega) + j(a_1 \sin \omega + \cdots + a_n \sin n\omega)}.$$
$$(8.69)$$

The frequency response curves of the system can then be determined by calculating the magnitude and angle of $H(e^{j\omega})$ given by (8.69) for values of ω ranging from 0 to 2π. A program for carrying out this computation is given in Figure A.11. Given a positive integer Q, the program computes $|H(e^{j\omega})|$ and $\angle H(e^{j\omega})$ for

$$\omega = k\frac{2\pi}{Q}, \qquad k = 0, 1, 2, \ldots, Q.$$

The general shape of the frequency response curves can be determined by considering the vector representations of the factors comprising $H(e^{j\omega})$. This approach is similar to the use of vector representations to compute the frequency response curves in the continuous-time case (see Section 8.5). The process is illustrated in the following example.

EXAMPLE 8.10. Suppose that

$$H(z) = \frac{1}{z - 0.5}.$$

Setting $z = e^{j\omega}$ in $H(z)$, we have

$$H(e^{j\omega}) = \frac{1}{e^{j\omega} - 0.5}.$$

The vector representation of the factor $e^{j\omega} - 0.5$ is shown in Figure 8.22. We see that the magnitude $|e^{j\omega} - 0.5|$ starts with value 0.5 when $\omega = 0$ and increases to the value 1.5 as ω is varied from 0 to π. Thus the magnitude function $|H(e^{j\omega})|$ rolls off as ω is varied from 0 to π. As ω is varied from π to 2π, $|H(e^{j\omega})|$ increases; in fact, from Figure 8.22 we see that

$$|e^{j\omega} - 0.5| = |e^{j(\omega - \pi)} - 0.5| \qquad \text{for } \pi \le \omega \le 2\pi$$

and thus

$$|H(e^{j\omega})| = |H(e^{j(\omega - \pi)})| \qquad \text{for } \pi \le \omega \le 2\pi.$$

So the plot of $|H(e^{j\omega})|$ is symmetrical about the point $\omega = \pi$. From Figure 8.22 we also see that the angle $\angle(e^{j\omega} - 0.5)$ varies from 0 to 360° as ω is

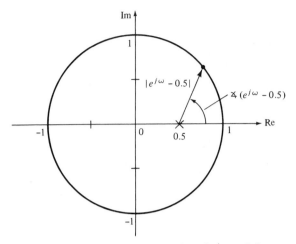

Figure 8.22. Vector representation of $e^{j\omega} - 0.5$.

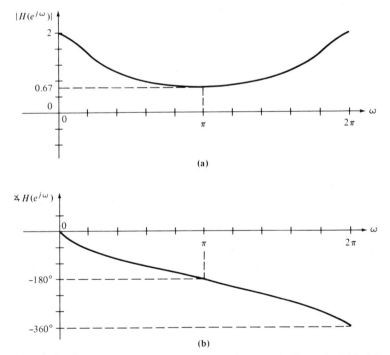

Figure 8.23. Frequency response curves of system in Example 8.10: (a) magnitude curve; (b) phase curve.

varied from 0 to 2π. Thus the phase function $\angle H(e^{j\omega})$ varies from 0 to $-360°$ as ω is varied from 0 to 2π. We can get a precise plot of the frequency response curves by using the program in Figure A.11. The results are displayed in Figure 8.23. Note that if the frequency ω of the input sinusoid $x(kT) = A \cos \omega k$ does not exceed π rad/sec, the system has the frequency characteristic of a lowpass filter. We will study discrete-time lowpass filters in Chapters 11 and 12.

PROBLEMS

Chapter 8

8.1 For the following linear time-invariant continuous-time systems, determine if the system is asymptotically stable, marginally stable, or unstable.

(a) $H(s) = \dfrac{s - 4}{s^2 + 7s}$.

(b) $H(s) = \dfrac{s + 3}{s^2 + 3}$.

(c) $H(s) = \dfrac{2s^2 + 3s + 1}{s^3 + 2s^2 + 4}$.

(d) $H(s) = \dfrac{3s^3 - 2s + 6}{s^3 + s^2 + s + 1}$.

8.2 A linear time-invariant continuous-time system has transfer function

$$H(s) = \frac{s + 5}{s^2 + s - 20}.$$

Assuming that $x(t) = 0$, $t \geq 0$, determine all initial conditions $y(0^-)$, $\dot{y}(0^-)$ for which the resulting output response $y(t)$ converges to zero as $t \to \infty$.

8.3 Consider the field-controlled dc motor given by the input/output differential equation

$$JL_f \frac{d^3 y(t)}{dt^3} + (fL_f + R_f J) \frac{d^2 y(t)}{dt^2} + R_f f \frac{dy(t)}{dt} = kx(t).$$

Assume that all the parameters J, L_f, f, R_f, and k are strictly positive (> 0). Determine if the motor is asymptotically stable, marginally stable, or unstable.

8.4 For the linear time-invariant continuous-time systems with impulse response $h(t)$ given below, determine if the system is BIBO stable.

(a) $h(t) = [2t^3 - 2t^2 + 3t - 2][u(t) - u(t - 10)]$.

(b) $h(t) = \dfrac{1}{t}$ for $t \geq 1$, $h(t) = 0$ for all $t < 1$.

(c) $h(t) = \sin 2t$ for $t \geq 0$.

(d) $h(t) = e^{-t} \sin 2t$ for $t \geq 0$.

(e) $h(t) = e^{-t^2}$ for $t \geq 0$.

8.5   Using the Routh test, determine all values of the parameter k for which the following systems are asymptotically stable.

(a) $H(s) = \dfrac{s^2 + 60s + 800}{s^3 + 30s^2 + (k + 200)s + 40k}$.

(b) $H(s) = \dfrac{2s^3 - 3s + 4}{s^4 + s^3 + ks^2 + 2s + 3}$.

(c) $H(s) = \dfrac{s^2 + 3s - 2}{s^3 + s^2 + (k + 3)s + 3k - 5}$.

(d) $H(s) = \dfrac{s^4 - 3s^2 + 4s + 6}{s^5 + 10s^4 + (9 + k)s^3 + (90 + 2k)s^2 + 12ks + 10k}$.

8.6   A linear time-invariant continuous-time system has transfer function

$$H(s) = \frac{s^2 + 16}{s^2 + 7s + 12}.$$

Compute the steady-state and transient responses resulting from the input $x(t) = 2 \cos 4t$, $t \geq 0$, with no initial energy at time $t = 0$.

8.7   A linear time-invariant continuous-time system has transfer function $H(s)$ with all poles in the open left-half plane and where

$$|H(\omega)| = \begin{cases} 1, & 0 \leq \omega \leq 2 \\ -\tfrac{1}{2}\omega^2 + \tfrac{5}{2}\omega - 2, & 2 \leq \omega \leq 4 \\ 0, & \omega > 4 \end{cases}$$

$$\sphericalangle H(\omega) = \begin{cases} 180°, & 0 \leq \omega \leq 2 \\ -90°\,\omega + 360°, & 2 < \omega \leq 4 \\ 0, & \omega > 4. \end{cases}$$

Compute the steady-state response $y_{ss}(t)$ due to the inputs given below assuming zero initial energy at time $t = 0$.

(a) $x(t) = u(t)$.
(b) $x(t) = \cos 5t$, $t \geq 0$.
(c) $x(t) = 2 \cos (3t - 45°)$, $t \geq 0$.
(d) $x(t) = 2 \cos t + \cos 2t$, $t \geq 0$.

8.8   A linear time-invariant continuous-time system has transfer function

$$H(s) = \frac{s^2 + 1}{(s + 1)(s^2 + 2s + 17)}.$$

Compute both the steady-state response $y_{ss}(t)$ and the transient response $y_{tr}(t)$ when the input $x(t)$ is

(a) $x(t) = u(t)$ with no initial energy.
(b) $x(t) = \cos t$, $t \geq 0$, with no initial energy.
(c) $x(t) = \cos 4t$, $t \geq 0$, with no initial energy.

8.9 A linear time-invariant continuous-time system has transfer function $H(s) = 2/(s + 1)$. Compute the transient response $y_{tr}(t)$ resulting from the input $x(t) = 3 \cos 2t - 4 \sin t$, $t \geq 0$, with no initial energy in the system.

8.10 A linear time-invariant continuous-time system has frequency function

$$H(\omega) = 6e^{-j3\omega}, \qquad -\infty < \omega < \infty.$$

Show that the steady-state response $y_{ss}(t)$ resulting from the input $x(t) = \cos \omega_0 t$, $t \geq 0$, can be expressed in the form

$$y_{ss}(t) = A \cos [\omega_0(t - c)], \qquad t \geq 0,$$

for some constants A and c. Compute A and c.

8.11 A linear time-invariant continuous-time system has transfer function

$$H(s) = \frac{s + 2}{(s + 1)^2 + 4}.$$

The input $x(t) = A \cos (\omega_0 t + \theta)$ is applied to the system for $t \geq 0$ with zero initial energy at time $t = 0$. The resulting steady-state response $y_{ss}(t)$ is

$$y_{ss}(t) = 6 \cos (t + 45°), \qquad t \geq 0.$$

(a) Compute A, ω_0, and θ.
(b) Compute the Laplace transform $Y_{tr}(s)$ of the transient response $y_{tr}(t)$ resulting from this input.

8.12 A linear time-invariant continuous-time system has transfer function $H(s)$ with $H(0) = 3$. The transient response $y_{tr}(t)$ resulting from the step-function input $x(t) = u(t)$ with the system at rest at time $t = 0$ has been determined to be

$$y_{tr}(t) = -2e^{-t} + 4e^{-3t}, \qquad t \geq 0.$$

(a) Compute the system's transfer function $H(s)$.
(b) Compute the steady-state response $y_{ss}(t)$ when the system's input $x(t)$ is equal to $2 \cos (3t + 60°)$, $t \geq 0$, with the system at rest at time $t = 0$.

8.13 A linear time-invariant continuous-time system has transfer function $H(s)$. The input $x(t) = 3 \cos t + 2 \cos (2t - 30°)$, $t \geq 0$, produces the steady-state response $y_{ss}(t) = 6 \cos (t - 45°) + 8 \cos (2t - 90°)$, $t \geq 0$, with no initial energy. Compute $H(1)$ and $H(2)$.

8.14 Sketch the magnitude and phase plots for the following systems. In each case, compute $|H(\omega)|$ and $\angle H(\omega)$ for $\omega = 0$, $\omega = $ 3-dB points, $\omega = \omega_p$, and $\omega \to \infty$. Here ω_p is the value of ω for which $|H(\omega)|$ is maximum.

(a) $H(s) = \dfrac{4}{(s + 2)^2}$.

(b) $H(s) = \dfrac{4s}{(s + 2)^2}.$

(c) $H(s) = \dfrac{s^2 + 2}{(s + 2)^2}.$

(d) $H(s) = \dfrac{4}{s^2 + \sqrt{2}(2s) + 4}.$

8.15   Sketch the magnitude and phase plots for the circuits shown in Figure P8.15. In each case, compute $|H(\omega)|$ and $\sphericalangle H(\omega)$ for $\omega = 0$, $\omega = $ 3-dB points, $\omega = \omega_p$, and $\omega \to \infty$.

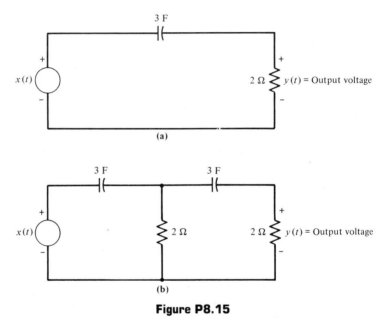

(a)

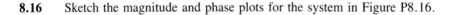

(b)

Figure P8.15

8.16   Sketch the magnitude and phase plots for the system in Figure P8.16.

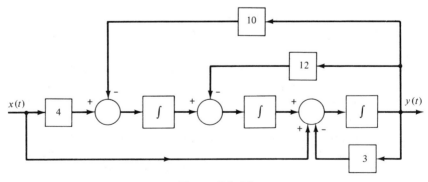

Figure P8.16

8.17 Consider the *RLC* circuit shown in Figure P8.17.

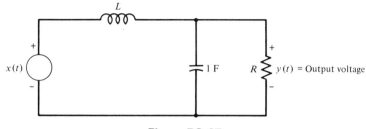

Figure P8.17

(a) Choose values for R and L such that the circuit is a lowpass filter with approximate 3-dB bandwidth equal to 20 rad/sec; that is, $|H(\omega)| \geq (0.707)|H(0)|$ for $0 \leq \omega \leq 20$.

(b) Choose values for R and L such that the circuit is a bandpass filter with approximate center frequency 40 rad/sec and with approximate passband 35 to 45 rad/sec; that is, $|H(\omega)| \geq (0.707) |H(40)|$ for $35 \leq \omega \leq 45$.

8.18 The transfer function of a linear time-invariant continuous-time system is given by

$$H(s) = \frac{9}{s^2 + 3.6s + 9}.$$

(a) Find the resonant frequency ω_r, peak magnitude M_p, damping ratio ζ, and 3-dB bandwidth of the system.

(b) Sketch the frequency response curves.

8.19 A linear time-invariant continuous-time system has transfer function $H(s)$. It is known that $H(0) = 1$ and that $H(s)$ has two poles and no zeros. In addition, the magnitude function $|H(\omega)|$ is shown in Figure P8.19. Determine $H(s)$.

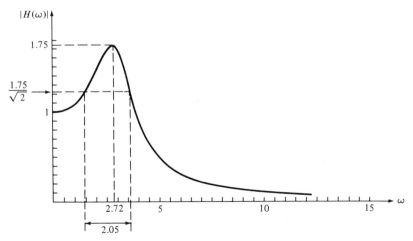

Figure P8.19

8.20 A linear time-invariant discrete-time system has transfer function $H(s) = K/(s + a)$, where $K > 0$ and $a > 0$ are unknown. The steady-state response to $x(t) = 2 \cos t$, $t \geq 0$, is $y_{ss}(t) = 20 \cos (t + \phi_1)$, $t \geq 0$. The steady-state response to $x(t) = 5 \cos 4t$, $t \geq 0$, is $y_{ss}(t) = 10 \cos (4t + \phi_2)$, $t \geq 0$. Here ϕ_1, ϕ_2 are unmeasurable phase shifts. Find K and a.

8.21 Using the Jury test, determine whether or not the following linear time-invariant discrete-time systems are asymptotically stable.

(a) $H(z) = \dfrac{z - 4}{z^2 + 1.5z + 0.5}$.

(b) $H(z) = \dfrac{z^2 - 3z + 1}{z^3 + z^2 - 0.5z + 0.5}$.

(c) $H(z) = \dfrac{1}{z^3 + 0.5z + 0.1}$.

8.22 Consider the discretization of the car on a level surface given by (see Problem 7.11)

$$y(kT + 2T) + \left(\frac{Tk_f}{M} - 2\right) y(kT + T) + \left(1 - \frac{Tk_f}{M}\right) y(kT) = \frac{T^2}{M} x(kT).$$

Determine all values of the discretization interval T for which the discretization is asymptotically stable.

8.23 Consider the discretization of the field-controlled dc motor computed in Problem 7.13. Determine all values of the discretization interval T for which the discretization is marginally stable or asymptotically stable.

8.24 A linear time-invariant discrete-time system has transfer function

$$H(z) = \frac{z}{(z^2 + 0.25)(z - 0.5)^2}.$$

Assuming that the system is at rest at time $k = 0$, find both the transient and steady-state response to $x(kT) = 12 \cos (\pi k/2)$, $k = 0, 1, 2, \ldots$.

8.25 A linear time-invariant discrete-time system has transfer function

$$H(z) = \frac{z}{z + 0.5}.$$

(a) Find the transient response and steady-state response resulting from the input $x(kT) = 5 \cos 3k$, $k = 0, 1, 2, \ldots$, with no initial energy in the system at time $k = 0$.

(b) Sketch the frequency response curves.

8.26 Consider the discrete-time system shown in Figure P8.26. Compute the steady-state output response $y_{ss}(k)$ and the transient output response $y_{tr}(k)$ when $y(-1) = x(-1) = 0$ and $x(k) = 2 \cos (\pi k/2)$, $k = 0, 1, 2, \ldots$.

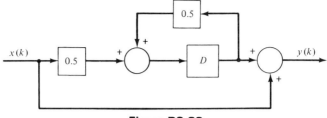

Figure P8.26

8.27 Consider the discrete-time system given by the input/output difference equation

$$y(k) = \frac{x(k) + x(k - 1)}{2}.$$

This system is called an *averager*. Sketch the frequency response curves of the averager.

8.28 The *differencer* is the discrete-time system with the input/output difference equation

$$y(k) = x(k) - x(k - 1).$$

Sketch the frequency response curves of the differencer.

The Fourier Series and Fourier Transform

In this chapter we introduce the fundamental notion of the frequency spectrum of a continuous-time signal. As will be seen, the frequency spectrum displays the various sinusoidal components that comprise a given continuous-time signal. In general, the frequency spectrum is a complex-valued function of the frequency variable, and thus it is usually specified in terms of an amplitude spectrum and a phase spectrum.

We begin by considering periodic signals. In this case the frequency spectrum of the signal can be generated by representing the signal as a sum of sinusoids, called a *Fourier series*. The Fourier series is named after the French physicist Jean Baptiste Fourier (1768–1830), who was the first to propose that periodic waveforms could be represented by a sum of sinusoids (or complex exponentials). It is interesting to note that in addition to his contributions to science and mathematics, Fourier was also very active in the politics of his time. For example, he played an important role in Napoleon's expeditions to Egypt during the late 1790s.

We define the trigonometric form of the Fourier series of a periodic signal in Section 9.1. In the first part of Section 9.2 we show that the Fourier series simplifies if the given periodic signal has even or odd symmetry. Then we study the complex exponential form of the Fourier series. In the last part of Section

9.2 we define the amplitude and phase spectra of a periodic signal in terms of the coefficients of the exponential form of the Fourier series.

The notion of the amplitude and phase spectra of a periodic signal gives new insight into how a linear time-invariant system processes a periodic input. In particular, in Section 9.3 we show that the response to a periodic input is also periodic and that the frequency spectrum of the response can be expressed in terms of the spectrum of the input and the frequency function of the system. This characterization of the output spectrum gives a procedure for computing the response to a periodic input. In Section 9.4 we apply this framework to the design of a rectifier.

Using the Fourier series representation, in the first part of Section 9.5 we show that one can generate a frequency-domain representation of a nonperiodic signal. This representation is defined in terms of the Fourier transform, which is studied in Sections 9.5 to 9.7. In contrast to a periodic signal, the spectra of a nonperiodic signal consist of a continuum of frequencies. In Chapter 10 we will see that the characterization of nonperiodic signals in terms of their spectra is very useful in determining how a linear time-invariant system processes non-periodic inputs.

9.1

FOURIER SERIES REPRESENTATION
OF PERIODIC SIGNALS

Let T be a fixed positive real number. A continuous-time signal $x(t)$ is said to be *periodic* with period T if

$$x(t + T) = x(t) \qquad \text{for all } t, \quad -\infty < t < \infty. \qquad (9.1)$$

The period T is the smallest positive number for which (9.1) is satisfied.

By the property (9.1), a periodic signal repeats every T seconds. For example, $x(t) = A \cos \omega t$ and $x(t) = A \sin \omega t$ are periodic signals with period $2\pi/\omega$. The waveform displayed in Figure 9.1 is periodic with period $T = 4$.

Let $x(t)$ be a periodic signal with period T. Then by Fourier's theorem, $x(t)$ can be expressed as a (in general infinite) sum of sinusoids

$$x(t) = c_0 + \sum_{n=1}^{\infty} 2|c_n| \cos (n\omega_0 t + \angle c_n), \qquad -\infty < t < \infty. \qquad (9.2)$$

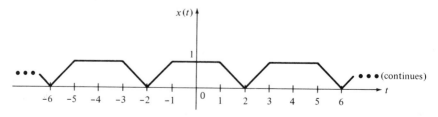

FIGURE 9.1. Periodic signal with $T = 4$.

The expression (9.2) for $x(t)$ is called the *trigonometric Fourier series* of the periodic signal $x(t)$.

In the representation (9.2), ω_0 is the fundamental frequency (in rad/sec) of $x(t)$ given by $\omega_0 = 2\pi/T$, where T is the period. The number c_0 is the dc component of $x(t)$ given by

$$c_0 = \frac{1}{T} \int_{-T/2}^{T/2} x(t) \, dt. \tag{9.3}$$

For $n \geq 1$, $|c_n|$ is the magnitude and $\angle\, c_n$ is the angle of the complex number $c_n = |c_n| \exp(j \angle\, c_n)$. For $n \geq 1$, c_n is given by

$$c_n = \frac{1}{T} \int_{-T/2}^{T/2} x(t) \exp(-jn\omega_0 t) \, dt. \tag{9.4}$$

The c_n can be computed by integrating $x(t) \exp(-jn\omega_0 t))$ over any full period; that is, for any real number h,

$$c_n = \frac{1}{T} \int_{h}^{T+h} x(t) \exp(-jn\omega_0 t) \, dt.$$

The component $2 |c_n| \cos(n\omega_0 t + \angle\, c_n)$ in the representation (9.2) is called the nth *harmonic* of the periodic signal $x(t)$. Here the term "nth harmonic" refers to the fact that the frequency $n\omega_0$ of this component is n times the fundamental frequency ω_0. The number $2 |c_n|$ is the peak value of the nth harmonic.

Using the trigonometric identity (5.68), we can rewrite the Fourier series representation (9.2) in the form

$$x(t) = c_0 + \sum_{n=1}^{\infty} [a_n \cos(n\omega_0 t) + b_n \sin(n\omega_0 t)]. \tag{9.5}$$

In (9.5),

$$a_n = 2 \operatorname{Re} c_n \quad \text{and} \quad b_n = -2 \operatorname{Im} c_n, \tag{9.6}$$

where $\operatorname{Re} c_n$ is the real part of c_n and $\operatorname{Im} c_n$ is the imaginary part of c_n. In the following development, we work mainly with the form (9.2).

The Fourier series representation of a periodic signal is a remarkable result. In particular, it says that a periodic signal such as the waveform with "corners" in Figure 9.1 can be expressed as a sum of sinusoids. Since sinusoids are infinitely smooth functions (i.e, they have ordinary derivatives of arbitrarily high order), it is difficult to believe that signals with corners can be expressed as a sum of sinusoids. Of course, the key here is that the sum is an infinite sum. It is not surprising that Fourier had a difficult time convincing his peers (in this case the members of the French Academy of Science) that his theorem was valid.

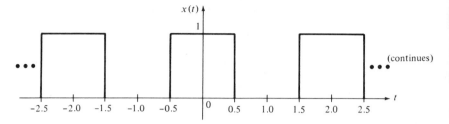

FIGURE 9.2. Signal in Example 9.1.

Fourier believed that any periodic signal could be expressed as a sum of sinusoids. However, this turned out not to be the case, although virtually all periodic signals arising in engineering do have a Fourier series representation. In particular, a periodic signal $x(t)$ has a Fourier series if it satisfies the *Dirichlet conditions* given by

1. $x(t)$ is absolutely integrable over any period; that is,

$$\int_{h}^{h+T} |x(t)|\, dt < \infty.$$

2. $x(t)$ has only a finite number of maxima and minima over any period.
3. $x(t)$ has only a finite number of discontinuities over any period.

EXAMPLE 9.1.   Consider the signal shown in Figure 9.2. This signal is periodic with period $T = 2$ and thus the fundamental frequency is $\omega_0 = 2\pi/2 = \pi$ rad/sec. The signal obviously satisfies the Dirichlet conditions, and thus it has a Fourier series representation. Evaluating (9.3), we have that the dc component of $x(t)$ is

$$c_0 = \tfrac{1}{2} \int_{-1}^{1} x(t)\, dt = \tfrac{1}{2} \int_{-0.5}^{0.5} (1)\, dt = \tfrac{1}{2}.$$

Evaluating (9.4), we obtain

$$\begin{aligned}
c_n &= \frac{1}{2} \int_{-1}^{1} x(t) \exp\left(-jn\pi t\right) dt \\
&= \frac{1}{2} \int_{-0.5}^{0.5} \exp\left(-jn\pi t\right) dt \\
&= \frac{-1}{j2n\pi} \exp\left(-jn\pi t\right) \Big]_{t=-0.5}^{t=0.5} \\
&= \frac{-1}{j2n\pi} \left[\exp\left(-j\frac{n\pi}{2}\right) - \exp\left(j\frac{n\pi}{2}\right) \right] \\
&= \frac{-1}{j2n\pi} (-j2) \sin\left(\frac{n\pi}{2}\right) = \frac{\sin(n\pi/2)}{n\pi}.
\end{aligned}$$

Hence

$$|c_n| = \begin{cases} \dfrac{1}{n\pi}, & n \text{ odd} \\ 0, & n \text{ even} \end{cases} \qquad \sphericalangle\, c_n = \begin{cases} \pi, & n = 3, 7, 11, \ldots \\ 0, & \text{all other } n. \end{cases}$$

Inserting the values of c_0 and $|c_n|$ into (9.2), we have that the Fourier series representation of the signal $x(t)$ shown in Figure 9.2 is

$$x(t) = \frac{1}{2} + \sum_{\substack{n=1 \\ n \text{ odd}}}^{\infty} \frac{2}{n\pi} \cos(n\pi t + \sphericalangle\, c_n), \qquad -\infty < t < \infty. \qquad (9.7)$$

Note that this signal does not contain any even harmonics.

GIBBS PHENOMENON   Again consider the signal $x(t)$ with the Fourier series representation (9.7). Given an odd positive integer N, let $x_N(t)$ denote the finite sum

$$x_N(t) = \frac{1}{2} + \sum_{\substack{n=1 \\ n \text{ odd}}}^{N} \frac{2}{n\pi} \cos(n\pi t + \sphericalangle\, c_n), \qquad -\infty < t < \infty.$$

By Fourier's theorem, $x_N(t)$ should converge to $x(t)$ as $N \to \infty$. In particular, for a suitably large value of N, $x_N(t)$ should be a close approximation to $x(t)$. To see if this is the case, we can simply plot $x_N(t)$ for various values of N. Let us start with $N = 3$, in which case

$$x_3(t) = \frac{1}{2} + \frac{2}{\pi} \cos \pi t + \frac{2}{3\pi} \cos(3\pi t + \pi), \qquad -\infty < t < \infty.$$

A plot of $x_3(t)$ is given in Figure 9.3. Note that even though $x_3(t)$ consists of the dc component and only two harmonics (the first and third), $x_3(t)$ does resemble the pulse train in Figure 9.2.

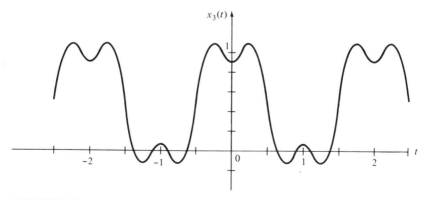

FIGURE 9.3.   Plot of $x_N(t)$ when $N = 3$.

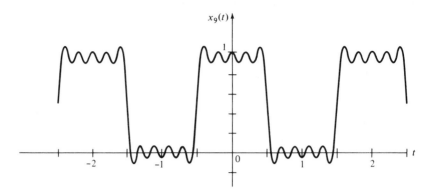

FIGURE 9.4. Approximation $x_9(t)$.

Increasing N to 9, we get the result shown in Figure 9.4. Comparing Figures 9.3 and 9.4, we see that $x_9(t)$ is a much closer approximation to the pulse train $x(t)$ than $x_3(t)$. Of course, $x_9(t)$ contains the dc component and the first, third, fifth, seventh, and ninth harmonics of $x(t)$, and thus we would expect it to be a much closer approximation than $x_3(t)$.

Setting $N = 21$, we obtain the result in Figure 9.5. Except for the overshoot at the corners of the pulses, the waveform in Figure 9.5 is a much better approximation to $x(t)$ than $x_9(t)$. From a careful examination of the plot in Figure 9.5, one can see that the magnitude of the overshoot is approximately equal to 9 percent.

Taking $N = 41$, we get the result displayed in Figure 9.6. Note that the 9 percent overshoot at the corners is still present. In fact, the 9 percent overshoot is present even in the limit as N approaches ∞. This characteristic was first discovered by Josiah Willard Gibbs (1839–1903), and thus the overshoot is referred to as the *Gibbs phenomenon*. Gibbs demonstrated the existence of the overshoot from mathematical properties rather than by direct computation.

Now let $x(t)$ be an arbitrary periodic signal. As a consequence of the Gibbs phenomenon, the Fourier series representation of $x(t)$ is not actually equal to

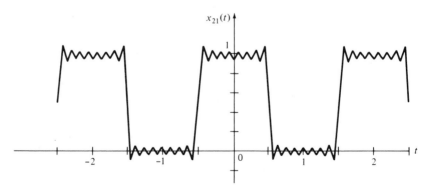

FIGURE 9.5. Signal $x_{21}(t)$.

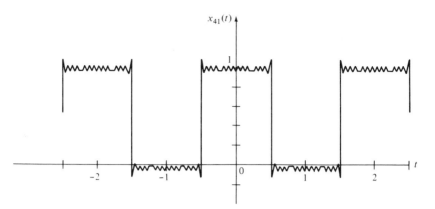

FIGURE 9.6. Signal $x_{41}(t)$.

the true value of $x(t)$ at any points where $x(t)$ is discontinuous. In fact, the value is off by approximately 9 percent.

9.2

SYMMETRY AND THE EXPONENTIAL FORM OF THE FOURIER SERIES

Again let $x(t)$ be a periodic signal. In the first part of this section, we show that when $x(t)$ possesses even or odd symmetry, we can simplify the expressions (9.3) and (9.4) for the coefficients c_n of the Fourier series for $x(t)$. We begin with the definition of even and odd symmetry.

A periodic signal $x(t)$ is said to be even if

$$x(-t) = x(t) \qquad \text{for all } t > 0.$$

This condition can be checked by folding the plot of $x(t)$ versus t about the vertical axis: If the plot "falls back on itself," the signal is even. Examples of even periodic signals are $x(t) = A \cos \omega t$ and the signals displayed in Figures 9.1 and 9.2.

If $x(t)$ and $v(t)$ are even periodic signals with period T, the product $x(t)v(t)$ is an even periodic signal with period T. To see that the product is even, let

$$w(t) = x(t)v(t).$$

Then

$$w(-t) = x(-t)v(-t) = x(t)v(t) = w(t).$$

A periodic signal $x(t)$ is odd if

$$x(-t) = -x(t) \qquad \text{for all } t > 0.$$

If the plot of $x(t)$ versus t falls back on itself after folding about both the vertical and horizontal axes, the signal is odd. An example of an odd periodic signal is $x(t) = A \sin \omega t$.

The product of two odd periodic signals is an even periodic signal, and the product of an even and odd periodic signal is an odd periodic signal. To prove the latter statement, again let

$$w(t) = x(t)v(t).$$

Then if $x(t)$ is even and $v(t)$ is odd,

$$w(-t) = x(-t)v(-t) = x(t)\,[-v(t)] = -x(t)v(t) = -w(t).$$

Now let $x(t)$ be an even periodic signal with period T. The coefficients of the Fourier series of $x(t)$ are given by (9.3) and (9.4). As a result of the even symmetry, the dc component c_0 of $x(t)$ can be expressed in the form

$$c_0 = \frac{2}{T} \int_0^{T/2} x(t)\,dt. \tag{9.8}$$

Using Euler's formula, the coefficients c_n for $n \geq 1$ are given by

$$c_n = \frac{1}{T} \int_{-T/2}^{T/2} x(t)[\cos\,(n\omega_0 t) - j\sin\,(n\omega_0 t)]\,dt. \tag{9.9}$$

Since $x(t)$ is even and $\sin\,(n\omega_0 t)$ is odd, the product $x(t)\sin\,(n\omega_0 t)$ is odd. As a result, the integral of $x(t)\sin\,(n\omega_0 t)$ over any T-second interval is equal to zero, and thus (9.9) reduces to

$$c_n = \frac{1}{T} \int_{-T/2}^{T/2} x(t)\cos\,(n\omega_0 t)\,dt, \qquad n = 1, 2, \ldots. \tag{9.10}$$

In addition, since $x(t)\cos\,(n\omega_0 t)$ is even, the integral in (9.10) is equal to two times the integral from 0 to $T/2$. Hence we have

$$c_n = \frac{2}{T} \int_0^{T/2} x(t)\cos\,(n\omega_0 t)\,dt, \qquad n = 1, 2, \ldots. \tag{9.11}$$

We can then compute the coefficients of the Fourier series of an even periodic signal $x(t)$ by using (9.8) and (9.11). It is important to note that by (9.11), the coefficients c_n of an even periodic signal are always real numbers. Since the c_n are real, Re $c_n = c_n$, Im $c_n = 0$, and thus by (9.5) and (9.6), we can write the Fourier series of an even periodic signal $x(t)$ in the form

$$x(t) = c_0 + \sum_{n=1}^{\infty} 2c_n \cos\,(n\omega_0 t), \qquad -\infty < t < \infty. \tag{9.12}$$

If $x(t)$ is odd, by an argument similar to the one above, the expressions (9.3) and (9.4) reduce to

$$c_0 = 0$$

$$c_n = \frac{-j2}{T} \int_0^{T/2} x(t) \sin (n\omega_0 t) \, dt, \qquad n = 1, 2, \ldots . \qquad (9.13)$$

Note that the dc component of an odd periodic signal is always equal to zero. By (9.13) we see that the coefficients c_n, $n \geq 1$, of an odd periodic signal are purely imaginary numbers. Since the c_n are purely imaginary,

$$\text{Re } c_n = 0, \qquad \text{Im } c_n = -jc_n,$$

and thus by (9.5) and (9.6), we can write the Fourier series of an odd periodic signal in the form

$$x(t) = \sum_{n=1}^{\infty} (j2c_n) \sin (n\omega_0 t), \qquad -\infty < t < \infty. \qquad (9.14)$$

EXAMPLE 9.2. Consider the periodic signal in Figure 9.1. As noted above, the signal is even, and thus we can use (9.8) and (9.11) to compute the Fourier series. Here $T = 4$, so $\omega_0 = 2\pi/T = \pi/2$. Then

$$c_0 = \frac{2}{4} \int_0^2 x(t) \, dt = \frac{1}{2} \int_0^1 (1) \, dt + \frac{1}{2} \int_1^2 (-t + 2) \, dt$$

$$= \frac{1}{2} + \frac{1}{2} \left(-\frac{t^2}{2} + 2t \right) \Big]_{t=1}^{t=2}$$

$$= \frac{3}{4}.$$

For $n \geq 1$,

$$c_n = \frac{2}{4} \int_0^2 x(t) \cos \left(\frac{n\pi}{2} t \right) dt$$

$$= \frac{1}{2} \int_0^1 \cos \left(\frac{n\pi}{2} t \right) dt + \frac{1}{2} \int_1^2 (-t + 2) \cos \left(\frac{n\pi}{2} t \right) dt$$

$$= \frac{1}{n\pi} \sin \left(\frac{n\pi}{2} t \right) \Big|_{t=0}^{t=1} - \frac{2}{n^2\pi^2} \left[\cos \left(\frac{n\pi}{2} t \right) + \frac{n\pi}{2} t \sin \left(\frac{n\pi}{2} t \right) \right]_{t=1}^{t=2}$$

$$+ \frac{2}{n\pi} \sin \left(\frac{n\pi}{2} t \right) \Big|_{t=1}^{t=2}$$

$$= \frac{1}{n\pi} \sin\left(\frac{n\pi}{2}\right) - \frac{2}{n^2\pi^2}\left[\cos(n\pi) - \cos\left(\frac{n\pi}{2}\right) - \frac{n\pi}{2}\sin\left(\frac{n\pi}{2}\right)\right]$$

$$- \frac{2}{n\pi}\sin\left(\frac{n\pi}{2}\right)$$

$$= \frac{2}{n^2\pi^2}\left[-\cos(n\pi) + \cos\left(\frac{n\pi}{2}\right)\right].$$

Inserting the c_n into (9.12), we have that the Fourier series representation of $x(t)$ is

$$x(t) = \frac{3}{4} + \sum_{n=1}^{\infty} \frac{4}{n^2\pi^2}\left[-\cos(n\pi) + \cos\left(\frac{n\pi}{2}\right)\right]\cos\left(\frac{n\pi}{2}t\right),$$

$$-\infty < t < \infty.$$

COMPLEX EXPONENTIAL FORM OF FOURIER SERIES A periodic signal $x(t)$ can also be expressed in terms of a sum of complex exponentials. The complex exponential form of the Fourier series can be derived from the trigonometric Fourier series given by (9.2). The steps are as follows. In the following derivation, we are assuming that the signal $x(t)$ is real valued; that is, $x(t)$ is a real number for every value of t.

Using the trigonometric identity (5.70) with $p_1 = jn\omega_0 t$, we can write (9.2) in the form

$$x(t) = c_0 + \sum_{n=1}^{\infty} [c_n \exp(jn\omega_0 t) + \bar{c}_n \exp(-jn\omega_0 t)]. \tag{9.15}$$

By defining

$$c_{-n} = \bar{c}_n \qquad \text{for } n \geq 1, \tag{9.16}$$

we have

$$\sum_{n=1}^{\infty} \bar{c}_n \exp(-jn\omega_0 t) = \sum_{n=-\infty}^{-1} c_n \exp(jn\omega_0 t).$$

Hence we can write the infinite sum in (9.15) as the bi-infinite sum

$$x(t) = c_0 + \sum_{\substack{n=-\infty \\ n\neq 0}}^{\infty} c_n \exp(jn\omega_0 t)$$

$$= \sum_{n=-\infty}^{\infty} c_n \exp(jn\omega_0 t). \tag{9.17}$$

Equation (9.17) is the complex exponential form of the Fourier series representation of $x(t)$. Note that for $n > 0$, the coefficients c_n in (9.17) are the same as the coefficients c_n in the trigonometric Fourier series given by (9.2). In either the trigonometeric form or the complex exponential form, the coefficients c_n for $n \geq 0$ are given by (9.3) and (9.4).

LINE SPECTRA Given a periodic signal $x(t)$ with period T, let c_n denote the coefficients of the exponential form of the Fourier series of $x(t)$. The coefficients c_n define a complex-valued function of the discrete frequencies $n\omega_0$, where $n = 0, \pm 1, \pm 2, \ldots$. We shall denote this function by $c(n\omega_0)$. By definition, we have

$$c(n\omega_0) = c_n \qquad \text{for } n = 0, \pm 1, \pm 2, \ldots.$$

The plot of $|c(n\omega_0)|$ versus $n\omega_0$ is called the *amplitude spectrum* of the periodic signal $x(t)$. Since the dc component of $x(t)$ is equal to $c(0)$ and the amplitudes of the sinusoidal components of $x(t)$ are equal to $2|c(n\omega_0)|$, the plot of $|c(n\omega_0)|$ versus $n\omega_0$ displays the amplitudes (scaled by a factor of 2) of the various frequency components comprising $x(t)$. This is why the plot is called the amplitude spectrum of $x(t)$.

The plot of $\measuredangle c(n\omega_0)$ versus $n\omega_0$ is called the *phase spectrum* of $x(t)$. Since the phase of the sinusoidal components comprising $x(t)$ is equal to $\measuredangle c(n\omega_0)$ [see (9.2)], the phase spectrum displays the phase of the sinusoidal components of $x(t)$.

Since the amplitude and phase spectra of the given periodic signal $x(t)$ are defined in terms of the magnitude and phase of $c(n\omega_0)$, the function $c(n\omega_0)$ is called the *frequency spectrum* of the signal $x(t)$.

Note that by (9.16),

$$c(-n\omega_0) = \overline{c(n\omega_0)} \qquad \text{for } n \geq 1,$$

and thus

$$|c(-n\omega_0)| = |c(n\omega_0)| \qquad \text{for } n \geq 1$$

and

$$\measuredangle c(-n\omega_0) = -\measuredangle c(n\omega_0) \qquad \text{for } n \geq 1.$$

In other words, the amplitude spectrum $|c(n\omega_0)|$ is an even function of $n\omega_0$, and the phase spectrum $\measuredangle c(n\omega_0)$ is an odd function of $n\omega_0$.

The amplitude and phase spectra of a periodic signal $x(t)$ are usually drawn with vertical lines connecting the values of $|c(n\omega_0)|$ and $\measuredangle c(n\omega_0)$ with the points $n\omega_0$. As a result, these spectra are often referred to as *line spectra*. Each line in the spectra corresponds to a particular frequency component of the signal $x(t)$.

| **EXAMPLE 9.3.** The line spectra for the periodic signal in Example 9.1 are displayed in Figure 9.7.

In Section 9.3 we will see that the notion of the amplitude and phase spectra can be utilized to give a very nice characterization of how a linear time-invariant system processes a periodic input signal.

PARSEVAL'S THEOREM Let $x(t)$ be a periodic signal with period T. We define the average power P of the signal by

$$P = \frac{1}{T} \int_{-T/2}^{T/2} x^2(t)\ dt. \tag{9.18}$$

If $x(t)$ is the voltage across a 1-Ω resistor or the current in a 1-Ω resistor, the average power is given by (9.18). So the expression (9.18) is a generalization of the notion of average power to arbitrary signals.

Again let $x(t)$ be an arbitrary periodic signal with period T. As discussed above, the exponential form of the Fourier series of $x(t)$ is

$$x(t) = \sum_{n=-\infty}^{\infty} c_n \exp\ (jn\omega_0 t).$$

Now by Parseval's theorem, the average power P of the signal $x(t)$ is given by

$$P = \sum_{n=-\infty}^{\infty} |c_n|^2. \tag{9.19}$$

The relationship (9.19) is useful since it relates the average power of a periodic signal to the coefficients of the Fourier series of the signal. The proof of Parseval's theorem will not be given.

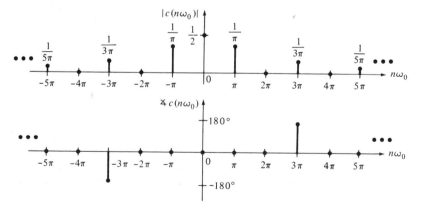

FIGURE 9.7. Line spectra for periodic signal in Example 9.1.

RESPONSE TO PERIODIC INPUTS

Consider a linear time-invariant continuous-time system with impulse response $h(t)$. By the results in Chapter 4, the response $y(t)$ resulting from an input $x(t)$ is given by

$$y(t) = \int_{-\infty}^{\infty} h(\lambda)x(t - \lambda) \, d\lambda. \tag{9.20}$$

Recall that $y(t)$ is the response with no initial energy in the system prior to the application of the input $x(t)$. If the input is applied at time $t = -\infty$, zero initial energy means that there is no energy stored in the system at time $t = -\infty$. Note that we cannot take the lower limit of the integral in (9.20) to be zero unless the system is causal [i.e., $h(t) = 0$ for $t < 0$]. In this section we do not require that the given system be causal.

Now suppose that $x(t)$ is periodic so that it has the Fourier series representation

$$x(t) = \sum_{n=-\infty}^{\infty} c_n \exp(jn\omega_0 t), \qquad -\infty < t < \infty. \tag{9.21}$$

We want to derive an expression for the output response resulting from this periodic input. First, we shall compute the output response resulting from the complex exponential

$$x(t) = e^{j\omega t}, \qquad -\infty < t < \infty.$$

Then using linearity, we will able to determine the response to the periodic input given by (9.21).

Inserting $x(t) = e^{j\omega t}$ into (9.20) gives

$$y(t) = \int_{-\infty}^{\infty} h(\lambda)e^{j\omega t} e^{-j\omega\lambda} \, d\lambda. \tag{9.22}$$

Let

$$H(\omega) = \int_{-\infty}^{\infty} h(\lambda)e^{-j\omega\lambda} \, d\lambda.$$

Then we can write (9.22) in the form

$$y(t) = H(\omega)e^{j\omega t} = H(\omega)x(t), \qquad -\infty < t < \infty. \tag{9.23}$$

From (9.23) we see that the system processes the complex exponential $x(t) = e^{j\omega t}$ by scaling it by the quantity $H(\omega)$. Thus the response to a complex exponential is also a complex exponential with the same frequency ω. Since the response to the input $x(t) = e^{j\omega t}$ can be determined directly from $H(\omega)$, the function $H(\omega)$ is called the *frequency function* (or the *system function*) of the system. The magnitude $|H(\omega)|$ is called the *magnitude function* of the system, and $\angle H(\omega)$ is called the *phase function* of the system.

If the given system is finite dimensional, causal, and asymptotically stable (as defined in Section 8.1), it turns out that $H(\omega)$ is identical to the frequency function defined in Chapter 8. In other words, $H(\omega)$ is equal to the transfer function $H(s)$ with $s = j\omega$. It is worth noting that the frequency function $H(\omega)$ can be determined without having to consider the transfer function. To see this, suppose that the system is given by the input/output differential equation

$$y^{(n)}(t) + \sum_{i=0}^{n-1} a_i y^{(i)}(t) = \sum_{i=0}^{m} b_i x^{(i)}(t). \qquad (9.24)$$

Setting $x(t) = e^{j\omega t}$ and $y(t) = H(\omega)e^{j\omega t}$ in (9.24), we obtain

$$\left[(j\omega)^n H(\omega) + \sum_{i=0}^{n-1} a_i (j\omega)^i H(\omega) \right] e^{j\omega t} = \sum_{i=0}^{m} b_i (j\omega)^i e^{j\omega t}. \qquad (9.25)$$

Solving (9.25) for $H(\omega)$ gives

$$H(\omega) = \frac{\displaystyle\sum_{i=0}^{m} b_i (j\omega)^i}{(j\omega)^n + \displaystyle\sum_{i=0}^{n-1} a_i (j\omega)^i}. \qquad (9.26)$$

DETERMINING THE RESPONSE TO PERIODIC INPUTS   Let us return to the problem of determining the response $y(t)$ to the periodic input $x(t)$ given by the Fourier series (9.21). Using linearity and (9.23), we have that

$$y(t) = \sum_{n=-\infty}^{\infty} H(n\omega_0)c_n \exp(jn\omega_0 t), \qquad -\infty < t < \infty. \qquad (9.27)$$

Since the right side of (9.27) is the complex exponential form of a Fourier series, we see that the response $y(t)$ is periodic. In addition, since the fundamental frequency of $y(t)$ is ω_0, which is the fundamental frequency of the input $x(t)$, we also have that the period of $y(t)$ is equal to the period of $x(t)$. Hence the response to a periodic input with period T is periodic with period T.

The response $y(t)$ to the periodic input $x(t)$ can also be expressed in the form of a trigonometric Fourier series. The result is

$$y(t) = c_0 H(0) + \sum_{n=1}^{\infty} 2|c_n||H(n\omega_0)|\cos [n\omega_0 t + \angle c_n + \angle H(n\omega_0)],$$
$$-\infty < t < \infty. \qquad (9.28)$$

Now let c_n^x denote the coefficients of the Fourier series for $x(t)$ and let c_n^y denote the coefficients of the Fourier series for the resulting output $y(t)$. From (9.27) we have

$$c_n^y = c_n^x H(n\omega_0), \qquad n = 0, \pm 1, \pm 2, \ldots \qquad (9.29)$$

From (9.29) we see that the Fourier coefficients of the response $y(t)$ are equal to the product of the coefficients of the periodic input with the frequency function $H(\omega)$ evaluated at $\omega = n\omega_0$. Taking the magnitude of both sides of (9.29), we have

$$|c_n^y| = |c_n^x||H(n\omega_0)|, \qquad n = 0, \pm 1, \pm 2, \ldots, \qquad (9.30)$$

so the output amplitude spectrum is the product of the input amplitude spectrum and the system's magnitude function $|H(\omega)|$ with $\omega = n\omega_0$. Taking the angle of both sides of (9.29), we get

$$\sphericalangle c_n^y = \sphericalangle c_n^x + \sphericalangle H(n\omega_0), \qquad n = 0, \pm 1, \pm 2, \ldots, \qquad (9.31)$$

so the output phase spectrum is the sum of the input phase spectrum and the system's phase function $\sphericalangle H(\omega)$ with $\omega = n\omega_0$.

The relationships (9.30) and (9.31) are very important. In particular, they describe how the system processes the various sinusoidal components comprising the periodic input signal. From (9.30) we can determine if the system will pass or will attenuate a given sinusoidal component of the input. From (9.31) we can determine how much of a phase shift the system will give to a particular sinusoidal component of the input.

Using (9.28) and the relationships (9.30) and (9.31), we can compute the trigonometric Fourier series of the output directly from the coefficients of the input Fourier series. The process is illustrated by the following example.

EXAMPLE 9.4. Consider the periodic signal displayed in Figure 9.8. This signal is the voltage input applied to the RC circuit shown in Figure 9.9. We want to determine the voltage $y(t)$ on the capacitor resulting from the periodic input $x(t)$ shown in Figure 9.8. The first step is to compute the coefficients c_n^x for the periodic input $x(t)$. The signal $x(t)$ is not even

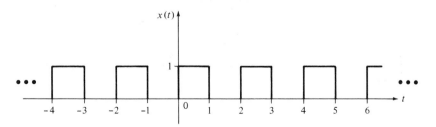

FIGURE 9.8. Periodic signal in Example 9.4.

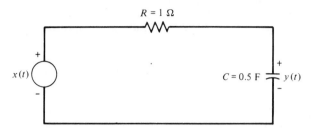

FIGURE 9.9.　*RC* circuit in Example 9.4.

or odd. However, the signal $x(t) - 0.5$ is odd and the Fourier coefficients of $x(t) - 0.5$ are the same as the coefficients of $x(t)$, except for the dc component c_0. Thus we can first compute the c_n^x for $x(t) - 0.5$. Using (9.13), we have

$$
\begin{aligned}
c_n^x &= \frac{-j2}{2} \int_0^1 (0.5) \sin (n\pi t)\, dt, \qquad n = 1, 2, \ldots \\
&= \frac{j}{2n\pi} \left[\cos (n\pi t) \right]_{t=0}^{t=1} \\
&= \begin{cases} 0, & n \text{ even} \\[2mm] -\dfrac{j}{n\pi}, & n \text{ odd.} \end{cases}
\end{aligned}
$$

In addition, $c_0 = 0$ for $x(t) - 0.5$, and thus $c_0 = 0.5$ for $x(t)$. The resulting trigonometric Fourier series for $x(t)$ is

$$
x(t) = 0.5 + \sum_{\substack{n=1 \\ n \text{ odd}}}^{\infty} \frac{2}{n\pi} \cos \left(n\pi t - \frac{\pi}{2} \right), \qquad t \geq 0.
$$

Now the *RC* circuit has the input/output differential equation

$$
(0.5) \frac{dy(t)}{dt} + y(t) = x(t).
$$

Thus, by (9.26), the frequency function is

$$
H(\omega) = \frac{1}{j\,0.5\omega + 1}.
$$

Hence

$$
H(n\omega_0) = H(n\pi) = \frac{1}{j\,0.5n\pi + 1}
$$

and

$$|H(n\pi)| = \frac{1}{\sqrt{0.25n^2\pi^2 + 1}}$$

$$\measuredangle H(n\pi) = -\tan^{-1}(0.5n\pi).$$

Note that the magnitude function $|H(n\pi)|$ rolls off as the integer n is increased. This is expected since the RC circuit is a lowpass filter (see Section 8.5). As a result of this lowpass filter characteristic, the system will not pass the high-frequency sinusoidal components of the input. Now inserting the expressions for c_n^x, $|H(n\pi)|$, and $\measuredangle H(n\pi)$ into (9.30) and (9.31), we have

$$c_0^y = (0.5)(1) = 0.5$$

$$|c_n^y| = \begin{cases} 0, & n \text{ even} \\ \dfrac{1}{n\pi} \dfrac{1}{\sqrt{0.25n^2\pi^2 + 1}} & n \text{ odd} \end{cases}$$

$$\measuredangle c_n^y = \frac{-\pi}{2} - \tan^{-1}(0.5n\pi), \qquad n \text{ odd}.$$

The trigonometric Fourier series of the capacitor voltage is

$$y(t) = 0.5 + \sum_{\substack{n=1 \\ n \text{ odd}}}^{\infty} \frac{2}{n\pi} \frac{1}{\sqrt{0.25n^2\pi^2 + 1}} \cos\left[n\pi t - \frac{\pi}{2} - \tan^{-1}(0.5n\pi) \right],$$

$$-\infty < t < \infty. \tag{9.32}$$

Since the coefficients of the series (9.32) for $y(t)$ are getting very small as n increases, we can determine the values of $y(t)$ by evaluating a suitable number of the terms comprising (9.32). Using a computer, we evaluated the first 25 terms, which gave the result shown in Figure 9.10. The waveform in Figure

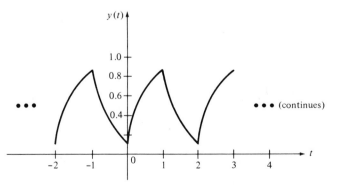

FIGURE 9.10. Capacitor voltage in Example 9.4.

9.10 shows that the capacitor charges up when the input voltage is on, and then discharges when the input voltage is off. This is the type of behavior we would expect from the physics of the circuit, so the Fourier theory is giving us a consistent result.

9.4

APPLICATION TO RECTIFIER DESIGN

The characterization of the spectra of the response in terms of the spectra of the input and the frequency response of the system is very useful in a number of applications. One such application is the design of a rectifier that produces dc (direct current) from ac (alternating current). In this section we consider the design of a half-wave rectifier. The full-wave rectifier is considered in the problems (see Problem 9.10).

Consider the half-wave rectifier shown in Figure 9.11. The input $v(t)$ to the rectifier is a voltage given by

$$v(t) = A \cos \omega_0 t,$$

where A is a positive constant and ω_0 is a fixed frequency. For example, if $v(t)$ is the usual "house voltage," then $A = 110\sqrt{2}$ V, $\omega_0 = 120\pi$ rad/sec. We assume that the voltage $v(t)$ has been "on" for a long time, and thus we can assume that $v(t) = A \cos \omega_0 t$ for $-\infty < t < \infty$.

The resistance R_s in Figure 9.11 is the source resistance of the input voltage. If the input impedance of the lowpass filter is large in comparison with R_s, the voltage $x(t)$ applied to the lowpass filter is approximately equal to $v(t)$ when $v(t) > 0$ and is equal to zero when $v(t) \leq 0$. The voltage $x(t)$ is shown in Figure 9.12.

The objective is to design the lowpass filter so that the voltage $y(t)$ applied to the load resistance R_L is approximately constant. It is not possible to eliminate completely all the sinusoidal components in $y(t)$, and thus $y(t)$ will have some "ripple." We can control the amount of ripple in the output voltage $y(t)$ by designing the lowpass filter so that the peak value of the largest sinusoidal component of $y(t)$ is a fraction $1/a$ of the dc component of $y(t)$, where a is suitably large. For example, if $a > 30$ the output voltage $y(t)$ may look like the waveform shown in Figure 9.13.

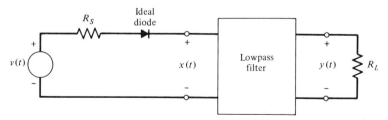

FIGURE 9.11.   Half-wave rectifier.

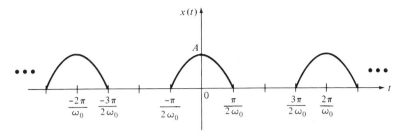

FIGURE 9.12. Voltage applied to lowpass filter.

In the following analysis, we shall take the lowpass filter to be the *RC* filter displayed in Figure 9.14. If we assume that $R_L >> R$, the input/output differential equation of the filter with load is approximately equal to

$$RC \frac{dy(t)}{dt} + y(t) = x(t). \tag{9.33}$$

To see this, from Figure 9.14 we have

$$- i_1(t) - i_2(t) = i_3(t)$$

Thus

$$C \frac{dy(t)}{dt} + \frac{1}{R_L} y(t) = \frac{1}{R} [x(t) - y(t)].$$

Multiplying both sides by R and rearranging terms gives

$$RC \frac{dy(t)}{dt} + \left(\frac{R}{R_L} + 1 \right) y(t) = x(t). \tag{9.34}$$

Now $R_L >> R$ implies that $(R/R_L) + 1 \approx 1$, and thus (9.33) is an approximation to (9.34). Then, by (9.26), the frequency function $H(\omega)$ of the filter with load is approximately equal to

$$H(\omega) = \frac{1}{jRC\omega + 1}.$$

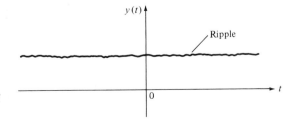

FIGURE 9.13. Output voltage of rectifier.

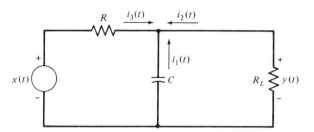

FIGURE 9.14. *RC* lowpass filter with load resistance.

The magnitude function $|H(\omega)|$ is plotted in Figure 9.15.

Now let us compute the Fourier coefficients c_n^x of the input voltage $x(t)$ of the lowpass filter. The period T of $x(t)$ is equal to $2\pi/\omega_0$, and thus the fundamental frequency ω_0 of $x(t)$ is equal to ω_0. Since $x(t)$ is an even function, from (9.8) we have

$$
c_0^x = \frac{1}{\pi/\omega_0} \int_0^{\pi/\omega_0} x(t)\, dt = \frac{\omega_0}{\pi} \int_0^{\pi/2\omega_0} A\cos(\omega_0 t)dt
$$

$$
= \frac{A}{\pi} \left[\sin \omega_0 t \right]_{t=0}^{t=\pi/2\omega_0} = \frac{A}{\pi}. \tag{9.35}
$$

Using (9.11), for $n \geq 1$ we have

$$
c_n^x = \frac{\omega_0}{\pi} \int_0^{\pi/\omega_0} x(t)\cos(n\omega_0 t)\, dt
$$

$$
= \frac{\omega_0}{\pi} \int_0^{\pi/2\omega_0} A\cos(\omega_0 t)\cos(n\omega_0 t)\, dt. \tag{9.36}
$$

The integral on the far right side of (9.36) can be evaluated using the identities

$$
\int \cos^2(\omega_0 t)\, dt = \frac{1}{\omega_0}\left[\left(\frac{1}{2}\omega_0 t + \frac{1}{4}\sin(2\omega_0 t)\right)\right] \tag{9.37}
$$

$$
\int \cos(\omega_0 t)\cos(n\omega_0 t)\, dt
$$

$$
= \frac{1}{\omega_0}\left[\frac{\sin[(n-1)\omega_0 t]}{2(n-1)} + \frac{\sin[(n+1)\omega_0 t]}{2(n+1)}\right], \quad n \geq 2. \tag{9.38}
$$

Using (9.37), we have

$$
c_1^x = \frac{A}{\pi}\left[\frac{1}{2}\omega_0 t + \frac{1}{4}\sin(2\omega_0 t)\right]_{t=0}^{t=\pi/2\omega_0} = \frac{A}{4}, \tag{9.39}
$$

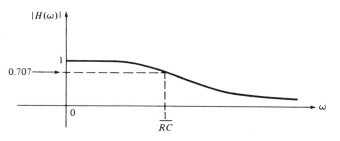

FIGURE 9.15. Magnitude function of lowpass filter.

and from (9.38),

$$c_n^x = \frac{A}{2\pi} \left[\frac{\sin[(n-1)\omega_0 t]}{n-1} + \frac{\sin[(n+1)\omega_0 t]}{n+1} \right]_{t=0}^{t=\pi/2\omega_0}, \qquad n \geq 2$$

$$= \frac{A}{2\pi} \left[\frac{\sin[(n-1)(\pi/2)]}{n-1} + \frac{\sin[(n+1)(\pi/2)]}{n+1} \right], \qquad n \geq 2$$

$$= \frac{A}{2\pi} \left[\frac{-2\cos(n\pi/2)}{n^2-1} \right], \qquad n \geq 2$$

$$c_n^x = \begin{cases} 0, & n = 3, 5, 7, 9, \ldots \\ \dfrac{A}{\pi} \dfrac{(-1)^{(n/2)+1}}{n^2-1}, & n = 2, 4, 6, 8, \ldots \end{cases} \qquad (9.40)$$

Using (9.35), (9.39), and (9.40), we have that the trigonometric Fourier series of $x(t)$ is

$$x(t) = \frac{A}{\pi} + \frac{A}{2}\cos\omega_0 t + \sum_{\substack{n=2 \\ n\ \text{even}}}^{\infty} \frac{2A}{\pi} \frac{(-1)^{(n/2)+1}}{n^2-1}\cos(n\omega_0 t), \qquad -\infty < t < \infty.$$

$$(9.41)$$

From (9.41) we see that the largest sinusoidal component of $x(t)$ is the first harmonic, whose peak value is equal to $A/2$. Also, for $n \geq 2$ we see that the magnitudes of the sinusoidal components of $x(t)$ roll off to zero at a rate that is proportional to $1/(n^2 - 1)$. The amplitude spectrum of $x(t)$ is displayed in Figure 9.16.

Now since $x(t)$ is periodic for $-\infty < t < \infty$ with fundamental frequency ω_0, from the results in Section 9.3 we know that the output $y(t)$ is also periodic with fundamental frequency ω_0. Hence $y(t)$ has the trigonometric Fourier series

$$y(t) = c_0^y + \sum_{n=1}^{\infty} 2|c_n^y| \cos(n\omega_0 t + \sphericalangle c_n^y), \qquad -\infty < t < \infty,$$

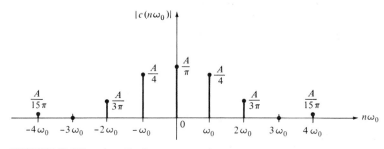

FIGURE 9.16. Amplitude spectrum of $x(t)$.

where

$$|c_n^y| = |c_n^x||H(n\omega_0)|, \qquad n = 0, 1, 2, \ldots \qquad (9.42)$$

Since the largest sinusoidal component of $x(t)$ is the first harmonic and since $|H(n\omega_0)|$ rolls off as n is increased, the largest sinusoidal component of $y(t)$ is also the first harmonic whose peak value is equal to $2|c_1^y|$. Then the largest sinusoidal component of $y(t)$ will be $1/a$ of the dc component of $y(t)$ if

$$2|c_1^y| = \frac{1}{a} |c_0^y|. \qquad (9.43)$$

So our design objective stated above will be met if we set the values of the resistor and capacitor in the RC lowpass filter so that (9.43) is satisfied. This can be accomplished as follows.

From (9.42) we have

$$|c_0^y| = |c_0^x||H(0)| = \frac{A}{\pi}(1) = \frac{A}{\pi}$$

and

$$|c_1^y| = |c_1^x||H(\omega_0)| = \frac{A}{4} \frac{1}{\sqrt{R^2C^2\omega_0^2 + 1}}.$$

Inserting these expressions for $|c_0^y|$ and $|c_1^y|$ into (9.43), we obtain

$$\frac{A}{2} \frac{1}{\sqrt{R^2C^2\omega_0^2 + 1}} = \frac{1}{a}\frac{A}{\pi}.$$

Thus

$$\frac{A}{2} = \frac{A}{a\pi} \sqrt{R^2 C^2 \omega_0^2 + 1}$$

$$\frac{1}{4} = \frac{1}{a^2 \pi^2} (R^2 C^2 \omega_0^2 + 1)$$

$$\approx \frac{R^2 C^2 \omega_0^2}{a^2 \pi^2}. \tag{9.44}$$

The value of R is chosen so that $R_L \gg R$. Then solving (9.44) for C, we get

$$C = \frac{a\pi}{2R\omega_0}. \tag{9.45}$$

As a specific example, suppose that $\omega_0 = 120\pi$ rad/sec and that $R = 100$ Ω. Let us take $a = 33.3$, so that the largest sinusoidal component of $y(t)$ is 3 percent of the dc component of $y(t)$. Inserting these values into (9.45), we have

$$C = \frac{(33.3\pi)}{2(100)(120\pi)} = 0.0014 \text{ F.}$$

Note that the value of C is rather large. This is a common characteristic of conventional dc power supplies.

9.5

FOURIER TRANSFORM

A key feature of the Fourier series theory of periodic signals is the definition of the amplitude and phase spectra of periodic signals. As we have seen, the characterization of a periodic signal in terms of its spectra is very useful in determining how a linear time-invariant continuous-time system processes a periodic input signal to produce the output response. More precisely, we know that the amplitude spectrum of the output is equal to the product of the amplitude spectrum of the periodic input signal and the system's magnitude function $|H(\omega)|$ with $\omega = n\omega_0$, and the phase spectrum of the output is equal to the sum of the phase spectrum of the periodic input signal and the system's phase function $\sphericalangle H(\omega)$ with $\omega = n\omega_0$.

The question then arises as to whether or not it is possible to define the notions of amplitude and phase spectra for nonperiodic signals, also called *aperiodic signals*. The answer is yes, and the analytical construct for doing this is the Fourier transform. As we shall see, the spectra associated with a nonperiodic signal are defined for all real values of ω, not just for discrete values of ω as in the case of a periodic signal. In other words, the spectra for a nonperiodic signal are not line spectra.

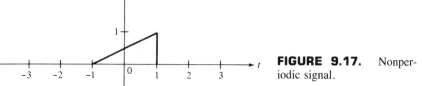

FIGURE 9.17. Nonperiodic signal.

We begin by showing that the Fourier transform can be generated by considering the Fourier series representation of a periodic signal. Let $x(t)$ be a continuous-time signal with $x(t) = 0$ for all $t > T_1$ and $x(t) = 0$ for all $t < -T_1$, where T_1 is some fixed positive number. Such a signal is said to be *time limited* or to be of *finite duration* in time. Clearly, a time-limited signal $x(t)$ cannot be periodic.

Given a positive number $T > 2T_1$, let $\tilde{x}_T(t)$ denote the periodic signal with period T which is equal to $x(t)$ for $-T/2 < t < T/2$; that is,

$$\tilde{x}_T(t + nT) = x(t), \qquad \frac{-T}{2} < t < \frac{T}{2}, \quad n = 0, \pm 1, \pm 2, \ldots$$

For example, suppose that $x(t)$ is the signal shown in Figure 9.17. Then with $T = 3$, $\tilde{x}_T(t)$ is the periodic signal displayed in Figure 9.18.

By definition of $\tilde{x}_T(t)$, we have that

$$x(t) = \lim_{T \to \infty} \tilde{x}_T(t).$$

Now since $\tilde{x}_T(t)$ is periodic, it has the exponential Fourier series

$$\tilde{x}_T(t) = \sum_{n=-\infty}^{\infty} c_n \exp(jn\omega_0 t), \tag{9.46}$$

where

$$c_n = \frac{1}{T} \int_{-T/2}^{T/2} x(t) \exp(-jn\omega_0 t) \, dt, \qquad n = 0, \pm 1, \pm 2, \ldots \tag{9.47}$$

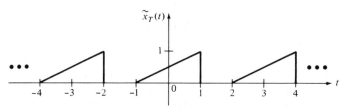

FIGURE 9.18. Signal $\tilde{x}_T(t)$ when $T = 3$.

Since $\tilde{x}_T(t) = x(t)$ for $-T/2 < t < T/2$ and $x(t) = 0$ for $t > T/2$ and $t < -T/2$, the expression (9.47) for c_n can be written in the form

$$c_n = \frac{1}{T} \int_{-\infty}^{\infty} x(t) \exp(-jn\omega_0 t) \, dt. \tag{9.48}$$

Now define

$$X(\omega) = \int_{-\infty}^{\infty} x(t) e^{-j\omega t} \, dt. \tag{9.49}$$

Then

$$X(n\omega_0) = \int_{-\infty}^{\infty} x(t) \exp(-jn\omega_0 t) \, dt, \qquad n = 0, \pm 1, \pm 2, \ldots,$$

and thus (9.48) reduces to

$$c_n = \frac{1}{T} X(n\omega_0), \qquad n = 0, \pm 1, \pm 2, \ldots. \tag{9.50}$$

Inserting the expression (9.50) for c_n into (9.46), we have

$$\tilde{x}_T(t) = \sum_{n=-\infty}^{\infty} \frac{1}{T} X(n\omega_0) \exp(jn\omega_0 t)$$

$$= \frac{1}{2\pi} \sum_{n=-\infty}^{\infty} X(n\omega_0) [\exp(jn\omega_0 t)]\omega_0 \qquad \text{since } T = \frac{2\pi}{\omega_0}. \tag{9.51}$$

Now as $T \to \infty$, we have that $n\omega_0 \to \omega$, $\omega_0 \to d\omega$, and the summation on the right side of (9.51) converges to the integral

$$\int_{-\infty}^{\infty} X(\omega) e^{j\omega t} \, d\omega.$$

Then since $\tilde{x}_T(t) \to x(t)$ as $T \to \infty$, from (9.51) we have that

$$x(t) = \frac{1}{2\pi} \int_{-\infty}^{\infty} X(\omega) e^{j\omega t} \, d\omega. \tag{9.52}$$

Equation (9.52) is the Fourier integral representation of the nonperiodic signal $x(t)$ and the function $X(\omega)$ defined by (9.49) is the Fourier transform of $x(t)$. The relationship (9.52) is referred to as the *inverse Fourier transform* since it shows how to reconstruct $x(t)$ from the transform $X(\omega)$.

As seen from the above derivation, the Fourier integral representation (9.52) of a nonperiodic signal can be viewed as a generalization of the Fourier series representation of a periodic signal. Note that in contrast to a periodic signal, a nonperiodic signal consists of a continuum of frequencies; that is, the frequency variable ω in (9.52) is a continuous variable. In the periodic case, the frequency variable is the discrete variable $n\omega_0$, $n = 0, 1, 2, \ldots$

It is also interesting to note that since $x(t) = \tilde{x}_T(t)$ for $-T/2 < t < T/2$, by (9.46) we have

$$x(t) = \sum_{n=-\infty}^{\infty} c_n \exp(jn\omega_0 t), \qquad \frac{-T}{2} < t < \frac{T}{2}. \qquad (9.53)$$

Thus the signal $x(t)$ has a Fourier series representation over the finite time interval $-T/2 < t < T/2$. The coefficients c_n of this Fourier series representation are related to the Fourier transform of $x(t)$ by equation (9.50). Hence we have a direct relationship between the Fourier transform (9.49) of $x(t)$ and the Fourier series representation (9.53) of $x(t)$.

EXISTENCE OF THE FOURIER TRANSFORM   The definition (9.49) of the Fourier transform can be extended to signals that are not time limited. In particular, given any signal $x(t)$, the Fourier transform $X(\omega)$ of $x(t)$ is defined to be the complex-valued function of the continuous frequency variable ω given by

$$X(\omega) = \int_{-\infty}^{\infty} x(t)e^{-j\omega t} \, dt. \qquad (9.54)$$

The signal $x(t)$ is said to have a Fourier transform in the ordinary sense if the integral in (9.54) converges (i.e., exists). The integral does converge if $x(t)$ is "well behaved" and if $x(t)$ is absolutely integrable. The latter condition means that

$$\int_{-\infty}^{\infty} |x(t)| \, dt < \infty. \qquad (9.55)$$

"Well behaved" means that the signal has a finite number of discontinuities, maxima, and minima within any finite interval of time. Except for impulses, most signals of interest are well behaved. All actual signals (i.e., signals that can be physically generated) are well behaved and satisfy (9.55).

Since any well-behaved signal of finite duration in time is absolutely integrable, any such signal has a Fourier transform in the ordinary sense. An example of a signal that does not have a Fourier transform in the ordinary sense follows.

EXAMPLE 9.5.   Consider the dc signal

$$x(t) = 1, \qquad -\infty < t < \infty.$$

Clearly, the dc signal is not an actual signal, but it plays a very important role in the theory of signals and systems. The Fourier transform of the dc signal is

$$X(\omega) = \int_{-\infty}^{\infty} (1)e^{-j\omega t}\, dt \qquad (9.56)$$

$$= \lim_{T\to\infty} \int_{-T/2}^{T/2} e^{-j\omega t}\, dt$$

$$= \lim_{T\to\infty} -\frac{1}{j\omega}\left[e^{-j\omega t} \right]_{t=-T/2}^{t=T/2}$$

$$= \lim_{T\to\infty} -\frac{1}{j\omega}\left[\exp\left(-\frac{j\omega T}{2}\right) - \exp\left(\frac{j\omega T}{2}\right) \right].$$

But $\exp(j\omega T/2)$ does not have a limit as $T \to \infty$, and thus the integral in (9.56) does not converge. Hence a dc signal does not have a Fourier transform in the ordinary sense. We could also see this by checking (9.55): The area under the dc signal is infinite, so the integral in (9.55) is not finite. Later, we will see that a dc signal has a Fourier transform in a generalized sense.

RELATIONSHIP BETWEEN THE FOURIER AND LAPLACE TRANS-FORMS

Let $x(t)$ be a continuous-time signal that is not necessarily zero for $t < 0$. In Chapter 5 we defined the two-sided Laplace transform $X(s)$ of $x(t)$ by

$$X(s) = \int_{-\infty}^{\infty} x(t)e^{-st}\, dt. \qquad (9.57)$$

Note that if the signal $x(t)$ is zero for all $t < 0$, the two-sided Laplace transform can be written in the form

$$X(s) = \int_{0}^{\infty} x(t)e^{-st}\, dt. \qquad (9.58)$$

The expression (9.58) for $X(s)$ is identical to the one-sided Laplace transform that we studied in Chapter 5.

Given a signal $x(t)$ that may be nonzero for $t < 0$, suppose that the integral in (9.57) converges when $s = j\omega$ and let $X(\omega)$ denote $X(s)$ with $s = j\omega$. Then

$$X(\omega) = \int_{-\infty}^{\infty} x(t)e^{-j\omega t}\, dt. \qquad (9.59)$$

Comparing (9.54) and (9.59), we see that $X(\omega)$ is equal to the Fourier transform of $x(t)$. Hence for any signal $x(t)$ whose two-sided Laplace transform $X(s)$ converges for $s = j\omega$, the Fourier transform $X(\omega)$ of $x(t)$ is given by

$$X(\omega) = X(s)\big|_{s=j\omega}.$$

So the Fourier transform can be computed directly from the two-sided Laplace transform. One must be careful in using this result since there are signals that have a two-sided Laplace transform, but not a Fourier transform in the ordinary sense. An example is given below.

EXAMPLE 9.6. Let $x(t)$ be the unit-step function $u(t)$. Since $u(t)$ is zero for all $t < 0$, the two-sided Laplace transform of $u(t)$ is the same as the one-sided Laplace transform. In Example 5.1 we showed that the Laplace transform $U(s)$ of $u(t)$ is equal to $1/s$. We also showed that the integral defining the transform $U(s)$ converges if and only if Re $s > 0$. Since Re $s = 0$ when $s = j\omega$, $U(\omega)$ does not exist and therefore the unit-step function does not have a Fourier transform in the ordinary sense. As in the dc-signal example, the nonexistence of the Fourier transform is a result of condition (9.55) not being satisfied.

EXAMPLE 9.7. Now suppose that $x(t) = e^{-bt}u(t)$, where b is an arbitrary real number. From the results in Example 5.3, we know that the Laplace transform $X(s)$ of $x(t)$ is equal to $1/(s + b)$ and that the integral in the definition of $X(s)$ converges if and only if Re $s > -b$. Thus, if $b > 0$, $X(\omega)$ exists and $x(t)$ has a Fourier transform given by

$$X(\omega) = X(s)\big|_{s=j\omega} = \frac{1}{j\omega + b}.$$

If $b \leq 0$, $X(\omega)$ does not exist and therefore in this case $x(t)$ does not have a Fourier transform in the ordinary sense.

Let $x(t)$ be a signal that is zero for all $t < 0$ and has one-sided Laplace transform $X(s)$. If $X(s)$ is a strictly proper rational function of s and if the poles of $X(s)$ are in the open left-half plane, it follows from the results in Chapter 8 that $x(t)$ has an ordinary Fourier transform $X(\omega)$ given by

$$X(\omega) = X(s)\big|_{s=j\omega}.$$

RECTANGULAR AND POLAR FORM OF THE FOURIER TRANSFORM
Consider the signal $x(t)$ with Fourier transform

$$X(\omega) = \int_{-\infty}^{\infty} x(t)e^{-j\omega t}\, dt.$$

Using Euler's formula, we can write $X(\omega)$ in the form

$$X(\omega) = \int_{-\infty}^{\infty} x(t)\cos(\omega t)\, dt - j\int_{-\infty}^{\infty} x(t)\sin(\omega t)\, dt.$$

Now let $R(\omega)$ and $I(\omega)$ denote the real-valued functions of ω defined by

$$R(\omega) = \int_{-\infty}^{\infty} x(t) \cos (\omega t) \, dt$$

$$I(\omega) = -\int_{-\infty}^{\infty} x(t) \sin (\omega t) \, dt.$$

Then we can express $X(\omega)$ in the rectangular form

$$X(\omega) = R(\omega) + jI(\omega). \tag{9.60}$$

The function $R(\omega)$ is the real part of $X(\omega)$ and the function $I(\omega)$ is the imaginary part of $X(\omega)$. Note that $R(\omega)$ and $I(\omega)$ could be computed first, and then $X(\omega)$ can be found using (9.60).

The Fourier transform $X(\omega)$ can also be written in the polar form

$$X(\omega) = |X(\omega)| \exp [j \sphericalangle X(\omega)], \tag{9.61}$$

where $|X(\omega)|$ is the magnitude of $X(\omega)$ and $\sphericalangle X(\omega)$ is the angle of $X(\omega)$. We can go from the rectangular form to the polar form by using the relationships

$$|X(\omega)| = \sqrt{R^2(\omega) + I^2(\omega)}$$

$$\sphericalangle X(\omega) = \begin{cases} \tan^{-1} \dfrac{I(\omega)}{R(\omega)} & \text{when } R(\omega) > 0 \\[2ex] 180° + \tan^{-1} \dfrac{I(\omega)}{R(\omega)} & \text{when } R(\omega) < 0. \end{cases}$$

Note that, by (9.54),

$$X(-\omega) = \overline{X(\omega)} = \text{complex conjugate of } X(\omega).$$

Then taking the complex conjugate of the polar form (9.61), we have that

$$X(-\omega) = |X(\omega)| \exp [-j \sphericalangle X(\omega)].$$

Thus

$$|X(-\omega)| = |X(\omega)|$$

and

$$\sphericalangle X(-\omega) = -\sphericalangle X(\omega).$$

These results show that the magnitude $|X(\omega)|$ is an even function of ω, while the angle $\sphericalangle X(\omega)$ is an odd function of ω.

SIGNALS WITH EVEN OR ODD SYMMETRY   Again suppose that $x(t)$ has Fourier transform $X(\omega)$ with $X(\omega)$ given in the rectangular form (9.60). We assume that the signal $x(t)$ is an even signal; that is, $x(t) = x(-t)$ for all $t > 0$. Then $x(t) \cos \omega t$ is an even function of t, and the real part $R(\omega)$ of the Fourier transform can be rewritten as

$$R(\omega) = 2 \int_0^\infty x(t) \cos (\omega t) \, dt.$$

Furthermore, since $x(t)$ is even, $x(t) \sin \omega t$ is odd, and the imaginary part $I(\omega)$ of the Fourier transform is zero. Hence the Fourier transform of an even signal $x(t)$ is a real-valued function of ω given by

$$X(\omega) = R(\omega) = 2 \int_0^\infty x(t) \cos (\omega t) \, dt. \tag{9.62}$$

If the signal $x(t)$ is odd, $x(t) \cos \omega t$ is odd and $x(t) \sin \omega t$ is even, in which case the Fourier transform of $x(t)$ is a purely imaginary function of ω given by

$$X(\omega) = jI(\omega) = -j2 \int_0^\infty x(t) \sin (\omega t) \, dt. \tag{9.63}$$

The expression (9.62) should be used to compute the Fourier transform of an even signal, and the expression (9.63) should be used to compute the Fourier transform of an odd signal.

EXAMPLE 9.8.   Given a fixed positive number τ, let $p_\tau(t)$ denote the rectangular pulse defined by

$$p_\tau(t) = \begin{cases} 1, & \dfrac{-\tau}{2} \le t < \dfrac{\tau}{2} \\[2mm] 0, & \text{all other } t. \end{cases}$$

The rectangular pulse is plotted in Figure 9.19. The rectangular function is even, and thus we can use (9.62) to compute the Fourier transform. Setting $x(t) = p_\tau(t)$ in (9.62), we have

$$X(\omega) = 2 \int_0^{\tau/2} (1) \cos (\omega t) \, dt$$

$$= \frac{2}{\omega} \left[\sin \omega t \right]_{t=0}^{t=\tau/2}$$

$$= \frac{2}{\omega} \sin \left(\frac{\omega \tau}{2} \right).$$

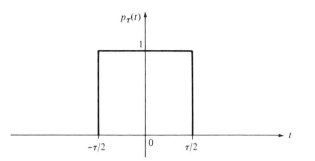

FIGURE 9.19. Rectangular pulse function.

We can express $X(\omega)$ in terms of the Sa function defined by

$$\text{Sa}\ (\lambda)\ =\ \frac{\sin \lambda}{\lambda}.$$

The Sa function is related to the sinc function $\text{sinc}\ \lambda\ =\ (\sin \pi\lambda)/\pi\lambda$ by

$$\text{Sa}\ (\lambda)\ =\ \text{sinc}\ \left(\frac{\lambda}{\pi}\right).$$

Now

$$\text{Sa}\ \left(\frac{\omega\tau}{2}\right)\ =\ \frac{\sin\ (\omega\tau/2)}{\omega\tau/2}\ =\ \frac{2}{\omega\tau}\ \sin\ \left(\frac{\omega\tau}{2}\right),$$

and thus

$$X(\omega)\ =\ \tau\ \text{Sa}\ \left(\frac{\omega\tau}{2}\right). \tag{9.64}$$

So the Fourier transform of the rectangular pulse function is a Sa function in frequency. Since $X(\omega)$ is real valued in this example, we can plot $X(\omega)$ versus ω. The result is displayed in Figure 9.20. Note that by l'Hôspital's rule, $X(0)\ =\ \tau$.

SPECTRUM OF A SIGNAL   Given a signal $x(t)$, let $X(\omega)$ denote the Fourier transform of $x(t)$. The function $X(\omega)$ is referred to as the frequency spectrum of the signal $x(t)$. The magnitude function $|X(\omega)|$ is called the amplitude spectrum of the signal, and the angle $\angle X(\omega)$ is called the phase spectrum of the signal. The amplitude and phase spectra are natural generalizations of the line spectra of periodic signals. As in the periodic case, the amplitude spectrum displays the frequency components comprising the signal, except that in the nonperiodic case, there is a continuum of frequency components that make up the signal.

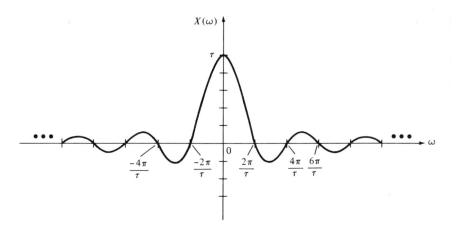

FIGURE 9.20.   Fourier transform of the rectangular pulse.

A signal $x(t)$ is said to be *bandlimited* if its spectrum $X(\omega)$ is zero for all $\omega > B$, where B is some positive number, called the bandwidth of the signal. If a signal is not bandlimited, it is said to have an *infinite bandwidth* or an *infinite spectrum*. It turns out that any bandlimited signal must be of infinite duration in time; in other words, bandlimited signals cannot be time limited. Conversely, any time-limited signal must have an infinite spectrum. The proof of these results is left to a more advanced treatment of Fourier theory.

Since all actual (physical) signals are time limited, we have the interesting result that the spectrum of an actual signal is infinite. However, for any well-behaved time-limited signal $x(t)$, it can be proved that $|X(\omega)|$ converges to zero as $\omega \to \infty$. Therefore, in practice one may always assume that $|X(\omega)| \approx 0$ for all $> B$, where B is chosen to be suitably large.

EXAMPLE 9.9.   Again consider the rectangular pulse function $x(t) = p_\tau(t)$. In Example 9.8 we showed that the Fourier transform $X(\omega)$ is equal to τ Sa $(\omega\tau/2)$. The plots of the amplitude and phase spectra for this example are given in Figure 9.21. The spectrum of the rectangular pulse is clearly infinite. However, from Figure 9.21a we see that most of the spectral content of the pulse is contained in the main lobe of the Sa function. Note that if we make the time duration τ of the rectangular pulse smaller, the main lobe of the amplitude spectrum is widened. This result shows that shorter time-duration signals have more spectral content at higher frequencies than longer time-duration signals.

INVERSE FOURIER TRANSFORM   Given a signal $x(t)$ with Fourier transform $X(\omega)$, we can recompute $x(t)$ from $X(\omega)$ by applying the inverse Fourier transform given by

$$x(t) = \frac{1}{2\pi} \int_{-\infty}^{\infty} X(\omega)\, e^{j\omega t}\, d\omega. \qquad (9.65)$$

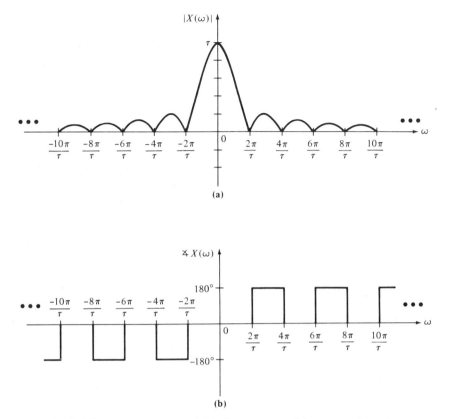

FIGURE 9.21. (a) Amplitude and (b) phase spectra of the rectangular pulse.

As in the Laplace transform theory, we shall use the transform pair notation

$$x(t) \leftrightarrow X(\omega)$$

to denote the fact that $X(\omega)$ is the Fourier transform of $x(t)$ and that $x(t)$ is the inverse Fourier transform of $X(\omega)$. One of the most fundamental transform pairs in the Fourier theory is the pair

$$p_\tau(t) \leftrightarrow \tau \, \mathrm{Sa} \left(\frac{\omega\tau}{2} \right). \tag{9.66}$$

It is often possible to compute the Fourier transform or the inverse Fourier transform without having to evaluate the integrals in (9.54) and (9.65). In particular, as in the Laplace transform theory, it is possible to derive many new transform pairs from a small set of "basic" transform pairs [such as (9.66)] by using the properties of the Fourier transform. These properties are given in the next section.

PROPERTIES OF THE FOURIER TRANSFORM

In this section we give the properties of the Fourier transform. The majority of these properties correspond directly to the properties of the Laplace transform that were studied in Section 5.2. Most of the properties of the Fourier transform that do correspond to properties of the Laplace transform can be proved simply by replacing s by $j\omega$ in the proof of the Laplace transform property. Thus the proof of these properties follows easily from the constructions given in Section 5.2. The details are left to the interested reader.

The Fourier transform does enjoy some properties for which there is no version in the Laplace transform theory. Two examples are the duality property and Parseval's theorem. These properties are proved in this section.

LINEARITY   The Fourier transform is a linear operation, as is the Laplace transform. In other words, if $x(t) \leftrightarrow X(\omega)$ and $v(t) \leftrightarrow V(\omega)$, then for any real or complex scalars a, b,

$$ax(t) + bv(t) \leftrightarrow aX(\omega) + bV(\omega). \tag{9.67}$$

> **EXAMPLE 9.10.**   Consider the signal $x(t)$ shown in Figure 9.22. As illustrated in Figure 9.22, this signal is equal to the sum of two rectangular pulse functions. From Figure 9.22 we have
>
> $$x(t) = p_4(t) + p_2(t).$$
>
> Then using linearity and the transform pair (9.66), we have that the Fourier transform $X(\omega)$ is
>
> $$X(\omega) = 4 \text{ Sa } (2\omega) + 2 \text{ Sa } (\omega).$$

LEFT OR RIGHT SHIFT IN TIME   If $x(t) \leftrightarrow X(\omega)$, then for any positive or negative real number c, we have

$$x(t - c) \leftrightarrow X(\omega)e^{-j\omega c}. \tag{9.68}$$

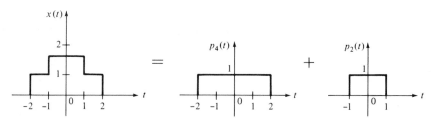

FIGURE 9.22.   Signal in Example 9.10.

Note that if $c > 0$, $x(t - c)$ is a c-second right shift of $x(t)$, and if $c < 0$, $x(t - c)$ is a c-second left shift of $x(t)$. Thus the transform pair (9.68) is valid for both left and right shifts of $x(t)$. This is in contrast to the one-sided Laplace transform, for which there is a version of (9.68) in the right-shift case only. There is a left-shift and right-shift property in the Fourier transform theory since the Fourier transform is a two-sided transform.

EXAMPLE 9.11. Consider the signal shown in Figure 9.23. The signal $x(t)$ is equal to a 1-second right shift of the rectangular pulse function $p_2(t)$; that is,

$$x(t) = p_2(t - 1).$$

Then using the time-shift property (9.68) and the transform pair (9.66), we have that

$$X(\omega) = 2 \text{ Sa } (\omega)e^{-j\omega}.$$

Note that since

$$|e^{-j\omega}| = 1 \text{ for all } \omega,$$

the amplitude spectrum $|X(\omega)|$ of $x(t) = p_2(t - 1)$ is the same as the amplitude spectrum of $p_2(t)$.

TIME SCALING   If $x(t) \leftrightarrow X(\omega)$, for any positive real number a,

$$x(at) \leftrightarrow \frac{1}{a} X\left(\frac{\omega}{a}\right). \tag{9.69}$$

The transform pair (9.69) follows from the corresponding transform pair in the Laplace transform theory [set $s = j\omega$ in the transform pair (5.21)].

TIME REVERSAL   Given the signal $x(t)$, consider the time-reversed signal $x(-t)$. The signal $x(-t)$ is equal to $x(t)$ folded about the vertical axis. Now if

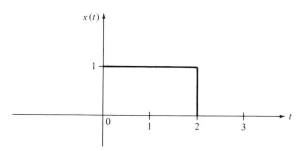

FIGURE 9.23.   Signal in Example 9.11.

$x(t) \leftrightarrow X(\omega)$, then $x(-t) \leftrightarrow X(-\omega)$. To prove this, simply replace t by $-t$ in the definition of the Fourier transform of $x(t)$. By definition of $X(\omega)$, we have that $X(-\omega) = \overline{X(\omega)}$, where $\overline{X(\omega)}$ is the complex conjugate of $X(\omega)$. Thus we have the transform pair $x(-t) \leftrightarrow \overline{X(\omega)}$. This result says that time reversal in the time domain corresponds to conjugation in the frequency domain.

MULTIPLICATION BY A POWER OF t If $x(t) \leftrightarrow X(\omega)$, for any positive integer n,

$$t^n x(t) \leftrightarrow (j)^n \frac{d^n}{d\omega^n} X(\omega). \qquad (9.70)$$

Setting $n = 1$ in (9.70), we see that multiplication by t in the time domain corresponds to differentiation with respect to ω in the Fourier transform domain (plus multiplication by j). The transform pair (9.70) can be proved in the same way that we proved the corresponding property in the Laplace transform theory.

EXAMPLE 9.12. Let $x(t) = tp_2(t)$. This signal is plotted in Figure 9.24. From the multiplication by t property (9.70) and the transform pair (9.66), we have that the transform $X(\omega)$ is given by

$$X(\omega) = j \frac{d}{d\omega} 2 \, \text{Sa} \, (\omega)$$

$$= j2 \frac{d}{d\omega} \frac{\sin \omega}{\omega}$$

$$= j2 \frac{\omega \cos \omega - \sin \omega}{\omega^2}.$$

The amplitude spectrum $|X(\omega)|$ is plotted in Figure 9.25.

MULTIPLICATION BY A COMPLEX EXPONENTIAL If $x(t) \leftrightarrow X(\omega)$, then for any real number ω_0,

$$x(t) \exp (j\omega_0 t) \leftrightarrow X(\omega - \omega_0). \qquad (9.71)$$

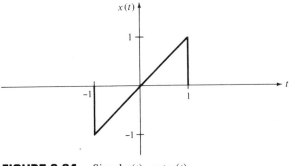

FIGURE 9.24. Signal $x(t) = tp_2(t)$.

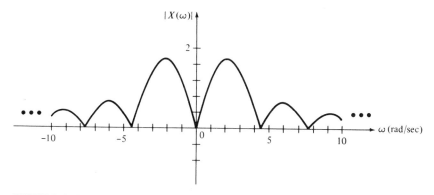

FIGURE 9.25. Amplitude spectrum of the signal in Figure 9.24.

So multiplication by a complex exponential in the time domain corresponds to a frequency shift in the Fourier transform domain.

MULTIPLICATION BY SIN $\omega_0 t$ OR COS $\omega_0 t$. If $x(t) \leftrightarrow X(\omega)$, then for any real number ω_0,

$$x(t) \sin \omega_0 t \leftrightarrow \frac{j}{2} [X(\omega + \omega_0) - X(\omega - \omega_0)] \qquad (9.72)$$

and

$$x(t) \cos \omega_0 t \leftrightarrow \tfrac{1}{2} [X(\omega + \omega_0) + X(\omega - \omega_0)]. \qquad (9.73)$$

As we shall discuss in more detail in Chapter 10, the signals $x(t) \sin \omega_0 t$ and $x(t) \cos \omega_0 t$ can be viewed as amplitude-modulated signals. More precisely, in forming the signal $x(t) \sin \omega_0 t$, the carrier $\sin \omega_0 t$ is modulated by the signal $x(t)$. As a result of this characterization of $x(t) \sin \omega_0 t$ [and $x(t) \cos \omega_0 t$], the relationships (9.72) and (9.73) are called the *modulation theorems* of the Fourier transform. The relationships (9.72) and (9.73) show that modulation of a carrier by a signal $x(t)$ results in the frequency translations $X(\omega + \omega_0)$, $X(\omega - \omega_0)$ of the spectrum $X(\omega)$ of $x(t)$.

EXAMPLE 9.13. Consider the signal $x(t) = p_\tau(t) \cos \omega_0 t$, which can be interpreted as a sinusoidal burst. For the case when $\tau = 0.5$ and $\omega_0 = 60$ rad/sec, the signal is plotted in Figure 9.26. By the modulation property (9.73) and the transform pair (9.66), the Fourier transform of the sinusoidal burst is equal to

$$\frac{1}{2} \left[\tau \, \text{Sa} \left(\frac{(\omega + \omega_0)\tau}{2} \right) + \tau \, \text{Sa} \left(\frac{(\omega - \omega_0)\tau}{2} \right) \right].$$

For the case $\tau = 0.5$ and $\omega_0 = 60$ rad/sec, the transform of the sinusoidal burst is plotted in Figure 9.27.

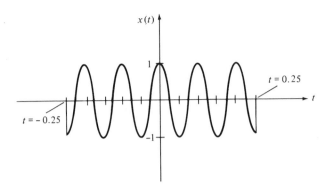

FIGURE 9.26. Sinusoidal burst.

DIFFERENTIATION IN THE TIME DOMAIN If $x(t) \leftrightarrow X(\omega)$, then for any positive integer n,

$$\frac{d^n}{dt^n} x(t) \leftrightarrow (j\omega)^n X(\omega). \tag{9.74}$$

The transform pair (9.74) looks like the corresponding pair in the Laplace transform theory, except that there are no initial conditions in (9.74) since the transform is two-sided.

CONVOLUTION IN THE TIME DOMAIN Given two signals $x(t)$ and $v(t)$ with Fourier transforms $X(\omega)$ and $V(\omega)$, the Fourier transform of the convolution $(x * v)(t)$ is equal to $X(\omega)V(\omega)$; that is, we have the transform pair

$$(x * v)(t) \leftrightarrow X(\omega)V(\omega). \tag{9.75}$$

The transform pair (9.75) follows by setting $s = j\omega$ in the corresponding transform pair in the Laplace transform theory.

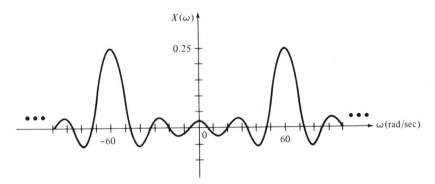

FIGURE 9.27. Fourier transform of the sinusoidal burst $x(t) = p_{0.5}(t) \cos 60t$.

INTEGRATION IN THE TIME DOMAIN   Given a signal $x(t)$, the integral of $x(t)$ is the function

$$x^{(-1)}(t) = \int_{-\infty}^{t} x(\lambda) \, d\lambda.$$

Suppose that $x(t)$ has Fourier transform $X(\omega)$. In general, the integral $x^{(-1)}(t)$ does not have a Fourier transform in the ordinary sense. But $x^{(-1)}(t)$ does have the generalized transform

$$\frac{1}{j\omega} X(\omega) + \pi X(0)\delta(\omega),$$

where $\delta(\omega)$ is the impulse function in the frequency domain. So we have the transform pair

$$x^{(-1)}(t) \leftrightarrow \frac{1}{j\omega} X(\omega) + \pi X(0)\delta(\omega). \tag{9.76}$$

We shall verify the transform pair (9.76) in Section 9.7 when we consider the generalized Fourier transform.

EXAMPLE 9.14.   Consider the triangular pulse function $v(t)$ displayed in Figure 9.28. As first noted in Chapter 1, we can write

$$v(t) = \left(1 - \frac{2|t|}{\tau}\right) p_\tau(t),$$

where $p_\tau(t)$ is the rectangular pulse of duration τ seconds. Let $x(t)$ denote the derivative of $v(t)$. The derivative is shown in Figure 9.29. From the plot we have

$$x(t) = \frac{2}{\tau} p_{\tau/2}\left(t + \frac{\tau}{4}\right) - \frac{2}{\tau} p_{\tau/2}\left(t - \frac{\tau}{4}\right).$$

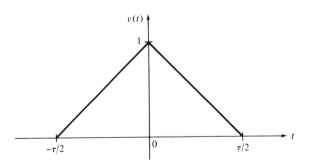

FIGURE 9.28.   Triangular pulse.

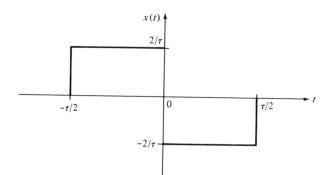

FIGURE 9.29. Derivative of the triangular pulse.

Using the transform pair (9.66) and the shift property (9.68), we have that the Fourier transform $X(\omega)$ of $x(t)$ is

$$X(\omega) = \text{Sa}\left(\frac{\omega\tau}{4}\right)\left[\exp\left(\frac{j\omega\tau}{4}\right) - \exp\left(\frac{-j\omega\tau}{4}\right)\right]$$

$$= \text{Sa}\left(\frac{\omega\tau}{4}\right)\left[j2\sin\left(\frac{\omega\tau}{4}\right)\right].$$

Now by definition of $x(t)$,

$$v(t) = \int_{-\infty}^{t} x(\lambda)\, d\lambda.$$

Thus by the integration property, the Fourier transform of $v(t)$ is equal to

$$\frac{1}{j\omega}\,\text{Sa}\left(\frac{\omega\tau}{4}\right)\left[j2\sin\left(\frac{\omega\tau}{4}\right)\right] + \pi X(0)\delta(\omega),$$

which reduces to $(\tau/2)\text{Sa}^2\left(\dfrac{\omega\tau}{4}\right)$. Thus we have the transform pair

$$\left(1 - \frac{2|t|}{\tau}\right)p_\tau(t) \leftrightarrow \frac{\tau}{2}\,\text{Sa}^2\left(\frac{\omega\tau}{4}\right). \qquad (9.77)$$

By (9.77) we see that the triangular function in the time domain corresponds to a Sa-squared function in the Fourier transform domain.

MULTIPLICATION IN THE TIME DOMAIN   If $x(t) \leftrightarrow X(\omega)$ and $v(t) \leftrightarrow V(\omega)$, then

$$x(t)v(t) \leftrightarrow \frac{1}{2\pi}\,(X * V)\,(\omega) = \frac{1}{2\pi}\int_{-\infty}^{\infty} X(\omega - \lambda)V(\lambda)\, d\lambda. \qquad (9.78)$$

The transform pair (9.78) says that multiplication in the time domain corresponds to convolution in the Fourier transform domain. The proof of (9.78) follows from the definition of the Fourier transform and the manipulation of integrals. We omit the details.

It is worth noting that there is no simple version of (9.78) in the Laplace transform theory. This is a result of the difficulty in defining the convolution of functions of complex variables.

PARSEVAL'S THEOREM   Again suppose that we are given the transform pairs $x(t) \leftrightarrow X(\omega)$ and $v(t) \leftrightarrow V(\omega)$. Then

$$\int_{-\infty}^{\infty} x(t)v(t) \, dt = \frac{1}{2\pi} \int_{-\infty}^{\infty} \overline{X(\omega)}V(\omega) \, d\omega, \tag{9.79}$$

where $\overline{X(\omega)}$ is the complex conjugate of $X(\omega)$. The relationship (9.79), which is called *Parseval's theorem*, follows directly from the transform pair (9.78). To see this, first note that the Fourier transform of the product $x(t)v(t)$ is equal to

$$\int_{-\infty}^{\infty} x(t)v(t)e^{-j\omega t} \, dt.$$

But by the transform pair (9.78), the Fourier transform of $x(t)v(t)$ is equal to

$$\frac{1}{2\pi} \int_{-\infty}^{\infty} X(\omega - \lambda)V(\lambda) \, d\lambda.$$

Thus

$$\int_{-\infty}^{\infty} x(t)v(t)e^{-j\omega t} \, dt = \frac{1}{2\pi} \int_{-\infty}^{\infty} X(\omega - \lambda)V(\lambda) \, d\lambda. \tag{9.80}$$

The relationship (9.80) must hold for all real values of ω. If we take $\omega = 0$, we get

$$\int_{-\infty}^{\infty} x(t)v(t) \, dt = \frac{1}{2\pi} \int_{-\infty}^{\infty} X(-\lambda)V(\lambda) \, d\lambda. \tag{9.81}$$

By definition of the Fourier transform, $X(-\omega) = \overline{X(\omega)}$, and thus changing the variable of integration from λ to ω on the right side of (9.81), we get (9.79). Note that if $v(t) = x(t)$, Parseval's theorem becomes

$$\int_{-\infty}^{\infty} x^2(t) \, dt = \frac{1}{2\pi} \int_{-\infty}^{\infty} \overline{X(\omega)}X(\omega) \, d\omega. \tag{9.82}$$

From the properties of complex numbers,

$$\overline{X(\omega)}X(\omega) = |X(\omega)|^2,$$

and thus (9.82) can be written in the form

$$\int_{-\infty}^{\infty} x^2(t)\ dt = \frac{1}{2\pi} \int_{-\infty}^{\infty} |X(\omega)|^2\ d\omega. \tag{9.83}$$

The left side of (9.83) can be interpreted as the energy of the signal $x(t)$. Thus (9.83) relates the energy of the signal and the integral of the square of the magnitude of the Fourier transform of the signal.

DUALITY   Suppose that we have the transform pair $x(t) \leftrightarrow X(\omega)$. We can define a new continuous-time signal by setting $\omega = t$ in $X(\omega)$. This results in the continuous-time signal $X(t)$. The duality property states that the Fourier transform of $X(t)$ is equal to $2\pi x(-\omega)$; that is, we have the transform pair

$$X(t) \leftrightarrow 2\pi x(-\omega). \tag{9.84}$$

In (9.84), $x(-\omega)$ is the frequency function constructed by setting $t = -\omega$ in the expression for $x(t)$.

For any given transform pair $x(t) \leftrightarrow X(\omega)$, by using duality we can construct the new transform pair (9.84). For example, applying the duality property to the pair (9.66), we get the transform pair

$$\tau\ \mathrm{Sa}\left(\frac{t\tau}{2}\right) \leftrightarrow 2\pi p_\tau(-\omega). \tag{9.85}$$

Since $p_\tau(\omega)$ is an even function of ω, (9.85) can be rewritten as

$$\tau\ \mathrm{Sa}\left(\frac{t\tau}{2}\right) \leftrightarrow 2\pi p_\tau(\omega). \tag{9.86}$$

The result (9.86) says that an Sa function in time corresponds to a rectangular pulse function in frequency.

If we apply the duality property to the transform pair (9.77), we get

$$\frac{\tau}{2}\ \mathrm{Sa}^2\left(\frac{t\tau}{4}\right) \leftrightarrow 2\pi \left(1 - \frac{2|\omega|}{\tau}\right) p_\tau(\omega). \tag{9.87}$$

Thus the Sa-squared time function has a Fourier transform equal to the triangular pulse function in frequency.

The duality property is easy to prove. First, by definition of the Fourier transform

$$X(\omega) = \int_{-\infty}^{\infty} x(t)e^{-j\omega t}\, dt. \tag{9.88}$$

Setting $\omega = t$ and $t = -\omega$ in (9.88), we have

$$X(t) = \int_{-\infty}^{\infty} x(-\omega)e^{j\omega t}\, d\omega$$

$$= \frac{1}{2\pi} \int_{-\infty}^{\infty} 2\pi x(-\omega)e^{j\omega t}\, d\omega.$$

Thus $X(t)$ is the inverse Fourier transform of the frequency function $2\pi x(-\omega)$, which proves (9.84).

For the convenience of the reader, the properties of the Fourier transform are summarized in Table 9.1.

TABLE 9.1. Properties of the Fourier transform

Property	Transform Pair/Property
Linearity	$ax(t) + bv(t) \leftrightarrow aX(\omega) + bV(\omega)$
Right or left shift in time	$x(t - c) \leftrightarrow X(\omega)e^{-j\omega c}$
Time scaling	$x(at) \leftrightarrow \dfrac{1}{a} X\left(\dfrac{\omega}{a}\right), \quad a > 0$
Time reversal	$x(-t) \leftrightarrow X(-\omega) = \overline{X(\omega)}$
Multiplication by a power of t	$t^n x(t) \leftrightarrow j^n \dfrac{d^n}{d\omega^n} X(\omega), \quad n = 1, 2, \ldots$
Multiplication by a complex exponential	$x(t)e^{j\omega_0 t} \leftrightarrow X(\omega - \omega_0), \quad \omega_0 \text{ real}$
Multiplication by $\sin \omega_0 t$	$x(t) \sin \omega_0 t \leftrightarrow \dfrac{j}{2} [X(\omega + \omega_0) - X(\omega - \omega_0)]$
Multiplication by $\cos \omega_0 t$	$x(t) \cos \omega_0 t \leftrightarrow \frac{1}{2} [X(\omega + \omega_0) + X(\omega - \omega_0)]$
Differentiation in the time domain	$\dfrac{d^n}{dt^n} x(t) \leftrightarrow (j\omega)^n X(\omega), \quad n = 1, 2, \ldots$
Convolution in the time domain	$(x * v)(t) \leftrightarrow X(\omega)V(\omega)$
Integration	$x^{-1}(t) \leftrightarrow \dfrac{1}{j\omega} X(\omega) + \pi X(0)\delta(\omega)$
Multiplication in the time domain	$x(t)v(t) \leftrightarrow \dfrac{1}{2\pi} (X * V)(\omega)$
Parseval's theorem	$\displaystyle\int_{-\infty}^{\infty} x(t)v(t)\, dt = \dfrac{1}{2\pi} \int_{-\infty}^{\infty} \overline{X(\omega)}V(\omega)\, d\omega$
Special case of Parseval's theorem	$\displaystyle\int_{-\infty}^{\infty} x^2(t)\, dt = \dfrac{1}{2\pi} \int_{-\infty}^{\infty} |X(\omega)|^2\, d\omega$
Duality	$X(t) \leftrightarrow 2\pi x(-\omega)$

GENERALIZED FOURIER TRANSFORM

In Example 9.6 we saw that the unit-step function $u(t)$ does not have a Fourier transform in the ordinary sense. It is also easy to see that $\cos \omega_0 t$ and $\sin \omega_0 t$ do not have a Fourier transform in the ordinary sense. Since the step function and sinusoidal functions often arise in the study of signals and systems, it is very desirable to be able to define the Fourier transform of these signals. This can be done by defining the notion of the generalized Fourier transform. We consider this concept in this section.

We shall first compute the Fourier transform of the unit impulse $\delta(t)$. Recall that $\delta(t)$ is defined by

$$\delta(t) = 0, \qquad t \neq 0$$

$$\int_{-\epsilon}^{\epsilon} \delta(\lambda) \, d\lambda = 1, \qquad \text{all } \epsilon > 0. \tag{9.89}$$

The Fourier transform of $\delta(t)$ is given by

$$\int_{-\infty}^{\infty} \delta(t) e^{-j\omega t} \, dt.$$

Since $\delta(t) = 0$ for all $t \neq 0$,

$$\delta(t) e^{-j\omega t} = \delta(t),$$

and the Fourier transform integral reduces to

$$\int_{-\infty}^{\infty} \delta(t) \, dt.$$

By (9.89), this integral is equal to 1. Thus we have the transform pair

$$\delta(t) \leftrightarrow 1. \tag{9.90}$$

The transform pair (9.90) is not surprising since the Laplace transform of $\delta(t)$ is also equal to 1.

Now applying the duality property to (9.90), we get the transform pair

$$x(t) = 1, \ -\infty < t < \infty \leftrightarrow 2\pi\delta(\omega). \tag{9.91}$$

This result says that the Fourier transform of a dc signal of amplitude 1 is equal to an impulse in frequency with area 2π. But from the results in Example 9.5, we know that the dc signal does not have a Fourier transform in the ordinary sense. The frequency function $2\pi\delta(\omega)$ is called the *generalized Fourier transform* of the dc signal $x(t) = 1$, $-\infty < t < \infty$.

Now consider the signal $x(t) = \cos \omega_0 t$, $-\infty < t < \infty$, where ω_0 is a fixed but arbitrary real number. Using the transform pair (9.91) and the modulation property (9.73), we have that $x(t)$ has the generalized Fourier transform

$$\pi[\delta(\omega + \omega_0) + \delta(\omega - \omega_0)].$$

Hence we have the transform pair

$$\cos \omega_0 t \leftrightarrow \pi[\delta(\omega + \omega_0) + \delta(\omega - \omega_0)]. \qquad (9.92)$$

In a similar manner, we can show that $\sin \omega_0 t$ has the generalized transform

$$j\pi[\delta(\omega + \omega_0) - \delta(\omega - \omega_0)],$$

so we have the transform pair

$$\sin \omega_0 t \leftrightarrow j\pi[\delta(\omega + \omega_0) - \delta(\omega - \omega_0)]. \qquad (9.93)$$

The plot of the Fourier transform of $\cos \omega_0 t$ is given in Figure 9.30. Note that the spectrum consists of two impulses located at $\pm\omega_0$ with each impulse having area π.

FOURIER TRANSFORM OF A PERIODIC SIGNAL   Using the transform pair (9.91) and the property (9.71), we obtain the transform pair

$$\exp (j\omega_0 t) \leftrightarrow 2\pi\delta(\omega - \omega_0). \qquad (9.94)$$

The transform pair (9.94) can be used to compute the generalized Fourier transform of a periodic signal. Let $x(t)$ be periodic for $-\infty < t < \infty$ with period T. Then $x(t)$ has the complex exponential Fourier series

$$x(t) = \sum_{n=-\infty}^{\infty} c_n \exp (jn\omega_0 t), \qquad (9.95)$$

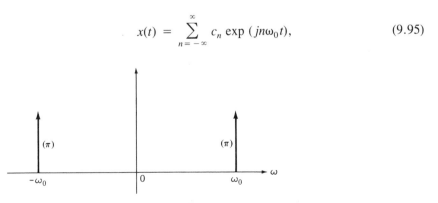

FIGURE 9.30.   Fourier transform of $\cos \omega_0 t$.

where $\omega_0 = 2\pi/T$. We can take the Fourier transform of both sides of (9.95). By linearity and the transform pair (9.94), we have

$$X(\omega) = \sum_{n=-\infty}^{\infty} 2\pi c_n \delta(\omega - n\omega_0).$$

So the Fourier transform of a periodic signal is a train of impulse functions located at $\omega = n\omega_0$, $n = 0, \pm 1, \pm 2, \ldots$.

TRANSFORM OF THE UNIT-STEP FUNCTION To compute the (generalized) Fourier transform of the unit step $u(t)$, we will first compute the transform of the signal $x(t)$, where

$$x(t) = -0.5 + u(t).$$

The signal $x(t)$ is plotted in Figure 9.31. Since $x(t)$ is constant except for a jump in value of 1 from $t = 0^-$ to $t = 0^+$, the (generalized) derivative of $x(t)$ is equal to the unit impulse $\delta(t)$. Then by the differentiation property, the Fourier transform of $x(t)$ must be equal to $1/j\omega$. Thus we have the transform pair

$$-0.5 + u(t) \leftrightarrow \frac{1}{j\omega}. \qquad (9.96)$$

Now the Fourier transform of the dc signal -0.5 is equal to $-\pi\delta(\omega)$. Then using (9.96) and linearity, we obtain the transform pair

$$u(t) \leftrightarrow \pi\delta(\omega) + \frac{1}{j\omega}. \qquad (9.97)$$

Using the transform pair (9.97), we can prove the integration property given in Table 9.1. Let $x(t)$ be a signal with Fourier transform $X(\omega)$. By definition of the convolution operation, the integral $x^{(-1)}(t)$ of $x(t)$ can be expressed in the form

$$x^{(-1)}(t) = (u * x)(t).$$

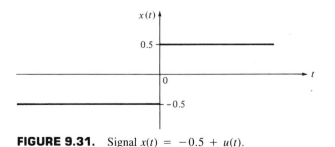

FIGURE 9.31. Signal $x(t) = -0.5 + u(t)$.

Then by the convolution property, the Fourier transform of $x^{(-1)}(t)$ is equal to

$$U(\omega)X(\omega).$$

By (9.97)

$$U(\omega) = \pi\delta(\omega) + \frac{1}{j\omega}.$$

and thus the Fourier transform of $x^{(-1)}(t)$ is equal to

$$\left[\pi\delta(\omega) + \frac{1}{j\omega}\right] X(\omega) = \pi X(0)\delta(\omega) + \frac{1}{j\omega} X(\omega).$$

In Table 9.2 we give a list of common Fourier transform pairs which includes the pairs that were derived in this chapter.

TABLE 9.2. Common Transform Pairs

$1, \quad -\infty < t < \infty \leftrightarrow 2\pi\delta(\omega)$

$-0.5 + u(t) \leftrightarrow \dfrac{1}{j\omega}$

$u(t) \leftrightarrow \pi\delta(\omega) + \dfrac{1}{j\omega}$

$\delta(t) \leftrightarrow 1$

$\delta(t - c) \leftrightarrow e^{-j\omega c}, \quad c$ any real number

$e^{-bt}u(t) \leftrightarrow \dfrac{1}{j\omega + b}, \quad b > 0$

$\exp(j\omega_0 t) \leftrightarrow 2\pi\delta(\omega - \omega_0), \quad \omega_0$ any real number

$p_\tau(t) \leftrightarrow \tau \, \text{Sa}\left(\dfrac{\omega\tau}{2}\right)$

$\tau \, \text{Sa}\left(\dfrac{t\tau}{2}\right) \leftrightarrow 2\pi p_\tau(\omega)$

$\left[1 - \dfrac{2|t|}{\tau}\right] p_\tau(t) \leftrightarrow \dfrac{\tau}{2} \, \text{Sa}^2\left(\dfrac{\omega\tau}{4}\right)$

$\dfrac{\tau}{2} \, \text{Sa}^2\left(\dfrac{t\tau}{4}\right) \leftrightarrow 2\pi\left(1 - \dfrac{2|\omega|}{\tau}\right)p_\tau(\omega)$

$\cos \omega_0 t \leftrightarrow \pi[\delta(\omega + \omega_0) + \delta(\omega - \omega_0)]$

$\cos(\omega_0 t + \theta) \leftrightarrow \pi[e^{-j\theta}\delta(\omega + \omega_0) + e^{j\theta}\delta(\omega - \omega_0)]$

$\sin \omega_0 t \leftrightarrow j\pi[\delta(\omega + \omega_0) - \delta(\omega - \omega_0)]$

$\sin(\omega_0 t + \theta) \leftrightarrow j\pi[e^{-j\theta}\delta(\omega + \omega_0) - e^{j\theta}\delta(\omega - \omega_0)]$

PROBLEMS

Chapter 9

9.1   For each of the periodic signals shown in Figure P9.1, compute the trigonometric Fourier series, the complex-exponential Fourier series, and sketch the amplitude and phase spectra for $n = 0, \pm1, \pm2, \pm3, \pm4, \pm5$.

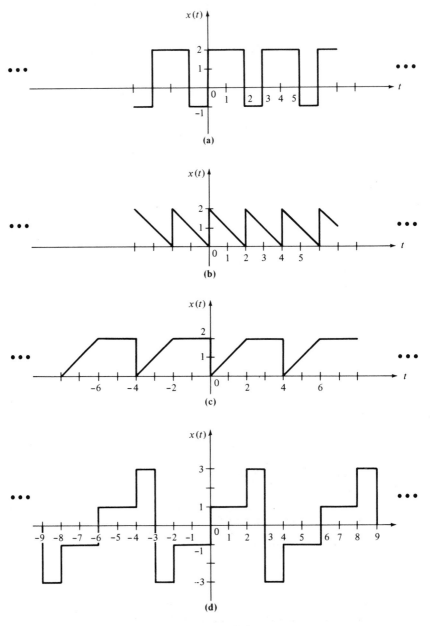

FIGURE P9.1

9.2 Repeat Problem 9.1 for the signals in Figure P9.2.

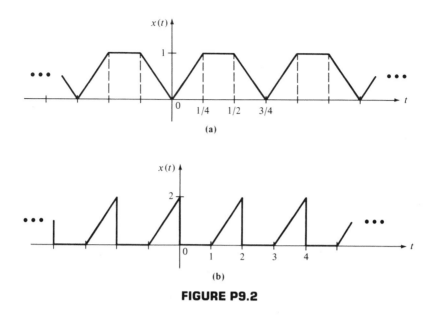

FIGURE P9.2

9.3 For each of the following signals, compute the complex-exponential Fourier series by using trigonometric identities, and then sketch the amplitude and phase spectra for all values of n.

(a) $x(t) = \cos(5t + \theta)$.

(b) $x(t) = \sin t + \cos t$.

(c) $x(t) = \sin^2 4t$.

(d) $x(t) = \cos 2t \sin 3t$.

(e) $x(t) = \cos^2 5t$.

(f) $x(t) = \cos 3t + \cos 5t$.

9.4 A periodic signal with period T has Fourier coefficients c_n^x; that is,

$$x(t) = \sum_{n=-\infty}^{\infty} c_n^x \exp(jn\omega_0 t), \qquad \omega_0 = \frac{2\pi}{T}, \qquad -\infty < t < \infty.$$

Compute the Fourier coefficients c_n^v for the periodic signal $v(t)$ where:

(a) $v(t) = x(t - 1)$.

(b) $v(t) = \dfrac{dx(t)}{dt}$.

(c) $v(t) = x(t) \exp[j(2\pi/T)t]$.

(d) $v(t) = x(t) \cos\left(\dfrac{2\pi}{T} t\right)$.

9.5 A linear time-invariant continuous-time system has frequency function

$$H(\omega) = \begin{cases} 2 \exp{(-|6 - \omega|)}\exp(-j3\omega), & 4 \le |\omega| \le 12 \\ 0, & \text{all other } \omega. \end{cases}$$

Compute the output response $y(t)$ resulting from the input $x(t)$ shown in Figure P9.5. Express $y(t)$ in trigonometric form.

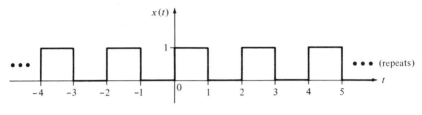

FIGURE P9.5

9.6 For the system defined in Problem 9.5, compute the output response $y(t)$ when the input $x(t)$ is given by:
(a) $x(t) = 2 \cos{(3.5t)} - 4 \sin{(6t)}, \quad -\infty < t < \infty$.
(b) $x(t) = 3 \cos{(6t + 30°)} + \sin{(10t - 30°)}, \quad -\infty < t < \infty$.
(c) $x(t) = 1 + \sum_{n=1}^{\infty} \frac{1}{n} \cos{(3nt)}, \quad -\infty < t < \infty$.

9.7 A linear time-invariant continuous-time system receives the periodic signal $x(t)$ shown in Figure P9.2b. The transfer function of the system is $H(s) = s/(s + 2)$. Compute the complex-exponential Fourier series of the output response $y(t)$, and then sketch the amplitude and phase spectra for $n = 0, \pm 1, \pm 2, \pm 3, \pm 4, \pm 5$.

9.8 A periodic signal $x(t)$ with period T has the dc component $c_0^x = 2$. The signal $x(t)$ is applied to a linear time-invariant continuous-time system with frequency function

$$H(\omega) = \begin{cases} 10e^{-j5\omega}, & \omega > \dfrac{\pi}{T}, \omega < \dfrac{-\pi}{T} \\ 0, & \text{all other } \omega. \end{cases}$$

Show that the resulting output response $y(t)$ can be expressed in the form

$$y(t) = ax(t - b) + c.$$

Compute the constants a, b, and c.

9.9 The voltage $x(t)$ shown in Figure P9.9b is applied to the RL circuit shown in Figure P9.9a. Find the value of L so that the peak value of the largest ac component (harmonic) in the output response $y(t)$ is $\frac{1}{30}$ of the dc component of the output.

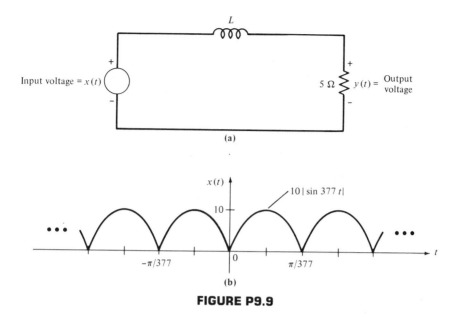

(a)

(b)

FIGURE P9.9

9.10 Consider the full-wave rectifier shown in Figure P9.10. The input voltage $v(t)$ is equal to $156 \cos (120\pi t)$, $-\infty < t < \infty$. The voltage $x(t)$ is equal to $|v(t)|$. Choose values for R and C such that the following two criteria are satisfied:

1. The dc component of $y(t)$ is equal to 90% of the dc component of the input $x(t)$.
2. The peak value of the largest harmonic in $y(t)$ is $\frac{1}{30}$ of the dc component of $y(t)$.

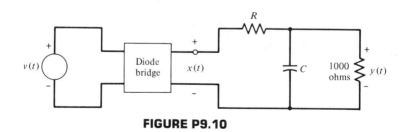

FIGURE P9.10

9.11 The input

$$x(t) = 1.5 + \sum_{n=1}^{\infty} \left(\frac{1}{n\pi} \sin (n\pi t) + \frac{2}{n\pi} \cos (n\pi t) \right), \qquad -\infty < t < \infty$$

is applied to a linear time-invariant system with frequency function $H(\omega)$. This input produces the output response $y(t)$ shown in Figure P9.11. Compute $H(n\pi)$ for $n = 1, 2, 3, \ldots$.

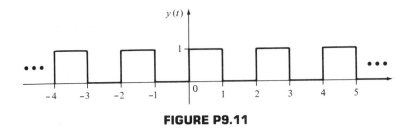

FIGURE P9.11

9.12 A linear time-invariant continuous-time system has frequency function $H(\omega)$ shown in Figure P9.12a. It is known that the system converts the sawtooth waveform in Figure P9.12b into the square waveform in Figure P9.12c; that is, the response to the sawtooth waveform is a square waveform. Compute the constants a and b in the plot of $H(\omega)$.

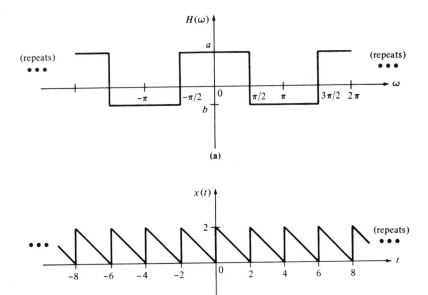

(a)

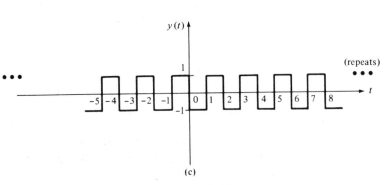

(b)

(c)

FIGURE P9.12

9.13   A linear time-invariant continuous-time system has frequency function

$$H(\omega) = b - ae^{j\omega c}, \qquad -\infty < \omega < \infty$$

where a, b, and c are constants (real numbers). The input $x(t)$ shown in Figure P9.13a is applied to the system. Determine the constants a, b, and c so that the output response $y(t)$ resulting from $x(t)$ is given by the plot in Figure P9.13b.

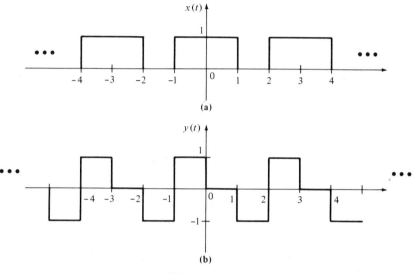

(a)

(b)

FIGURE P9.13

9.14   A continuous-time signal $x(t)$ has Fourier transform

$$X(\omega) = \frac{1}{j\omega + b},$$

where b is a constant. Determine the Fourier transform $V(\omega)$ of the following signals.

(a) $v(t) = x(5t - 4)$.
(b) $v(t) = t^2 x(t)$.
(c) $v(t) = x(t)e^{j2t}$.
(d) $v(t) = x(t) \cos 4t$.
(e) $v(t) = \dfrac{d^2 x(t)}{dt^2}$.
(f) $v(t) = (x * x)(t)$.
(g) $v(t) = x^2(t)$.
(h) $v(t) = \dfrac{1}{jt - b}$.

9.15   By first expressing $x(t)$ in terms of rectangular pulse functions and triangular pulse functions, compute the Fourier transform of the signals in Figure P9.15.

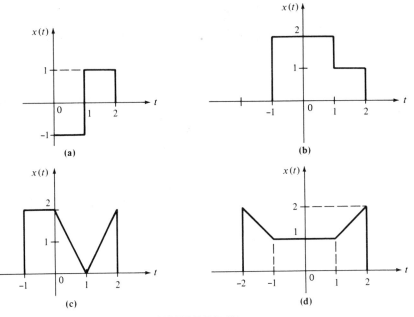

FIGURE P9.15

9.16 By using equation (9.50) and the Fourier transforms computed in Problem 9.15, determine the complex-exponential Fourier series of the periodic signals in Figure P9.16.

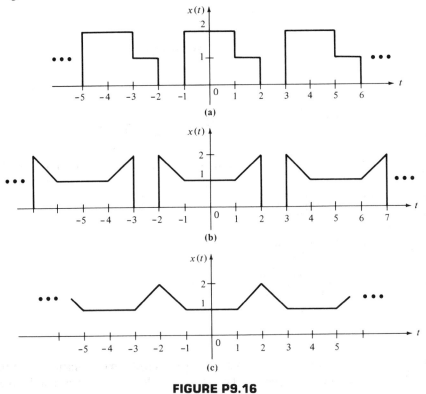

FIGURE P9.16

9.17 Compute the Fourier transform of the signals in Figure P9.17.

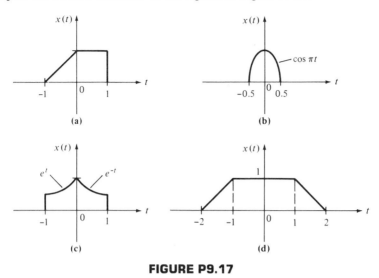

FIGURE P9.17

9.18 Compute the inverse Fourier transforms of the frequency functions $X(\omega)$ shown in Figure P9.18.

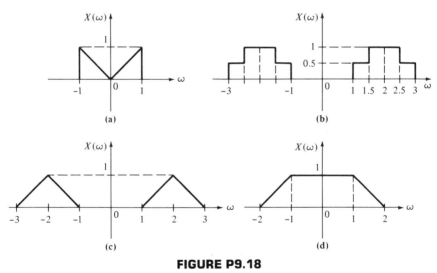

FIGURE P9.18

9.19 Compute the inverse Fourier transform of the following frequency functions.
(a) $X(\omega) = \cos 4\omega$, $-\infty < \omega < \infty$.
(b) $X(\omega) = \sin^2 3\omega$, $-\infty < \omega < \infty$.
(c) $X(\omega) = p_4(\omega) \cos\left(\dfrac{\pi\omega}{2}\right)$.
(d) $X(\omega) = \dfrac{\sin(\omega/2)}{j\omega + 2} e^{-j\omega 2}$, $-\infty < \omega < \infty$.

9.20 A signal $x(t)$ has Fourier transform

$$X(\omega) = \frac{1}{j} \left[Sa \left(2\omega - \frac{\pi}{2} \right) - Sa \left(2\omega + \frac{\pi}{2} \right) \right].$$

(a) Compute $x(t)$.

(b) Let $x_p(t)$ denote the periodic signal defined by

$$x_p(t) = \sum_{n=-\infty}^{\infty} x(t - 16n).$$

Compute the Fourier transform $X_p(\omega)$ of $x_p(t)$.

9.21 Compute the Fourier transform of the following signals.

(a) $x(t) = (e^{-t} \cos 4t) u(t)$.

(b) $x(t) = (\cos 4t) u(t)$.

(c) $x(t) = e^{-|t|}, \quad -\infty < t < \infty$.

(d) $x(t) = e^{-t^2}, \quad -\infty < t < \infty$.

System Analysis via the Fourier Transform

In this chapter we apply the Fourier transform to the analysis of systems, including simple types of systems employing amplitude modulation. We begin in Section 10.1 by showing that the Fourier transform can be used to find the output response resulting from an input with the system initially at rest. We shall see that the spectrum of the output response is equal to the product of the spectrum of the input and the frequency function of the system. This frequency-domain description of system behavior gives a great deal of insight into how a system processes a given input. The theory is illustrated in Section 10.2 when we consider ideal linear-phase filters.

The Fourier transform is the primary analytical tool in the study of communications and analog signal processing. In Sections 10.3 and 10.4 we consider applications of the Fourier theory to the study of amplitude modulation. We begin in Section 10.3 with sine-wave amplitude modulation. Then we consider the simultaneous transmission of several signals over the same channel using frequency-division multiplexing. In Section 10.4 we study pulse amplitude modulation, which can be viewed as a sampling operation. Using the theory of Fourier series and the Fourier transform, we prove the sampling theorem. This result states that a bandlimited continuous-time signal can be completely reconstructed from a sampled version of the signal if the sampling frequency is suitably fast. In the last part of Section 10.4, we consider time-division multiplexing, which is another technique for sending a large number of signals over the same channel at the same time.

10.1

COMPUTATION OF OUTPUT RESPONSE VIA THE FOURIER TRANSFORM

Consider a linear time-invariant continuous-time system with impulse response $h(t)$. Let $y(t)$ denote the output response resulting from the application of an input $x(t)$ with no initial energy in the system prior to the application of $x(t)$. In this section we show that $y(t)$ can be computed for $-\infty < t < \infty$ by using the Fourier transform. As will be seen, the computation of the output response via the Fourier transform is very similar to the approach we developed in Chapter 5 using the one-sided Laplace transform.

From the results in Chapter 4, the output resulting from the input $x(t)$ is given by

$$y(t) = \int_{-\infty}^{\infty} h(\lambda)x(t - \lambda)\,d\lambda. \tag{10.1}$$

As in Chapter 9, we are not requiring that the system be causal, nor are we assuming that the input $x(t)$ is necessarily zero for $t < 0$. Thus the limits of the integral in (10.1) are $t = -\infty$ and $t = \infty$. As will be seen in Section 10.2, we cannot require causality if we wish to study ideal filters.

Now let $H(\omega)$ denote the Fourier transform of the impulse response $h(t)$ of the given system, that is,

$$H(\omega) = \int_{-\infty}^{\infty} h(t)\, e^{-j\omega t}\,dt. \tag{10.2}$$

Then if we take the Fourier transform of both sides of (10.1), using the convolution property of the transform we have

$$Y(\omega) = H(\omega)X(\omega), \tag{10.3}$$

where $Y(\omega)$ is the Fourier transform of the output $y(t)$ and $X(\omega)$ is the Fourier transform of the input $x(t)$. Equation (10.3) is the Fourier transform domain (or ω-domain) representation of the given system.

The function $H(\omega)$ defined by (10.2) is identical to the frequency function that we defined in Section 9.3. Thus the Fourier transform of the impulse response is equal to the frequency function of the system. As defined in Chapter 9, the magnitude $|H(\omega)|$ is the magnitude function of the system and the angle $\sphericalangle H(\omega)$ is the phase function of the system.

From the ω-domain representation (10.3), we have that the amplitude spectrum $|Y(\omega)|$ of the output response $y(t)$ is given by

$$|Y(\omega)| = |H(\omega)||X(\omega)|, \tag{10.4}$$

and the phase spectrum $\sphericalangle\, Y(\omega)$ is given by

$$\sphericalangle\, Y(\omega) = \sphericalangle\, H(\omega) + \sphericalangle\, X(\omega). \tag{10.5}$$

From (10.4) we see that the amplitude spectrum of the output is equal to the product of the amplitude spectrum of the input with the system's magnitude function. From (10.5) we see that the phase spectrum of the output is equal to the sum of the phase spectrum of the input and the phase function of the system. These relationships provide a good deal of insight into how a system processes inputs. Examples are given in Section 10.2.

The relationships (10.4) and (10.5) are generalizations of the results derived in Section 9.3 on the spectrum of the response to a periodic input signal.

COMPARISON WITH THE TRANSFER FUNCTION REPRESENTA-TION Now suppose that the given system is causal so that $h(t) = 0$ for $t <$ 0. Then we can take the lower limit of the integral in (10.2) to be zero, and thus

$$H(\omega) = \int_0^\infty h(t)e^{-j\omega t}\, dt. \tag{10.6}$$

Letting $H(s)$ denote the transfer function of the system as defined in Chapter 5, we know that $H(s)$ is equal to the one-sided Laplace transform of $h(t)$, and thus

$$H(s) = \int_0^\infty h(t)e^{-st}\, dt. \tag{10.7}$$

Let us suppose that the integral in (10.7) exists when $s = j\omega$. This will be the case if the given system is asymptotically stable. Then setting $s = j\omega$ in (10.7) and comparing the result with (10.6), we see that

$$H(s)|_{s=j\omega} = H(\omega). \tag{10.8}$$

Therefore, in a causal stable system the Fourier transform of the impulse response $h(t)$ is equal to the transfer function $H(s)$ evaluated at $s = j\omega$. In addition, if $x(t)$ is zero for $t < 0$, the ω-domain representation (10.3) can be viewed as a special case of the transfer function model

$$Y(s) = H(s)X(s), \tag{10.9}$$

where $Y(s)$ and $X(s)$ are the (one-sided) Laplace transforms of $y(t)$ and $x(t)$, respectively. More precisely, if the transforms $Y(s)$ and $X(s)$ exist when

$s = j\omega$, by setting $s = j\omega$ in (10.9) we obtain the ω-domain representation (10.3).

A very interesting consequence of the relationship (10.8) is that the system's transfer function $H(s)$ can be determined from knowledge of the system's frequency function $H(\omega)$ (e.g., the system's Bode diagrams). In particular, if we know $H(\omega)$, then by taking the inverse Fourier transform of $H(\omega)$, we get the impulse response $h(t)$. We can then take the Laplace transform of $h(t)$ to get the transfer function $H(s)$.

COMPUTATION OF THE OUTPUT RESPONSE   Applying the inverse Fourier transform to both sides of the ω-domain representation (10.3), we obtain

$$y(t) = \frac{1}{2\pi} \int_{-\infty}^{\infty} H(\omega)X(\omega)e^{j\omega t}\, d\omega. \qquad (10.10)$$

The expression (10.10) for $y(t)$ is the output response resulting from input $x(t)$ with no initial energy in the system prior to the application of $x(t)$.

The expression (10.10) for $y(t)$ can be used to compute $y(t)$. If possible, one should determine $y(t)$ by working with transform pairs from a table rather than by computing the integral in (10.10). Examples are given in Section 10.2.

As we know, the output response $y(t)$ could also be computed by using the one-sided or two-sided Laplace transform. The question then arises as to whether it is easier to use the Fourier transform or the Laplace transform to compute the output response. Which approach is easier to use depends on the application. For example, if the Laplace transform $X(s)$ of the input $x(t)$ and the transfer function $H(s)$ are rational functions of s, we know that the output response can be determined by first expanding $H(s)X(s)$ by partial fractions. So in this case, it is easier to use the Laplace transform. If $X(s)$ and $H(s)$ are not rational functions in s, it may be easier to use the Fourier transform to compute the output. We illustrate this point in the next section.

10.2

ANALYSIS OF IDEAL LINEAR-PHASE FILTERS

In this section we use the ω-domain representation to study the properties of ideal linear-phase filters. We begin with the lowpass filter.

Consider the system with the frequency function

$$H(\omega) = \begin{cases} K \exp(-j\omega t_d), & -B \leq \omega \leq B \\ 0, & \omega < -B, \quad \omega > B \end{cases} \qquad (10.11)$$

where K and t_d are positive real numbers. Equation (10.11) is the polar-form representation of $H(\omega)$. From (10.11) we have

$$|H(\omega)| = \begin{cases} K, & -B \leq \omega \leq B \\ 0, & \omega < -B, \quad \omega > B \end{cases}$$

and

$$\sphericalangle H(\omega) = \begin{cases} -\omega t_d, & -B \le \omega \le B \\ 0, & \omega < -B, \quad \omega > B. \end{cases}$$

The magnitude function $|H(\omega)|$ and phase function $\sphericalangle H(\omega)$ of the system are plotted in Figure 10.1. From Figure 10.1b we see that over the frequency range 0 to B, the phase function of the system is linear with slope equal to $-t_d$. We show below that t_d is the time delay through the system.

Now suppose that the input $x(t) = A \cos \omega_0 t$, $-\infty < t < \infty$, is applied to the system. Then from Table 9.2,

$$X(\omega) = A\pi[\delta(\omega + \omega_0) + \delta(\omega - \omega_0)]. \tag{10.12}$$

Inserting (10.11) and (10.12) into (10.3), we have that the Fourier transform of the output response is given by

$$Y(\omega) = \begin{cases} AK\pi \exp(-j\omega t_d)[\delta(\omega + \omega_0) + \delta(\omega - \omega_0)], & \omega_0 \le B \\ 0, & \omega_0 > B. \end{cases} \tag{10.13}$$

Taking the inverse Fourier transform of (10.13), we obtain

$$y(t) = \begin{cases} AK \cos[\omega_0(t - t_d)], & \omega_0 \le B \\ 0, & \omega_0 > B. \end{cases} \tag{10.14}$$

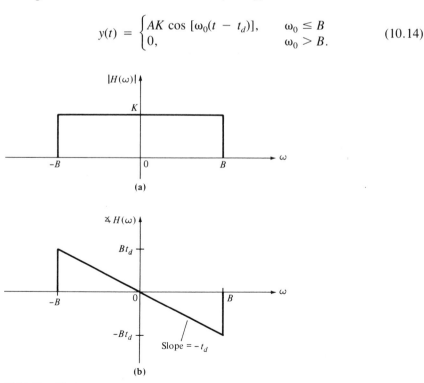

FIGURE 10.1. (a) Magnitude and (b) phase functions of given system.

From (10.14) we see that the system passes (with a time delay of t_d seconds) all input sinusoids whose frequency is between 0 and B, while it completely stops all input sinusoids whose frequency is above B. Thus the system is a linear-phase lowpass filter with bandwidth B (in rad/sec). The system is called an *ideal lowpass filter* as a result of the "sharp corner" of the magnitude function between the passband and the stopband. Note that if $K > 1$, the system is an ideal linear-phase lowpass amplifier.

It is interesting to note that if the phase function of the filter were not linear, there would be phase distortion in the output of the filter. To see this, suppose that the phase function $\angle H(\omega)$ is equal to some nonzero constant C and the input to the filter is

$$x(t) = A_0 \cos \omega_0 t + A_1 \cos \omega_1 t, \qquad -\infty < t < \infty,$$

where $\omega_0 < B$ and $\omega_1 < B$. Then using (10.3), we have that the resulting response is

$$y(t) = A_0 K \cos (\omega_0 t + C) + A_1 K \cos (\omega_1 t + C), \qquad -\infty < t < \infty.$$

This output is not a time-delayed version of the input, so there is distortion in the filtering process. If the filter has the linear-phase characteristic given above, the response is

$$y(t) = A_0 K \cos [\omega_0(t - t_d)] + A_1 K \cos [\omega_1(t - t_d)].$$

This output is equal to a t_d-second time delay of the input, and hence in this case there is no distortion. Therefore, for distortionless filtering the phase function of the filter should be as close to linear as possible over the passband of the filter.

COMPUTATION OF THE IMPULSE RESPONSE   We shall now calculate the impulse response $h(t)$ of the filter with the frequency function (10.11). This can be computed by taking the inverse Fourier transform of $H(\omega)$. First, using the definition of the rectangular pulse, we can express $H(\omega)$ in the form

$$H(\omega) = K p_{2B}(\omega) \exp (-j\omega t_d), \qquad -\infty < \omega < \infty. \tag{10.15}$$

From Table 9.2 we have the transform pair

$$\frac{\tau}{2\pi} \operatorname{Sa}\left(\frac{\pi t}{2}\right) \leftrightarrow p_\tau(\omega). \tag{10.16}$$

Setting $\tau = 2B$ in (10.16) and multiplying both sides by K, we obtain

$$\frac{KB}{\pi} \operatorname{Sa}(Bt) \leftrightarrow K p_{2B}(\omega). \tag{10.17}$$

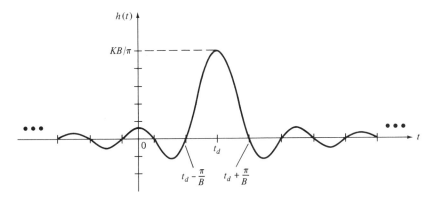

FIGURE 10.2. Impulse response of ideal linear-phase lowpass filter.

Applying the time-shift property to the transform pair (10.17) gives

$$\frac{KB}{\pi} \text{ Sa } [B(t - t_d)] \leftrightarrow Kp_{2B}(\omega) \exp(-j\omega t_d). \qquad (10.18)$$

Since the right side of the transform pair (10.18) is equal to $H(\omega)$, the impulse response of the ideal lowpass filter is

$$h(t) = \frac{KB}{\pi} \text{ Sa } [B(t - t_d)], \qquad -\infty < t < \infty. \qquad (10.19)$$

The impulse response $h(t)$ is plotted in Figure 10.2.

In Figure 10.2 we see that the impulse response $h(t)$ is not zero for $t < 0$, and thus the filter is noncausal. As a result, it is not possible to build an ideal lowpass filter. However, we can build a good approximation to an ideal lowpass filter. To be a good approximation to an ideal lowpass filter, the impulse response of the filter should look like the time function in Figure 10.2 for $t > 0$.

EXAMPLE 10.1. Consider the three-pole Butterworth filter given by the transfer function

$$H(s) = \frac{\omega_c^3}{(s + \omega_c)(s^2 + \omega_c s + \omega_c^2)}.$$

Recall from Chapter 8 that ω_c is the 3-dB bandwidth of the filter. Setting $\omega_c = 1$, we have

$$H(s) = \frac{1}{(s + 1)(s^2 + s + 1)}.$$

We can compute the impulse response $h(t)$ of the filter by setting $s = j\omega$ in $H(s)$ and then computing the inverse Fourier transform of $H(\omega)$ using the

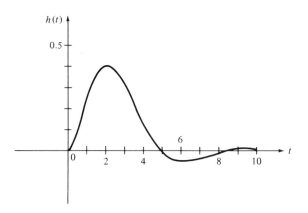

FIGURE 10.3. Impulse response of three-pole Butterworth filter.

integral expression for the inverse Fourier transform. To avoid having to evaluate this integral, we can compute $h(t)$ by first expanding $H(s)$ by partial fractions and then computing the inverse Laplace transform. Expanding $H(s)$, we have

$$H(s) = \frac{1}{s + 1} - \frac{s}{(s + 0.5)^2 + 0.75}$$

$$= \frac{1}{s + 1} - \frac{s + 0.5}{(s + 0.5)^2 + 0.75} + \frac{0.5}{(s + 0.5)^2 + 0.75}.$$

The impulse response is then

$$h(t) = e^{-t} + e^{-0.5t}\left[\frac{\sqrt{3}}{3}\sin\left(\frac{\sqrt{3}}{2}t\right) - \cos\left(\frac{\sqrt{3}}{2}t\right)\right], \qquad t \ge 0.$$

The impulse response is plotted in Figure 10.3. Note that there is some resemblance between the waveform in Figure 10.3 and the impulse response of the ideal filter shown in Figure 10.2.

RESPONSE TO NONSINUSOIDAL INPUTS   We now want to illustrate the use of the ω-domain representation $Y(\omega) = H(\omega)X(\omega)$ in computing the output response resulting from a nonsinusoidal input $x(t)$. In particular, suppose that

$$x(t) = \text{Sa}\,(t), \qquad -\infty < t < \infty.$$

Then from Table 9.2,

$$X(\omega) = \pi p_2(\omega).$$

Since

$$|X(\omega)| = 0 \qquad \text{for } \omega > 1,$$

the bandwidth of the signal $x(t) = $ Sa (t) is equal to 1.

Now using (10.11), we have

$$Y(\omega) = H(\omega)X(\omega) = Kp_{2B}(\omega) \exp(-j\omega t_d)\pi p_2(\omega).$$

If $2B > 2$ or $B > 1$,

$$p_{2B}(\omega)p_2(\omega) = p_2(\omega),$$

and thus

$$Y(\omega) = K\pi p_2(\omega) \exp(-j\omega t_d).$$

Using the time-shift property, we have

$$y(t) = K \text{ Sa } (t - t_d) = Kx(t - t_d), \qquad -\infty < t < \infty.$$

This result shows that if the bandwidth B of the filter is wide enough to pass all the frequency components of the input signal, the filter simply passes the input with a t_d-second time delay.

If $2B \leq 2$ or $B \leq 1$,

$$p_{2B}(\omega)p_2(\omega) = p_{2B}(\omega),$$

and thus

$$Y(\omega) = K\pi p_{2B}(\omega) \exp(-j\omega t_d) = \pi H(\omega).$$

Therefore,

$$y(t) = \pi h(t) = KB \text{ Sa } [B(t - t_d)], \qquad -\infty < t < \infty. \qquad (10.20)$$

In this case, the bandwidth B of the filter is not wide enough to pass all the frequency components of the input, and thus the output is not a time-delayed version of the input. In fact, by (10.20) the response to $x(t) = $ Sa (t) is equal to a scalar multiple of the impulse response when $B < 1$. This is a very interesting result since it implies that the response to the input $x(t) = \pi\delta(t)$ is the same as the response to the input $x(t) = $ Sa (t).

Now suppose that

$$x(t) = \text{Sa } (t) \cos 2t, \qquad -\infty < t < \infty.$$

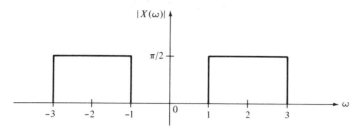

FIGURE 10.4. Amplitude spectrum of the signal Sa t cos 2t.

Then by the modulation property of the Fourier transform,

$$X(\omega) = \frac{\pi}{2}[p_2(\omega + 2) + p_2(\omega - 2)].$$

The amplitude spectrum $|X(\omega)|$ is plotted in Figure 10.4. In this case, $|X(\omega)| = 0$ for $\omega > 3$, and thus the bandwidth of $x(t)$ is equal to 3. The transform of the resulting output is

$$Y(\omega) = Kp_{2B}(\omega) \exp(-j\omega t_d) \frac{\pi}{2}[p_2(\omega + 2) + p_2(\omega - 2)].$$

If $B > 3$, the filter again passes all frequency components of the input, and thus

$$y(t) = Kx(t - t_d) = K \, \text{Sa} \, (t - t_d) \cos[2(t - t_d)].$$

If $B < 1$, the filter does not pass any of the frequency components of the input $x(t) = \text{Sa} \, (t) \cos 2t$, and thus

$$Y(\omega) = 0,$$

which implies that

$$y(t) = 0, \qquad -\infty < t < \infty.$$

If $1 < B < 3$, the filter passes the frequency components in the range from 1 to B. In this case, the product

$$p_{2B}(\omega)[p_2(\omega + 2) + p_2(\omega - 2)]$$

is computed in Figure 10.5.
From Figure 10.5 we have that

$$p_{2B}(\omega)[p_2(\omega + 2) + p_2(\omega - 2)]$$

$$= p_{B-1}\left(\omega + \frac{B + 1}{2}\right) + p_{B-1}\left(\omega - \frac{B + 1}{2}\right).$$

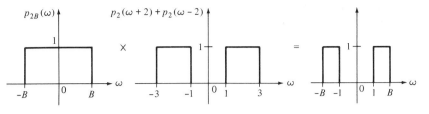

FIGURE 10.5. Computation of $p_{2B}(\omega)[p_2(\omega + 2) + p_2(\omega - 2)]$.

Let $V(\omega) = p_{B-1}(\omega)$. Then

$$
Y(\omega) = \frac{K\pi}{2} \left[V\left(\omega + \frac{B + 1}{2} \right) + V\left(\omega - \frac{B + 1}{2} \right) \right] \exp\left(-j\omega t_d\right).
$$

Using the modulation and shift properties, we get

$$
y(t) = K\pi v(t - t_d) \cos\left[\frac{B + 1}{2} (t - t_d) \right], \qquad -\infty < t < \infty.
$$

But

$$
v(t) = \frac{B - 1}{2\pi} \operatorname{Sa}\left(\frac{B - 1}{2} t \right), \qquad -\infty < t < \infty,
$$

and thus

$$
y(t) = \frac{K(B - 1)}{2} \operatorname{Sa}\left[\frac{B - 1}{2} (t - t_d) \right] \cos\left[\frac{B + 1}{2} (t - t_d) \right],
$$
$$
-\infty < t < \infty.
$$

Clearly, $y(t)$ is not a time-delayed version of the input $x(t) = \operatorname{Sa}(t) \cos(2t)$. So "cutting off" some of the frequency components of the input results in a significant distortion of the input.

Now let $x(t)$ be an arbitrary input with Fourier transform $X(\omega)$. We assume that $|X(\omega)| = 0$ for all $\omega > \Omega$; that is, Ω is the bandwidth of the input signal $x(t)$. If $\Omega < B$, the lowpass filter passes all the frequency components of the input; that is,

$$
Y(\omega) = KX(\omega) \exp\left(-j\omega t_d\right).
$$

Hence

$$
y(t) = Kx(t - t_d).
$$

If $\Omega > B$, the filter does not pass all the frequency components of the input, and thus the filter output $y(t)$ will be a distorted version of the input.

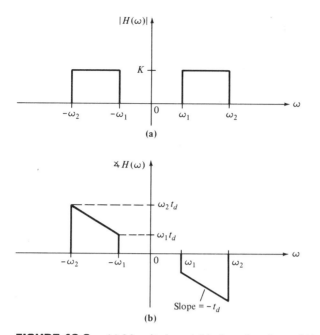

FIGURE 10.6. (a) Magnitude and (b) phase functions of ideal linear-phase bandpass filter.

IDEAL BANDPASS FILTERS   One can extend the analysis given above to ideal linear-phase bandpass filters. The frequency function of an ideal linear-phase bandpass filter is given by

$$H(\omega) = \begin{cases} K \exp\left(-j\omega t_d\right), & \omega_1 \leq |\omega| \leq \omega_2 \\ 0, & \text{all other } \omega, \end{cases}$$

where K, t_d, ω_1, and ω_2 are positive real numbers. The magnitude and phase functions are plotted in Figure 10.6. From Figure 10.6a, we see that the passband of the filter is from ω_1 to ω_2. Thus for any input signal $x(t)$ whose spectrum is contained in the region from ω_1 to ω_2, the filter will pass the signal with no distortion, although there will be a time delay of t_d seconds.

10.3

AMPLITUDE MODULATION

In this section we illustrate the use of the Fourier transform in the study of amplitude modulation. Let $x(t)$ be a signal such as an audio signal that we want to transmit through a cable or through the atmosphere. In AM (amplitude-modulated) transmission, the signal $x(t)$ is first put on a carrier signal $\cos \omega_c t$ of frequency ω_c (rad/sec). In other words, the amplitude of the carrier signal $\cos \omega_c t$ is modulated by the signal $x(t)$. In one form of AM transmission, the signal $x(t)$ and the carrier $\cos \omega_c t$ are simply multiplied together. The process

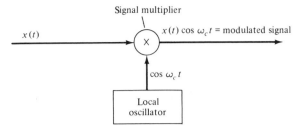

FIGURE 10.7. Amplitude modulation.

is illustrated in Figure 10.7. The local oscillator in Figure 10.7 is a device that produces the carrier signal $\cos \omega_c t$. The signal multiplier may be realized by using a nonlinear element such as a diode (see Problem 10.12).

> **EXAMPLE 10.2.** Suppose that $x(t)$ is the signal shown in Figure 10.8a. The modulated carrier $x(t) \cos \omega_c t$ is sketched in Figure 10.8b.

We shall determine the spectrum of the modulated carrier $x(t) \cos \omega_c t$. In the following analysis, we assume that the signal $x(t)$ is bandlimited with bandwidth B; that is,

$$|X(\omega)| = 0 \qquad \text{for all } \omega > B,$$

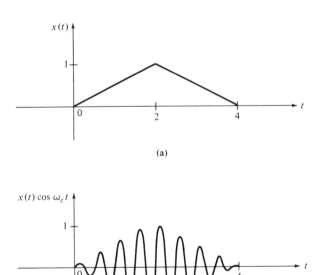

(a)

(b)

FIGURE 10.8. (a) Signal and (b) modulated carrier in Example 10.2.

where $X(\omega)$ is the Fourier transform of $x(t)$. We also assume that $\omega_c > B$; that is, the frequency ω_c of the carrier is greater than the bandwidth B of the signal.

If $x(t)$ is an audio signal such as a speech or music waveform, we can usually take the bandwidth B to be equal to 20 kHz. In other words, an audio signal is not likely to contain many frequency components above 20 kHz.

Now by the modulation property, the Fourier transform of the modulated carrier $x(t) \cos \omega_c t$ is equal to

$$\tfrac{1}{2}[X(\omega + \omega_c) + X(\omega - \omega_c)].$$

This result shows that the modulation process translates the spectrum $X(\omega)$ of $x(t)$ up to the frequency range from $\omega_c - B$ to $\omega_c + B$ (and to the negative-frequency range from $-\omega_c + B$ to $-\omega_c - B$). For example, suppose that the amplitude spectrum $|X(\omega)|$ has the shape shown in Figure 10.9a. The amplitude spectrum of the modulated carrier is shown in Figure 10.9b. As illustrated, the portion of $X(\omega - \omega_c)$ from $\omega_c - B$ to ω_c is called the *lower sideband* and the portion of $X(\omega - \omega_c)$ from ω_c to $\omega_c + B$ is called the *upper sideband*. Each sideband contains all the spectral components of the signal $x(t)$. As a result, the signal $x(t)$ can be reconstructed from either the upper or lower sideband. This will be clear from the results to be given later.

Since both the upper and lower sidebands are present in the spectrum of the modulated carrier, the modulation process we are considering is called *double-sideband transmission*.

In some types of AM transmission, such as AM radio, the modulated carrier is of the form

$$[K + x(t)] \cos \omega_c t,$$

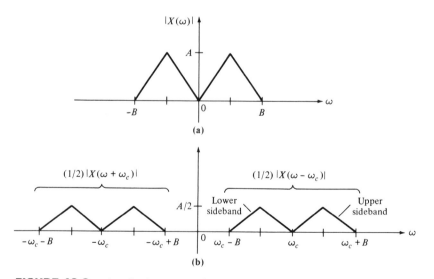

FIGURE 10.9. Amplitude spectra of (a) signal $x(t)$, (b) modulated carrier $x(t) \cos \omega_c t$.

where K is a constant which is chosen so that $K + x(t) > 0$ for all t. The condition $K + x(t) > 0$ ensures that the envelope of the modulated signal $[K + x(t)] \cos \omega_c t$ is a replica of the signal $x(t)$.

The spectrum of the modulated signal $[K + x(t)] \cos \omega_c t$ also contains the upper and lower sidebands. Thus this modulation process is another example of double-sideband transmission. The difference between this form of amplitude modulation and the one considered above is that $[K + x(t)] \cos \omega_c t$ contains the carrier $K \cos \omega_c t$; whereas $x(t) \cos \omega_c t$ does not. As a result, the latter process is referred to as *double-sideband suppressed-carrier transmission*.

The key property of amplitude modulation in the transmission of a signal $x(t)$ is the up conversion of the spectrum of $x(t)$. The higher frequency range of the modulated signal makes it possible to achieve good propagation properties in transmission through a cable or through the atmosphere. For instance, in AM radio transmission the frequency of the transmitted signal occupies a narrow bandwidth in the frequency range from 540 to 1600 kHz. In optical communication, a beam of light is modulated with the result that the spectrum of the signal $x(t)$ is up-converted to an optical-frequency range.

SINGLE-SIDEBAND TRANSMISSION Since both the upper and lower sidebands contain all the spectral components of $x(t)$, it is necessary to transmit only one of the sidebands. The resulting process is called *single-sideband transmission*. Single-sideband transmission can be realized by bandpass filtering the modulated signals considered above. For example, the upper sideband associated with the signal $x(t) \cos \omega_c t$ can be generated by applying $x(t) \cos \omega_c t$ to a bandpass filter with passband from ω_c to $\omega_c + B$. The process is illustrated in Figure 10.10. The frequency function $H(\omega)$ of the bandpass filter is given by

$$H(\omega) = \begin{cases} 1, & \omega_c \le |\omega| \le \omega_c + B \\ 0, & \text{all other } \omega. \end{cases}$$

Note that we are assuming that the phase of the filter is zero. A linear phase would result in a time delay that can be ignored since it does not cause any signal distortion.

DEMODULATION The process of reconstructing the signal $x(t)$ from the modulated signal $x(t) \cos \omega_c t$ or the signal $[K + x(t)] \cos \omega_c t$ is called *demodulation*. In the latter case, demodulation is often accomplished by using an envelope detector. We shall not consider this.

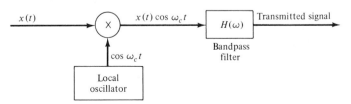

FIGURE 10.10. Single-sideband suppressed carrier transmission.

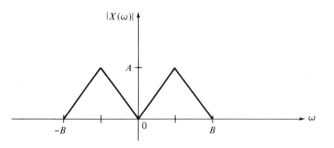

FIGURE 10.11. Amplitude spectrum of $x(t)$.

There is more than one way to reconstruct $x(t)$ from $x(t) \cos \omega_c t$. One technique is synchronous demodulation, which we consider here. In the following analysis, we assume that the amplitude spectrum $|X(\omega)|$ of $x(t)$ has the shape shown in Figure 10.11. We are choosing this shape for $|X(\omega)|$ for illustrative purposes only. The following analysis applies to any signal $x(t)$ with bandwidth $B < \omega_c$.

We begin by applying the signal $x(t) \cos \omega_c t$ to the process shown in Figure 10.12. Note here that the local oscillator is synchronized with the carrier signal $\cos \omega_c t$. The type of demodulator we are considering here requires synchronization.

Now using the trigonometric identity

$$x(t) \cos^2 \omega_c t = \tfrac{1}{2}(1 + \cos 2\omega_c t)x(t),$$

and the modulation property of the Fourier transform, we have that the transform of $x(t) \cos^2 \omega_c t$ is equal to

$$\tfrac{1}{2}X(\omega) + \tfrac{1}{4}[X(\omega + 2\omega_c) + X(\omega - 2\omega_c)].$$

The amplitude spectrum is displayed in Figure 10.13. We see that we can extract $x(t)$ from $x(t) \cos^2 \omega_c t$ by applying $x(t) \cos^2 \omega_c t$ to a lowpass amplifier with gain 2 and bandwidth B. In particular, if we take the frequency function $H(\omega)$ of the amplifier to be

$$H(\omega) = \begin{cases} 2, & -B \le \omega \le B \\ 0, & \text{all other } \omega \end{cases}$$

the output of the amplifier will be equal to $x(t)$.

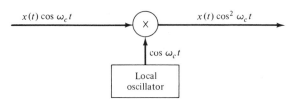

FIGURE 10.12. First stage of demodulation.

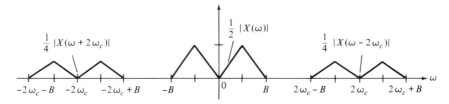

FIGURE 10.13. Amplitude spectrum of $x(t) \cos^2 \omega_c t$.

Combining the multiplication process in Figure 10.12 and the lowpass amplifier, we obtain the synchronous demodulator shown in Figure 10.14.

OTHER TYPES OF MODULATION In addition to amplitude modulation, information in the form of a signal $x(t)$ can be "put on" a carrier by modulating other features of the carrier. For example, in frequency modulation (FM), the frequency of the carrier $\cos \omega_c t$ is modulated by the signal $x(t)$. The resulting modulated carrier has the form

$$\cos [\omega_c t + B_f x(t) t],$$

where B_f is a constant.

In phase modulation (PM), the phase of the carrier is modulated, so that the modulated carrier has the form

$$\cos [\omega_c t + B_p x(t)],$$

where B_p is a constant. For details on these other types of modulation, see Schwartz [1980], Couch [1983], or Kanefsky [1985].

FREQUENCY-DIVISION MULTIPLEXING Suppose that we have N signals $x_1(t)$, $x_2(t)$, . . . , $x_N(t)$ that we want to send from one point to another. To keep the cost of the communication process as small as possible, we would like to be able to send the signals through the same channel (e.g., wire cable) at the same time. For example, if the signals are telephone calls, we would like to be able to send them over the same wire cable or optical-fiber cable at the same time. The process of transmitting several signals over the same channel simultaneously is called *multiplexing*. Multiplexing can be achieved by using frequency-division multiplexing (FDM) or time-division multiplexing (TDM). We consider FDM in this section, while TDM is studied in Section 10.4.

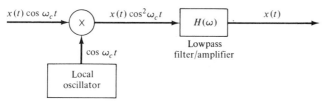

FIGURE 10.14. Synchronous demodulator.

We assume that the spectra of the signals $x_1(t), x_2(t), \ldots, x_N(t)$ are contained in the frequency range from 0 to B (rad/sec); that is,

$$|X_i(\omega)| = 0, \qquad \omega > B, \quad \text{all } i,$$

where $X_i(\omega)$ is the Fourier transform of $x_i(t)$. The idea of FDM is to up convert the spectra of the signals into adjacent (nonoverlapping) frequency "slots." If double-sideband transmission is used, the bandwidth of each slot is equal to $2B$; whereas the bandwidth is equal to B in the case of single-sideband transmission. In practice, single-sideband transmission is typically used since the bandwidth required for each signal is one half of that required in the double-sideband case. However, to keep the analysis as simple as possible, we shall consider double-sideband transmission only.

The up conversion of the spectra of the given signals is accomplished by putting the signals on carriers whose frequencies differ by an amount that is greater than or equal to $2B$. More precisely, given $\omega_c > B$, the frequencies of the carrier signals may be chosen to be $\omega_c, \omega_c + 2B, \omega_c + 4B, \ldots, \omega_c + 2(N - 1)B$. This modulation process is illustrated in the first part of the block diagram in Figure 10.15.

As shown in Figure 10.15, the outputs of the modulators are added together, which results in the signal

$$v(t) = \sum_{i=1}^{N} x_i(t) \cos [\omega_c + 2(i - 1)B]t.$$

The signal $v(t)$ is then applied to a channel which may be a cable, a waveguide, or free space. We assume that the channel is an ideal channel; in other words, the output of the channel is equal to the input.

The spectrum $V(\omega)$ of the output $v(t)$ of the channel consists of the up-converted spectra of the signals $x_1(t), x_2(t), \ldots, x_N(t)$. Since the up-converted spectra comprising $V(\omega)$ do not overlap, they can be separated by bandpass filtering. So as shown in Figure 10.15, we apply the output of the channel to a collection of N bandpass filters, with the frequency function $H_i(\omega)$ of the ith filter given by

$$H_i(\omega) = \begin{cases} 1, & \omega_c + 2(i - 1.5)B \leq |\omega| \leq \omega_c + 2(i - 0.5)B \\ 0, & \text{all other } \omega. \end{cases}$$

To simplify the analysis, we are assuming that the filters are ideal with zero phase.

Now the spectrum of the output of the ith bandpass filter contains only the up-converted spectrum of the ith signal $x_i(t)$. Thus $x_i(t)$ can be extracted from the output of the ith bandpass filter by demodulation. We can use the synchronous demodulator shown in Figure 10.14. The individual demodulators are shown in the last part of the block diagram in Figure 10.15. The lowpass filters in these

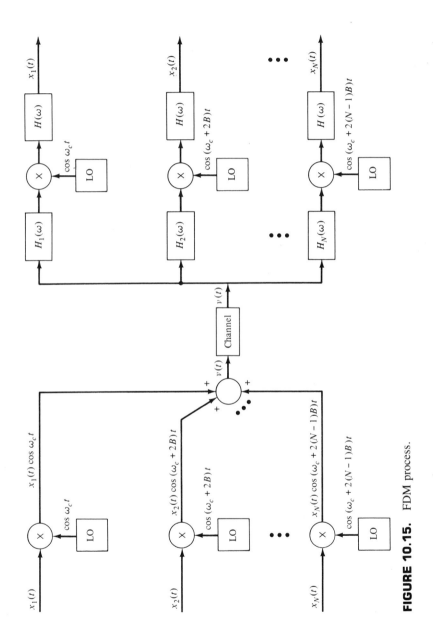

FIGURE 10.15. FDM process.

demodulators all have the same frequency function, given by

$$H(\omega) = \begin{cases} 2, & -B \leq \omega \leq B \\ 0, & \text{all other } \omega. \end{cases}$$

The number of signals that can be transmitted simultaneously using FDM depends on the base frequency ω_c of the carrier signals $\cos(\omega_c + 2iB)t$ and the bandwidth of the channel. For example, let us suppose that the channel is an optical-fiber cable and that we are transmitting telephone calls. In this case, the carrier signals are visible light, which ranges in frequency from 4.0×10^{14} Hz (red) to 7.5×10^{14} Hz (violet).

To have voice recognition in the transmission of phone calls, it is necessary to transmit only a 4-kHz bandwidth of the voice; that is, a person can recognize a filtered version of the voice of someone he or she knows if the bandwidth of the filter is at least 4 kHz. We can restrict telephone calls to a 4-kHz bandwidth by first lowpass filtering the voice waveform. Hence we can take $B = 4$ kHz for FDM transmission of telephone calls. This is done in practice.

Now the bandwidth of a channel consisting of an optical-fiber cable can be as large as several gigahertz, where 1 GHz $= 10^9$ Hz. In a frequency range of 1 GHz, the total number of 8-kHz frequency slots is equal to

$$\frac{10^9}{8 \times 10^3} = 125,000.$$

Thus we can transmit 125,000 phone calls over the same optical-fiber cable at the same time. If we were using single-sideband transmission, we could transmit 250,000 calls over the same cable at the same time.

The main problem in achieving such large capacities has been the difficulty in modulating and demodulating light at high frequencies. However, the capacity indicated above has been achieved in practice and is constantly being expanded as new optical techniques and optical devices are developed. For details on optical devices and optical communication systems, see Wilson and Hawkes [1983], Gowar [1984], or Palais [1984].

We conclude this section by noting that the communication processes discussed above are examples of analog communication systems. Today there is a great deal of emphasis on digital communication systems, in which analog signals are first sampled before they are transmitted. The sampling process is considered in the next section.

10.4

PULSE-AMPLITUDE MODULATION (SAMPLING)

In Section 10.3 we considered the modulation of a sinusoidal carrier signal. In many applications, nonsinusoidal signals are modulated. A very common example is the modulation of the pulse train shown in Figure 10.16. As seen from Figure 10.16, the pulse train $p(t)$ is periodic with period T.

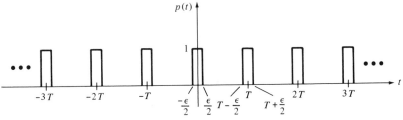

FIGURE 10.16. Pulse train with period T.

Now given a continuous-time signal $x(t)$, we can modulate the amplitude of the pulse train $p(t)$ by multiplying $p(t)$ and $x(t)$ as illustrated in Figure 10.17. This process is called *pulse-amplitude modulation* (PAM).

> **EXAMPLE 10.3.** Consider the signal $x(t)$ displayed in Figure 10.18. With $T = 0.2$ and $\epsilon \ll 0.2$, the PAM signal is shown in Figure 10.19.

A PAM signal can be generated by applying $x(t)$ to a switch which is closed for ϵ seconds every T seconds. If ϵ is much smaller than T, the modulated signal $x(t)p(t)$ is actually a sampled version of $x(t)$, where T is the sampling interval. Thus sampling and pulse-amplitude modulation are really the same process.

IDEALIZED SAMPLING To simplify the analysis of systems containing the sampling operation, many authors consider idealized sampling where the pulse train $p(t)$ is a train of impulses as shown in Figure 10.20. Since

$$x(t)\delta(t - kT) = x(kT)\delta(t - kT) \qquad \text{for any integer } k,$$

when $p(t)$ is the impulse train, the sampled signal $x(t)p(t)$ is given by

$$x(t)p(t) = \sum_{k=-\infty}^{\infty} x(kT)\delta(t - kT). \tag{10.21}$$

Thus, in idealized sampling, the sampled signal (10.21) is an impulse train whose weights (areas) are the instantaneous values of the signal $x(t)$ at the sample times $t = kT$.

In the following development we restrict our attention to (nonideal) sampling where $p(t)$ is given by the pulse train in Figure 10.16.

SIGNAL RECONSTRUCTION A key question is whether or not it is possible to reconstruct $x(t)$ from the sampled signal (or PAM signal) $x(t)p(t)$. The

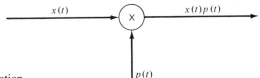

FIGURE 10.17. Pulse-amplitude modulation.

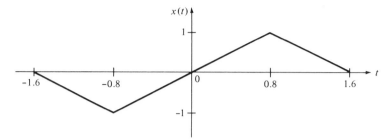

FIGURE 10.18. Signal in Example 10.3.

synchronous demodulator that we used in the case of a sinusoidal carrier signal will not work when the carrier is a pulse train. The demodulation of a PAM signal can be approached using the results in Chapter 9 on the Fourier series and Fourier transform.

First, since the pulse train $p(t)$ is periodic with period T, we can express $p(t)$ by its trigonometric Fourier series representation. Since $p(t)$ is an even periodic function,

$$c_0 = \frac{2}{T} \int_0^{T/2} p(t) \, dt \tag{10.22}$$

and

$$c_n = \frac{2}{T} \int_0^{T/2} p(t) \cos (\omega_s t) \, dt, \qquad n = 1, 2, \ldots, \tag{10.23}$$

where $\omega_s = 2\pi/T$ is the sampling frequency. Evaluating the integrals in (10.22) and (10.23), we obtain

$$c_0 = \frac{\epsilon}{T}$$

and

$$c_n = \frac{1}{n\pi} \sin \left(\frac{n\pi\epsilon}{T} \right) = \frac{\epsilon}{T} \operatorname{sinc} \left(\frac{n\pi\epsilon}{T} \right).$$

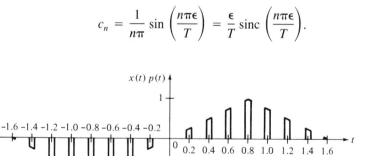

FIGURE 10.19. PAM signal in Example 10.3.

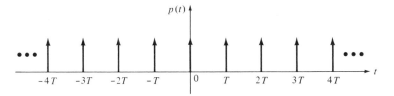

FIGURE 10.20. Pulse train $p(t)$ in the case of idealized sampling.

Therefore,

$$p(t) = \frac{\epsilon}{T} + \sum_{n=1}^{\infty} \frac{2\epsilon}{T} \text{sinc}\left(\frac{n\pi\epsilon}{T}\right) \cos(n\omega_s t)$$

$$p(t) = \frac{\epsilon}{T}\left[1 + \sum_{n=1}^{\infty} 2 \text{sinc}\left(\frac{n\pi\epsilon}{T}\right) \cos(n\omega_s t)\right].$$

To simplify the notation, let us set

$$e_n = 2 \text{sinc}\left(\frac{n\pi\epsilon}{T}\right), \qquad n = 1, 2, \ldots$$

Then

$$x(t)p(t) = \frac{\epsilon}{T}\left[x(t) + \sum_{n=1}^{\infty} e_n x(t) \cos(n\omega_s t)\right]. \tag{10.24}$$

Now using the expression (10.24), we shall compute the Fourier transform of the sampled signal $x(t)p(t)$. In the following development, we denote the sampled signal by $x_s(t)$; that is, $x_s(t) = x(t)p(t)$.

Letting $X(\omega)$ denote the Fourier transform of the given signal $x(t)$, by the modulation property, we have the transform pair

$$e_n x(t) \cos(n\omega_s t) \leftrightarrow \frac{e_n}{2}[X(\omega + n\omega_s) + X(\omega - n\omega_s)]. \tag{10.25}$$

Using (10.24), (10.25), and linearity, we have that the transform $X_s(\omega)$ of the sampled signal $x_s(t)$ is given by

$$X_s(\omega) = \frac{\epsilon}{T}\left\{X(\omega) + \sum_{n=1}^{\infty} \frac{e_n}{2}[X(\omega + n\omega_s) + X(\omega - n\omega_s)]\right\}. \tag{10.26}$$

From (10.26) we see that the Fourier transform $X_s(\omega)$ consists of a sum of magnitude-scaled replicas of $X(\omega)$ sitting at integer multiples $n\omega_s$ of ω_s for $n = 0, \pm 1, \pm 2, \ldots$.

Now suppose that $x(t)$ has a bandwidth equal to B. Then if $\omega_s > 2B$, the replicas of $X(\omega)$ in $X_s(\omega)$ do not overlap in frequency. Thus if we lowpass filter

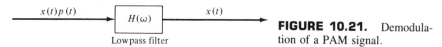

FIGURE 10.21. Demodulation of a PAM signal.

and amplify the sampled signal $x_s(t)$ with the filter cutoff frequency equal to B, we can recover $x(t)$ from $x_s(t)$. The frequency function $H(\omega)$ of the lowpass filter/amplifier is

$$H(\omega) = \begin{cases} T/\epsilon, & -B \leq \omega \leq B \\ 0, & \text{otherwise.} \end{cases}$$

So the demodulation of a PAM signal can be accomplished by simply lowpass filtering and amplifying the signal. The process is illustrated in Figure 10.21. The filter/amplifier in this figure is sometimes called an *interpolating filter* since it reproduces $x(t)$ from the values of $x(t)$ in a neighborhood of the time points $t = kT$.

Summarizing the above results, we have proved the famous sampling theorem: If the signal $x(t)$ has bandwidth B and the sampling frequency ω_s is greater than or equal to $2B$, then $x(t)$ can be completely and exactly reconstructed from the sampled signal $x_s(t) = x(t)p(t)$ by lowpass filtering/amplifying with cutoff frequency B. The minimum sampling frequency $\omega_s = 2B$ is called the *Nyquist sampling frequency*.

> **EXAMPLE 10.4.** The spectrum of a speech signal is essentially zero for all frequencies above 10 kHz, so we can take the bandwidth of a speech signal to be $2\pi \times 10^4$ rad/sec. Then the Nyquist sampling frequency for speech is
>
> $$\omega_s = 2B = 4\pi \times 10^4 \text{ rad/sec.}$$
>
> Since $\omega_s = 2\pi/T$, the sampling interval T is equal to $2\pi/\omega_s = 50$ μs. So the sampling interval corresponding to the Nyquist sampling rate is very small.

ALIASING In Chapter 9 we noted that a time-limited signal cannot be band-limited. Since all actual signals are time limited, they cannot be bandlimited. Therefore, if we sample a time-limited signal with sampling interval T, no matter how small T is, the replicas of $X(\omega)$ in (10.26) will overlap. As a result of the overlap of frequency components, it is not possible to reconstruct $x(t)$ exactly by lowpass filtering the sampled signal $x_s(t) = x(t)p(t)$.

Although time-limited signals are not bandlimited, the amplitude spectrum $|X(\omega)|$ of a time-limited signal $x(t)$ will be small for suitably large values of ω. Thus for some finite B, all the significant components of $X(\omega)$ will be in the range $-B \leq \omega \leq B$. For instance, the signal $x(t)$ may have the amplitude spectrum shown in Figure 10.22. If we take B to have the value indicated, and if we sample $x(t)$ with sampling frequency $\omega_s = 2B$, the amplitude spectrum of the resulting sampled signal $x_s(t)$ is shown in Figure 10.23.

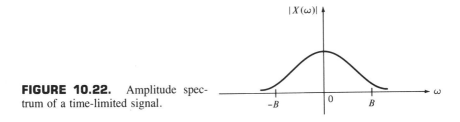

FIGURE 10.22. Amplitude spectrum of a time-limited signal.

Now if we lowpass filter the sampled signal $x_s(t)$ with cutoff frequency B, we see that the output spectrum of the filter will contain high-frequency components of $x(t)$ transposed to low-frequency components. This phenomenon is called *aliasing*.

Aliasing will result in a distorted version of the original signal $x(t)$. It can be eliminated (theoretically) by first lowpass filtering $x(t)$ before $x(t)$ is sampled: If we lowpass filter $x(t)$ so that all frequency components above B are removed, there will be no overlap of frequency components in $X_s(\omega)$ if we sample at the Nyquist rate $\omega_s = 2B$.

In practice, aliasing cannot be eliminated completely since we cannot build a lowpass filter that cuts off all frequency components above a certain frequency. However, we can reduce the magnitude of the aliased components if we lowpass filter the signal $x(t)$ before sampling. This approach is feasible as long as lowpass filtering $x(t)$ does not remove the "information content" of the signal $x(t)$.

EXAMPLE 10.5. Again suppose that $x(t)$ is a speech waveform. Although a speech waveform usually contains sizable frequency components above 4 kHz, in Section 10.3 we noted that voice recognition is possible for speech signals that have been filtered to a 4-kHz bandwidth. If we take $B =$ 4 kHz for filtered speech, the resulting Nyquist sampling frequency is

$$\omega_s = 2(2\pi)(4 \times 10^3) = 16\pi \times 10^3 \text{ rad/sec.}$$

For this sampling frequency, the sampling interval T is

$$T = \frac{2\pi}{\omega_s} = 0.125 \text{ ms.}$$

This is a much longer sampling interval than the 50-μs sampling interval

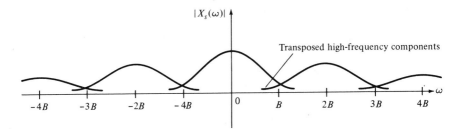

FIGURE 10.23. Amplitude spectrum of a sampled signal.

required to transmit a 10-kHz bandwidth of speech. In general, the wider the bandwidth of a signal, the more expensive it is to transmit the signal. So it is much "cheaper" to send filtered speech over phone lines. In particular, since the bandwidth of filtered speech is much smaller than that of unfiltered speech, we can simultaneously transmit many more filtered speech signals over the same line using FDM.

In many applications, the signal $x(t)$ cannot be lowpass filtered without removing information contained in $x(t)$. In such cases, we must take the bandwidth B of the signal to be sufficiently large so that the aliased components do not seriously distort the reconstructed signal. Equivalently, for a given value of B, we can sample at a rate higher than the Nyquist rate. For example, in applications to sampled-data control, the sampling frequency may be as large as eight or ten times B, where B is the bandwidth of the system being controlled.

TIME-DIVISION MULTIPLEXING   By interleaving the samples from several time signals, we can transmit many signals simultaneously over the same channel. This process is called *time-division multiplexing* (TDM). A brief description of TDM follows. For an in-depth treatment, see Couch [1983].

Suppose that we have N signals $x_1(t), x_2(t), \ldots, x_N(t)$ that we want to transmit over a cable. We assume that the bandwidth of each of the signals is equal to B, so that the Nyquist sampling frequency is $\omega_s = 2B$. Let $p(t)$ denote the pulse train in Figure 10.16 with $T = 2\pi/\omega_s = \pi/B$. We assume that $\epsilon < T/N$.

In the first part of the TDM process, we sample the signals $x_1(t), x_2(t), \ldots, x_N(t)$ by multiplying the ith signal $x_i(t)$ by $p[t - (i - 1)(T/N)]$. As shown in Figure 10.24, the outputs of the multipliers are added together, which results in the signal $v(t)$ given by

$$v(t) = \sum_{i=1}^{N} x_i(t)p\left[t - (i - 1)\frac{T}{N}\right].$$

The signal $v(t)$ consists of the interleaved samples of the signals $x_1(t), x_2(t), \ldots, x_N(t)$.

The signal $v(t)$ is then applied to a channel whose output is assumed to be equal to $v(t)$. As shown in Figure 10.24, we then recover the original signals from $v(t)$ by first multiplying by $p[t - (i - 1)(T/N)]$. These multipliers (samplers) must be synchronized with the samplers in the first part of the process. If they are not synchronized, the sample values of one signal will be mixed with the sample values of the other signals, resulting in distortion in the reconstruction (demodulation) process. In the final stage of the TDM process, the original signals are recovered from the sampled signals $x_i(t)p[t - (i - 1)(T/N)]$ by lowpass filtering. The cutoff frequency of each lowpass filter is equal to B.

We should note that in digital communication systems employing TDM, the sampled signals are quantized in amplitude and then encoded before the sample values are interleaved. The process of quantization and encoding is discussed in Chapter 11.

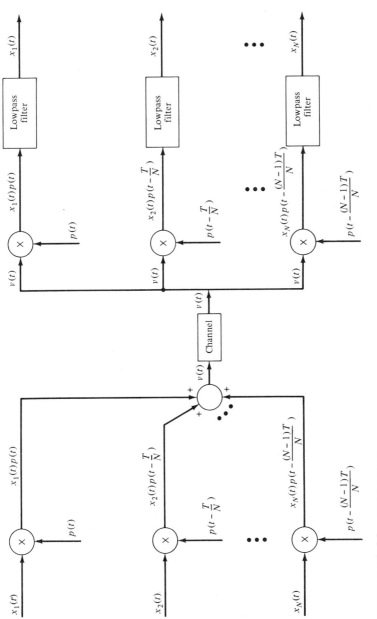

FIGURE 10.24. TDM process.

PROBLEMS

Chapter 10

10.1   An ideal linear-phase lowpass filter has the frequency function

$$H(\omega) = \begin{cases} e^{-j\omega}, & -2 < \omega < 2 \\ 0, & \text{all other } \omega. \end{cases}$$

Compute the filter's output response $y(t)$ when the input $x(t)$ is

(a) $x(t) = 5 \, \text{Sa} \left(\dfrac{3t}{2} \right), \quad -\infty < t < \infty.$

(b) $x(t) = 5 \, \text{Sa} \left(\dfrac{t}{2} \right) \cos 2t, \quad -\infty < t < \infty.$

(c) $x(t) = \text{Sa}^2 \left(\dfrac{t}{2} \right), \quad -\infty < t < \infty.$

(d) $x(t) = \displaystyle\sum_{n=1}^{\infty} \frac{1}{n} \cos \left(\frac{n\pi}{2} t + 30° \right), \quad -\infty < t < \infty.$

10.2   A lowpass filter has the frequency response curves shown in Figure P10.2.
   (a) Compute the impulse response $h(t)$ of the filter.
   (b) Compute the response $y(t)$ when $x(t) = 3 \, \text{Sa} \, (t) \cos 4t, \quad -\infty < t < \infty.$

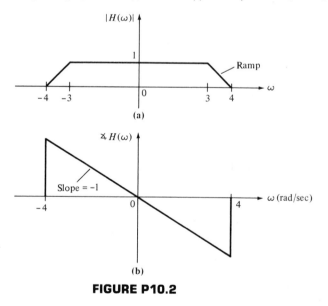

(a)

(b)

FIGURE P10.2

10.3   An ideal linear-phase highpass filter has frequency function

$$H(\omega) = \begin{cases} 6e^{-j2\omega}, & \omega > 3, \ \omega < -3 \\ 0, & -3 < \omega < 3. \end{cases}$$

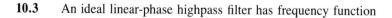

(a) Compute the impulse response $h(t)$ of the filter.

(b) Compute the output response $y(t)$ when the input $x(t)$ is given by $x(t) =$ Sa $(5t)$, $-\infty < t < \infty$.

(c) Compute the output response $y(t)$ when the input $x(t)$ is the periodic signal shown in Figure P10.3.

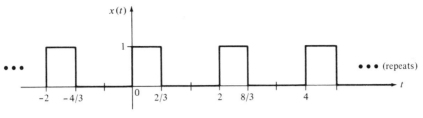

FIGURE P10.3

10.4   The input

$$x(t) = \text{Sa}\left(\frac{t}{2}\right)(\cos 3t)^2 + \text{Sa}\left(\frac{t}{2}\right)\cos t, \qquad -\infty < t < \infty$$

is applied to a linear time-invariant continuous-time system with the frequency function $H(\omega)$. Determine $H(\omega)$ so that the output response $y(t)$ resulting from this input is given by

$$y(t) = \text{Sa}\left(\frac{t}{2}\right), \qquad -\infty < t < \infty.$$

Express your answer by giving $H(\omega)$ in analytical form.

10.5   An ideal linear-phase bandpass filter has frequency function

$$H(\omega) = \begin{cases} 10e^{-j4\omega}, & -4 < \omega < -2, \ 2 < \omega < 4 \\ 0, & \text{all other } \omega. \end{cases}$$

Compute the output response $y(t)$ of the filter when the input $x(t)$ is

(a) $x(t) = \text{Sa}(2t)$, $-\infty < t < \infty$.

(b) $x(t) = \text{Sa}(3t)$, $-\infty < t < \infty$.

(c) $x(t) = \text{Sa}(4t)$, $-\infty < t < \infty$.

(d) $x(t) = \text{Sa}(2t)\cos t$, $-\infty < t < \infty$.

(e) $x(t) = \text{Sa}(2t)\cos 3t$, $-\infty < t < \infty$.

(f) $x(t) = \text{Sa}(2t)\cos 6t$, $-\infty < t < \infty$.

(g) $x(t) = \text{Sa}^2(t)\cos 2t$, $-\infty < t < \infty$.

10.6   A linear time-invariant continuous-time system has frequency function $H(\omega) =$ $5\cos 2\omega$, $-\infty < \omega < \infty$.

(a) Sketch the system's magnitude function $|H(\omega)|$ and phase function $\angle H(\omega)$.

(b) Compute the system's impulse response $h(t)$.

(c) Derive an expression for the output response $y(t)$ resulting from an arbitrary input $x(t)$ with the system at rest prior to the application of $x(t)$.

10.7　The input $x(t) = $ Sa $(t) \cos 2t$, $-\infty < t < \infty$, is applied to an ideal lowpass filter with frequency function $H(\omega) = 1$, $-a < \omega < a$, $H(\omega) = 0$ for all other ω. Determine the smallest possible value of a for which the resulting output response $y(t)$ is equal to the input $x(t) = $ Sa $(t) \cos 2t$.

10.8　A linear time-invariant continuous-time system has transfer function

$$H(s) = \frac{s + 1}{s^2 + 2s + 2}.$$

The input $x(t) = (\cos t)u(t)$ is applied to the system with no initial energy in the system at time $t = 0$.

(a) Compute the Fourier transform of the steady-state response $y_{ss}(t)$ resulting from the input $x(t)$ (see Section 8.4).

(b) Compute the Laplace transform of the transient response $y_{tr}(t)$ resulting from the input $x(t)$.

(c) Using your result in part (b), compute the Fourier transform of the transient response $y_{tr}(t)$.

(d) Show that the sum of your answers in parts (a) and (c) is equal to $H(\omega)X(\omega)$, where $H(\omega) = H(s)|_{s=j\omega}$ and $X(\omega)$ is the Fourier transform of $x(t)$.

10.9　A linear time-invariant continuous-time system has a rational transfer function $H(s)$ with two poles and two zeros. The frequency function $H(\omega)$ of the system is given by

$$H(\omega) = \frac{-\omega^2 + j3\omega}{8 + j12\omega - 4\omega^2}.$$

Determine $H(s)$.

10.10　Consider the system in Figure P10.10a. The frequency function $H(\omega)$ of the

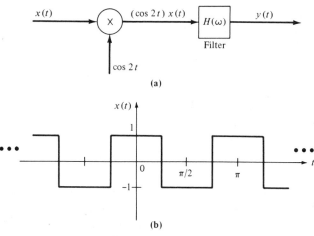

(a)

(b)

FIGURE P10.10

filter in Figure P10.10a is given by

$$H(\omega) = \begin{cases} 2e^{-j3\omega}, & -2 \le \omega \le 2 \\ 0, & \text{all other } \omega. \end{cases}$$

(a) Compute the output response $y(t)$ when $x(t) = \cos t$, $-\infty < t < \infty$.
(b) Compute $y(t)$ when $x(t) = \cos 2t$, $-\infty < t < \infty$.
(c) Compute $y(t)$ when $x(t) = \text{Sa}(t) \cos 3t$, $-\infty < t < \infty$.
(d) Compute $y(t)$ when $x(t) = \text{Sa}^2(t)$, $-\infty < t < \infty$.
(e) Compute the output response $y(t)$ when $x(t)$ is the periodic signal shown in Figure P10.10b.

 HINT: Sketch $(\cos 2t)x(t)$.

10.11 Consider the system in Figure P10.11a. The frequency functions of the filters in Figure P10.11a are given by

$$H_1(\omega) = \begin{cases} 3, & -2 \le \omega \le 2 \\ 0, & \text{all other } \omega \end{cases}, \quad H_2(\omega) = \begin{cases} e^{-j\omega}, & \omega < -2, \omega > 2 \\ 0, & -2 \le \omega \le 2. \end{cases}$$

(a) Compute the output response $y(t)$ when $x(t) = \text{Sa}(t)$, $-\infty < t < \infty$.
(b) Compute $y(t)$ when $x(t) = \text{Sa}(t) \cos 2t$, $-\infty < t < \infty$.
(c) Compute the output response $y(t)$ when $x(t)$ is the periodic signal shown in Figure P10.11b.

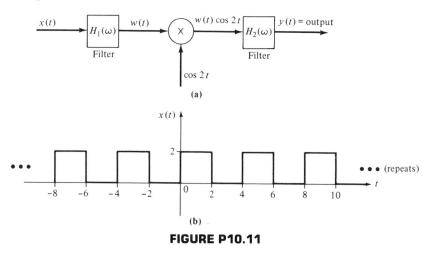

(a)

(b)

FIGURE P10.11

10.12 Consider the system in Figure P10.12a. The signal $z(t)$ into the square-law device is given by

$$z(t) = x(t) + \cos \omega_c t.$$

The amplitude spectrum $|X(\omega)|$ of the signal $x(t)$ applied to the system is shown in Figure P10.12b.

(a) Sketch the amplitude spectrum of the output $z^2(t)$ of the square-law device.

(b) Determine the bandwidth of $x^2(t)$.

(c) What condition on ω_c is required so that $y(t) = x(t) \cos \omega_c t$? When this condition on ω_c is satisfied, the resulting system is called a *square-law modulator*.

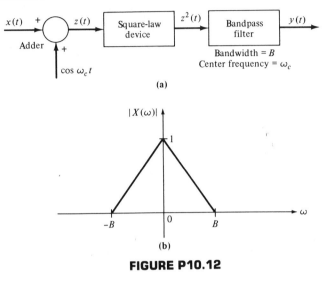

(a)

(b)

FIGURE P10.12

10.13 A signal $x(t)$ with bandwidth B is put on a carrier $\cos \omega_c t$ with $\omega_c \gg B$. The modulated signal $x(t) \cos \omega_c t$ is then applied to the system shown in Figure P10.13. The frequency function $H(\omega)$ of the filter in the system is given by

$$H(\omega) = \begin{cases} 2, & -B < \omega < B \\ 0, & \text{all other } \omega. \end{cases}$$

Compute $y(t)$ when
(a) $\theta = 90°$.
(b) $\theta = 180°$.

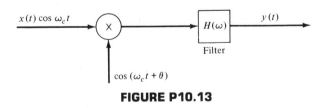

FIGURE P10.13

10.14 The input $x(t)$ to the demodulator in Figure P10.14a is an amplitude-modulated signal $s(t) \cos \omega_c t$ plus an interfering signal $\cos Bt \cos At$; that is,

$$x(t) = s(t) \cos \omega_c t + \cos Bt \cos At, \qquad B = \text{bandwidth of } s(t).$$

The amplitude spectrum $|S(\omega)|$ of $s(t)$ is shown in Figure P10.14b.

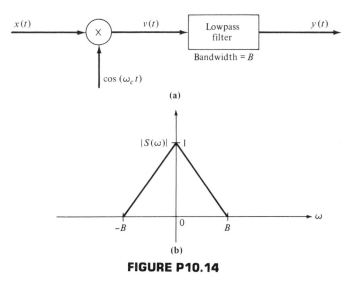

FIGURE P10.14

(a) Sketch the amplitude spectrum $|V(\omega)|$ of $v(t)$, where $v(t)$ is the signal applied to the lowpass filter in Figure P10.14a.

(b) For what range of values of A will the interfering signal cause the output $y(t)$ to be not equal to $s(t)$?

10.15 The objective of this problem is to show that the balanced modulator illustrated in Figure P10.15a can be used to recover the signal $x(t)$ from the amplitude-

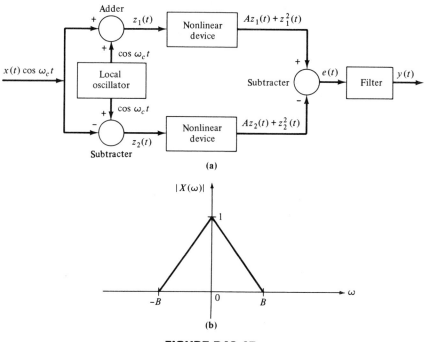

FIGURE P10.15

modulated waveform $x(t) \cos \omega_c t$. The signal $x(t)$ has the amplitude spectrum shown in Figure P10.15b.

(a) Determine the signal $e(t)$ in the output of the subtracter.

(b) Determine $E(\omega)$ and sketch the amplitude spectrum $|E(\omega)|$.

(c) What conditions on ω_c are necessary to ensure that $x(t)$ can be recovered from $x(t) \cos \omega_c t$ using the system in Figure P10.15a?

(d) Assuming that the conditions in part (c) are met, design the filter so that $y(t) = x(t)$. Give the filter type, bandwidth, gain, and so on.

10.16 A signal $x(t)$ with bandwidth B is applied to the system shown in Figure P10.16. Design a continuous-time system that reproduces $x(t)$ from the output of the system in Figure P10.16; that is, the response of the system to input $4x(t) + x(t) \cos Bt$ is equal to $x(t)$.

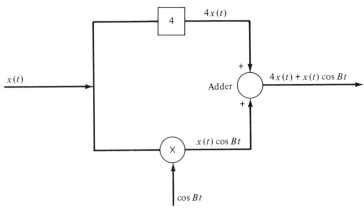

FIGURE P10.16

10.17 A speech signal $x(t)$ with bandwidth B is applied to the speech scrambler shown in Figure P10.17. Filter 1 is an ideal zero-phase highpass filter that stops all frequencies below ω_c (rad/sec), where $\omega_c \gg B$. The second filter is an ideal zero-phase lowpass filter with bandwidth ω_c. Design a descrambler that reproduces $x(t)$ from the output $y(t)$ of the scrambler.

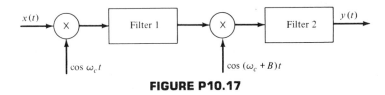

FIGURE P10.17

10.18 A special multiplexing technique for two signals $x_1(t)$ and $x_2(t)$ produces the signal

$$x(t) = x_1(t) \cos \omega_c t + x_2(t) \sin \omega_c t,$$

where ω_c is much greater than the bandwidth B_1 of $x_1(t)$ and the bandwidth B_2 of $x_2(t)$. The multiplexed signal $x(t)$ is applied to the system shown in Figure P10.18. Design the filters in this system so that the first output $y_1(t)$ is equal to $x_1(t)$ and the second output $y_2(t)$ is equal to $x_2(t)$. Give the filter types, bandwidths, and so on.

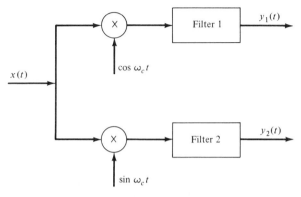

FIGURE P10.18

10.19 Compute the Nyquist sampling frequency (rad/sec) for the following signals.
(a) $x(t) = 3 \cos 4t$.
(b) $x(t) = 2 \cos 3t + \cos 5t$.
(c) $x(t) = 4 \operatorname{Sa}(3t)$.
(d) $x(t) = 6 \operatorname{Sa}^2(5t)$.
(e) $x(t) = \operatorname{Sa}(4t) + 2 \operatorname{Sa}^2(3t)$.
(f) $x(t) = 3 \operatorname{Sa}^2(2t)$.
(g) $x(t) = \operatorname{Sa}(3t) \cos 2t$.
(h) $x(t) = \operatorname{Sa}(3t) \cos 4t$.
(i) $x(t) = \operatorname{Sa}(2t) \cos^2 3t$.

10.20 Compute the Nyquist sampling frequency for the signals shown in Figure P10.20.

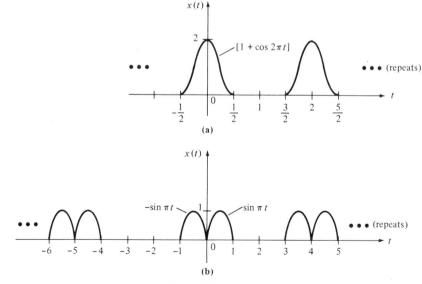

FIGURE P10.20

Fourier Analysis of Discrete-Time Signals and Systems

In this chapter we present the discrete-time counterpart to the theory developed in Chapters 9 and 10. We begin in Section 11.1 by considering the Fourier series representation of periodic discrete-time signals. In Sections 11.2 and 11.3 we define and study the discrete-time Fourier transform (DTFT). The DTFT is the discrete-time counterpart to the Fourier transform. (The DTFT should not be confused with the discrete Fourier transform, which is discussed below.) As in the continuous-time case, the DTFT is a function of a continuum of frequencies, but unlike the continuous-time case, the DTFT is always a periodic function with period 2π. The use of the DTFT in system analysis is considered in Section 11.4. The results in this section closely resemble those derived in the continuous-time case in Sections 10.1 and 10.2.

In Section 11.5 we define a transform that is a function of a finite number of frequencies. This transform is called the discrete Fourier transform (DFT). For time-limited discrete-time signals, we show that the DFT is equal to the DTFT with the frequency variable evaluated at a finite number of points. Since the DFT is a function of a finite number of frequencies, it is the transform that is often used in practice. In particular, the DFT is used extensively in digital signal processing. As discussed in Section 11.6, there is a fast method for computing the DFT, called the fast Fourier transform (FFT) algorithm. In Section 11.7 we apply the FFT algorithm to the computation of the Fourier transform of an analog signal and to the computation of a convolution sum (fast convolution).

★11.1

PERIODIC DISCRETE-TIME SIGNALS

In the first part of this section we study the Fourier series representation of a periodic discrete-time signal. Then we use the Fourier series representation to compute the output response resulting from a periodic input.

FOURIER-SERIES REPRESENTATION   Let N be a fixed positive integer. A discrete-time signal $x(kT)$ is periodic with period NT if

$$x[(k + N)T] = x(kT) \qquad \text{for } k = 0, \pm 1, \pm 2, \ldots . \qquad (11.1)$$

If $x(kT)$ is periodic with period NT, we always assume that N is the smallest positive integer for which (11.1) is satisfied. Note that by (11.1), a periodic signal repeats its values every NT seconds.

Examples of periodic signals follow: $x(kT) = A \cos (2\pi k/N)$, $x(kT) = A \sin (2\pi k/N)$, and $x(kT) = \exp (j2\pi k/N)$. All three of these signals are periodic with period NT. To show that $x(kT) = \exp (j2\pi k/N)$ is periodic, we have

$$x[(k + N)T] = \exp \left[\frac{j2\pi(k + N)}{N} \right] = \exp \left(\frac{j2\pi k}{N} \right) \exp (j2\pi).$$

Now

$$\exp (j2\pi) = \cos 2\pi + j \sin 2\pi = 1,$$

and thus $x[(k + N)T] = x(kT)$ for all integers k.

Let $x(kT)$ be a periodic signal with period NT. Then $x(kT)$ has the complex exponential Fourier series

$$x(kT) = \sum_{n=0}^{N-1} a_n \exp \left(\frac{j2\pi nk}{N} \right), \qquad k = 0, \pm 1, \pm 2, \ldots, \qquad (11.2)$$

where

$$a_n = \frac{1}{N} \sum_{k=0}^{N-1} x(kT) \exp \left(\frac{-j2\pi kn}{N} \right), \qquad n = 0, 1, \ldots, N - 1. \qquad (11.3)$$

Note that the Fourier series (11.2) consists of a finite number of complex exponentials. This is in contrast to the Fourier series of a periodic continuous-time signal, which usually consists of an infinite number of complex exponentials.

From the expression (11.2) for $x(kT)$, we see that a periodic discrete-time signal is comprised of a finite number of frequency components. In particular, there is the dc component given by a_0. The fundamental or first harmonic of $x(kT)$ is the term on the right side of (11.2) with frequency $2\pi/N$. The harmonics of $x(kT)$ are the terms in (11.2) with frequencies $2\pi n/N$, $n = 2, 3, \ldots,$ $N - 1$.

To simplify the notation, from here on we shall let w_N equal exp $(j2\pi/N)$. The complex number w_N is a Nth root of unity; that is,

$$w_N^N = e^{j2\pi} = 1.$$

We assume that $N > 1$, and thus $w_N \neq 1$.

In terms of w_N, the Fourier series (11.2) and the expression (11.3) for the coefficients can be rewritten in the form

$$x(kT) = \sum_{n=0}^{N-1} a_n w_N^{nk}, \qquad k = 0, \pm 1, \pm 2, \ldots \qquad (11.4)$$

$$a_n = \frac{1}{N} \sum_{k=0}^{N-1} x(kT) w_N^{-kn}, \qquad n = 0, 1, \ldots, N - 1. \qquad (11.5)$$

PROOF OF THE FOURIER SERIES REPRESENTATION Unlike the continuous-time case, every periodic discrete-time signal has a Fourier series of the form (11.2) or (11.4). We can verify that $x(kT)$ can be expressed in the form (11.4) by inserting the expression (11.5) for a_n into the right side of (11.4). If the result is equal to $x(kT)$, this will prove that the Fourier series representation is valid.

First, let us change the index of the summation in (11.5) to m, so that

$$a_n = \frac{1}{N} \sum_{m=0}^{N-1} x(mT) w_N^{-mn}.$$

Then inserting this into (11.4), we obtain

$$\sum_{n=0}^{N-1} a_n w_N^{nk} = \frac{1}{N} \sum_{n=0}^{N-1} \sum_{m=0}^{N-1} x(mT) w_N^{-mn} w_N^{nk}$$

$$= \frac{1}{N} \sum_{m=0}^{N-1} x(mT) \sum_{n=0}^{N-1} w_N^{n(k-m)}. \qquad (11.6)$$

Now when $k = m$,

$$\sum_{n=0}^{N-1} w_N^{n(k-m)} = N, \qquad (11.7)$$

and when $k \neq m$,

$$\sum_{n=0}^{N-1} w_N^{n(k-m)} = \frac{1 - w_N^{N(k-m)}}{1 - w_N^{k-m}}. \tag{11.8}$$

The relationship (11.8) can be checked by multiplying both sides by $1 - w_N^{k-m}$. Since $w_N^N = 1$, (11.8) reduces to

$$\sum_{n=0}^{N-1} w_N^{n(k-m)} = 0, \qquad k \neq m. \tag{11.9}$$

Using (11.7) and (11.9) in (11.6), we obtain

$$\sum_{n=0}^{N-1} a_n w_N^{nk} = \frac{1}{N} x(kT)N = x(kT).$$

Hence we have verified (11.4).

It is worth noting that to compute the coefficients a_n of the Fourier series (11.4), we can sum over any full period of $x(kT)$. Thus, for any integer k_0,

$$a_n = \frac{1}{N} \sum_{k=k_0}^{N+k_0-1} x(kT) w_N^{-kn}. \tag{11.10}$$

In particular, if N is an even integer, we can take $k_0 = -N/2$, which gives

$$a_n = \frac{1}{N} \sum_{k=-(N/2)}^{(N/2)-1} x(kT) w_N^{-kn}. \tag{11.11}$$

The computation of the Fourier series is illustrated by the following two examples.

EXAMPLE 11.1. Suppose that $x(kT) = A \cos(2\pi k/N)$. As noted above, this signal is periodic with period equal to NT. Using Euler's formula, we have

$$A \cos \frac{2\pi k}{N} = \frac{A}{2} \left[\exp\left(\frac{j2\pi k}{N}\right) + \exp\left(\frac{-j2\pi k}{N}\right) \right]$$

$$= \frac{A}{2} (w_N^k + w_N^{-k}).$$

Now

$$w_N^{(N-1)k} = w_N^{Nk} w_N^{-k} = w_N^{-k}.$$

Therefore,

$$A \cos\left(\frac{2\pi k}{N}\right) = \frac{A}{2}(w_N^k + w_N^{(N-1)k}). \tag{11.12}$$

Comparing (11.12) and (11.4), we see that the right side of (11.12) is the Fourier series of the signal $A \cos(2\pi k/N)$. In this example the coefficients of the series are

$$a_1 = \frac{A}{2} \qquad a_{N-1} = \frac{A}{2}, \qquad a_n = 0 \quad \text{all other } n.$$

EXAMPLE 11.2. In this example $x(kT)$ is a square wave with period NT, where $N \geq 2$ is an even integer. The signal is displayed in Figure 11.1. Evaluating a_n using (11.5), we have

$$a_n = \frac{1}{N}\sum_{k=0}^{N-1} x(kT)w_N^{-kn}$$

$$= \frac{1}{N}\sum_{k=0}^{(N/2)-1} (1)w_N^{-kn}.$$

Then

$$a_0 = \frac{1}{N}\frac{N}{2} = \frac{1}{2}.$$

When $n \neq 0$,

$$a_n = \frac{1}{N}\frac{1 - w_N^{-Nn/2}}{1 - w_N^{-n}}$$

$$= \begin{cases} \dfrac{2/N}{1 - w_N^{-n}} & n = 1, 3, 5, \ldots, N-1 \\ 0, & n = 2, 4, 6, \ldots, N-2. \end{cases}$$

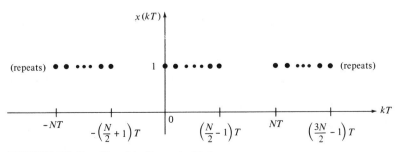

FIGURE 11.1. Signal in Example 11.2.

So the Fourier series for $x(kT)$ is

$$x(kT) = \frac{1}{2} + \sum_{\substack{n=1 \\ n \text{ odd}}}^{N-1} \frac{2/N}{1 - w_N^{-n}} w_N^{nk}, \qquad k = 0, \pm 1, \pm 2, \ldots$$

RESPONSE TO A PERIODIC INPUT   Now suppose that a periodic input signal $x(kT)$ is applied to a linear time-invariant discrete-time system with unit-pulse response $h(kT)$. We assume that the system has no initial energy at the initial time $kT = -\infty$. From the results in Chapter 4 we know that the resulting output response $y(kT)$ is given by

$$y(kT) = \sum_{i=-\infty}^{\infty} h(iT)x(kT - iT), \qquad k = 0, \pm 1, \pm 2, \ldots \quad (11.13)$$

We are not assuming that $h(kT)$ and $x(kT)$ are equal to zero for $k < 0$, and thus the sum in (11.13) must be from $i = -\infty$ to $i = \infty$. Inserting the Fourier series (11.4) for $x(kT)$ into (11.13), we obtain

$$y(kT) = \sum_{i=-\infty}^{\infty} h(iT) \sum_{n=0}^{N-1} a_n w_N^{n(k-i)}$$

$$= \sum_{n=0}^{N-1} \left[\sum_{i=-\infty}^{\infty} h(iT) w_N^{-ni} a_n \right] w_N^{nk}. \qquad (11.14)$$

Now let

$$H(\omega) = \sum_{i=-\infty}^{\infty} h(iT) e^{-j\omega i}. \qquad (11.15)$$

Then

$$H\left(\frac{2\pi n}{N}\right) = \sum_{i=-\infty}^{\infty} h(iT) \exp\left(-\frac{j2\pi ni}{N}\right) = \sum_{i=-\infty}^{\infty} h(iT) w_N^{-ni},$$

and thus we can write (11.14) in the form

$$y(kT) = \sum_{n=0}^{N-1} \left[H\left(\frac{2\pi n}{N}\right) a_n \right] w_N^{-nk}. \qquad (11.16)$$

The right side of (11.16) is in the form of a Fourier series, and hence the output response is also a periodic signal. Note that the period of $y(kT)$ is equal to NT, which is the period of $x(kT)$.

Now suppose that $x(kT)$ is the complex exponential $x(kT) = e^{j\omega k}$. Then inserting $x(kT)$ into (11.13) and using (11.15), we have

$$y(kT) = H(\omega)e^{j\omega k} = H(\omega)x(kT).$$

Since the output response $y(kT)$ can be expressed directly in terms of $H(\omega)$, the function $H(\omega)$ is called the frequency function or the system function. The magnitude $|H(\omega)|$ is the magnitude function of the system, and $\sphericalangle H(\omega)$ is the phase function of the system.

The frequency function $H(\omega)$ defined by (11.15) is the discrete-time counterpart of the frequency function of a linear time-invariant continuous-time system (as defined in Section 9.3). Observe that we are using the same notation for the frequency function in both the discrete-time and continuous-time cases. This should not result in any confusion since it will always be clear whether we are considering a discrete-time or a continuous-time system.

If the given discrete-time system is causal, finite-dimensional, and asymptotically stable, the frequency function $H(\omega)$ defined by (11.15) turns out to be identical to the frequency function defined in Section 8.9. The verification of this result is left to the reader.

If the discrete-time system under study is defined by the input/output difference equation

$$y(kT + nT) + \sum_{i=0}^{n-1} a_i y(kT + iT) = \sum_{i=0}^{m} b_i x(kT + iT), \qquad m \le n, \quad (11.17)$$

the frequency function $H(\omega)$ can be determined directly from the coefficients of (11.17). In particular, setting $x(kT) = e^{j\omega k}$ and $y(kT) = H(\omega)e^{j\omega k}$ in (11.17), we obtain

$$H(\omega) = \frac{\sum_{i=0}^{m} b_i e^{j\omega i}}{e^{j\omega n} + \sum_{i=0}^{n-1} a_i e^{j\omega i}}. \qquad (11.18)$$

Returning to the Fourier series representation (11.16) of the output $y(kT)$, let a_n^y denote the coefficients of the output and let a_n^x denote the coefficients of the input. Then from (11.16) we have that

$$a_n^y = H\left(\frac{2\pi n}{N}\right) a_n^x, \qquad n = 0, 1, \ldots, N - 1.$$

So the Fourier coefficients of the output response are equal to the coefficients of the input times the frequency function $H(\omega)$ of the system evaluated at $\omega = 2\pi n/N$. This result is an obvious discrete-time version of the result in the continuous-time case studied in Section 9.3.

DISCRETE-TIME FOURIER TRANSFORM

In this section we consider the discrete-time version of the Fourier transform. We begin by showing that the discrete-time Fourier transform can be viewed as a generalization of the Fourier series representation of a periodic discrete-time signal. The following derivation is directly analogous to the procedure we used to generate the Fourier transform representation from the Fourier series representation in the continuous-time case (see Section 9.5).

Let $x(kT)$ be a discrete-time signal with $x(kT) = 0$ for all integers $k > N_1$ and all integers $k < -N_1$, where N_1 is a fixed positive integer. The signal $x(kT)$ is a time-limited discrete-time signal. Given a positive even integer $N > 2N_1$, let $\tilde{x}_N(kT)$ denote the periodic signal with period NT, which is equal to $x(kT)$ for $k = -N/2, -(N/2) + 1, \ldots, -1, 0, 1, \ldots, (N/2) - 1$. By definition of $\tilde{x}_N(kT)$, we have that

$$x(kT) = \lim_{N \to \infty} \tilde{x}_N(kT).$$

Since $\tilde{x}_N(kT)$ is periodic with period NT, it has the Fourier series

$$\tilde{x}_N(kT) = \sum_{n=0}^{N-1} a_n \exp\left(\frac{j2\pi nk}{N}\right), \tag{11.19}$$

where

$$a_n = \frac{1}{N} \sum_{k=-N/2}^{(N/2)-1} \tilde{x}_N(kT) \exp\left(\frac{-j2\pi kn}{N}\right), \qquad n = 0, 1, \ldots, N - 1. \tag{11.20}$$

By definition of $\tilde{x}_N(kT)$, we can rewrite (11.20) in the form

$$a_n = \frac{1}{N} \sum_{k=-\infty}^{\infty} x(kT) \exp\left(\frac{-j2\pi kn}{N}\right), \qquad n = 0, 1, \ldots, N - 1. \tag{11.21}$$

Now let

$$X(\omega) = \sum_{k=-\infty}^{\infty} x(kT)e^{-jk\omega}, \tag{11.22}$$

where ω is a continuous (nondiscrete) variable. Then we can rewrite (11.21) in the form

$$a_n = \frac{1}{N} X\left(\frac{2\pi n}{N}\right), \qquad n = 0, 1, \ldots, N - 1. \tag{11.23}$$

Inserting the expression (11.23) for a_n into (11.19) gives

$$\tilde{x}_N(kT) = \sum_{n=0}^{N-1} \frac{1}{N} X\left(\frac{2\pi n}{N}\right) \exp\left(\frac{j2\pi nk}{N}\right)$$

$$\tilde{x}_N(kT) = \frac{1}{2\pi} \sum_{n=0}^{N-1} X\left(\frac{2\pi n}{N}\right) \frac{2\pi}{N} \exp\left(\frac{j2\pi nk}{N}\right). \tag{11.24}$$

Letting $\omega_0 = 2\pi/N$, we can express (11.24) in the form

$$\tilde{x}_N(kT) = \frac{1}{2\pi} \sum_{n\omega_0=0}^{2\pi(1-1/N)} X(n\omega_0)\, e^{jkn\omega_0}\, \omega_0. \tag{11.25}$$

Now as $\dot{N} \to \infty$, we have that $n\omega_0 \to \omega$, $\omega_0 \to d\omega$ and the summation on the right side of (11.25) converges to the integral

$$\int_0^{2\pi} X(\omega)e^{jk\omega}\, d\omega.$$

Since $\tilde{x}_N(kT)$ converges to $x(kT)$ as $N \to \infty$, we then have that

$$x(kT) = \frac{1}{2\pi} \int_0^{2\pi} X(\omega)e^{jk\omega}\, d\omega. \tag{11.26}$$

Equation (11.26) is the Fourier integral representation of the discrete-time signal $x(kT)$ and the function $X(\omega)$ defined by (11.22) is the discrete-time Fourier transform (DTFT) of $x(kT)$. The right side of (11.26) is the inverse Fourier transform of $X(\omega)$.

From (11.26) we see that the Fourier integral representation of a nonperiodic discrete-time signal consists of a continuum of frequencies in general. In analogy with the continuous-time case, the DTFT $X(\omega)$ of $x(kT)$ is called the frequency spectrum of $x(kT)$. The amplitude spectrum of $x(kT)$ is the function $|X(\omega)|$ and the phase spectrum of $x(kT)$ is the function $\sphericalangle X(\omega)$.

As in the Fourier transform theory of continuous-time signals, we shall use the transform pair notation

$$x(kT) \leftrightarrow X(\omega)$$

to denote the fact that $X(\omega)$ is the DTFT of $x(kT)$, and conversely, that $x(kT)$ is the inverse DTFT of $X(\omega)$.

DTFT AND INVERSE DTFT IN THE GENERAL CASE The definition of the DTFT given by (11.22) can be extended to signals that may be of infinite duration in time. In particular, let $x(kT)$ be a discrete-time signal that may be nonzero for all k, $k = 0, \pm 1, \pm 2, \ldots$. The DTFT of $x(kT)$ is the complex-valued function of the continuous variable ω given by

$$X(\omega) = \sum_{k=-\infty}^{\infty} x(kT)e^{-jk\omega}. \tag{11.27}$$

The signal $x(kT)$ is said to have a DTFT in the ordinary sense if the bi-infinite sum in (11.27) converges for all real values of ω. A sufficient condition for $x(kT)$ to have a DTFT in the ordinary sense is that $x(kT)$ be absolutely summable; that is,

$$\sum_{k=-\infty}^{\infty} |x(kT)| < \infty. \tag{11.28}$$

Since any signal $x(kT)$ that is of finite duration in time satisfies (11.28), any such signal has a DTFT in the ordinary sense.

The DTFT $X(\omega)$ of a signal $x(kT)$ is a periodic function of ω with period 2π; that is,

$$X(\omega + 2\pi) = X(\omega) \qquad \text{for all } \omega, \ -\infty < \omega < \infty.$$

To prove the periodicity property, note that

$$X(\omega + 2\pi) = \sum_{k=-\infty}^{\infty} x(kT) \, e^{-jk(\omega + 2\pi)}$$

$$= \sum_{k=-\infty}^{\infty} x(kT)e^{-jk\omega} \, e^{-jk2\pi}.$$

But

$$e^{-jk2\pi} = 1 \qquad \text{for all integers } k,$$

and thus

$$X(\omega + 2\pi) = X(\omega) \qquad \text{for all } \omega.$$

An important consequence of periodicity of $X(\omega)$ is that the spectrum $X(\omega)$ is completely determined by computing $X(\omega)$ over any 2π interval such as $0 \le \omega < 2\pi$ or $-\pi \le \omega < \pi$. Since $X(\omega)$ is periodic with period 2π, both the amplitude spectrum $|X(\omega)|$ and the phase spectrum $\sphericalangle X(\omega)$ are periodic with period 2π.

Given a signal $x(kT)$ with DTFT $X(\omega)$, we can recompute $x(kT)$ from $X(\omega)$ by applying the inverse DTFT to $X(\omega)$. The inverse DTFT is defined by

$$x(kT) = \frac{1}{2\pi} \int_{0}^{2\pi} X(\omega)e^{jk\omega} \, d\omega. \tag{11.29}$$

Since $X(\omega)$ and $e^{jk\omega}$ are both periodic functions of ω with period 2π, the product $X(\omega)e^{jk\omega}$ is also a periodic function of ω with period 2π. As a result, the integral in (11.29) can be evaluated over any interval of length 2π. For example, we have

$$x(kT) = \frac{1}{2\pi} \int_{-\pi}^{\pi} X(\omega)e^{jk\omega}\, d\omega. \qquad (11.30)$$

GENERALIZED DTFT   As in the continuous-time Fourier transform, there are discrete-time signals that do not have a DTFT in the ordinary sense, but do have a generalized DTFT. One such signal is given in the following example.

EXAMPLE 11.3.   Consider the dc signal $x(kT) = 1$ for all integers k. Since

$$\sum_{k=-\infty}^{\infty} x(kT) = \infty,$$

this signal does not have a DTFT in the ordinary sense. The dc signal does have a generalized DTFT that is defined to be the impulse train

$$X(\omega) = \sum_{n=-\infty}^{\infty} 2\pi\delta(\omega - 2\pi n).$$

This transform is displayed in Figure 11.2.

The justification for taking the transform in Figure 11.2 to be the generalized DTFT of the dc signal follows from the property that the inverse DTFT of $X(\omega)$ is equal to the dc signal. To see this, by (11.30) we have

$$x(kT) = \frac{1}{2\pi} \int_{-\pi}^{\pi} X(\omega)e^{jk\omega}\, d\omega$$

$$= \frac{1}{2\pi} \int_{-\pi}^{\pi} 2\pi\delta(\omega)e^{0}\, d\omega = \int_{-\pi}^{\pi} \delta(\omega)\, d\omega$$

$$= 1.$$

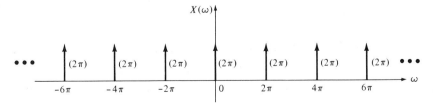

FIGURE 11.2.   DTFT of the discrete-time dc signal.

To simplify the terminology, from here on we shall refer to the ordinary or generalized DTFT as the "DTFT."

RECTANGULAR AND POLAR FORM OF THE DTFT Given the discrete-time signal $x(kT)$ with DTFT $X(\omega)$, by Euler's formula we can write $X(\omega)$ in the form

$$X(\omega) = R(\omega) + jI(\omega), \qquad (11.31)$$

where $R(\omega)$ and $I(\omega)$ are real-valued functions of ω given by

$$R(\omega) = \sum_{k=-\infty}^{\infty} x(kT) \cos k\omega$$

$$I(\omega) = -\sum_{k=-\infty}^{\infty} x(kT) \sin k\omega.$$

The function $R(\omega)$ is the real part of $X(\omega)$ and the function $I(\omega)$ is the imaginary part of $X(\omega)$. The expression (11.31) is the rectangular form of $X(\omega)$.

The DTFT $X(\omega)$ can also be written in the polar form

$$X(\omega) = |X(\omega)| \exp [j \sphericalangle X(\omega)], \qquad (11.32)$$

where $|X(\omega)|$ is the magnitude of $X(\omega)$ and $\sphericalangle X(\omega)$ is the angle of $X(\omega)$. We can go from the rectangular form to the polar form by using the relationships

$$|X(\omega)| = \sqrt{R^2(\omega) + I^2(\omega)}$$

$$\sphericalangle X(\omega) = \begin{cases} \tan^{-1} \dfrac{I(\omega)}{R(\omega)} & \text{when } R(\omega) > 0 \\[3mm] 180° + \tan^{-1} \dfrac{I(\omega)}{R(\omega)} & \text{when } R(\omega) < 0. \end{cases}$$

From (11.27) we have that $X(-\omega) = \overline{X(\omega)} =$ complex conjugate of $X(\omega)$. Then taking the complex conjugate of the polar form (11.32), we have that

$$X(-\omega) = |X(\omega)| \exp [-j\sphericalangle X(\omega)].$$

Hence $|X(-\omega)| = |X(\omega)|$ and $\sphericalangle X(-\omega) = -\sphericalangle X(\omega)$, which shows that $|X(\omega)|$ is an even function of ω and $\sphericalangle X(\omega)$ is an odd function of ω.

RELATIONSHIP BETWEEN THE DTFT AND THE z-TRANSFORM
As first defined in Chapter 7, the two-sided z-transform $X(z)$ of the discrete-time signal $x(kT)$ is given by

$$X(z) = \sum_{k=-\infty}^{\infty} x(kT)z^{-k}. \qquad (11.33)$$

If the signal $x(kT)$ is zero for all negative integers k, the two-sided z-transform is identical to the one-sided z-transform that we studied in Chapter 7.

If the bi-infinite sum in (11.33) converges when $z = e^{j\omega}$, then

$$X(\omega) = X(z)|_{z = e^{j\omega}}$$

exists, and is given by

$$X(\omega) = \sum_{k = -\infty}^{\infty} x(kT)e^{-jk\omega}. \qquad (11.34)$$

Comparing (11.34) and (11.27), we see that $X(\omega)$ is the DTFT of $x(kT)$. Thus when the two-sided z-transform $X(z)$ exists for $z = e^{j\omega}$, the signal has a DTFT in the ordinary sense, and the DTFT is equal to $X(z)$ with $z = e^{j\omega}$.

Now suppose that $x(kT) = 0$ for $k = -1, -2, \ldots$ and the z-transform $X(z)$ is a proper rational function of z. From the stability analysis in Chapter 8, it follows that $X(z)$ exists for $z = e^{j\omega}$ if all the poles of $X(z)$ are inside the open unit-disc of the complex plane. Hence in this case the DTFT exists and is equal to $X(z)$ with $z = e^{j\omega}$.

EXAMPLE 11.4. Let $x(kT) = a^k u(kT)$, where a is a nonzero real or complex number. From Table 7.3 the z-transform of this signal is equal to

$$\frac{z}{z - a}.$$

If $|a| < 1$, the only pole of this transform is inside the unit-disc, and thus in this case the signal has a DTFT given by

$$X(\omega) = \frac{z}{z - a}\bigg|_{z = e^{j\omega}} = \frac{e^{j\omega}}{e^{j\omega} - a}.$$

If $|a| > 1$, the given signal does not have a DTFT in the ordinary sense.

SIGNALS WITH EVEN OR ODD SYMMETRY Suppose that $x(kT)$ is a real-valued signal. If $x(kT)$ is an even signal, that is, $x(-kT) = x(kT)$ for all integers $k \geq 1$, the expression for the DTFT reduces to

$$X(\omega) = x(0) + 2 \sum_{k=1}^{\infty} x(kT) \cos k\omega. \qquad (11.35)$$

If $x(kT)$ is an odd signal $[x(-kT) = -x(kT)$ for all integers $k \geq 1]$, the DTFT is

$$X(\omega) = x(0) - j2 \sum_{k=1}^{\infty} x(kT) \sin k\omega. \qquad (11.36)$$

From (11.35) and (11.36), we see that the DTFT is a real-valued function of ω when $x(kT)$ is an even signal and the DTFT is a purely imaginary-valued function of ω when $x(kT)$ is an odd signal and $x(0) = 0$.

EXAMPLE 11.5. Given a fixed positive integer N, let $p_{2N}(k)$ denote the discrete-time rectangular pulse function defined by

$$p_{2N}(k) = \begin{cases} 1, & k = -N, -N + 1, \ldots, -1, 0, 1, \ldots, N \\ 0, & \text{all other } k. \end{cases}$$

This signal is even and thus by (11.35) the DTFT is

$$X(\omega) = 1 + 2 \sum_{k=1}^{N} \cos k\omega.$$

We can express $X(\omega)$ in a closed form as follows. Let $v(k)$ denote the signal defined by

$$v(k) = 2 \sum_{i=0}^{k} \cos i\omega - 1, \qquad k = 0, 1, \ldots .$$

By definition of $v(k)$,

$$v(N) = X(\omega). \tag{11.37}$$

We shall first express $v(k)$ in closed form by using the z-transform. Then applying (11.37), we obtain a closed-form expression for $X(\omega)$. Using the summation property of the z-transform and the transform pairs in Table 7.3, we have that the z-transform of $v(k)$ is

$$\begin{aligned} V(z) &= \frac{2z}{z - 1} \frac{z^2 - (\cos \omega)z}{z^2 - (2 \cos \omega)z + 1} - \frac{z}{z - 1} \\ &= \frac{z^2 + z}{z^2 - (2 \cos \omega) z + 1}. \end{aligned}$$

From equation (7.70) we have that the inverse z-transform of $V(z)$ is

$$v(k) = \cos k\omega + \frac{1 + \cos \omega}{\sin \omega} \sin k\omega.$$

Using the trigonometric identity

$$\frac{1 + \cos \omega}{\sin \omega} = \frac{\cos (\omega/2)}{\sin (\omega/2)},$$

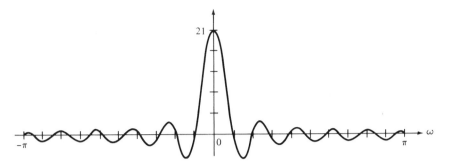

FIGURE 11.3. Fourier transform of rectangular pulse function $p_{20}(k)$.

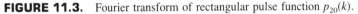

we have

$$v(k) = \cos k\omega + \frac{\cos (\omega/2)}{\sin (\omega/2)} \sin k\omega$$

$$= \frac{\sin (\omega/2) \cos k\omega + \cos (\omega/2) \sin k\omega}{\sin (\omega/2)}$$

$$= \frac{\sin [(k + \frac{1}{2})\omega]}{\sin (\omega/2)}.$$

Finally, by (11.37),

$$X(\omega) = \frac{\sin [(N + \frac{1}{2})\omega]}{\sin (\omega/2)}.$$

Thus we have the basic transform pair

$$p_{2N}(k) \leftrightarrow \frac{\sin [(N + \frac{1}{2})\omega]}{\sin (\omega/2)}. \tag{11.38}$$

As N is increased, the plot of $X(\omega)$ looks more and more like a Sa function. For example, $X(\omega)$ is plotted in Figure 11.3 in the case $N = 10$. The transform pair (11.38) is the discrete-time counterpart to the rectangle-Sa pair in the continuous-time case.

11.3

PROPERTIES OF THE DTFT

The DTFT has several properties, most of which are discrete-time versions of the properties of the continuous-time Fourier transform (CTFT). The properties of the DTFT are listed in Table 11.1. Except for the last property in Table 11.1, the proofs of these properties closely resemble the proofs of the corresponding properties of the CTFT. We omit the details.

TABLE 11.1. Properties of the DTFT

Property	Transform Pair/Property
Linearity	$ax(kT) + bv(kT) \leftrightarrow aX(\omega) + bV(\omega)$
Right or left shift in time	$x(kT - qT) \leftrightarrow X(\omega)e^{-jq\omega}$, q any integer
Time scaling	$x(akT) \leftrightarrow \dfrac{1}{a} X\left(\dfrac{\omega}{a}\right)$, $a > 0$
Time reversal	$x(-kT) \leftrightarrow X(-\omega) = \overline{X(\omega)}$
Multiplication by k	$kx(kT) \leftrightarrow j\dfrac{d}{d\omega} X(\omega)$
Multiplication by a complex exponential	$x(kT) \exp(jkT\omega_0) \leftrightarrow X(\omega - \omega_0 T)$, ω_0 real
Multiplication by $\sin \omega_0 kT$	$x(kT) \sin \omega_0 kT \leftrightarrow \dfrac{j}{2}[X(\omega + \omega_0 T) - X(\omega - \omega_0 T)]$
Multiplication by $\cos \omega_0 kT$	$x(kT) \cos \omega_0 kT \leftrightarrow \frac{1}{2}[X(\omega + \omega_0 T) + X(\omega - \omega_0 T)]$
Convolution in the time domain	$(x * v)(kT) \leftrightarrow X(\omega)V(\omega)$
Summation	$\displaystyle\sum_{i=0}^{k} x(iT) \leftrightarrow \dfrac{1}{1 - e^{-j\omega}} X(\omega) + \sum_{n=-\infty}^{\infty} \pi X(2\pi n)\delta(\omega - 2\pi n)$
Multiplication in the time domain	$x(kT)v(kT) \leftrightarrow \dfrac{1}{2\pi}\displaystyle\int_{-\pi}^{\pi} X(\omega - \lambda)V(\lambda)\, d\lambda$
Parseval's theorem	$\displaystyle\sum_{k=-\infty}^{\infty} x(kT)v(kT) = \dfrac{1}{2\pi}\int_{-\pi}^{\pi} \overline{X(\omega)}V(\omega)\, d\omega$
Special case of Parseval's theorem	$\displaystyle\sum_{k=-\infty}^{\infty} x^2(kT) = \dfrac{1}{2\pi}\int_{-\pi}^{\pi} |X(\omega)|^2\, d\omega$
Relationship to inverse CTFT	If $x(kT) \leftrightarrow X(\omega)$ and $\gamma(t) \leftrightarrow X(\omega)p_{2\pi}(\omega)$, then $x(kT) = \gamma(k)$

We should note that in contrast to the CTFT, there is no duality property for the DTFT. However, there is a relationship between the inverse DTFT and the inverse CTFT. This is the last property listed in Table 11.1. We state and then prove this property below.

Suppose that we have a discrete-time signal $x(kT)$ with DTFT $X(\omega)$. Let $p_{2\pi}(\omega)$ denote the rectangular frequency function with width equal to 2π. Then the product $X(\omega)p_{2\pi}(\omega)$ is equal to $X(\omega)$ for $-\pi \leq \omega < \pi$ and is equal to zero for all other values of ω. Let $\gamma(t)$ denote the inverse CTFT of $X(\omega)p_{2\pi}(\omega)$. Then the last property in Table 11.1 states that $x(kT) = \gamma(k)$. To prove this, first observe that by definition of the inverse CTFT,

$$\gamma(t) = \frac{1}{2\pi}\int_{-\infty}^{\infty} X(\omega)p_{2\pi}(\omega)e^{j\omega t}\, d\omega. \tag{11.39}$$

By definition of $X(\omega)p_{2\pi}(\omega)$, (11.39) reduces to

$$\gamma(t) = \frac{1}{2\pi}\int_{-\pi}^{\pi} X(\omega)e^{j\omega t}\, d\omega. \tag{11.40}$$

Replacing t by k in (11.40) gives

$$\gamma(k) = \frac{1}{2\pi} \int_{-\pi}^{\pi} X(\omega)e^{j\omega k}\, d\omega. \tag{11.41}$$

The right side of (11.41) is equal to the inverse DTFT of $X(\omega)$, and thus $\gamma(k)$ is equal to $x(kT)$.

The relationship between the inverse CTFT and inverse DTFT can be used to generate DTFT pairs from CTFT pairs, as illustrated in the following example.

EXAMPLE 11.6. Suppose that

$$X(\omega) = \sum_{n=-\infty}^{\infty} p_{2BT}(\omega + 2\pi n),$$

where $BT < \pi$. The transform $X(\omega)$ is plotted in Figure 11.4. We see that

$$X(\omega)p_{2\pi}(\omega) = p_{2BT}(\omega).$$

From Table 9.2 the inverse CTFT of $p_{2BT}(\omega)$ is equal to

$$\frac{BT}{\pi}\, \text{Sa}\,(BTt), \qquad -\infty < t < \infty.$$

Thus

$$x(kT) = \gamma(k) = \frac{BT}{\pi}\, \text{Sa}\,(BTk), \qquad k = 0,\, \pm 1,\, \pm 2,\, \ldots.$$

So we have the DTFT pair

$$\frac{BT}{\pi}\, \text{Sa}\,(BTk) \leftrightarrow \sum_{n=-\infty}^{\infty} p_{2BT}(\omega + 2\pi n). \tag{11.42}$$

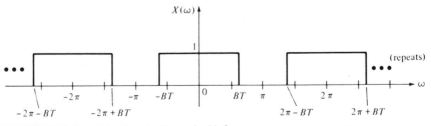

FIGURE 11.4. Transform in Example 11.6.

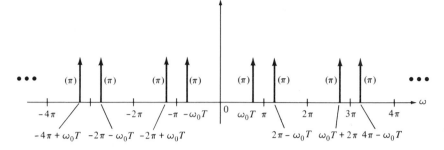

FIGURE 11.5. DTFT of $\cos \omega_0 kT$.

The properties in Table 11.1 can be used to generate many new transform pairs from a basic set of pairs. Three examples follow.

EXAMPLE 11.7. Consider the signal $x(kT) = \cos \omega_0 kT$, $k = 0, \pm 1, \pm 2, \ldots$, where $\omega_0 T < \pi$. In Section 11.2 we defined the DTFT of the dc signal $x(kT) = 1$ by

$$2\pi \sum_{n=-\infty}^{\infty} \delta(\omega - 2\pi n).$$

Using this transform and the multiplication by $\cos \omega_0 kT$ property, we obtain the transform pair

$$\cos \omega_0 kT \leftrightarrow \pi \sum_{n=-\infty}^{\infty} [\delta(\omega + \omega_0 T - 2\pi n) + \delta(\omega - \omega_0 T - 2\pi n)].$$

$$(11.43)$$

The DTFT of $\cos \omega_0 kT$ is displayed in Figure 11.5.

EXAMPLE 11.8. Now consider the discrete-time sgn function $x(kT) = \text{sgn } kT$, where

$$\text{sgn } (kT) = \begin{cases} -1, & k = -1, -2, \ldots \\ 1, & k = 0, 1, 2, \ldots \end{cases}$$

By definition of sgn (kT),

$$\text{sgn } (kT) - \text{sgn } (kT - T) = 2\Delta(kT),$$

where $\Delta(kT)$ is the unit-pulse function. The DTFT of $\Delta(kT)$ is easily seen to be the constant function 1, and thus the DTFT of sgn (kT) − sgn $(kT - T)$

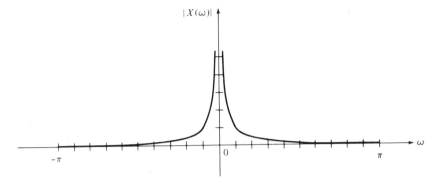

FIGURE 11.6. Magnitude of DTFT of sgn kT.

is the constant function 2. By linearity and the right-shift property, the DTFT of sgn $(kT) -$ sgn $(kT - T)$ is equal to

$$X(\omega) - X(\omega)e^{-j\omega},$$

where $X(\omega)$ is the DTFT of sgn (kT). Thus

$$X(\omega) - X(\omega)e^{-j\omega} = 2.$$

Solving for $X(\omega)$ gives

$$X(\omega) = \frac{2}{1 - e^{-j\omega}}.$$

We therefore have the transform pair

$$\text{sgn } (kT) \leftrightarrow \frac{2}{1 - e^{-j\omega}}. \tag{11.44}$$

The magnitude $|X(\omega)|$ of the transform of sgn (kT) is plotted in Figure 11.6.

EXAMPLE 11.9. Now let

$$x(kT) = \text{Sa } (Bk) \cos \omega_0 kT, \qquad k = 0, \pm 1, \pm 2, \ldots.$$

From the transform pair (11.42),

$$\text{Sa } (Bk) \leftrightarrow \frac{\pi}{B} \sum_{n=-\infty}^{\infty} p_{2B}(\omega + 2\pi n).$$

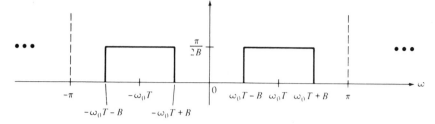

FIGURE 11.7. DTFT of Sa $(Bk) \cos \omega_0 kT$ when $2B < \omega_0 T + B < \pi$.

Then using the multiplication by $\cos \omega_0 kT$ property, we get the transform pair

$$\text{Sa }(Bk) \cos \omega_0 kT \leftrightarrow \frac{\pi}{2B} \sum_{n=-\infty}^{\infty} [p_{2B}(\omega + \omega_0 T + 2\pi n)$$
$$+ p_{2B}(\omega - \omega_0 T + 2\pi n)]. \qquad (11.45)$$

This transform is plotted in Figure 11.7 in the case when

$$2B < \omega_0 T + B < \pi.$$

TABLE 11.2. Common DTFT Pairs.

1, all $k \leftrightarrow \displaystyle\sum_{n=-\infty}^{\infty} 2\pi\delta(\omega - 2\pi n)$

$\text{sgn }(kT) \leftrightarrow \dfrac{2}{1 - e^{-j\omega}}$

$u(kT) \leftrightarrow \dfrac{1}{1 - e^{-j\omega}} + \displaystyle\sum_{n=-\infty}^{\infty} \pi\delta(\omega - 2\pi n)$

$\Delta(kT) \leftrightarrow 1$

$\Delta(kT - NT) \leftrightarrow e^{-jN\omega}$, N any positive integer

$a^k u(kT) \leftrightarrow \dfrac{e^{j\omega}}{e^{j\omega} - a}$, $|a| < 1$

$\exp(jkT\omega_0) \leftrightarrow \displaystyle\sum_{n=-\infty}^{\infty} 2\pi\delta(\omega - \omega_0 T - 2\pi n)$

$p_{2N}(k) \leftrightarrow \dfrac{\sin [(N + \frac{1}{2})\omega]}{\sin (\omega/2)}$

$\dfrac{BT}{\pi} \text{Sa }(BTk) \leftrightarrow \displaystyle\sum_{n=-\infty}^{\infty} p_{2BT}(\omega + 2\pi n)$

$\cos \omega_0 KT \leftrightarrow \displaystyle\sum_{n=-\infty}^{\infty} \pi[\delta(\omega + \omega_0 T - 2\pi n) + \delta(\omega - \omega_0 T - 2\pi n)]$

$\sin \omega_0 kT \leftrightarrow \displaystyle\sum_{n=-\infty}^{\infty} j\pi[\delta(\omega + \omega_0 T - 2\pi n) - \delta(\omega - \omega_0 T - 2\pi n)]$

$\cos (\omega_0 KT + \theta) \leftrightarrow \displaystyle\sum_{n=-\infty}^{\infty} \pi[e^{-j\theta}\delta(\omega + \omega_0 T - 2\pi n) + e^{j\theta}\delta(\omega - \omega_0 T - 2\pi n)]$

For the convenience of the reader, a list of common DTFT pairs is given in Table 11.2. This table includes the transform pairs that were generated in this section and Section 11.2.

11.4

SYSTEM ANALYSIS VIA THE DTFT

In this section we develop the DTFT-domain representation of a linear time-invariant discrete-time system. This mathematical representation is the discrete-time counterpart of the Fourier transform representation of a linear time-invariant continuous-time system.

Consider a linear time-invariant discrete-time system with unit-pulse response $h(kT)$. We do not assume that the system is stable or causal. By the results in Chapter 4, the output response $y(kT)$ resulting from the input $x(kT)$ is given by

$$y(kT) = \sum_{i=-\infty}^{\infty} h(iT)x(kT - iT). \tag{11.46}$$

Let $H(\omega)$ denote the DTFT of the unit-pulse response $h(kT)$. Then taking the DTFT of both sides of (11.46), we obtain

$$Y(\omega) = H(\omega)X(\omega), \tag{11.47}$$

where $Y(\omega)$ is the DTFT of the output $y(kT)$ and $X(\omega)$ is the DTFT of the input $x(kT)$. Equation (11.47) is the DTFT-domain (or ω-domain) representation of the given discrete-time system.

The function $H(\omega)$ in (11.47) is identical to the frequency function that we defined in Section 11.1. Thus the DTFT of the unit-pulse response $h(kT)$ is equal to the frequency function of the system.

Taking the magnitude and angle of both sides of (11.47), we get

$$|Y(\omega)| = |H(\omega)||X(\omega)| \tag{11.48}$$

$$\sphericalangle Y(\omega) = \sphericalangle H(\omega) + \sphericalangle X(\omega). \tag{11.49}$$

By (11.48) the amplitude spectrum $|Y(\omega)|$ of the output is the product of the amplitude spectrum $|X(\omega)|$ of the input and the system's magnitude function $|H(\omega)|$. By (11.49) the phase spectrum $\sphericalangle Y(\omega)$ of the output is the sum of the phase spectrum $\sphericalangle X(\omega)$ of the input and the system's phase function $\sphericalangle H(\omega)$.

ANALYSIS OF AN IDEAL LOWPASS FILTER As an illustration of the use of the DTFT representation, we shall apply it to the study of an ideal lowpass filter.

Consider the discrete-time system with the frequency function

$$H(\omega) = \sum_{n=-\infty}^{\infty} p_{2BT}(\omega + 2\pi n), \tag{11.50}$$

where $BT < \pi$. The function $H(\omega)$ is plotted in Figure 11.4. We shall first compute the output response $y(kT)$ resulting from the input

$$x(kT) = A \cos \omega_0 kT, \qquad k = 0, \pm 1, \pm 2, \ldots$$

From Table 11.2 we have the transform pair

$$A \cos \omega_0 kT \leftrightarrow \sum_{n=-\infty}^{\infty} A\pi[\delta(\omega + \omega_0 T - 2\pi n) + \delta(\omega - \omega_0 T - 2\pi n)].$$

Then, using (11.47), we have that the DTFT $Y(\omega)$ of the output response is given by

$$Y(\omega) = \begin{cases} \sum_{n=-\infty}^{\infty} A\pi[\delta(\omega + \omega_0 T - 2\pi n) + \delta(\omega - \omega_0 T - 2\pi n)], & \omega_0 T \le BT \\ 0, & BT < \omega_0 T < \pi. \end{cases}$$

Taking the inverse Fourier transform gives

$$y(kT) = \begin{cases} A \cos \omega_0 kT, & \omega_0 T \le BT \\ 0, & BT < \omega_0 T < \pi. \end{cases} \tag{11.51}$$

As a result of the periodicity of $X(\omega)$ and $H(\omega)$, the output response $y(kT)$ is equal to $A \cos \omega_0 kT$ when

$$2\pi n - BT \le \omega_0 T \le 2\pi n + BT, \qquad n = 0, 1, 2, \ldots \tag{11.52}$$

For all other positive values of $\omega_0 T$, the response $y(kT)$ is zero.

From (11.51) we see that the system passes all input sinusoids $A \cos \omega_0 kT$ with $\omega_0 \le B$, while it completely stops all such inputs with $B < \omega_0 < \pi/T$. However, as a result of periodicity of the frequency function, this discrete-time filter is not a "true" lowpass filter since it passes input sinusoids $A \cos \omega_0 kT$ with $\omega_0 T$ belonging to the intervals given by (11.52). If we restrict $\omega_0 T$ to lie in the range $0 \le \omega_0 T < \pi$, the filter can be viewed as an ideal lowpass filter with bandwidth B.

DIGITAL-FILTER REALIZATION OF AN IDEAL ANALOG LOWPASS FILTER

The filter with frequency function (11.50) can be used as a digital-filter realization (or discrete-time realization) of an ideal lowpass zero-phase analog filter with bandwidth B. To see this, suppose that the input

$$x(t) = A \cos \omega_0 t, \qquad -\infty < t < \infty$$

is applied to an analog filter with the frequency function $p_{2B}(\omega)$. From the results in Chapter 10, we know that the output of the filter is equal to $x(t)$ when $\omega_0 \le B$ and is equal to zero when $\omega_0 > B$.

Now suppose that the sampled version

$$x(kT) = A \cos \omega_0 kT$$

of the analog-filter input is applied to the discrete-time filter with frequency function (11.50). Then by (11.51), the output is equal to $x(kT)$ when $\omega_0 T \leq BT$ and is equal to zero when $BT < \omega_0 T < \pi$. Thus as long as $\omega_0 T < \pi$ or $\omega_0 < \pi/T$, the output $y(kT)$ of the discrete-time filter will be equal to a sampled version of the output of the analog filter. We can then generate an analog signal from the sampled output using a hold circuit as discussed in Chapter 12. Hence we have a digital-filter realization of the given analog filter.

The requirement that the frequency ω_0 of the input sinusoid $A \cos \omega_0 t$ be less than π/T is not an insurmountable constraint since we can increase π/T by decreasing the sampling interval T, which is equivalent to increasing the sampling frequency $\omega_s = 2\pi/T$ (see Section 10.4). Therefore, as long as we can sample suitably fast, the upper bound π/T on the input frequency is not a problem.

UNIT-PULSE RESPONSE OF IDEAL LOWPASS FILTER From the transform pairs in Table 11.2 we have that the unit-pulse response $h(kT)$ of the filter with the frequency function (11.50) is given by

$$h(kT) = \frac{BT}{\pi} \text{ Sa } (kBT), \qquad k = 0, \pm 1, \pm 2, \dots . \qquad (11.53)$$

The unit-pulse response is displayed in Figure 11.8. Note that the Sa function form of the unit-pulse response is very similar to the form of the impulse response of an ideal analog lowpass filter (see Section 10.2).

From Figure 11.8 we see that $h(kT)$ is not zero for $k < 0$, and thus the filter is noncausal. Therefore, the filter cannot be implemented on-line (in real time); but it can be implemented off-line. By an off-line implementation, we mean that the filtering process is applied to the values of signals that have been stored

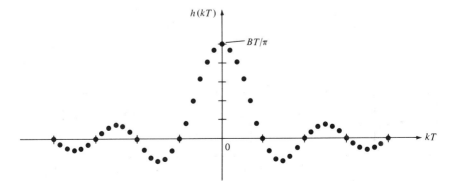

FIGURE 11.8. Unit-pulse response of ideal lowpass discrete-time filter.

in the memory of a digital computer or stored by some other means. So in the discrete-time case, ideal filters can be used in practice, as long as the filtering process is carried out off-line.

RESPONSE TO NONSINUSOIDAL INPUTS   We want to illustrate the use of the DTFT representation in computing the output response resulting from a nonsinusoidal input. We continue to assume that the given system is the discrete-time filter with the frequency function (11.50).

Suppose that the input applied to the filter is

$$x(kT) = \text{Sa}\left(\frac{\pi k}{2}\right), \qquad k = 0, \pm 1, \pm 2, \ldots .$$

From Table 11.2 the DTFT of $x(kT)$ is

$$X(\omega) = 2 \sum_{n=-\infty}^{\infty} P_{\pi}(\omega + 2\pi n).$$

The transform $X(\omega)$ is shown in Figure 11.9.

Now the DTFT of the response to the input $x(kT) = \text{Sa}\,(\pi k/2)$ is

$$Y(\omega) = H(\omega)X(\omega) = \left[\sum_{n=-\infty}^{\infty} P_{2BT}(\omega + 2\pi n)\right]\left[2 \sum_{n=-\infty}^{\infty} P_{\pi}(\omega + 2\pi n)\right].$$

If $BT > \pi/2$, then $Y(\omega) = X(\omega)$, and thus the response is

$$y(kT) = x(kT) = \text{Sa}\left(\frac{\pi k}{2}\right), \qquad k = 0, \pm 1, \pm 2, \ldots .$$

This result shows that if the filter bandwidth is wide enough to pass all the frequency components of the input over the interval $-\pi < \omega < \pi$, then the output of the filter is equal to the input.

If $BT < \pi/2$,

$$Y(\omega) = 2H(\omega),$$

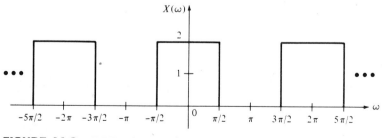

FIGURE 11.9.   DTFT of Sa $(\pi k/2)$.

and thus

$$y(kT) = 2h(kT) = \frac{2BT}{\pi} \text{ Sa } (kBT), \qquad k = 0, \pm 1, \pm 2, \ldots$$

In this case the filter does not pass all the frequency components of the input, so the output is a distorted version of the input.

Finally, let us consider the input

$$x(kT) = \text{Sa}\left(\frac{\pi k}{2}\right) \cos\left(\frac{\pi k}{4}\right), \qquad k = 0, \pm 1, \pm 2, \ldots$$

Using the multiplication-by-cos $\omega_0 kT$ property, we find that the DTFT of this input is

$$X(\omega) = \sum_{n=-\infty}^{\infty} \left[p_\pi\left(\omega - \frac{\pi}{4} + 2\pi n\right) + p_\pi\left(\omega + \frac{\pi}{4} + 2\pi n\right) \right].$$

Over the interval $-\pi < \omega < \pi$, $X(\omega)$ is equal to

$$p_\pi\left(\omega - \frac{\pi}{4}\right) + p_\pi\left(\omega + \frac{\pi}{4}\right),$$

which can be rewritten in the form

$$p_{\pi/2}(\omega) + p_{3\pi/2}(\omega).$$

Since $X(\omega)$ is periodic with period 2π, we have that

$$X(\omega) = \sum_{n=-\infty}^{\infty} [p_{\pi/2}(\omega + 2\pi n) + p_{3\pi/2}(\omega + 2\pi n)].$$

The transform $X(\omega)$ is plotted in Figure 11.10.

If $BT > 3\pi/4$, the filter again passes all the frequency components of the input, and thus $y(kT) = x(kT)$. If $BT < 3\pi/4$, the filter will distort the input.

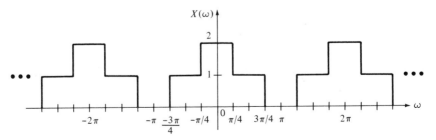

FIGURE 11.10.   DTFT of Sa $(\pi k/2)$ cos $(\pi k/4)$.

For example, if $BT = \pi/4$,

$$Y(\omega) = 2 \sum_{n=-\infty}^{\infty} P_{\pi/2}(\omega + 2\pi n).$$

Taking the inverse DTFT of $Y(\omega)$, we have that

$$y(kT) = \frac{1}{2} \operatorname{Sa}\left(\frac{\pi k}{4}\right), \qquad k = 0, \pm 1, \pm 2, \ldots.$$

CAUSAL LOWPASS FILTER As noted above, an ideal lowpass filter cannot be implemented in real time since the filter is noncausal. For "real-time filtering" we need to consider a causal lowpass filter. One very simple example is the averager, which is defined by the input/output difference equation

$$y(kT) = \tfrac{1}{2}[x(kT) + x(kT - T)] \tag{11.54}$$

or

$$y(kT + T) = \tfrac{1}{2}[x(kT + T) + x(kT)].$$

Since the system (11.54) averages the past two values of the input, the time variation of the input should be smoothed out to some extent in the output; in other words, the system should behave like a lowpass filter. To verify this, we will determine the magnitude function $|H(\omega)|$ of the system. First, from (11.18) we have

$$H(\omega) = \tfrac{1}{2}[1 + e^{-j\omega}] = \tfrac{1}{2}[(1 + \cos \omega) - j \sin \omega].$$

Taking the magnitude gives

$$|H(\omega)| = \frac{1}{\sqrt{2}} \sqrt{1 + \cos \omega}.$$

The magnitude function $|H(\omega)|$ is plotted in Figure 11.11. We see that the

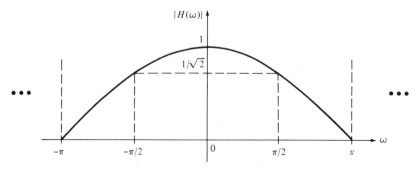

FIGURE 11.11. Magnitude function of the averager.

averager is a (nonideal) lowpass filter with $|H(\omega)|$ down by 3 dB when $\omega = \pi/2$. If the input $x(kT)$ to the averager is the sampled sinusoid

$$x(kT) = A \cos \omega_0 kT, \qquad k = 0, \pm 1, \pm 2, \ldots,$$

the averager will pass the sinusoid if $\omega_0 T < \pi/2$ or $\omega_0 < \pi/2T$; otherwise, the averager will attenuate the sinusoid. Thus we can set the effective bandwidth of the filter by choosing the sampling interval T.

In this example the separation between the passband and the stopband is not very sharp. A filter with a sharper frequency cutoff is the filter with the input/output difference equation

$$y(kT) = \frac{1}{N} \sum_{i=0}^{N-1} x(kT - iT), \qquad (11.55)$$

where $N \geq 3$. The filter (11.55) is called the *mean filter*. The mean filter is an example of a causal lowpass discrete-time filter.

11.5

DISCRETE FOURIER TRANSFORM

Let $x(kT)$ be a discrete-time signal with DTFT $X(\omega)$. Since $X(\omega)$ is a function of the continuous variable ω, we cannot place $X(\omega)$ into the memory of a digital computer unless $X(\omega)$ can be expressed in a closed form. To implement DTFT techniques on a digital computer, we need to discretize in frequency. This leads to the concept of the discrete Fourier transform, which we define below.

Suppose that the discrete-time signal $x(kT)$ is zero for all integers $k < 0$ and all integers $k \geq N$, where N is a fixed positive integer. The integer N could be very large, for example $N = 2^{10} = 1024$.

The discrete Fourier transform (DFT) X_n of $x(kT)$ is defined by

$$X_n = \sum_{k=0}^{N-1} x(kT) \exp\left(\frac{-j2\pi kn}{N}\right), \qquad n = 0, 1, \ldots, N - 1. \quad (11.56)$$

From (11.56) we see that the DFT X_n is a function of the discrete (integer) variable n. Also note that

$$X_n = X(\omega)|_{\omega = 2\pi n/N} = X\left(\frac{2\pi n}{N}\right), \qquad (11.57)$$

where $X(\omega)$ is the DTFT of $x(kT)$ defined by (11.27). A program for computing the DFT is given in Figure A.12.

The inverse DFT is defined by

$$x(kT) = \frac{1}{N} \sum_{n=0}^{N-1} X_n \exp\left(\frac{j2\pi kn}{N}\right), \qquad k = 0, 1, 2, \ldots, N - 1. \quad (11.58)$$

Note that the right side of (11.56) is equal to N times the right side of the expression (11.3) for the coefficients of a periodic discrete-time signal with period NT. In addition, the right side of (11.58) is equal to $1/N$ times the Fourier series representation (11.2) of a periodic discrete-time signal. We can establish a precise interconnection between the DFT and the Fourier series as follows.

Given the discrete-time signal $x(kT)$ as defined above, let $\tilde{x}_N(kT)$ denote the periodic signal with period NT, which is equal to $x(kT)$ for $k = 0, 1, \ldots, N - 1$. Since $\tilde{x}_N(kT)$ is periodic with period NT, it has the Fourier series representation

$$\tilde{x}_N(kT) = \sum_{n=0}^{N-1} a_n \exp\left(\frac{j2\pi kn}{N}\right), \tag{11.59}$$

where

$$a_n = \frac{1}{N} \sum_{k=0}^{N-1} \tilde{x}_N(kT) \exp\left(\frac{-j2\pi kn}{N}\right), \qquad n = 0, 1, \ldots, N - 1.$$

Since $x(kT) = \tilde{x}_N(kT)$ for $k = 0, 1, \ldots, N - 1$, from (11.59) we have that

$$x(kT) = \sum_{n=0}^{N-1} a_n \exp\left(\frac{j2\pi kn}{N}\right), \qquad k = 0, 1, \ldots, N - 1. \tag{11.60}$$

Comparing (11.60) and (11.58) gives

$$X_n = Na_n, \qquad n = 0, 1, \ldots, N - 1. \tag{11.61}$$

Thus for each value of n ranging from 0 to $N - 1$, the DFT X_n of $x(kT)$ is equal to N times the coefficient a_n of the Fourier series of $\tilde{x}_N(kT)$.

As in the theory of the discrete-time Fourier series, we shall simplify the notation by letting w_N denote the Nth root of unity given by $w_N = \exp(j2\pi/N)$. We can then write the DFT and inverse DFT in the forms

$$X_n = \sum_{k=0}^{N-1} x(kT)w_N^{-kn}, \qquad n = 0, 1, \ldots, N - 1 \tag{11.62}$$

$$x(kT) = \frac{1}{N} \sum_{n=0}^{N-1} X_n w_N^{kn}, \qquad k = 0, 1, \ldots, N - 1. \tag{11.63}$$

SYSTEM ANALYSIS VIA THE DFT Now suppose that the signal $x(kT)$ is applied to a linear time-invariant discrete-time system with unit-pulse response $h(kT)$. We continue to assume that $x(kT) = 0$ for $k < 0$ and $k \geq N$. We also assume that $h(kT) = 0$ for $k < 0$ and $k > Q$, where Q is some positive integer. The resulting output response $y(kT)$ with the system at rest at time $kT = 0$ is given by

$$y(kT) = \sum_{i=0}^{\infty} h(iT)x(kT - iT), \qquad k = 0, 1, 2, \ldots . \qquad (11.64)$$

Since $h(iT) = 0$ for $i > Q$, we have

$$y(kT) = \sum_{i=0}^{Q} h(iT)x(kT - iT), \qquad k = 0, 1, 2, \ldots . \qquad (11.65)$$

Now since

$$x(kT - iT) = 0 \qquad \text{when } k - i \geq N,$$

from (11.65) we see that

$$y(kT) = 0 \qquad \text{for all integers } k \geq N + Q.$$

So the output response is also of finite duration in time.

We now want to show that the output $y(kT)$ can be computed by using the DFT. First, we must "pad" $x(kT)$ and $h(kT)$ with zeros so that the DFT of each of these signals is evaluated with respect to $N + Q$ time points; that is, the DFTs of $x(kT)$ and $h(kT)$ are defined by

$$X_n = \sum_{k=0}^{N+Q-1} x(kT)w_{N+Q}^{-kn}, \qquad n = 0, 1, \ldots, N + Q - 1$$

$$H_n = \sum_{k=0}^{N+Q-1} h(kT)w_{N+Q}^{-kn}, \qquad n = 0, 1, \ldots, N + Q - 1.$$

We claim that the response $y(kT)$ is equal to the inverse DFT of H_nX_n; that is,

$$y(kT) = \frac{1}{N + Q} \sum_{n=0}^{N+Q-1} H_nX_nw_{N+Q}^{kn}, \qquad k = 0, 1, \ldots, N + Q - 1. \qquad (11.66)$$

The relationship (11.66) can be proved using the results derived in Section 11.4 on the DTFT-domain representation. Let $X(\omega)$, $Y(\omega)$, and $H(\omega)$ denote the DTFTs of $x(kT)$, $y(kT)$, and $h(kT)$, respectively. Then

$$Y(\omega) = H(\omega)X(\omega). \qquad (11.67)$$

Setting $\omega = 2\pi n/N$ in both sides of (11.67), we get

$$Y\left(\frac{2\pi n}{N}\right) = H\left(\frac{2\pi n}{N}\right)X\left(\frac{2\pi n}{N}\right), \qquad n = 0, 1, \ldots, N + Q - 1. \qquad (11.68)$$

Let Y_n denote the DFT of $y(kT)$ taken with respect to $N + Q$ time points. Then by the relationship (11.57) between the DTFT and DFT, (11.68) implies that

$$Y_n = H_n X_n, \qquad n = 0, 1, \ldots, N + Q - 1.$$

Therefore, $y(kT)$ is equal to the inverse DFT of $H_n X_n$, which verifies (11.66).

The expression (11.66) for $y(kT)$ can be computed using a "fast method." The issue of computation is considered in the next section.

★11.6

FFT ALGORITHM

Again let $x(kT)$ be a discrete-time signal with $x(kT) = 0$ for $k < 0$ and $k \geq N$. In Section 11.5 we defined the DFT and inverse DFT by

$$X_n = \sum_{k=0}^{N-1} x(kT) w_N^{-kn}, \qquad n = 0, 1, \ldots, N - 1 \qquad (11.69)$$

$$x(kT) = \frac{1}{N} \sum_{n=0}^{N-1} X_n w_N^{kn}, \qquad k = 0, 1, \ldots, N - 1. \qquad (11.70)$$

From (11.69) we see that for each value of n, the computation of X_n from $x(kT)$ requires N multiplications. Thus the computation of X_n for $n = 0, 1, \ldots, N - 1$ requires N^2 multiplications. Similarly, from (11.70) we have that the computation of $x(kT)$ from X_n also requires N^2 multiplications.

It should be mentioned that the multiplications in (11.69) and (11.70) are complex multiplications in general; that is, the numbers being multiplied are complex numbers. The multiplication of two complex numbers requires four real multiplications. In the following analysis, we shall count the number of complex multiplications. The number of additions required to compute the DFT or inverse DFT will not be considered.

Since the direct evaluation of (11.69) or (11.70) requires N^2 multiplications, this can result in a great deal of computation if N is large. It turns out that (11.69) or (11.70) can be computed using a fast Fourier transform (FFT) algorithm, which requires on the order of $(N \log_2 N)/2$ multiplications. This is a significant decrease in the N^2 multiplications required in the direct evaluation of (11.69) or (11.70). For instance, if $N = 1024$, the direct evaluation requires $N^2 = 1,048,576$ multiplications. In contrast, the FFT algorithm requires

$$\frac{1024(\log_2 1024)}{2} = 5120 \text{ multiplications.}$$

There are different versions of the FFT algorithm. Here we shall limit our attention to one particular approach based on decimation in time. For an in-depth treatment of the FFT algorithm, we refer the reader to Brigham (1974) or Rabiner and Gold [1975].

The basic idea of the decimation-in-time approach is to subdivide the time interval into intervals having a smaller number of points. We illustrate this by first showing that the computation of X_n can be broken up into two parts.

Let N be an even integer, so that $N/2$ is an integer. Given the signal $x(kT)$ with $x(kT) = 0$ for $k < 0$ and $k \geq N$, define the signals

$$a(kT) = x(2kT), \qquad k = 0, 1, 2, \ldots, \frac{N}{2} - 1$$

$$b(kT) = x(2kT + T), \qquad k = 0, 1, 2, \ldots, \frac{N}{2} - 1.$$

Note that the signal $a(kT)$ consists of the values of $x(kT)$ at the even values of the time index k, while $b(kT)$ consists of the values at the odd time points.

Let A_n and B_n denote the DFTs of $a(kT)$ and $b(kT)$; that is,

$$A_n = \sum_{k=0}^{(N/2)-1} a(kT)w_{N/2}^{-kn}, \qquad n = 0, 1, \ldots, \frac{N}{2} - 1$$

$$B_n = \sum_{k=0}^{(N/2)-1} b(kT)w_{N/2}^{-kn}, \qquad n = 0, 1, \ldots, \frac{N}{2} - 1.$$

Let X_n denote the DFT of $x(kT)$. Then we claim that

$$X_n = A_n + w_N^{-n}B_n, \qquad n = 0, 1, \ldots, \frac{N}{2} - 1 \qquad (11.71)$$

$$X_{N+n} = A_n - w_N^{-n}B_n, \qquad n = 0, 1, \ldots, \frac{N}{2} - 1. \qquad (11.72)$$

To verify (11.71), let us insert the expressions for A_n and B_n into the right side of (11.71). This gives

$$A_n + w_N^{-n}B_n = \sum_{k=0}^{(N/2)-1} a(kT)w_{N/2}^{-kn} + \sum_{k=0}^{(N/2)-1} b(kT)w_N^{-n}w_{N/2}^{-kn}.$$

Now $a(kT) = x(2kT)$ and $b(kT) = x(2kT + T)$, and thus

$$A_n + w_N^{-n}B_n = \sum_{k=0}^{(N/2)-1} x(2kT)w_{N/2}^{-kn} + \sum_{k=0}^{(N/2)-1} x(2kT + T)w_N^{-n}w_{N/2}^{-kn}.$$

Using the properties

$$w_{N/2}^{-kn} = w_N^{-2kn}, \qquad w_N^{-n}w_{N/2}^{-kn} = w_N^{-(1+2k)n},$$

we obtain

$$A_n + w_N^{-n} B_n = \sum_{k=0}^{(N/2)-1} x(2kT) w_N^{-2kn} + \sum_{k=0}^{(N/2)-1} x(2kT + T) w_N^{-(1+2k)n}.$$

$$(11.73)$$

Defining the change of index $\bar{k} = 2k$ in the first sum of the right side of (11.73) and the change of index $\bar{k} = 2k + 1$ in the second sum, we have

$$A_n + w_N^{-n} B_n = \sum_{\substack{\bar{k}=0 \\ \bar{k} \text{ even}}}^{N-2} x(\bar{k}T) w_N^{-\bar{k}n} + \sum_{\substack{\bar{k}=0 \\ \bar{k} \text{ odd}}}^{N-1} x(\bar{k}T) w_N^{-\bar{k}n}$$

$$A_n + w_N^{-n} B_n = \sum_{\bar{k}=0}^{N-1} x(\bar{k}T) w_N^{-\bar{k}n}$$

$$A_n + w_N^{-n} B_n = X_n.$$

Hence we have verified (11.71). The proof of (11.72) is similar and is therefore omitted.

The computation of X_n using (11.71) and (11.72) requires $N^2/2 + N/2$ multiplications. To see this, first note that the computation of A_n requires $(N/2)^2 = N^2/4$ multiplications, as does the computation of B_n. The computation of the products $w_N^{-n} B_n$ in (11.71) and (11.72) requires $N/2$ multiplications. So the total number of multiplications is equal to $N^2/2 + N/2$. This is $N^2/2 - N/2$ multiplications less than N^2 multiplications. Therefore, when N is large, the computation of X_n using (11.71) and (11.72) requires significantly fewer multiplications than the computation of X_n using (11.69).

If $N/2$ is even, we can express each of the signals $a(kT)$ and $b(kT)$ in two parts, and then we can repeat the process described above. If $N = 2^q$ for some positive integer q, the subdivision process can be continued until we obtain signals with only one nonzero value [with each value equal to one of the values of the given signal $x(kT)$].

In the case $N = 8$, a block diagram of the FFT algorithm is given in Figure 11.12. On the far left side of the diagram, we have the values of the given signal $x(kT)$. Note the order (in terms of row position) in which the signal values $x(kT)$ are applied to the process. The order can be determined by a process called *bit reversing*. Suppose that $N = 2^q$. Given an integer k ranging from 0 to $N - 1$, we represent the time index k by the q-bit binary word for the integer k. Reversing the bits comprising this word, we have that the integer corresponding to the reversed-bit word is the row at which the signal value $x(kT)$ is applied to the FFT algorithm. For example, when $N = 8$ the binary words and bit-reversed words corresponding to the time index k are shown in Table 11.3. The last column in Table 11.3 gives the order for which the signal values are applied to the FFT algorithm shown in Figure 11.12.

A computer program for the FFT algorithm is given in Figure A.13. Since the program is written in Basic, the complex multiplications and additions in the FFT algorithm are expressed in terms of real multiplications and additions.

TABLE 11.3. Bit Reversing in the Case $N = 8$.

Time Point (k)	Binary Word	Reversed-Bit Word	Order
0	000	000	$x(0)$
1	001	100	$x(4T)$
2	010	010	$x(2T)$
3	011	110	$x(6T)$
4	100	001	$x(T)$
5	101	101	$x(5T)$
6	110	011	$x(3T)$
7	111	111	$x(7T)$

As a result, the program runs slower than a program written in a language (such as Fortran) for which complex multiplications and additions are commands in the language.

★11.7

APPLICATIONS OF THE FFT ALGORITHM

The FFT algorithm is very useful in a wide range of applications involving digital signal processing and digital communications. In this section we first show that the FFT algorithm can be used to compute the Fourier transform of

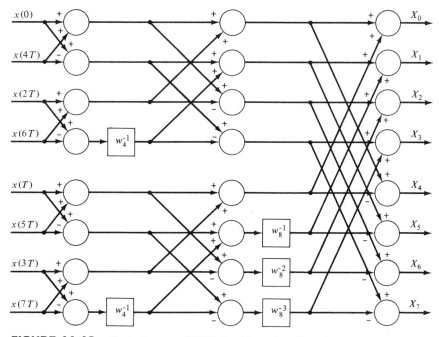

FIGURE 11.12 Block diagram of FFT algorithm when $N = 8$.

a continuous-time signal. Then we apply the FFT algorithm to the computation of the output response of a linear time-invariant system.

COMPUTATION OF THE FOURIER TRANSFORM VIA THE FFT   Let $x(t)$ be a continuous-time signal with Fourier transform $X(\omega)$. We assume that $x(t) = 0$ for all $t < 0$. Then the Fourier transform $X(\omega)$ of $x(t)$ is given by

$$X(\omega) = \int_0^\infty x(t)e^{-j\omega t}\,dt. \qquad (11.74)$$

Let Ω be a fixed positive real number and let N be a fixed positive integer. We shall show that by using the FFT algorithm, we can compute $X(\omega)$ for $\omega = n\Omega$, $n = 0, 1, 2, \ldots, N - 1$.

Given a fixed positive number T, we can rewrite the integral in (11.74) in the form

$$X(\omega) = \sum_{k=0}^{\infty} \int_{kT}^{kT+T} x(t)e^{-j\omega t}\,dt. \qquad (11.75)$$

Suppose that we have chosen T small enough so that the variation in $x(t)$ is small over each T-second interval $kT \leq t < kT + T$. Then we can approximate the sum in (11.75) by

$$X(\omega) = \sum_{k=0}^{\infty} \left(\int_{kT}^{kT+T} e^{-j\omega t}\,dt \right) x(kT)$$

$$= \sum_{k=0}^{\infty} \left[\left(\frac{-1}{j\omega} e^{-j\omega t} \right)_{t=kT}^{t=kT+T} \right] x(kT)$$

$$= \frac{e^{-j\omega T} - 1}{j\omega} \sum_{k=0}^{\infty} \exp(-j\omega kT)x(kT). \qquad (11.76)$$

Now suppose that for some sufficiently large positive integer N, the magnitude $|x(kT)|$ is small for all integers $k \geq N$. Then (11.76) becomes

$$X(\omega) = \frac{e^{-j\omega T} - 1}{j\omega} \sum_{k=0}^{N-1} \exp(-j\omega kT)x(kT). \qquad (11.77)$$

Evaluating both sides of (11.77) at $\omega = 2\pi n/NT$, we obtain

$$X\left(\frac{2\pi n}{NT}\right) = \frac{\exp(-j2\pi n/N) - 1}{j2\pi n/NT} \sum_{k=0}^{N-1} \exp\left(\frac{-j2\pi nk}{N}\right) x(kT). \qquad (11.78)$$

Now let X_n denote the DFT of the signal $x(kT)$ with the number of time points equal to N. By definition of the DFT,

$$X_n = \sum_{k=0}^{N-1} x(kT) \exp\left(\frac{-j2\pi kn}{N}\right), \qquad n = 0, 1, \ldots, N - 1. \quad (11.79)$$

Comparing (11.78) and (11.79), we have that

$$X\left(\frac{2\pi n}{NT}\right) = \frac{\exp(-j2\pi n/N) - 1}{j2\pi n/NT} X_n. \quad (11.80)$$

Finally, letting $\Omega = 2\pi/NT$, we can rewrite (11.80) in the form

$$X(n\Omega) = \frac{\exp(-jn\Omega T) - 1}{jn\Omega} X_n, \qquad n = 0, 1, 2, \ldots, N - 1. \quad (11.81)$$

By first calculating X_n via the FFT algorithm and then using (11.81), we can compute $X(n\Omega)$ for $n = 0, 1, \ldots, N - 1$. A program for carrying out this computation is given in Figure A.14.

We should stress that the relationship (11.81) is an approximation, so the values of $X(\omega)$ computed using (11.81) are only approximate values. We can obtain better accuracy by taking a smaller value for the sampling interval T and/or by taking a larger value for N. If we know that the amplitude spectrum $|X(\omega)|$ is small for $\omega > B$, a good choice for T is the sampling interval π/B corresponding to the Nyquist sampling frequency $\omega_s = 2B$.

EXAMPLE 11.10. Consider the continuous-time signal shown in Figure 11.13. Let $T = 0.2$ and $N = 2^5 = 32$. Then $\Omega = 2\pi/NT = 0.9817$ and $x(0.2k) = 0$ for all $k > 10$. Computing the Fourier transform of $x(t)$ using the program in Figure A.14, we obtain the amplitude spectrum plotted in Figure 11.14. The exact amplitude spectrum is plotted in Figure 9.25.

FAST CONVOLUTION The FFT algorithm can be used to perform a fast version of convolution. Since the output is the convolution of the input and the unit-pulse response, this can be used to compute the output response as follows.

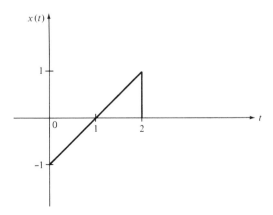

FIGURE 11.13. Continuous-time signal in Example 11.10.

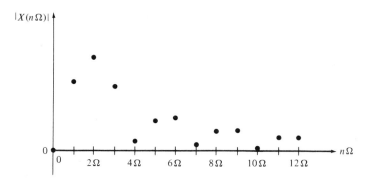

FIGURE 11.14. Amplitude spectrum of signal in Example 11.10.

Given signals $x(kT)$ and $h(kT)$, with $x(kT) = 0$ for $k \geq N$ and $h(kT) = 0$ for $k > Q$, we pad $x(kT)$ and $h(kT)$ with zeros as discussed in Section 11.5. Let X_n and H_n denote the DFTs computed using the FFT algorithm with respect to $N + Q$ points. Then the convolution $(x * h)(kT)$ is equal to the inverse DFT of the product $H_n X_n$, which also can be computed using the FFT algorithm.

This approach requires on the order of

$$1.5(N + Q) \log_2 (N + Q) + N + Q \text{ multiplications.}$$

In contrast, the computation of $y(kT)$ using the convolution sum requires on the order of

$$0.5(N + Q)^2 + 1.5(N + Q) \text{ multiplications.}$$

PROBLEMS

Chapter 11

11.1 For each of the periodic discrete-time signals shown in Figure P11.1, compute the complex-exponential form of the Fourier series.

11.2 A periodic discrete-time signal $x(kT)$ with period NT has the Fourier series representation

$$x(kT) = \sum_{n=0}^{N-1} a_n \exp \left(\frac{j2\pi nk}{N} \right), \qquad k = 0, \pm 1, \pm 2, \ldots$$

Compute the Fourier series representation of the periodic signal $v(kT)$ where:
(a) $v(kT) = x(kT - T)$.
(b) $v(kT) = x(kT) - x(kT - T)$.

(c) $v(kT) = x(kT) \exp \left(\frac{j2\pi k}{N} \right)$.

(d) $v(kT) = x(kT) \cos \left(\frac{j2\pi k}{N} \right)$.

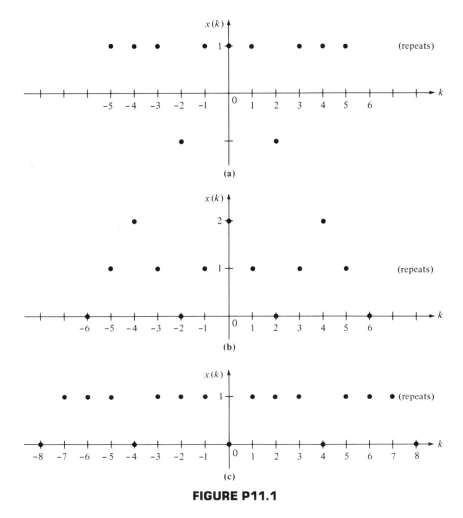

FIGURE P11.1

11.3 Compute the Fourier series for the following periodic discrete-time signals.
 (a) $x(k) = \cos(\pi k + \theta)$, where θ is an arbitrary constant.
 (b) $x(k) = \sin \pi k + \cos \pi k$.
 (c) $x(k) = \sin^2 \dfrac{\pi k}{2}$.
 (d) $x(k) = \cos \pi k \sin \dfrac{\pi k}{2}$.
 (e) $x(k) = \cos \pi k + \cos \dfrac{\pi k}{3}$.

11.4 A linear time-invariant discrete-time system has the frequency function $H(\omega)$ shown in Figure P11.4. Compute the output response $y(k)$ when the input $x(k)$ is the periodic signal:
 (a) $x(k) = \cos \dfrac{\pi k}{2}$.

(b) $x(k) = \cos \dfrac{\pi k}{4}$.

(c) $x(k) = \sin \left(\dfrac{\pi k}{4} + \theta \right)$, where θ in an arbitrary constant.

(d) $x(k) = 1 + \displaystyle\sum_{n=1}^{4} \dfrac{1}{n} \cos \left(\dfrac{2\pi k}{n} \right)$.

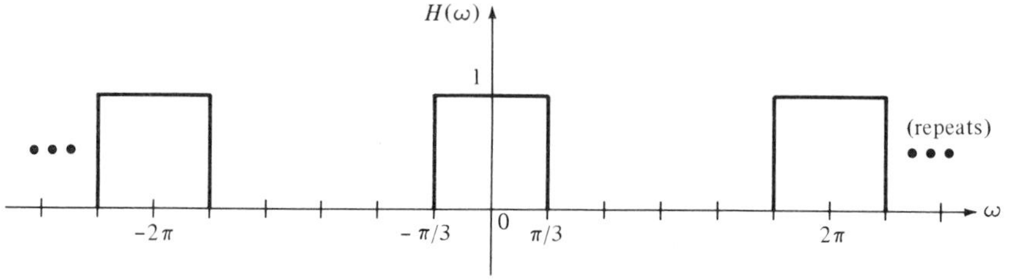

FIGURE P11.4

11.5 Consider the discrete-time system given by the input/output difference equation

$$y(k + 1) + 0.9y(k) = 1.9x(k + 1).$$

(a) Sketch the magnitude function $|H(\omega)|$ for $\omega \in [0, 2\pi]$.
(b) Compute the output response $y(k)$ resulting from the application of the periodic signal $x(k)$ shown in Figure P11.1a.
(c) For your answer in part (b), compute $y(0)$, $y(1)$, $y(2)$, and $y(3)$. Compare these values with the values of $x(0)$, $x(1)$, $x(2)$, and $x(3)$. Are the results expected? Explain.

11.6 Repeat Problem 11.5 for the system given by

$$y(k + 1) - 0.9y(k) = 0.1x(k + 1).$$

11.7 A linear time-invariant discrete-time system has frequency function

$$H(\omega) = b - ae^{j\omega c}, \qquad -\infty < \omega < \infty,$$

where a, b, and c are constants. The periodic input $x(k)$ shown in Figure P11.7a is applied to the system. Determine the constants a, b, and c so that the resulting output response $y(k)$ is as shown in Figure P11.7b.

11.8 A discrete-time signal $x(k)$ has DTFT

$$X(\omega) = \frac{1}{e^{j\omega} + b},$$

where b is an arbitrary constant. Determine the DTFT $V(\omega)$ of the following signals.

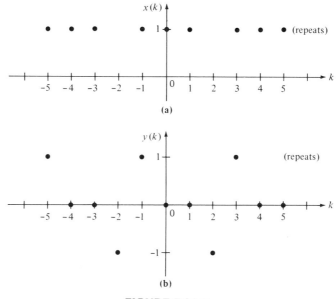

FIGURE P11.7

(a) $v(k) = x(3k - 5)$.
(b) $v(k) = x(-k)$.
(c) $v(k) = kx(k)$.
(d) $v(k) = x(k) - x(k - 1)$.
(e) $v(k) = (x * x)(k)$.
(f) $v(k) = x(k) \cos 3k$.
(g) $v(k) = x^2(k)$
(h) $v(k) = x(k)e^{j2k}$.

11.9 Compute the DTFT of the discrete-time signals shown in Figure P11.9. Express the DTFT in the simplest possible form.

11.10 Determine the inverse DTFT of the frequency functions $X(\omega)$ shown in Figure P11.10.

11.11 Determine the inverse DTFT of the following frequency functions.
(a) $X(\omega) = \sin \omega$.
(b) $X(\omega) = \cos \omega$.
(c) $X(\omega) = \cos^2 \omega$.
(d) $X(\omega) = \sin \omega \cos \omega$.

11.12 An ideal linear-phase highpass discrete-time filter has frequency function $H(\omega)$, where for one period, $H(\omega)$ is given by

$$H(\omega) = \begin{cases} e^{-j3\omega}, & \dfrac{\pi}{2} \le |\omega| \le \pi \\ 0, & 0 < |\omega| < \dfrac{\pi}{2}. \end{cases}$$

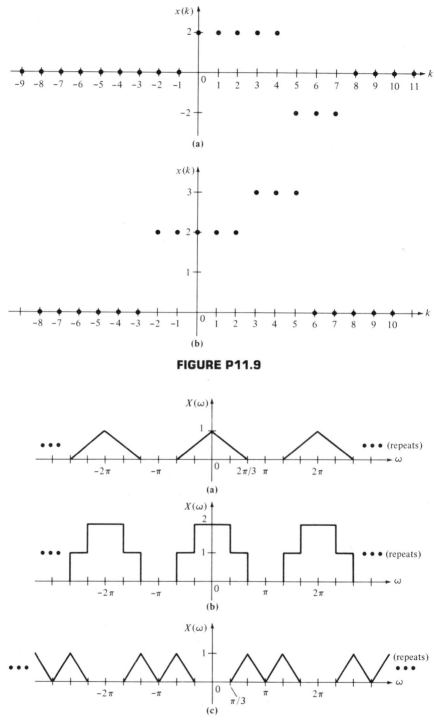

FIGURE P11.9

FIGURE P11.10

(a) Determine the unit-pulse response $h(kT)$ of the filter.

(b) Compute the output response $y(kT)$ of the filter when the input $x(kT)$ is given by

(i) $x(kT) = \cos \dfrac{\pi k}{4}, \quad k = 0, \pm 1, \pm 2, \ldots$

(ii) $x(kT) = \cos \dfrac{3\pi k}{4}, \quad k = 0, \pm 1, \pm 2, \ldots$

(iii) $x(kT) = \text{Sa}\left(\dfrac{\pi k}{2}\right), \quad k = 0, \pm 1, \pm 2, \ldots$

(iv) $x(kT) = \text{Sa}\left(\dfrac{\pi k}{4}\right), \quad k = 0, \pm 1, \pm 2, \ldots$

(v) $x(kT) = \text{Sa}\left(\dfrac{\pi k}{4}\right) \cos \dfrac{\pi k}{8}, \quad k = 0, \pm 1, \pm 2, \ldots$

(vi) $x(kT) = \text{Sa}\left(\dfrac{\pi k}{2}\right) \cos \dfrac{\pi k}{8}, \quad k = 0, \pm 1, \pm 2, \ldots$

11.13 A linear time-invariant discrete-time system has the frequency function $H(\omega)$ shown in Figure P11.13.

(a) Determine the unit-pulse response $h(kT)$ of the system.

(b) Compute the output response $y(kT)$ when the input $x(kT)$ is equal to $\Delta(kT) - \Delta(kT - T)$.

(c) Compute the output response $y(kT)$ when $x(kT) = \text{Sa}\,(\pi k/4), k = 0, \pm 1, \pm 2, \ldots$

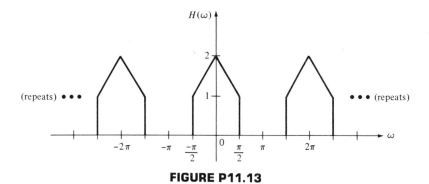

FIGURE P11.13

11.14 Consider the mean filter given by the input/output difference equation

$$y(kT) = \frac{1}{N} \sum_{i=0}^{N-1} x(kT - iT).$$

(a) Determine the unit-pulse response $h(kT)$ of the filter.

(b) Show that the frequency function $H(\omega)$ of the filter can be expressed in the form

$$H(\omega) = \begin{cases} 1, & \omega = 0 \\ \dfrac{1 - \cos N\omega + j \sin N\omega}{N(1 - \cos \omega + j \sin \omega)}, & 0 < \omega < 2\pi. \end{cases}$$

(c) Sketch $|H(\omega)|$ for $0 \leq \omega \leq 2\pi$ in the case when $N = 3$ and $N = 4$.

(d) Sketch $\not{\angle} H(\omega)$ for $0 \leq \omega \leq 2\pi$ in the case when $N = 3$ and $N = 4$.

(e) Compute the output response $y(kT)$ when $N = 3$ and

　　(i)　$x(kT) = 1, k = 0, \pm 1, \pm 2, \ldots$

　　(ii)　$x(kT) = \cos \dfrac{\pi k}{4}, k = 0, \pm 1, \pm 2, \ldots$

　　(iii)　$x(kT) = \cos \dfrac{\pi k}{2}, k = 0, \pm 1, \pm 2, \ldots$

　　(iv)　$x(kT) = \cos \dfrac{\pi k}{2} \sin \dfrac{\pi k}{4}, k = 0, \pm 1, \pm 2, \ldots$

11.15　Using the program in Figure A.12 or A.13, compute the magnitude of the DFT X_n of the following signals.

(a) $x(k) = \begin{cases} 1, & k = 0 \\ \dfrac{1}{k}, & k = 1, 2, 3, \ldots, 31 \\ 0, & k = 32, 33, \ldots \end{cases}$

(b) $x(k) = \begin{cases} 1, & k = 0 \\ \dfrac{1}{k^2}, & k = 1, 2, 3, \ldots, 31 \\ 0, & k = 32, 33, \ldots \end{cases}$

(c) $x(k) = \begin{cases} 1, & k = 0 \\ \dfrac{1}{k!}, & k = 1, 2, 3, \ldots, 31 \\ 0, & k = 32, 33, \ldots \end{cases}$

(d) Compare the results obtained for parts (a), (b), and (c). Explain the differences in the results.

11.16　Consider the discrete-time signal

$$x(k) = \begin{cases} \text{RND } (k) - 0.5, & k = 0, 1, 2, \ldots, 31 \\ 0, & \text{all other } k. \end{cases}$$

where RND (k) is the random number generator (in Basic) which produces a sequence of random numbers whose values lie between zero and 1. The signal $x(k)$ can be interpreted as random noise. Using the program in Figure A.12 or

A.13, compute the magnitude of the DFT of $x(k)$. What frequencies would you expect to see in the amplitude spectrum of $x(k)$? Explain.

11.17 Consider the ideal lowpass discrete-time filter with frequency function $H(\omega)$ defined by

$$
H(\omega) = \begin{cases} 1, & 0 < \omega < \dfrac{\pi}{4} \\[3mm] 0, & \dfrac{\pi}{4} < \omega < 2\pi. \end{cases}
$$

(a) The input $x(k) = u(k) - u(k - 20)$ is applied to the filter.
  (i) Using the FFT algorithm in Figure A.13, compute the DFT of the resulting output response. Take $N = 32$.
  (ii) Using the inverse FFT algorithm in Figure A.13, compute the output response $y(k)$ for $k = 0, 1, 2, \ldots, 31$.
(b) Repeat part (a) for the input $u(k) = u(k) - u(k - 10)$.
(c) Compare the output response obtained in parts (a) and (b). In what respects do the responses differ? Explain.
(d) Repeat part (a) for the input

$$
x(k) = \begin{cases} \text{RND } (k) - 0.5, & k = 0, 1, 2, \ldots, 15 \\ 0, & \text{all other } k. \end{cases}
$$

How does the magnitude of the response compare with the magnitude of the input? Explain.

11.18 Repeat Problem 11.17 for the mean filter given by the input/output difference equation

$$
y(k) = \frac{x(k) + x(k - 1) + x(k - 2)}{3}.
$$

Note: To calculate the DFT Y_n of the output, write a program that carries out the multiplication of the DFT X_n of the input with $H(2\pi n/N)$, where $H(\omega)$ is the frequency function of the filter. Take $N = 32$.

Discretization and the Design of Digital Filters

In many applications the system under study is a hybrid system consisting of an interconnection of continuous-time and discrete-time components. An example is a continuous-time system such as the engine of an automobile that is controlled by a microprocessor-implemented digital controller. Continuous-time and discrete-time (or digital) components are connected together through analog-to-digital (A/D) and digital-to-analog (D/A) converters. These devices are studied in Sections 12.1 and 12.2. As pointed out in Section 12.1, the A/D conversion process is the same as pulse-code modulation, which is used extensively in digital communications.

In Section 12.3 we consider the design of digital filters by discretizing in time a given continuous-time system. Design procedures are developed based on the matching of responses to piecewise-constant inputs and sinusoidal inputs. We also consider a design method based on a bilinear transformation between the s-domain and the z-domain. In Section 12.5 we give an introduction to digital control.

12.1

ANALOG-TO-DIGITAL CONVERTERS

Let $x(t)$ be a continuous-time signal. As discussed in Section 10.4, the signal $x(t)$ can be sampled by multiplying $x(t)$ times the pulse train $p(t)$ shown in

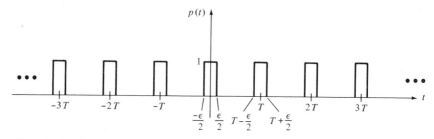

FIGURE 12.1.   Pulse train with period T.

Figure 12.1. Recall that the period T of the pulse train is equal to $2\pi/\omega_s$, where ω_s is the sampling frequency.

From the results in Chapter 10 we know that if the signal $x(t)$ is bandlimited with bandwidth B, then $x(t)$ can be recovered from the sampled signal $x_s(t) = x(t)p(t)$ by applying $x_s(t)$ to a lowpass filter/amplifier.

In analog-to-digital (A/D) conversion, the sampled signal $x_s(t)$ is converted into a signal consisting of a string of pulses with amplitude equal to 0 or 1. The process of sampling an analog signal $x(t)$ and then converting the sampled signal $x_s(t)$ into a binary-amplitude signal is also called *pulse-code modulation*. The binary-amplitude signal is constructed by quantizing and then encoding the sampled signal $x_s(t)$. We describe this process below.

Given a fixed positive integer r, let $Q = 2^r$. We quantize the amplitudes of the pulses comprising $x_s(t)$ so that the pulse amplitudes are equal to the sample values $x(kT)$ rounded off to one of Q levels, called the *quantization levels*. This process results in a *quantized sampled signal*, which we shall denote by $x_{sq}(t)$. We then represent each pulse in $x_{sq}(t)$ by a r-bit binary word. This process is called *encoding*. We illustrate how one encodes in the case when the number Q of quantization levels is equal to eight.

Let $a_0, a_1, \ldots, a_7$ denote the eight quantization levels. For the integer i ranging from 0 to 7, we write i with respect to base 2; that is,

$$i = b_2 \times 2^2 + b_1 \times 2^1 + b_0 \times 2^0,$$

TABLE 12.1.   Encoded Quantization Levels
When $Q = 8$.

Quantization Level	Encoded Representation
a_0	000
a_1	001
a_2	010
a_3	011
a_4	100
a_5	101
a_6	110
a_7	111

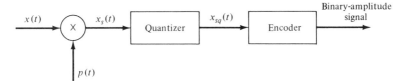

FIGURE 12.2. A/D converter.

where b_2, b_1, b_0 are binary numbers (0 or 1). Then the quantization level a_i is encoded by the binary word $b_2b_1b_0$. The encoding of the eight quantization levels is displayed in Table 12.1.

Let us return to the general case where $Q = 2^r$. In the last stage of the A/D converter, each pulse in $x_{sq}(t)$ is converted into r binary-amplitude pulses corresponding to the encoded representation of the quantized value. Combining the sampling, quantization, and encoding operations, we obtain an A/D converter. A block diagram of the A/D converter is given in Figure 12.2.

Since the output of the A/D converter is a binary signal, it can be applied directly to a digital computer for processing. In this way, analog signals can be processed in real time (on-line) using digital computers.

In Section 12.3 we will see how the original analog signal $x(t)$ can be reconstructed from the binary signal. Actually, even if the signal $x(t)$ is bandlimited, it is not possible to reconstruct $x(t)$ exactly from the binary signal due to the quantization process. The effect of quantization can be characterized as a noise signal that appears in the reconstructed signal. The magnitude of the noise, which can be expressed in terms of a signal-to-noise ratio, can be reduced by increasing the number of quantization levels. This results in an increase in the number of bits required to represent the quantization levels. For an in-depth treatment of this aspect of A/D conversion, the reader may consult Oppenheim and Schafer [1975].

In telephone communications, sampled voice signals are usually quantized to 256 levels, so that quantized sample values are represented by 8-bit binary words. In compact disc digital audio players, signal reconstruction is based on a 16-bit system. This means that the system is capable of recognizing $2^{16} = 65,536$ quantization levels. With this number of levels, it is possible to achieve signal-to-noise ratios exceeding 90 dB; that is, the magnitude of the noise is 90 dB below the magnitude of the signal. The A/D process is illustrated by the following example.

EXAMPLE 12.1. Consider the analog signal shown in Figure 12.3. Since $x(t)$ is time limited, it is not bandlimited. However, most of the spectral content of $x(t)$ is contained in the frequency range from 0 to 6π rad/sec. This can be verified by taking the Fourier transform of $x(t)$. (We leave this computation to the interested reader.) We can then compute the Nyquist sampling frequency based on a lowpass-filtered version of $x(t)$ with the cutoff frequency of the filter equal to 6π. Although the filtered signal will be a "rounded" version of the original signal, we assume that the difference between the signals is very small and thus can be neglected. The Nyquist

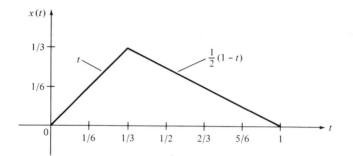

FIGURE 12.3. Signal in Example 12.1.

sampling frequency is therefore equal to $2(6\pi) = 12\pi$ rad/sec, which results in a sampling interval T equal to $2\pi/12\pi = \frac{1}{6}$ sec. The sampled signal $x(t)p(t)$ is displayed in Figure 12.4.

The sample values $x(kT)$ are given by

$$x(0) = 0$$
$$x(\tfrac{1}{6}) = \tfrac{1}{6} = 0.1667$$
$$x(\tfrac{1}{3}) = \tfrac{1}{3} = 0.3333$$
$$x(\tfrac{1}{2}) = \tfrac{1}{4} = 0.2500$$
$$x(\tfrac{2}{3}) = \tfrac{1}{6} = 0.1667$$
$$x(\tfrac{5}{6}) = \tfrac{1}{12} = 0.0833$$
$$x(1) = 0.$$

Now let us take eight quantization levels given by

$$a_i = \frac{i}{21}, \qquad i = 0, 1, \ldots, 7.$$

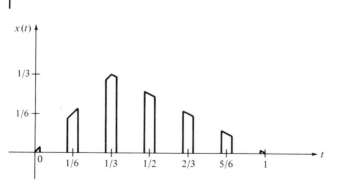

FIGURE 12.4. Sampled signal.

The values of the quantization levels are

$$a_0 = 0$$
$$a_1 = \tfrac{1}{21} = 0.0476$$
$$a_2 = \tfrac{2}{21} = 0.0952$$
$$a_3 = \tfrac{3}{21} = 0.1429$$
$$a_4 = \tfrac{4}{21} = 0.1905$$
$$a_5 = \tfrac{5}{21} = 0.2381$$
$$a_6 = \tfrac{6}{21} = 0.2857$$
$$a_7 = \tfrac{7}{21} = 0.3333.$$

Then the quantized sampled signal $x_{sq}(t)$ has the values

$$x_{sq}(0) = a_0$$
$$x_{sq}(\tfrac{1}{6}) = a_3$$
$$x_{sq}(\tfrac{1}{3}) = a_7$$
$$x_{sq}(\tfrac{1}{2}) = a_5$$
$$x_{sq}(\tfrac{2}{3}) = a_3$$
$$x_{sq}(\tfrac{5}{6}) = a_2$$
$$x_{sq}(1) = a_0.$$

Note that we could have taken the quantized level for $x(\tfrac{1}{6}) = \tfrac{1}{6}$ to be a_4, since $\tfrac{1}{6}$ is equidistant between a_3 and a_4. Finally, for each quantized value we generate three binary pulses corresponding to the encoded representation of the quantized value. The resulting binary signal is shown in Figure 12.5.

PCM TRANSMISSION   As noted above, the A/D conversion process is the same as pulse-code modulation (PCM). PCM is a very common technique used in digital communications. A block diagram of a PCM transmission system is given in Figure 12.6.

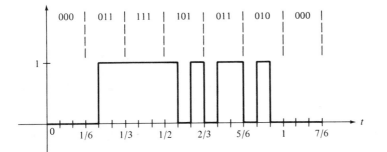

FIGURE 12.5.   Binary-amplitude signal resulting from A/D conversion.

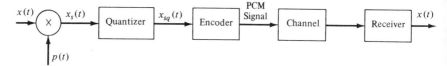

FIGURE 12.6. PCM transmission.

The primary advantage of PCM transmission over analog transmission (e.g., AM or PAM transmission) is in noise suppression. Since only pulses of amplitude 1 or 0 are transmitted in a PCM system, the receiver only has to distinguish between 0 and 1. Even if the pulses are seriously distorted during transmission, the receiver has a good chance of deciding whether a received pulse is 0 or 1. In contrast, if the received pulse can have any particular amplitude (as in PAM transmission), the receiver must be able to determine the correct value of the amplitude. This may be impossible to do if the signal is distorted during transmission.

Even if zeros look like ones and conversely, the receiver in a PCM system may still be able to reconstruct the transmitted signal by using error-correcting techniques. These are ingenious techniques involving the encoding and decoding of sampled signals. For an introductory treatment of error-correcting coding, see Couch [1983]. For an in-depth treatment, see Gallager [1968].

A disadvantage of PCM transmission over analog transmission is that PCM requires a wider bandwidth for transmission. To see this, we shall compute the bandwidth required to transmit a PCM signal.

Let $x(t)$ be an analog signal with bandwidth B. For example, $x(t)$ could be a speech signal that has been filtered to a 4-kHz bandwidth. To transmit such a signal, we can use single-sideband sine-wave amplitude modulation. The bandwidth necessary for this type of transmission is 4 kHz.

Again let $x(t)$ be an arbitrary signal with bandwidth B, and suppose that we want to transmit the signal using PCM. Let $Q = 2^r$ denote the number of quantization levels. If we sample the signal at the Nyquist frequency $\omega_s = 2B$, the duration τ of each of the binary pulses (bits) comprising the PCM signal is given by

$$\tau = \frac{T}{r} = \frac{\pi}{rB}.$$

The bit rate $1/\tau$ for PCM transmission is therefore equal to $(rB)/\pi$ bits/sec.

Much of the spectral content of a pulse of duration τ seconds is contained in the frequency range 0 to π/τ rad/sec (see Figure 9.21a). Hence to transmit a pulse of time duration τ, we will need a bandwidth of π/τ rad/sec. Setting $\tau = \pi/(rB)$, the bandwidth required to transmit the PCM signal is therefore rB rad/sec. So the bandwidth is r times the bandwidth needed for direct (analog) transmission, where r is the number of bits in the binary representation of the quantized sample values.

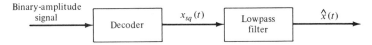

FIGURE 12.7. D/A converter.

EXAMPLE 12.2. Again suppose that $x(t)$ is a filtered speech signal with a 4-kHz bandwidth. Then the Nyquist sampling frequency is

$$\omega_s = 2B = 8 \text{ kHz} = 16\pi \times 10^3 \text{ rad/sec.}$$

As noted previously, in phone communication the number of quantization levels is 256, which corresponds to a 8-bit PCM signal. The resulting bit rate is

$$\frac{rB}{\pi} = \frac{(8)(8\pi \times 10^3)}{\pi} = 64 \text{ kilobits/sec.}$$

The bandwidth required to transmit this bit rate is

$$rB = 64\pi \times 10^3 \text{ rad/sec} = 32 \text{ kHz.}$$

12.2

DIGITAL-TO-ANALOG CONVERTERS

To recover the analog signal $x(t)$ applied to a A/D converter, we apply the binary-amplitude signal of the A/D converter to a digital-to-analog (D/A) converter. A block diagram of a D/A converter is given in Figure 12.7.

The decoder in the D/A converter reconstructs the quantized sampled signal $x_{sq}(t)$ from the binary-amplitude signal. The output of the decoder is then lowpass filtered, which gives an approximation $\hat{x}(t)$ to the original analog signal $x(t)$. As noted in Section 12.1, $\hat{x}(t)$ is not exactly equal to $x(t)$ as a result of the quantization process.

HOLD OPERATION Instead of lowpass filtering, the quantized sampled signal $x_{sq}(t)$ can be applied to a hold device as illustrated in Figure 12.8. The output $\tilde{x}(t)$ of the hold device is also an approximation to the original analog signal $x(t)$.

The output $\tilde{x}(t)$ of the hold device is given by

$$x(t) = x_{sq}(kT), \qquad kT \leq t < kT + T.$$

FIGURE 12.8. Hold operation.

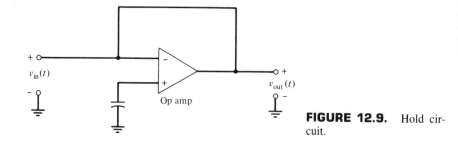

FIGURE 12.9. Hold circuit.

In other words, the hold device receives a pulse with amplitude $x_{sq}(kT)$ and "holds" this amplitude until it receives the next pulse with amplitude $x_{sq}(kT + T)$. The output $\tilde{x}(t)$ of the hold device is a piecewise-constant analog signal; that is, the value of $\tilde{x}(t)$ is constant over each T-second interval $kT \le t < kT + T$. Since the amplitude of $\tilde{x}(t)$ is constant over each T-second interval, the device is sometimes called a *zero-order hold*.

An op-amp realization of a hold device is shown in Figure 12.9. In this realization, an input voltage pulse charges the capacitor so that the voltage on the capacitor is equal to the amplitude of the input pulse. The capacitor then holds this voltage until the next pulse is received. The output of the circuit is equal to the voltage on the capacitor.

TRANSFER FUNCTION OF THE HOLD DEVICE   A hold device can be characterized in terms of a transfer function model that is useful in the analysis of systems containing hold circuits. The transfer function representation is constructed as follows.

Given an analog signal $x(t)$ with $x(t) = 0$ for $t < 0$, we shall apply the sampled signal $x_s(t) = x(t)p(t)$ to a hold device. The function $p(t)$ is the pulse train displayed in Figure 12.1. In the following analysis, we do not assume that $x_s(t)$ has been quantized in amplitude.

We let $\tilde{x}(t)$ denote the output of the hold device in response to the input $x_s(t)$. By definition of the hold device, $\tilde{x}(t)$ is given by

$$\tilde{x}(t) = x(kT), \qquad kT \le t < kT + T.$$

Then the Laplace transform $\tilde{X}(s)$ of $\tilde{x}(t)$ is given by

$$\tilde{X}(s) = \int_0^\infty \tilde{x}(t)e^{-st}\, dt$$

$$= \sum_{k=0}^\infty \int_{kT}^{kT+T} x(kT)e^{-st}\, dt$$

$$= \sum_{k=0}^\infty \left(\int_{kT}^{kT+T} e^{-st}\, dt \right) x(kT)$$

$$\tilde{X}(s) = \sum_{k=0}^{\infty} \left[-\frac{1}{s} e^{-st} \right]_{t=kT}^{t=kT+T} x(kT)$$

$$= \frac{1 - e^{-sT}}{s} \sum_{k=0}^{\infty} e^{-skT} x(kT). \tag{12.1}$$

Now let $X_s(s)$ denote the Laplace transform of the sampled signal $x_s(t) = x(t)p(t)$. Since the time duration ϵ of the pulses comprising $p(t)$ is assumed to be very small, we can approximate $X_s(s)$ by

$$X_s(s) = \sum_{k=0}^{\infty} \epsilon \, x(kT) e^{-skT}.$$

Then rewriting (12.1) in terms of $X_s(s)$, we obtain

$$\tilde{X}(s) = \frac{1 - e^{-sT}}{s} \frac{X_s(s)}{\epsilon}. \tag{12.2}$$

From (12.2) we have that the transfer function of the hold device is equal to

$$\frac{1 - e^{-sT}}{s} \frac{1}{\epsilon}.$$

We should mention that in most textbooks on sampling, the transfer function of the hold device is defined to be

$$\frac{1 - e^{-sT}}{s}.$$

This is the transfer function if we take the input to the hold device to be the idealized sampled signal given by

$$\sum_{k=0}^{\infty} x(kT)\delta(t - kT),$$

where $\delta(t - kT)$ is the impulse concentrated at the point $t = kT$. The Laplace transform of the idealized sampled signal is equal to

$$\sum_{k=0}^{\infty} x(kT)e^{-skT},$$

which in turn is equal to $(1/\epsilon)$ times the transform $X_s(s)$ of the sampled signal $x_s(t) = x(t)p(t)$. So the $(1/\epsilon)$-factor difference in the two transfer functions is a result of the difference in the definition of the sampled signal applied to the hold device.

There is also an interesting relationship between the Laplace transform of the output of the hold device and the z-transform of the sample values $x(kT)$. Let $X_d(z)$ denote the z-transform of the discrete-time signal $x(kT)$, that is,

$$X_d(z) = \sum_{k=0}^{\infty} x(kT)z^{-k}.$$

Then, from (12.1), we have that

$$\tilde{X}(s) = \frac{1 - e^{-sT}}{s} X_d(e^{sT}). \tag{12.3}$$

So the Laplace transform of the hold output is equal to the transfer function of the hold device times the z-transform of $x(kT)$ evaluated at $z = e^{sT}$.

12.3

DESIGN OF DIGITAL FILTERS BY RESPONSE MATCHING

Suppose that we have a causal linear time-invariant continuous-time system with input $x(t)$ and output $y(t)$. The system is specified by the transfer function representation

$$Y(s) = H(s)X(s),$$

where $Y(s)$, $X(s)$ are the Laplace transforms of the output and input and $H(s)$ is the transfer function of the system.

We can consider realizing (or simulating) the given continuous-time system by using a discrete-time system with sample and hold operations as illustrated in Figure 12.10. In Figure 12.10 the sampling operation is represented by a switch that closes briefly every T seconds. We assume that the time during which the switch is closed is so small that we can take the sampled signal to be the discrete-time signal $x(kT)$. The sampled signal $x(kT)$ is then applied to a causal linear time-invariant discrete-time system with transfer function $H_d(z)$. The discrete-time system has the transfer function representation

$$\Gamma(z) = H_d(z)X_d(z), \tag{12.4}$$

where $\Gamma(z)$ is the z-transform of $\gamma(kT)$ and $X_d(z)$ is the z-transform of the sampled input $x(kT)$. The output $\gamma(kT)$ of the discrete-time system is then applied to a hold device which produces the analog signal $\tilde{\gamma}(t)$ given by

$$\tilde{\gamma}(t) = \gamma(kT), \qquad kT \le t < kT + T.$$

To keep the analysis as simple as possible, we are not including quantization and encoding/decoding operations in the discrete-time realization. However, if

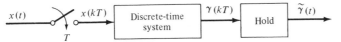

FIGURE 12.10. Realization of given continuous-time system.

the discrete-time system in Figure 12.10 is to be implemented in real time by a digital computer or microprocessor, it is necessary to use A/D and D/A converters. This aspect of the design problem is left to a course on the application of microprocessors (see Andrews [1982]).

Now the objective is to design the discrete-time system in Figure 12.10 so that this system with the sample-and-hold operations "behaves" as much as possible like the given continuous-time system. One way to obtain corresponding behavior is to require that for a specified collection of inputs $x(t)$, the output response of the given system matches the output response of the system in Figure 12.10; that is,

$$\tilde{y}(t) \approx y(t) \qquad \text{for all } t. \tag{12.5}$$

If the sampling interval T is suitably small, the matching condition (12.5) can be replaced by the requirement

$$\gamma(kT) = y(kT) \qquad \text{for all integers } k. \tag{12.6}$$

So given a specified collection of inputs $x(t)$, the problem is to design the transfer function $H_d(z)$ of the discrete-time system so that the matching condition (12.6) is satisfied. This process is called *design by response matching*.

If the given continuous-time system is a filter, the discrete-time system in Figure 12.10 is called a *digital filter*. This terminology is somewhat misleading since a discrete-time system is not a digital system unless we include the quantization of signal amplitudes. Nevertheless, the term "digital filter" is firmly entrenched in the literature.

There are two important points that one should be aware of in solving the design-by-response-matching problem:

1. The transfer function $H_d(z)$ of the discrete-time system is not equal to $H(s)$ with s replaced by z. In other words, matching cannot be achieved by simply replacing s by z in the transfer function $H(s)$ of the given continuous-time system.
2. There is no unique answer for $H_d(z)$. The solution depends on the particular choice of input (or inputs) for which matching of the resulting output responses is desired.

MATCHING THE RESPONSES TO A SINGLE INPUT Given a continuous-time signal $x(t)$ with $x(t) = 0$ for $t < 0$, let $y(t)$ denote the output response of the continuous-time system resulting from input $x(t)$ with the system at rest at time $t = 0$. The sampled input $x(kT)$ is applied to the discrete-time system in Figure 12.10. Then by (12.6), we have matching if the output $\gamma(kT)$

of the discrete-time system is equal to $y(kT)$. Taking the z-transform of both sides of (12.6), this is equivalent to requiring that

$$\Gamma(z) = Y_d(z), \tag{12.7}$$

where $Y_d(z)$ is the z-transform of $y(kT)$. Inserting the expression (12.4) for $\Gamma(z)$ into (12.7) gives

$$H_d(z)X_d(z) = Y_d(z). \tag{12.8}$$

Solving (12.8) for $H_d(z)$, we obtain

$$H_d(z) = \frac{Y_d(z)}{X_d(z)}. \tag{12.9}$$

So the transfer function $H_d(z)$ of the discrete-time system is equal to the z-transform of the discretized output $y(kT)$ divided by the z-transform of the discretized input $x(kT)$.

STEP-REPONSE MATCHING   In some applications we would like to have matching when $x(t)$ is the unit-step function $u(t)$. If $x(t)$ is the unit-step function, the response $y(t)$ of the continuous-time system is the step response of the system, and the response of the discrete-time system to $u(kT)$ is the step response of the discrete-time system. Therefore, when $x(t) = u(t)$, response matching means that we are matching the step responses of the given continuous-time system and the discrete-time realization.

Letting $g(t)$ denote the step response of the continuous-time system and letting $G_d(z)$ denote the z-transform of the $g(kT)$, from (12.9) we have

$$H_d(z) = \frac{G_d(z)}{z/(z-1)} = \frac{z-1}{z} G_d(z). \tag{12.10}$$

Thus to achieve step-response matching, the transfer function $H_d(z)$ of the discrete-time system must be equal to $(z - 1)/z$ times the z-transform of the discretized step response of the given continuous-time system.

EXAMPLE 12.3.   Consider the two-pole Butterworth lowpass filter with transfer function

$$H(s) = \frac{\omega_n^2}{s^2 + \sqrt{2}\,\omega_n s + \omega_n^2}.$$

Recall from Chapter 8 that ω_n is the 3-dB bandwidth of the filter. To compute the step response $g(t)$ of the filter, we first write $H(s)$ in the form

$$H(s) = \frac{2\omega_p^2}{(s + \omega_p)^2 + \omega_p^2}, \qquad \text{where } \omega_p = \frac{\omega_n}{\sqrt{2}}.$$

The Laplace transform $G(s)$ of the step response is given by

$$G(s) = H(s)\frac{1}{s} = \frac{1}{s} - \frac{s + \omega_p}{(s + \omega_p)^2 + \omega_p^2} - \frac{\omega_p}{(s + \omega_p)^2 + \omega_p^2}.$$

Using the transform pairs in Table 5.3, we obtain

$$g(t) = 1 - e^{-\omega_p T}\cos\omega_p t - e^{-\omega_p t}\sin\omega_p t, \qquad t \geq 0. \qquad (12.11)$$

Setting $t = kT$ in (12.11) gives

$$g(kT) = 1 - (e^{-\omega_p T})^k \cos\omega_p kT - (e^{-\omega_p T})^k \sin\omega_p kT,$$
$$k = 0, 1, \ldots.$$

Taking the z-transform of $g(kT)$, we get

$$G_d(z) = \frac{z}{z - 1} - \frac{z(z - e^{-\omega_p T}\cos\omega_p T) + (e^{-\omega_p T}\sin\omega_p T)z}{z^2 - (2e^{-\omega_p T}\cos\omega_p T)z + e^{-2\omega_p T}}.$$

Therefore, the transfer function of the digital filter (discrete-time realization) is

$$H_d(z)$$

$$= \frac{z - 1}{z}\, G_d(z)$$

$$= 1 - \frac{(z - 1)[z - e^{-\omega_p T}\cos\omega_p T + e^{-\omega_p T}\sin\omega_p T]}{z^2 - (2e^{-\omega_p T}\cos\omega_p T)z + e^{-2\omega_p T}}$$

$$= \frac{[-e^{-\omega_p T}(\cos\omega_p T + \sin\omega_p T) + 1]z + e^{-\omega_p T}[e^{-\omega_p T} + \sin\omega_p T - \cos\omega_p T]}{z^2 - (2e^{-\omega_p T}\cos\omega_p T)z + e^{-2\omega_p T}}$$

For the case when $\omega_n = 2$ and $T = 0.2$, the magnitude function $|H_d(\omega)|$ of the digital filter is plotted in Figure 12.11.

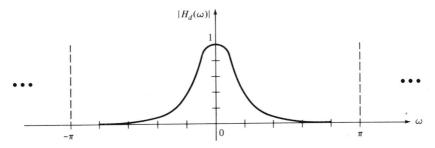

FIGURE 12.11. Magnitude function of digital-filter realization of two-pole Butterworth filter.

RESPONSE TO PIECEWISE-CONSTANT INPUTS An important characteristic of the design by step-response matching is that the output of the discrete-time system matches the samples values $y(kT)$ of the output $y(t)$ of the continuous-time system for any input $x(t)$ which is constant over the T-second intervals $kT \le t < kT + T$. To see this, first observe that any such input can be written in the form

$$x(t) = \sum_{i=0}^{\infty} c_i u(t - iT),$$

where $c_0, c_1, \ldots$ are constants. By linearity and time invariance, the output response $y(t)$ of the given continuous-time system is

$$y(t) = \sum_{i=0}^{\infty} c_i g(t - iT), \qquad (12.12)$$

where again $g(t)$ is the step response of the system. The output response $\gamma(kT)$ of the discrete-time system resulting from input $x(kT)$ is

$$\gamma(kT) = \sum_{i=0}^{\infty} c_i g_d(kT - iT), \qquad (12.13)$$

where $g_d(kT)$ is the step response of the discrete-time system. Now if the discrete-time system has been designed by step-response matching, then

$$g_d(kT) = g(kT).$$

Thus by (12.12) and (12.13), we have that

$$\gamma(kT) = y(kT) \qquad \text{for all } k.$$

So the output of the discrete-time system does match the sample values of $y(t)$.

The matching of responses to piecewise-constant inputs is a useful property in the design of digital control systems. We consider this in Section 12.5.

12.4

MATCHING OF RESPONSES TO SINUSOIDS

Again consider a linear time-invariant continuous-time system with the transfer function representation

$$Y(s) = H(s)X(s) = \frac{N(s)}{D(s)} X(s).$$

Throughout this section we assume that the system is asymptotically stable.

Recall that asymptotic stability is equivalent to requiring that all the poles of the system lie in the open left-half plane (see Section 8.1).

Let ω_{max} be a fixed positive number. With the input $x(t)$ equal to $A \cos \omega_0 t$, $t \geq 0$, where $0 \leq \omega_0 \leq \omega_{max}$, we want to select the transfer function $H_d(z)$ of the discrete-time system in Figure 12.10 so that its steady-state output $\gamma_{ss}(kT)$ matches the values $y_{ss}(kT)$ of the steady-state output $y_{ss}(t)$ of the given continuous-time system. This type of matching is often used in the case when the given continuous-time system is a filter.

We shall first determine the steady-state output response $y_{ss}(t)$ of the continuous-time system resulting from the input $x(t) = A \cos \omega_0 t$, $t \geq 0$. Since the system is asymptotically stable, from the results in Section 8.4 $y_{ss}(t)$ is given by

$$y_{ss}(t) = A|H(\omega_0)| \cos [\omega_0 t + \sphericalangle H(\omega_0)], \qquad t \geq 0. \qquad (12.14)$$

Now let us choose the sampling interval T so that it corresponds to the Nyquist sampling frequency ω_s for signals with bandwidth ω_{max}. That is, with $\omega_s = 2\omega_{max}$, take

$$T = \frac{2\pi}{\omega_s} = \frac{\pi}{\omega_{max}}.$$

The input to the discrete-time system is then equal to $A \cos \omega_0 kT$, $k = 0, 1, 2, \ldots$, where $T = \pi/\omega_{max}$. If the discrete-time system is asymptotically stable, from the results in Section 8.7 the steady-state response to this input is

$$\gamma_{ss}(kT) = A|H_d(\omega_0 T)|\cos [\omega_0 kT + \sphericalangle H_d(\omega_0 T)], \qquad k = 0, 1, \ldots, \qquad (12.15)$$

where

$$H_d(\omega_0 T) = H_d(z)\big|_{z = e^{j\omega_0 T}}.$$

Comparing (12.14) and (12.15), we see that in order to have $y_{ss}(kT) = \gamma_{ss}(kT)$ for $0 \leq |\omega_0| \leq \omega_{max}$, it is necessary and sufficient that

$$H_d(\omega_0 T) = H(\omega_0) \qquad \text{for } 0 \leq |\omega_0| \leq \omega_{max}. \qquad (12.16)$$

Using the change of variable $\overline{\omega}_0 = \omega_0 T$, we have that the condition (12.16) is equivalent to

$$H_d(\overline{\omega}_0) = H\left(\frac{\overline{\omega}_0}{T}\right) \qquad \text{for } 0 \leq |\overline{\omega}_0| < \pi. \qquad (12.17)$$

From (12.17) we see that the frequency function $H_d(\omega)$ of the discrete-time system over the interval $-\pi < \omega < \pi$ is equal to the scaled frequency function $H(\omega/T)$ of the given continuous-time system.

EXAMPLE 12.4. Suppose that the continuous-time system is the ideal linear-phase lowpass filter with frequency function

$$H(\omega) = K \exp(-j\omega t_d) \, p_{2B}(\omega).$$

Recall that t_d is the time delay through the filter. If we choose T so that $BT < \pi$, then by (12.17) the frequency function of the discrete-time realization is

$$H_d(\omega) = K \exp\left(\frac{-j\omega t_d}{T}\right) p_{2BT}(\omega), \qquad -\pi < \omega < \pi.$$

This discrete-time system is an ideal linear-phase lowpass filter. In particular, the system passes all inputs $x(kT) = A \cos \omega_0 kT$ with $0 \le \omega_0 < B$, while it completely stops all such inputs with $B < \omega_0 < \pi/T$. Note that if the time delay t_d is equal to zero, this filter is the same as the one considered in Section 11.4.

CHARACTERIZATION OF MATCHING IN TERMS OF THE IM-PULSE RESPONSE

The expression (12.17) for the frequency function of the discrete-time system is useful if the frequency function of the continuous-time system is known. In some cases, the continuous-time system may be specified by the impulse response or the transfer function. In such cases, the discrete-time system can be determined without having to compute the frequency function. We first show that the unit-pulse response of the discrete-time system can be calculated directly from the impulse response $h(t)$ of the continuous-time system.

Again let $H(\omega)$ denote the frequency function of the given continuous-time system. The impulse response $h(t)$ of the system is equal to the inverse Fourier transform of $H(\omega)$; that is,

$$h(t) = \frac{1}{2\pi} \int_{-\infty}^{\infty} H(\omega) e^{j\omega t} \, d\omega. \tag{12.18}$$

Let us suppose that

$$|H(\omega)| \approx 0 \qquad \text{for all } \omega > \omega_{\max},$$

where again $\omega_{\max}$ is the maximum frequency of the sinusoidal inputs. Then since $T = \pi/\omega_{\max}$,

$$|H(\omega)| \approx 0 \qquad \text{for all } \omega > \frac{\pi}{T},$$

and we can rewrite (12.18) in the form

$$h(t) = \frac{1}{2\pi} \int_{-\pi/T}^{\pi/T} H(\omega) e^{j\omega t} \, d\omega. \tag{12.19}$$

Now consider the change of variable $\overline{\omega} = T\omega$ in the integral in (12.19). Then $\omega = \overline{\omega}/T$, $d\overline{\omega} = T d\omega$, and (12.19) becomes

$$h(t) = \frac{1}{2\pi T} \int_{-\pi}^{\pi} H\left(\frac{\overline{\omega}}{T}\right) \exp\left(\frac{j\overline{\omega}t}{T}\right) d\overline{\omega}. \qquad (12.20)$$

Setting $t = kT$ in (12.20) and using (12.17), we obtain

$$h(kT) = \frac{1}{2\pi T} \int_{-\pi}^{\pi} H_d(\overline{\omega}) e^{j\overline{\omega}k} \, d\overline{\omega}. \qquad (12.21)$$

The right side of (12.21) is equal to $1/T$ times the inverse DTFT of $H_d(\omega)$, which in turn is equal to the unit-pulse response $h_d(kT)$ of the discrete-time system. Therefore, we have that

$$h(kT) = \frac{1}{T} h_d(kT)$$

or

$$h_d(kT) = Th(kT). \qquad (12.22)$$

So the unit-pulse response of the discrete-time system is equal to T times the sample values of the impulse response of the given continuous-time system.

The condition (12.22) is sometimes referred to as *impulse-response matching* since the unit-pulse response of the discrete-time system matches a scaled version of the sampled impulse response of the continuous-time system. Hence if $|H(\omega)| \approx 0$ for $\omega > \omega_{\max}$, sinusoidal-response matching is equivalent to impulse-response matching.

It is also interesting to note that the discrete-time system with unit-pulse response defined by (12.22) is the same as the discrete-time simulation constructed in Chapter 4 by discretizing the convolution integral. Therefore, sinusoidal-response matching, impulse-response matching, and the discretization of the convolution integral all yield the same discrete-time realization of the given continuous-time system.

TRANSFER FUNCTION OF THE DISCRETE-TIME REALIZATION

The transfer function $H_d(z)$ of the discrete-time realization can be computed by taking the z-transform of both sides of (12.22). In the case when $H(s)$ is rational in s, $H_d(z)$ can be computed from $H(s)$ via the following steps.

Step 1. By first expanding $H(s)$ using partial fractions, compute the impulse response $h(t)$ of the continuous-time system. For example, suppose that the poles $p_1, p_2, \ldots, p_n$ of $H(s)$ are distinct and the expansion for $H(s)$ has the form

$$H(s) = \sum_{i=1}^{n} \frac{c_i}{s - p_i},$$

where the c_i are real or complex numbers. Taking the inverse Laplace transform, we have that the impulse response is

$$h(t) = \sum_{i=1}^{n} c_i e^{p_i t}, \qquad t \geq 0. \tag{12.23}$$

Step 2. Set $t = kT$ in the expression for $h(t)$. For example, if $h(t)$ is given by (12.23), then

$$h(kT) = \sum_{i=1}^{n} c_i e^{p_i kT}, \qquad k = 0, 1, 2, \ldots. \tag{12.24}$$

Step 3. Take the z-transform of $Th(kT)$. The result is equal to the transfer function $H_d(z)$ of the discrete-time realization. For instance, if $h(kT)$ is given by (12.24), then

$$H_d(z) = T \sum_{k=0}^{\infty} h(kT) z^{-k}$$

$$= T \sum_{k=0}^{\infty} \left[\sum_{i=1}^{n} c_i e^{p_i kT} \right] z^{-k}$$

$$= T \sum_{i=1}^{n} c_i \left[\sum_{k=0}^{\infty} (e^{p_i T})^k z^{-k} \right]$$

$$= T \sum_{i=1}^{n} \frac{c_i z}{z - e^{p_i T}}.$$

EXAMPLE 12.5. Again consider the two-pole Butterworth lowpass filter with transfer function

$$H(s) = \frac{2\omega_p^2}{(s + \omega_p)^2 + \omega_p^2},$$

where $\omega_p = \omega_n / \sqrt{2}$. Using Table 5.3, we have that the impulse response of the filter is

$$h(t) = 2\omega_p e^{-\omega_p t} \sin \omega_p t, \qquad t \geq 0.$$

Thus

$$h(kT) = 2\omega_p (e^{-\omega_p T})^k \sin (\omega_p kT), \qquad k = 0, 1, 2, \ldots.$$

Taking the z-transform of $Th(kT)$ using Table 7.2, we have that

$$H_d(z) = \frac{(2\omega_p T e^{-\omega_p T} \sin \omega_p T) z}{z^2 - (2e^{-\omega_p T} \cos \omega_p T) z + e^{-2\omega_p T}}.$$

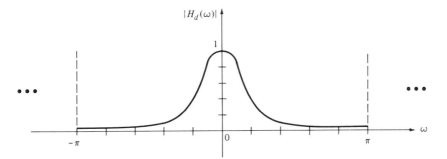

FIGURE 12.12. Magnitude function of discretization based on impulse-response matching.

Note that the denominator of $H_d(z)$ is the same as the denominator of the discretization computed in Example 12.3 using step-response matching. However, the numerators of the two discretized transfer functions are different, and thus the designs are not identical. In the case when $\omega_n = 2$ and $T = 0.2$, the magnitude function for the discretization based on impulse-response matching is shown in Figure 12.12. Comparing with Figure 12.11, we see that the magnitude function is almost identical to that obtained using step-response matching.

DESIGN BASED ON THE BILINEAR TRANSFORMATION In design by sinusoidal-response matching, it is not possible to compute the transfer function $H_d(z)$ of the discretization directly from the transfer function $H(s)$ of the given continuous-time system. However, there is an approximation to this design procedure that allows us to compute $H_d(z)$ directly from $H(s)$. The steps are as follows.

First observe that the matching condition (12.16) is satisfied if

$$H_d(z)\big|_{z=e^{sT}} = H(s). \tag{12.25}$$

To verify that (12.25) does imply (12.16), simply set $s = j\omega_0$ on both sides of (12.25). Because of the nature of the transformation

$$z = e^{sT},$$

it is not possible to derive an expression for $H_d(z)$ from (12.25). However, we can derive an expression for $H_d(z)$ if we approximate the transformation $z = e^{sT}$ by

$$z = \frac{1 + (T/2)s}{1 - (T/2)s} = 1 + sT + s^2\frac{T^2}{2} + s^3\frac{T^3}{4} + \cdots. \tag{12.26}$$

The transformation (12.26) is a *bilinear transformation* from the complex plane into itself. The term *bilinear* means that the relationship (12.26) is a bilinear function of z and s. For a detailed development of bilinear transformations in complex function theory, we refer the reader to Churchill et. al. [1976].

The transformation (12.26) has the property that it maps the open left-half plane into the open unit-disc. In other words, if Re $s < 0$, then z given by (12.26) is located in the open unit-disc of the complex plane; that is, $|z| < 1$.

Solving (12.26) for s in terms of z, we get the *inverse transformation* given by

$$s = \frac{2}{T} \frac{z - 1}{z + 1}.$$

By definition of the inverse transformation, we have that

$$\frac{1 + (T/2)s}{1 - (T/2)s}\bigg|_{s = \frac{2}{T}\frac{z-1}{z+1}} = z. \tag{12.27}$$

Now using the approximation (12.26) for e^{sT}, from (12.25) we obtain

$$H_d\left(\frac{1 + (T/2)s}{1 - (T/2)s}\right) \approx H(s). \tag{12.28}$$

Then setting

$$s = \frac{2}{T} \frac{z - 1}{z + 1}$$

on both sides of (12.28), we get

$$H_d\left(\frac{1 + (T/2)s}{1 - (T/2)s}\right)\bigg|_{s = \frac{2}{T}\frac{z-1}{z+1}} = H\left(\frac{2}{T} \frac{z - 1}{z + 1}\right).$$

Using (12.27), we then have that

$$H_d(z) = H\left(\frac{2}{T} \frac{z - 1}{z + 1}\right). \tag{12.29}$$

By (12.29) the transfer function of the discretization is (approximately) equal to the transfer function of the given continuous-time system with s replaced by $(2/T)[(z - 1)/(z + 1)]$.

Note that since the bilinear transformation maps the open left-half plane into the open unit-disc, the poles of the discretization $H_d(z)$ are in the open unit-disc if and only if the poles of $H(s)$ are in the open left-half plane. Thus the discretization is stable if and only if the given continuous-time system is stable. In other words, the discretization process preserves stability. This is a very desirable property for any discretization process.

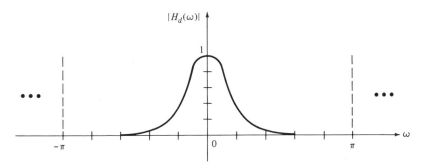

FIGURE 12.13. Magnitude function of discretization based on the bilinear transformation.

EXAMPLE 12.6. For the two-pole Butterworth filter, we have

$$H_d(z) = H\left(\frac{2}{T}\frac{z-1}{z+1}\right)$$

$$= \frac{\omega_n^2}{\left(\dfrac{2}{T}\dfrac{z-1}{z+1}\right)^2 + \sqrt{2}\,\omega_n\left(\dfrac{2}{T}\dfrac{z-1}{z+1}\right) + \omega_n^2}$$

$$= \frac{(T^2/4)(z+1)^2\,\omega_n^2}{(z-1)^2 + (T/\sqrt{2})\omega_n(z+1)(z-1) + (T^2/4)(z+1)^2\,\omega_n^2}$$

$$= \frac{(T^2\omega_n^2/4)(z^2 + 2z + 1)}{\left(1 + \dfrac{\omega_n T}{\sqrt{2}} + \dfrac{T^2\omega_n^2}{4}\right)z^2 + \left(\dfrac{T^2\omega_n^2}{2} - 2\right)z + \left(1 - \dfrac{\omega_n T}{\sqrt{2}} + \dfrac{T^2\omega_n^2}{4}\right)}.$$

For $\omega_n = 2$ and $T = 0.2$, the magnitude function $|H_d(\omega)|$ is plotted in Figure 12.13. The magnitude response of this discretization closely resembles the magnitude responses of the other two designs (see Figures 12.11 and 12.12).

12.5

DIGITAL CONTROL

In Section 6.5 we considered a common type of control problem involving the tracking of a reference signal in the presence of a disturbance. The controller that we used to achieve tracking was a continuous-time system. To obtain a high degree of accuracy and flexibility, controllers are often implemented in digital form. We give a brief introduction to digital control in this section. For a comprehensive treatment of digital control, we refer the reader to Kuo [1980], Franklin and Powell [1980], and Vanlandingham [1985].

Given a linear time-invariant continuous-time system with transfer function $H(s)$, consider the feedback control system shown in Figure 12.14. As illustrated in Figure 12.14, the output $y(t)$ of the plant (the given system) is fed back and

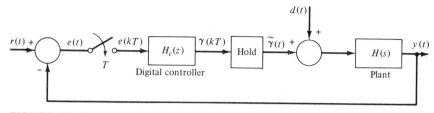

FIGURE 12.14. Feedback control system with digital controller.

compared with a reference signal $r(t)$. The error signal $e(t) = r(t) - y(t)$ is sampled and then the sampled error signal $e(kT)$ is applied to a linear time-invariant discrete-time system with transfer function $H_c(z)$. This discrete-time system is the digital controller. Again, the term "digital" is somewhat misleading since we are not considering the quantization of signal amplitudes.

The continuous-time signal $\tilde{\gamma}(t)$ in Figure 12.14 is the control signal applied to the plant. Since $\tilde{\gamma}(t)$ is the output of the hold device, the signal $\tilde{\gamma}(t)$ is constant over the T-second time intervals $kT \leq t < kT + T$. The signal $d(t)$ in Figure 12.14 is a disturbance signal applied to the plant.

In practice, the discrete-time system with sample-and-hold operations would be implemented using D/A and A/D converters and a microprocessor. We shall not consider this aspect of the control process.

The objective of the control process is to design the transfer function $H_c(z)$ of the digital controller so that the plant's output $y(t)$ tracks the reference signal $r(t)$. This will be the case if

$$e(t) \rightarrow 0 \qquad \text{as } t \rightarrow \infty. \tag{12.30}$$

In Section 6.5 we considered the solution of this tracking problem using an analog controller. We now want to solve it using a digital controller.

In order that (12.30) be satisfied, it is necessary that the sampled error converges to zero; that is,

$$e(kT) \rightarrow 0 \qquad \text{as } k \rightarrow \infty. \tag{12.31}$$

We can first consider the design of $H_c(z)$ so that (12.31) is satisfied. Given such a design, we can then check to see if (12.30) is also satisfied. The tracking condition (12.30) will be satisfied for the design procedure given below.

Since we want to achieve the discrete-time tracking condition (12.31), we can restrict our attention to the behavior of the feedback system at the sample times $t = kT$ only. In particular, we can work with a discretization in time of the plant. Since the control input $\tilde{\gamma}(t)$ applied to the plant is constant over the T-second intervals $kT \leq t < kT + T$, we should discrete the plant using the step-response matching technique. If the disturbance $d(t)$ is approximately constant over the T-second intervals $kT \leq t < kT + T$, the output values of the discretization will match the sample values $y(kT)$ of the plant's output $y(t)$.

Letting $H_d(z)$ denote the transfer function of the discretization of the plant, we can represent the feedback system at the sample times $t = kT$ by the discrete-time representation given in Figure 12.15. Note that for $t = kT$, the output

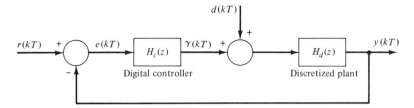

FIGURE 12.15. Discrete-time representation of feedback system.

$e(kT)$ of the sampler is equal to the input $e(t)$ and the output $\tilde{\gamma}(t)$ of the hold operation is equal to the input $\gamma(kT)$, and thus these operations do not appear in the discretized representation.

Let $R(z)$, $D(z)$, $Y(z)$ denote the z-transforms of $r(kT)$, $d(kT)$, $y(kT)$, respectively. Then the feedback system in Figure 12.15 has the transfer function representation

$$Y(z) = \frac{H_c(z)H_d(z)}{1 + H_c(z)H_d(z)} R(z) + \frac{H_d(z)}{1 + H_c(z)H_d(z)} D(z). \qquad (12.32)$$

We assume that $H_d(z)$ and $H_c(z)$ are proper rational functions of z given by

$$H_d(z) = \frac{N_d(z)}{D_d(z)} \qquad \text{and} \qquad H_c(z) = \frac{N_c(z)}{D_c(z)},$$

where $N_d(z)$, $D_d(z)$, $N_c(z)$, and $D_c(z)$ are polynomials in z with real coefficients. Then we can write (12.32) in the form

$$Y(z) = \frac{N_d(z)[N_c(z)R(z) + D_c(z)D(z)]}{D_d(z)D_c(z) + N_d(z)N_c(z)}. \qquad (12.33)$$

Now suppose that both $r(t)$ and $d(t)$ are step functions given by

$$r(t) = r_0, \quad t \geq 0 \qquad \text{and} \qquad d(t) = d_0, \quad t \geq 0.$$

Then

$$R(z) = \frac{r_0 z}{z - 1} \qquad \text{and} \qquad D(z) = \frac{d_0 z}{z - 1}.$$

Inserting these expressions for $R(z)$ and $D(z)$ into (12.33), we obtain

$$
\begin{aligned}
Y(z) &= \frac{N_d(z)\left[N_c(z) \dfrac{r_0 z}{z - 1} + D_c(z) \dfrac{d_0 z}{z - 1} \right]}{D_d(z)D_c(z) + N_d(z)N_c(z)} \\
&= \frac{N_d(z)[N_c(z) r_0 z + D_c(z) d_0 z]}{[D_d(z)D_c(z) + N_d(z)N_c(z)](z - 1)}. \qquad (12.34)
\end{aligned}
$$

Since $r(kT) = r_0$, $k = 0, 1, \ldots$, the sampled error signal is

$$e(kT) = r_0 - y(kT).$$

Therefore, the tracking condition (12.31) will be satisfied if

$$y(kT) \to r_0 \qquad \text{as } k \to \infty. \tag{12.35}$$

The condition (12.35) can be satisfied by choosing $H_c(z) = N_c(z)/D_c(z)$ so that

1. $D_c(z) = (z - 1)\overline{D}_c(z)$; that is, $D_c(z)$ has a zero at $z = 1$.
2. The zeros of $D_d(z)D_c(z) + N_d(z)N_c(z)$ are all located inside the unit-disc of the complex plane.

The proof that these two conditions imply that (12.35) is satisfied follows from the final-value theorem for the z-transform (see Section 7.2). We omit the details.

EXAMPLE 12.7 (Digital Speed Controller). Consider the automobile given by the transfer function

$$H(s) = \frac{1/M}{s - k_f/M}.$$

Here the input is the force applied to the car and the output $y(t)$ is the velocity $v(t)$ of the car at time t. We want to design a digital controller that keeps the car at a fixed velocity v_0. The feedback system with the digital controller is shown in Figure 12.16. The disturbance force $d(t)$ applied to the car is a step function with magnitude d_0. Now the first step in the design of the digital controller $H_c(z)$ is to compute the discretization of the given system. Using the step-response matching method, the transfer function $H_d(z)$ of the discretized plant is

$$H_d(z) = \frac{z - 1}{z} G_d(z),$$

where $G_d(z)$ is the z-transform of the discretized step response. From previous results [see (1.31)], the step response of the car is

$$g(t) = \frac{1}{k_f}\left[1 - \exp\left(\frac{-k_f}{M} t\right)\right], \qquad t \geq 0.$$

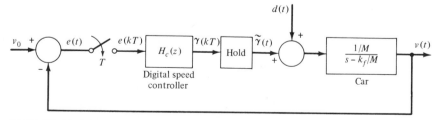

FIGURE 12.16. Car with digital controller.

Then

$$g(kT) = \frac{1}{k_f}\left\{1 - \left[\exp\left(-\frac{k_f T}{M}\right)\right]^k\right\}, \qquad k = 0, 1, \ldots,$$

and thus

$$G_d(z) = \frac{1}{k_f}\left[\frac{z}{z - 1} - \frac{z}{z - \exp(-k_f T/M)}\right]$$

$$= \frac{(1/k_f)[1 - \exp(k_f T/M)]z}{(z - 1)[z - \exp(-k_f T/M)]}.$$

Finally,

$$H_d(z) = \frac{(1/k_f)[1 - \exp(-k_f T/M)]}{z - \exp(-k_f T/M)}.$$

Now using (12.34), we have that the z-transform of the discretized velocity $v(kT)$ is

$$V(z) = \frac{(1/k_f)[1 - \exp(-k_f T/M)][N_c(z)v_0 z + D_c(z)\, d_0 z]}{\{[z - \exp(-k_f T/M)]D_c(z) + (1/k_f)\,[1 - \exp(-k_f T/M)]N_c(t)\}(z - 1)}.$$

where

$$H_c(z) = \frac{N_c(z)}{D_c(z)}.$$

If we choose

$$D_c(z) = z - 1 \qquad \text{and} \qquad N_c(z) = k_f z,$$

then

$$D_d(z)D_c(z) + N_d(z)N_c(z) = \left[z - \exp\left(\frac{-k_f T}{M}\right)\right](z - 1)$$

$$+ \frac{1}{k_f}\left[1 - \exp\left(-\frac{k_f T}{M}\right)\right]k_f z$$

$$= z^2 - 2\left[\exp\left(-\frac{k_f T}{M}\right)\right]z + \exp\left(-\frac{k_f T}{M}\right).$$

Since (if $k_f > 0$)

$$0 < \exp\left(-\frac{k_f T}{M}\right) < 1 \quad \text{and} \quad 2 \exp\left(-\frac{k_f T}{M}\right) < 1 + \exp\left(-\frac{k_f T}{M}\right),$$

by the stability conditions (8.62), the zeros of $D_p(z)D_c(z) + N_p(z)N_c(z)$ are located inside the unit-disc, and thus condition (2) above is satisfied. In addition, since $D_c(z) = z - 1$, condition 1 is also satisfied. Therefore, $v(kT) \to v_0$ as $k \to \infty$. Although we shall not prove this, it is also true that $v(t) \to v_0$ as $t \to \infty$.

PROBLEMS

Chapter 12

12.1 Consider the signal $x(t) = 3 \, \mathrm{Sa}^2 (t/2)$, $-\infty < t < \infty$.
(a) Compute the Nyquist sampling frequency (in rad/sec).
(b) For the time interval $-\pi \le t \le 3\pi$, compute the PCM signal with sampling interval $T = \pi$ and quantization levels $a_i = 3[-1 + (i/3)]$, $i = 0, 1, 2, \ldots, 7$. Sketch the PCM signal.

12.2 For the signal $x(t)$ in Figure P12.2, sketch the PCM signal with $T = 0.5$ and quantization levels $a_i = i/7$, $i = 0, 1, 2, \ldots, 7$.

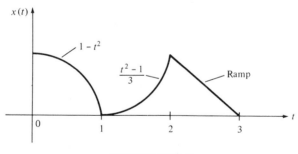

FIGURE P12.2

12.3 For the signal $x(t) = 4 \, \mathrm{Sa} \, (t) \cos 2t$, $-\infty < t < \infty$:
(a) Compute the Nyquist sampling frequency in rad/sec.
(b) For the time interval $0 \le t \le 2\pi$, sketch the PCM signal with sampling interval $T = \pi/2$ and quantization levels $a_0 = -3$, $a_1 = -2$, $a_2 = -1$, $a_3 = 0$, $a_4 = 1$, $a_5 = 2$, $a_6 = 3$, $a_7 = 4$.

12.4 Consider the hybrid system shown in Figure P12.4. Compute exact values for $y(1)$, $y(2)$, and $y(3)$ when
(a) $w(0) = -1$, $y(0) = 1$, and $x(t) = \delta(t) = $ unit impulse.
(b) $w(0) = y(0) = 0$ and $x(t) = e^{-2t} u(t)$.
(c) $w(0) = 2$, $y(0) = -1$, and $x(t) = u(t)$.

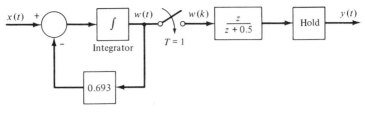

FIGURE P12.4

12.5 Consider the system shown in Figure P12.5.

(a) Compute $Y(z)/X(z)$, where $Y(z)$ is the z-transform of $y(k)$ and $X(z)$ is the z-transform of $x(k)$. Assume that there is no initial energy in the system at time $t = 0^-$.

(b) Using your answer in part (a), compute $y(t)$ for $t = 1, 2, \ldots$ when $y(0) = 0$, $\dot{y}(0) = 0$, and $x(t) = u(t)$.

(c) Compute *exact* values for $y(1)$, $y(2)$, and $y(3)$ when $y(0) = 2$, $\dot{y}(0) = -2$, and $x(t) = (t + 1)u(t)$.

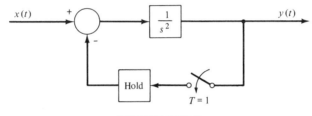

FIGURE P12.5

12.6 For the system in Figure P12.6,

(a) Compute $Y(z)/X(z)$ assuming that there is no initial energy in the system at time $t = 0^-$.

(b) Compute $y(t)$ for $t = 1, 2, \ldots$ when $w(0) = -2$, $y(0) = 3$, $e(0) = -1$, and $x(t) = -4u(t)$.

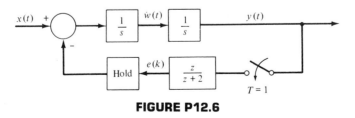

FIGURE P12.6

12.7 Compute $Y(z)/X(z)$ for the system in Figure P12.7. Assume that there is no initial energy in the system at time $t = 0^-$.

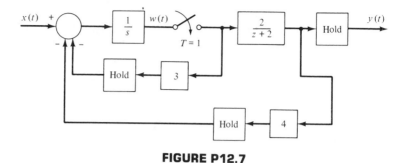

FIGURE P12.7

12.8 Consider the one-pole lowpass filter given by the transfer function

$$H(s) = \frac{B}{s + B}.$$

Using the step-response method, design a discrete-time system that realizes the lowpass filter. Let the sampling interval T be arbitrary. Express your answer by giving the transfer function $H_d(z)$ of the discretization.

12.9 Repeat Problem 12.8 using the sinusoidal-response matching method.

12.10 Repeat Problem 12.8 using the discretization procedure based on the bilinear transformation.

12.11 A two-pole bandpass filter is given by the transfer function

$$H(s) = \frac{100}{s^2 + 2s + 101}.$$

Using the step-response method, design a discrete-time system that realizes the bandpass filter. Take the sampling interval T to be 0.02. Express your answer by giving the transfer function $H_d(z)$ of the discretization.

12.12 Repeat Problem 12.11 using the sinusoidal-response matching method.

12.13 Repeat Problem 12.11 using the discretization procedure based on the bilinear transformation.

12.14 Using the Program in Figure A.11, sketch the magnitude function $|H_d(\omega)|$ and the phase function $\measuredangle H_d(\omega)$ of the discretizations computed in Problems 12.11, 12.12, and 12.13.

12.15 Suppose that the sampled sinusoid $x(kT) = \cos \omega_0 kT$, $k = 0, \pm 1, \pm 2, \ldots$, is applied (with $T = 0.02$) to each of the discretizations constructed in Problems 12.11, 12.12, and 12.13. For each discretization, determine the range of values of ω_0 for which the peak magnitude of the resulting output response is greater than or equal to 0.707. In other words, we are determining the "effective" 3-dB bandwidth of the discretizations.

12.16 A linear time-invariant continuous-time system has the transfer function

$$H(s) = \frac{3s}{s^2 + 4s + 13} = \frac{3s}{(s + 2)^2 + 3^2}.$$

Design a discrete-time system that results in response matching for *all* inputs $x(t) = A \cos \omega_0 t$, $-\infty < t < \infty$, where $0 \le \omega_0 \le 10\pi$. Express your answer by giving the transfer function $H_d(z)$ of the discretization.

12.17 Consider the single-car system given by the input/output differential equation

$$\frac{d^2 y(t)}{dt^2} + \frac{k_f}{M}\frac{dy(t)}{dt} = \frac{1}{M} x(t),$$

where $y(t)$ is the position of the car at time t. Construct discrete-time simulations of the car by using the three discretization methods presented in Chapter 12. Express your answer by giving the transfer functions of the discretizations. Take $M = 1$, $k_f = 0.1$, and $T = 0.1$.

12.18 Consider a digital controller $H_c(z) = N_c(z)/D_c(z)$ whose purpose is to force the output $y(kT)$ of a discretized plant $H_d(z) = N_d(z)/D_d(z)$ to track a constant reference r_0 in the presence of a constant disturbance d_0 (see Figure 12.15). Suppose that $H_c(z)$ is chosen so that $D_c(z) = (z - 1)\overline{D}_c(z)$ and

$$D_d(z)D_c(z) + N_d(z)N_c(z) = z^q,$$

where q is some positive integer. Show that the error $e(kT) = r_0 - y(kT)$ is identically zero for $k = q + 1$, $q + 2$, $\ldots$. This type of performance is sometimes referred to as *dead-beat tracking*.

12.19 Using the analysis given in Problem 12.18, design a dead-beat tracker for speed control of the automobile described in Example 12.7. Take $M = 1$, $k_f = 0.1$, and $T = 0.1$. Design the dead-beat tracker so that the integer q is as small as possible.

12.20 For the system in Figure P12.20, determine the feedback gains k_1, k_2 so that for any initial conditions $y(0)$ and $w(0)$, both $w(t)$ and $y(t)$ are identically zero for all values of $t \ge 2$. Here it is assumed that $x(t) = 0$ for all $t \ge 0$.

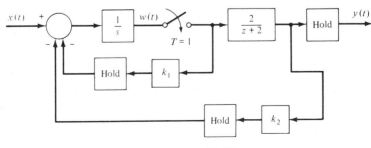

FIGURE P12.20

State Representation

The models that we have considered so far in this book are mathematical representations of the input/output behavior of the system under study. In this chapter we define a new type of model, which is specified in terms of a collection of variables that describe the internal behavior of the system. These variables are called the state variables of the system. The model that we will define in terms of the state variables is called the state or state-variable representation. Our objective is to define the state model and to study the basic properties of this model for both continuous-time and discrete-time systems. An in-depth development of the state approach to systems can be found in a number of textbooks. For example, we mention Friedland [1986], Chen [1984], Brogan [1985], and Kailath [1980].

Since the state model is given in terms of a matrix equation, the reader should have some familiarity with matrix algebra. A brief review is given in Appendix C.

We begin in Section 13.1 with the notion of state and then we define the state equations for a continuous-time system. The construction of state models from input/output differential equations is considered in Section 13.2. The solution of the state equations is studied in Section 13.3. Then in Section 13.4, we give the discrete-time version of the state-variable theory. The notion of equivalent state representations and the discretization of state models are studied in Sections 13.5 and 13.6.

STATE MODEL

Consider a single-input single-output causal continuous-time system given by the input/output relationship

$$y(t) = (Fx)(t), \qquad t \geq t_0. \tag{13.1}$$

Recall that $y(t)$ is the output response of the system resulting from input $x(t)$ with no initial energy in the system at time t_0 (or t_0^-).

Let t_1 be a fixed but arbitrary value of time with $t_1 > t_0$. To compute the output response $y(t)$ for all $t \geq t_1$, we need to know more than the input $x(t)$ for $t \geq t_1$. In other words, it is not possible in general to compute $y(t)$ for $t \geq t_1$ from knowledge of $x(t)$ for $t \geq t_1$. The reason for this is that the application of the input for $t_0 \leq t \leq t_1$ may put energy into the system which affects the output response for $t \geq t_1$. For example, a voltage or current applied to an RLC circuit for $t \geq t_0$ may result in voltages on the capacitors and/or currents in the inductors at time t_1 $(t_1 > t_0)$. These voltages and currents at time t_1 can then affect the output of the RLC circuit for $t \geq t_1$.

Let us return to the general system given by (13.1). Given $t_1 > t_0$, the state $q(t_1)$ of the system at time $t = t_1$ is defined to be that portion of the past history $(t_0 \leq t \leq t_1)$ of the system required to determine the output response $y(t)$ for all $t \geq t_1$ given the input $x(t)$ for $t \geq t_1$. A nonzero state $q(t_1)$ at time t_1 indicates the presence of energy in the system at time t_1. In particular, the system is in the zero state at time t_1 if and only if there is no energy in the system at time t_1. If the system is in the zero state at time t_1, the response $y(t)$ for $t \geq t_1$ can be computed from knowledge of the input $x(t)$ for $t \geq t_1$; in fact, we have

$$y(t) = (Fx)(t), \qquad t \geq t_1. \tag{13.2}$$

If the state at time t_1 is not zero, the relationship (13.2) must be modified to include the effect of the initial state on the output response for $t \geq t_1$. We will consider this later.

If the given system is finite dimensional, the state $q(t)$ of the system at time t is an n-element column vector given by

$$q(t) = \begin{bmatrix} q_1(t) \\ q_2(t) \\ \cdot \\ \cdot \\ \cdot \\ q_n(t) \end{bmatrix}.$$

The components $q_1(t), q_2(t), \ldots, q_n(t)$ are called the *states* or the *state variables* of the system. For example, suppose that the given system is an RLC circuit. From circuit theory, we know that any energy in the circuit at time t is completely characterized by the voltages across the capacitors at time t and the currents in the inductors at time t. Thus we can define the state of the circuit at time t to be a vector whose components are the voltages across the capacitors at time t and the currents in the inductors at time t. If the number of capacitors in the circuit is equal to n_C and the number of inductors in the circuit is equal to n_L, the total number of state variables is equal to $n_C + n_L$.

Now suppose that the system is an integrator with the input/output relationship

$$y(t) = \int_{t_0}^{t} x(\lambda) \, d\lambda, \qquad t > t_0, \tag{13.3}$$

where $y(t_0) = 0$. We want to show that the state $q(t)$ of the integrator can be chosen to be the output $y(t)$ of the integrator at time t. To see this, let t_1 be an arbitrary value of time with $t_1 > t_0$. Then rewriting (13.3), we have

$$y(t) = \int_{t_0}^{t_1} x(\lambda) \, d\lambda + \int_{t_1}^{t} x(\lambda) \, d\lambda, \qquad t \geq t_1. \tag{13.4}$$

By (13.3) we see that the first term on the right side of (13.4) is equal to $y(t_1)$. Therefore,

$$y(t) = y(t_1) + \int_{t_1}^{t} x(\lambda) \, d\lambda, \qquad t \geq t_1. \tag{13.5}$$

The relationship (13.5) shows that $y(t_1)$ characterizes the energy in the system at time t_1. More precisely, from (13.5) we see that $y(t)$ can be computed for all $t \geq t_1$ from knowledge of $x(t)$ for $t \geq t_1$ and knowledge of $y(t_1)$. Thus we can take the state at time t_1 to be $y(t_1)$.

Now suppose that we have an interconnection of integrators, adders, subtracters, and scalar multipliers. Since adders, subtracters, and scalar multipliers are memoryless devices, the energy in the interconnection is completely characterized by the values of the outputs of the integrators. Thus we can define the state at time t to be a vector whose components are the outputs of the integrators at time t.

STATE EQUATIONS Consider a single-input single-output finite-dimensional continuous-time system with state $q(t)$ given by

$$q(t) = \begin{bmatrix} q_1(t) \\ q_2(t) \\ \cdot \\ \cdot \\ \cdot \\ q_n(t) \end{bmatrix}$$

The state $q(t)$ is a vector-valued function of time t. In other words, for any particular value of t, $q(t)$ is an n-element column vector. The vector-valued function $q(t)$ is called the *state trajectory* of the system.

The system with state $q(t)$ can be modeled by the state equations given by

$$\dot{q}(t) = f(q(t), x(t), t) \tag{13.6}$$

$$y(t) = g(q(t), x(t), t). \tag{13.7}$$

In (13.6), $\dot{q}(t)$ is the derivative of the state vector with the derivative taken component by component; that is,

$$\dot{q}(t) = \begin{bmatrix} \dot{q}_1(t) \\ \dot{q}_2(t) \\ \cdot \\ \cdot \\ \cdot \\ \dot{q}_n(t) \end{bmatrix}.$$

On the right side of (13.6), f is a function of the state $q(t)$ at time t, the input $x(t)$ at time t, and time t. Hence, by (13.6), the derivative of the state at time t is a function of time t and the state and input at time t.

The function f in (13.6) is a vector-valued function of several variables. That is, if we insert values for t, $x(t)$, and the components of $q(t)$ in $f(q(t), x(t), t)$, we obtain an n-element column vector that is equal to $\dot{q}(t)$. Since $q(t)$ and $\dot{q}(t)$ are n-element column vectors, (13.6) is a vector differential equation. In particular, (13.6) is a first-order vector differential equation.

In the second equation (13.7), g is another function of the state $q(t)$ at time t, the input $x(t)$ at time t, and time t. So the output $y(t)$ of the system at time t is a function of $q(t)$ and $x(t)$ at time t and time t. Equation (13.7) is called the *output equation* of the system.

Equations (13.6) and (13.7) comprise the state model of the system. This representation is a time-domain model of the system since the equations are in terms of functions of time. Note that the state model is specified in two parts; (13.6) describes the state response resulting from the application of an input $x(t)$ with initial state $q(t_0) = q_0$, while (13.7) gives the output response as a function of the state and input. The two parts of the state model correspond to a cascade decomposition of the system as illustrated in Figure 13.1. The double line for $q(t)$ in Figure 13.1 indicates that $q(t)$ is a vector signal.

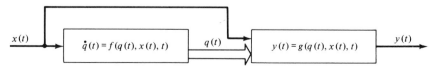

FIGURE 13.1. Cascade structure corresponding to state model.

From Figure 13.1 we see that the system state $q(t)$ is an "internal" vector variable of the system; that is, the state variables [the components of $q(t)$] are signals within the system. Since the state model is specified in terms of the internal vector variable $q(t)$, the representation is an internal model of the system. The form of this model is quite different from that of the external or input/output models that we studied in Chapters 2 and 4.

The functions f and g in the state equations (13.6) and (13.7) may be nonlinear functions of their arguments, in which case the given system is nonlinear. The system is linear if and only if the functions f and g are both linear. If f and g are linear, we can write the state equations in the form

$$\dot{q}(t) = A(t)q(t) + b(t)x(t) \tag{13.8}$$

$$y(t) = c(t)q(t) + d(t)x(t). \tag{13.9}$$

In (13.8), $A(t)$ is a $n \times n$ matrix whose entries are functions of time t, and $b(t)$ is an n-element column vector whose components are functions of t. In (13.9), $c(t)$ is a n-element row vector with time-varying components and $d(t)$ is a real-valued function of time. The number n of state variables is called the *dimension* of the state model (or system).

The system is time invariant if and only if $A(t)$, $b(t)$, $c(t)$, and $d(t)$ are constant, that is, they do not vary with t. If this is the case, the state model is given by

$$\dot{q}(t) = Aq(t) + bx(t) \tag{13.10}$$

$$y(t) = cq(t) + dx(t). \tag{13.11}$$

Here the components of the matrix A and the vectors b and c are real numbers. The element d in (13.11) is a real number. The term $dx(t)$ is a "direct feed" between the input $x(t)$ and output $y(t)$. If $d = 0$, there is no direct connection between $x(t)$ and $y(t)$.

With a_{ij} equal to the ij entry of A and b_i equal to the ith component of b, we can write (13.10) in the expanded form

$$\dot{q}_1(t) = a_{11}q_1(t) + a_{12}q_2(t) + \cdots + a_{1n}q_n(t) + b_1x(t)$$

$$\dot{q}_2(t) = a_{21}q_1(t) + a_{22}q_2(t) + \cdots + a_{2n}q_n(t) + b_2x(t)$$

$$\vdots \qquad\qquad \vdots$$

$$\dot{q}_n(t) = a_{n1}q_1(t) + a_{n2}q_2(t) + \cdots + a_{nn}q_n(t) + b_nx(t).$$

With $c = [c_1 \quad c_2 \quad \cdots \quad c_n]$, the expanded form of (13.11) is

$$y(t) = c_1q_1(t) + c_2q_2(t) + \cdots + c_nq_n(t) + dx(t).$$

From the expanded form of the state equations, we see that the derivative $\dot{q}_i(t)$ of the ith state and the output $y(t)$ are equal to linear combinations of all the states and the input.

13.2

CONSTRUCTION OF STATE MODELS

In the first part of this section we show how to construct a state model from the input/output differential equation of the system. We begin with the first-order case.

Consider the single-input single-output continuous-time system given by the first-order input/output differential equation

$$\dot{y}(t) = f(y(t), x(t), t). \qquad (13.12)$$

If we define the state $q(t)$ of the system to be equal to $y(t)$, we obtain the state model

$$\dot{q}(t) = f(q(t), x(t), t)$$

$$y(t) = x(t).$$

Thus we can easily construct a state model from a first-order input/output differential equation. If the given system is linear and time invariant so that

$$\dot{y}(t) = -ay(t) + bx(t)$$

for some constants a and b, then the state model is

$$\dot{q}(t) = -aq(t) + bx(t)$$

$$y(t) = q(t).$$

In terms of the notation of (13.10) and (13.11), we have that the coefficients A, b, c, and d of this state model are

$$A = -a, \qquad b = b, \qquad c = 1, \qquad d = 0.$$

Now suppose that the given system has the second-order input/output differential equation

$$\ddot{y}(t) = f(y(t), \dot{y}(t), x(t), t). \qquad (13.13)$$

If we define the states by

$$q_1(t) = y(t), \qquad q_2(t) = \dot{y}(t),$$

we obtain the state equations

$$\dot{q}_1(t) = q_2(t)$$

$$\dot{q}_2(t) = f(q_1(t), q_2(t), x(t), t)$$

$$y(t) = q_1(t).$$

If the system with input/output differential equation (13.13) is linear and time invariant, (13.13) can be written in the form

$$\ddot{y}(t) = -a_1\dot{y}(t) - a_0y(t) + b_0x(t).$$

Then with $q_1(t) = y(t)$ and $q_2(t) = \dot{y}(t)$, the state model is

$$\begin{bmatrix} \dot{q}_1(t) \\ \dot{q}_2(t) \end{bmatrix} = \begin{bmatrix} 0 & 1 \\ -a_0 & -a_1 \end{bmatrix} \begin{bmatrix} q_1(t) \\ q_2(t) \end{bmatrix} + \begin{bmatrix} 0 \\ b_0 \end{bmatrix} x(t)$$

$$y(t) = \begin{bmatrix} 1 & 0 \end{bmatrix} \begin{bmatrix} q_1(t) \\ q_2(t) \end{bmatrix}.$$

The definition of states in terms of the output and derivatives of the output extends to any system given by the nth-order input/output differential equation

$$y^{(n)}(t) = f(y(t), y^{(1)}(t), \dots, y^{(n-1)}(t), x(t), t). \tag{13.14}$$

If we define the states by

$$q_i(t) = y^{(i-1)}(t), \qquad i = 1, 2, \dots, n,$$

the resulting state equations are

$$\dot{q}_1(t) = q_2(t)$$

$$\dot{q}_2(t) = q_3(t)$$

$$\vdots$$

$$\dot{q}_{n-1}(t) = q_n(t)$$

$$\dot{q}_n(t) = f(q_1(t), q_2(t), \dots, q_n(t), x(t), t)$$

$$y(t) = q_1(t).$$

From the above constructions, it is tempting to conclude that we can always define the states of a system to be equal to the output $y(t)$ and derivatives of

$y(t)$. Unfortunately, this is not the case. For instance, suppose that the system is given by the linear second-order input/output differential equation

$$\ddot{y}(t) + a_1\dot{y}(t) + a_0y(t) = b_1\dot{x}(t) + b_0x(t), \tag{13.15}$$

where $b_1 \neq 0$. Note that (13.15) is not a special case of (13.14) since $\ddot{y}(t)$ depends on $\dot{x}(t)$.

If we define $q_1(t) = y(t)$ and $q_2(t) = \dot{y}(t)$, we cannot eliminate the term $b_1\dot{x}(t)$ in (13.15). Thus there is no state model with respect to this definition of states. But the system does have the following state model:

$$\begin{bmatrix} \dot{q}_1(t) \\ \dot{q}_2(t) \end{bmatrix} = \begin{bmatrix} 0 & 1 \\ -a_0 & -a_1 \end{bmatrix}\begin{bmatrix} q_1(t) \\ q_2(t) \end{bmatrix} + \begin{bmatrix} 0 \\ 1 \end{bmatrix}x(t), \quad y(t) = [b_0 \ \ b_1]\begin{bmatrix} q_1(t) \\ q_2(t) \end{bmatrix}. \tag{13.16}$$

To verify that (13.16) is a state model, we must show that the input/output differential equation corresponding to (13.16) is the same as (13.15). Expanding (13.16), we have

$$\dot{q}_1(t) = q_2(t) \tag{13.17}$$

$$\dot{q}_2(t) = -a_0q_1(t) - a_1q_2(t) + x(t) \tag{13.18}$$

$$y(t) = b_0q_1(t) + b_1q_2(t). \tag{13.19}$$

Differentiating both sides of (13.19) and using (13.17) and (13.18), we obtain

$$\begin{aligned} \dot{y}(t) &= b_0q_2(t) + b_1[-a_0q_1(t) - a_1q_2(t) + x(t)] \\ &= -a_1y(t) + (a_1b_0 - a_0b_1)q_1(t) + b_0q_2(t) + b_1x(t). \end{aligned} \tag{13.20}$$

Differentiating both sides of (13.20) and again using (13.17) and (13.18), we get

$$\begin{aligned} \ddot{y}(t) &= -a_1\dot{y}(t) + (a_1b_0 - a_0b_1)q_2(t) \\ &\quad + b_0[-a_0q_1(t) - a_1q_2(t) + x(t)] + b_1\dot{x}(t) \\ &= -a_1\dot{y}(t) - a_0y(t) + b_0x(t) + b_1\dot{x}(t). \end{aligned}$$

This is the same as the input/output differential equation (13.15) of the given system, and thus (13.16) is a state model.

The states $q_1(t)$ and $q_2(t)$ in the state model (13.16) can be expressed in terms of $x(t)$, $y(t)$, and $\dot{y}(t)$. We leave the derivation of these expressions to the interested reader (see Problem 13.9).

Now consider the linear time-invariant system given by the nth-order input/output differential equation

$$y^{(n)}(t) + \sum_{i=0}^{n-1} a_iy^{(i)}(t) = \sum_{i=0}^{n-1} b_ix^{(i)}(t).$$

This system has the n-dimensional state model $\dot{q}(t) = Aq(t) + bx(t)$, $y(t) = cq(t)$, where

$$A = \begin{bmatrix} 0 & 1 & 0 & \cdots & & 0 \\ 0 & 0 & 1 & & & 0 \\ \cdot & & & & & \cdot \\ \cdot & & & \cdot & & \cdot \\ \cdot & & & & \cdot & \cdot \\ 0 & 0 & 0 & & & 1 \\ -a_0 & -a_1 & -a_2 & \cdots & & -a_{n-1} \end{bmatrix}, \quad b = \begin{bmatrix} 0 \\ 0 \\ \cdot \\ \cdot \\ \cdot \\ 0 \\ 1 \end{bmatrix},$$

$$c = [b_0 \quad b_1 \quad b_2 \quad \cdots \quad b_{n-1}].$$

The verification that this is a state model is omitted.

INTEGRATOR REALIZATIONS Any linear time-invariant system given by the n-dimensional state model

$$\dot{q}(t) = Aq(t) + bx(t)$$

$$y(t) = cq(t) + dx(t)$$

can be realized by an interconnection of n integrators and combinations of adders, subtracters, and scalar multipliers. The steps of the realization process are as follows.

Step 1. For each state variable $q_i(t)$, construct an integrator and define the output of the integrator to be $q_i(t)$. The input to the ith integrator will then be equal to $\dot{q}_i(t)$. Note that if there are n state variables, the integrator realization will contain n integrators.

Step 2. Put an adder/subtracter in front of each integrator. Feed into the adders/subtracters scalar multiples of the states and input according to the vector equation $\dot{q}(t) = Aq(t) + bx(t)$.

Step 3. Put scalar multiples of the states and input into an adder/subtracter to realize the output $y(t)$ in accordance with the equation $y(t) = cq(t) + dx(t)$.

EXAMPLE 13.1. Consider a two-dimensional state model with arbitrary coefficients; that is,

$$\begin{bmatrix} \dot{q}_1(t) \\ \dot{q}_2(t) \end{bmatrix} = \begin{bmatrix} a_{11} & a_{12} \\ a_{21} & a_{22} \end{bmatrix} \begin{bmatrix} q_1(t) \\ q_2(t) \end{bmatrix} + \begin{bmatrix} b_1 \\ b_2 \end{bmatrix} x(t),$$

$$y(t) = [c_1 \quad c_2] \begin{bmatrix} q_1(t) \\ q_2(t) \end{bmatrix}.$$

Following the steps above, we obtain the realization shown in Figure 13.2.

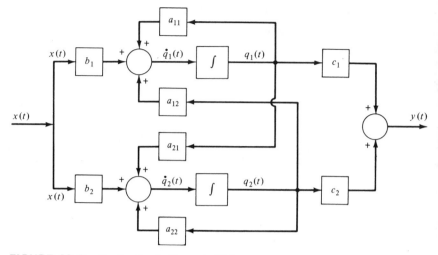

FIGURE 13.2. Realization in Example 13.1.

There is a converse to the result that any linear time-invariant system given by a state model has an integrator realization. Namely, any system specified by an interconnection consisting of n integrators and combinations of adders, subtracters, and scalar multipliers has a state model of dimension n. A state model can be computed directly from the interconnection by employing the following steps.

Step 1. Define the output of each integrator in the interconnection to be a state variable. Then if the output of the ith integrator is $q_i(t)$, the input to this integrator is $\dot{q}_i(t)$.

Step 2. By looking at the interconnection, express each $\dot{q}_i(t)$ in terms of a sum of scalar multiples of the states and input. Writing these relationships in matrix form yields the vector equation $\dot{q}(t) = Aq(t) + bx(t)$.

Step 3. Again looking at the interconnection, express the output $y(t)$ in terms of scalar multiples of the states and input. Writing this in vector form yields the output equation $y(t) = cq(t) + dx(t)$.

EXAMPLE 13.2. Consider the system shown in Figure 13.3. Denoting the output of the first integrator by $q_1(t)$ and the output of the second integrator by $q_2(t)$, from Figure 13.3 we have

$$\dot{q}_1(t) = -q_1(t) - 3q_2(t) + x(t)$$

$$\dot{q}_2(t) = q_1(t) + 2x(t).$$

Also from Figure 13.3, we have

$$y(t) = q_1(t) + q_2(t) + 2x(t).$$

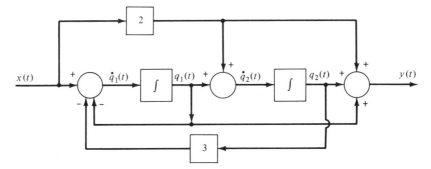

FIGURE 13.3.   System in Example 13.2.

Thus the coefficient matrices of the state model are

$$A = \begin{bmatrix} -1 & -3 \\ 1 & 0 \end{bmatrix}, \qquad b = \begin{bmatrix} 1 \\ 2 \end{bmatrix}, \qquad c = [1 \ \ 1], \qquad d = 2.$$

From the above results we see that there is a one-to-one correspondence between integrator realizations and state models.

MULTI-INPUT MULTI-OUTPUT SYSTEMS   The state model generalizes very easily to multi-input multi-output systems. In particular, the state model of a p-input r-output linear time-invariant finite-dimensional continuous-time system is given by

$$\dot{q}(t) = Aq(t) + Bx(t)$$

$$y(t) = Cq(t) + Dx(t),$$

where now B is a $n \times p$ matrix of real numbers, C is a $r \times n$ matrix of real numbers, and D is a $r \times p$ matrix of real numbers. The matrix A is still $n \times n$, as in the single-input single-output case.

If a p-input r-output system is specified by a collection of coupled input/output differential equations, a state model can be constructed by generalizing the procedure given above in the single-input single-output case. The process is illustrated by the following example.

EXAMPLE 13.3.   Again consider the two-car system that was studied in Examples 1.20, 2.9, and 6.9. The system is given by the input/output differential equations

$$\dot{v}_1(t) + \frac{k_f}{M} v_1(t) = \frac{1}{M} x_1(t)$$

$$\dot{v}_2(t) + \frac{k_f}{M} v_2(t) = \frac{1}{M} x_2(t)$$

$$\dot{w}(t) = v_2(t) - v_1(t).$$

Here $v_1(t)$ and $v_2(t)$ are the velocities of the two cars, and $w(t)$ is the distance between the cars. The inputs to the system are the forces $x_1(t)$ and $x_2(t)$ applied to the cars. So the system is a two-input system. We shall take the output to be

$$y(t) = \begin{bmatrix} v_1(t) \\ v_2(t) \\ w(t) \end{bmatrix}.$$

With this definition of the output, the system is a three-output system. Now we define the states by

$$q_1(t) = v_1(t)$$

$$q_2(t) = v_2(t)$$

$$q_3(t) = w(t).$$

Then the state model of the system is

$$\begin{bmatrix} \dot{q}_1(t) \\ \dot{q}_2(t) \\ \dot{q}_3(t) \end{bmatrix} = \begin{bmatrix} \dfrac{-k_f}{M} & 0 & 0 \\ 0 & \dfrac{-k_f}{M} & 0 \\ -1 & 1 & 0 \end{bmatrix} \begin{bmatrix} q_1(t) \\ q_2(t) \\ q_3(t) \end{bmatrix} + \begin{bmatrix} \dfrac{1}{M} & 0 \\ 0 & \dfrac{1}{M} \\ 0 & 0 \end{bmatrix} \begin{bmatrix} x_1(t) \\ x_2(t) \end{bmatrix}$$

$$y(t) = \begin{bmatrix} 1 & 0 & 0 \\ 0 & 1 & 0 \\ 0 & 0 & 1 \end{bmatrix} \begin{bmatrix} q_1(t) \\ q_2(t) \\ q_3(t) \end{bmatrix}.$$

If a p-input r-output system is given by an interconnection of integrators, adders, subtracters, and scalar multipliers, a state model can be constructed directly from the interconnection. The process is very similar to the steps given above in the single-input single-output case.

EXAMPLE 13.4. Consider the two-input two-output system shown in Figure 13.4. The transfer function matrix of this system was computed in Example 6.10. From Figure 13.4 we have

$$\dot{q}_1(t) = -3y_1(t) + x_1(t)$$

$$\dot{q}_2(t) = x_2(t)$$

$$y_1(t) = q_1(t) + q_2(t)$$

$$y_2(t) = q_2(t).$$

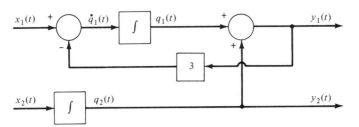

FIGURE 13.4.   System in Example 13.4.

Inserting the expression for $y_1(t)$ into the expression for $\dot{q}_1(t)$, we obtain

$$\dot{q}_1(t) = -3[q_1(t) + q_2(t)] + x_1(t).$$

Putting these equations in matrix form, we obtain the state model

$$\begin{bmatrix} \dot{q}_1(t) \\ \dot{q}_2(t) \end{bmatrix} = \begin{bmatrix} -3 & -3 \\ 0 & 0 \end{bmatrix} \begin{bmatrix} q_1(t) \\ q_2(t) \end{bmatrix} + \begin{bmatrix} 1 & 0 \\ 0 & 1 \end{bmatrix} \begin{bmatrix} x_1(t) \\ x_2(t) \end{bmatrix}$$

$$\begin{bmatrix} y_1(t) \\ y_2(t) \end{bmatrix} = \begin{bmatrix} 1 & 1 \\ 0 & 1 \end{bmatrix} \begin{bmatrix} q_1(t) \\ q_2(t) \end{bmatrix}.$$

13.3

SOLUTION OF STATE EQUATIONS

Consider the p-input r-output linear time-invariant continuous-time system given by the state model

$$\dot{q}(t) = Aq(t) + Bx(t) \tag{13.21}$$

$$y(t) = Cq(t) + Dx(t). \tag{13.22}$$

Recall that the matrix A is $n \times n$, B is $n \times p$, C is $r \times n$, and D is $r \times p$. Given an initial state $x(0)$ at initial time $t = 0$ and an input $x(t)$, $t \geq 0$, our objective is to solve (13.21) for the resulting state response $x(t)$. From this we can then compute the output response $y(t)$ by using (13.22).

We begin by considering the free (unforced) vector differential equation

$$\dot{q}(t) = Aq(t), \qquad t > 0, \tag{13.23}$$

with initial state $q(0)$. To solve (13.23), we need to define the *matrix exponential* e^{At}, which is a natural generalization of the scalar exponential e^{at}. For each real value of t, e^{At} is defined by the matrix power series

$$e^{At} = I_n + At + \frac{A^2 t^2}{2!} + \frac{A^3 t^3}{3!} + \frac{A^4 t^4}{4!} + \cdots, \tag{13.24}$$

where I_n is the $n \times n$ identity matrix. The matrix exponential e^{At} is a $n \times n$ matrix of time functions. Later we will show how the elements of e^{At} can be computed by using the Laplace transform.

There are a couple of properties of e^{At} that we shall need. First, for any real numbers t and λ, we have that

$$e^{A(t+\lambda)} = e^{At}e^{A\lambda}. \tag{13.25}$$

The relationship (13.25) can be checked by setting $t = t + \lambda$ in (13.24).

Taking $\lambda = -t$ in (13.25), we obtain

$$e^{At}e^{-At} = e^{A(t-t)} = I_n. \tag{13.26}$$

The relationship (13.26) shows that the matrix e^{At} always has an inverse, which is equal to the matrix e^{-At}.

We define the derivative $d/dt\ (e^{At})$ of the matrix exponential e^{At} to be the matrix formed by taking the derivative of the components of e^{At}. The derivative $d/dt\ (e^{At})$ can be computed by taking the derivative of the terms comprising the matrix power series in (13.24). The result is

$$\frac{d}{dt}\, e^{At} = A + A^2 t + \frac{A^3 t^2}{2!} + \frac{A^4 t^3}{3!} + \cdots$$

$$= A\left(I_n + At + \frac{A^2 t^2}{2!} + \frac{A^3 t^3}{3!} + \cdots\right)$$

$$= Ae^{At} = e^{At}A. \tag{13.27}$$

Returning to the problem of solving (13.23), we claim that the solution is

$$q(t) = e^{At}q(0), \qquad t \geq 0. \tag{13.28}$$

To verify that the expression (13.28) for $q(t)$ is the solution, take the derivative of both sides of (13.28). This gives

$$\frac{d}{dt}\, q(t) = \frac{d}{dt}\, [e^{At}q(0)]$$

$$= \left[\frac{d}{dt}\, e^{At}\right] q(0)$$

$$= Ae^{At}q(0) = Aq(t).$$

Thus the expression (13.28) for $q(t)$ does satisfy the vector differential equation (13.23).

From (13.28) we see that the state $q(t)$ at time t resulting from state $q(0)$ at time $t = 0$ with no input applied for $t \geq 0$ can be computed by multiplying $q(0)$ by the matrix e^{At}. As a result of this property, the matrix e^{At} is called the *state-transition matrix* of the system.

SOLUTION TO FORCED EQUATION We shall now derive an expression for the solution to the forced equation (13.21). The solution can be computed by using a matrix version of the integrating factor method. In Chapter 1 we used this solution technique to solve a first-order scalar differential equation. The following steps are simply a matrix version of the derivation given in Section 1.5.

Multiplying both sides of (13.21) on the left by e^{-At} and rearranging terms, we get

$$e^{-At}[\dot{q}(t) - Aq(t)] = e^{-At}Bx(t). \tag{13.29}$$

Using (13.27), we have that the left side of (13.29) is equal to the derivative of $e^{-At}q(t)$. Thus

$$\frac{d}{dt}[e^{-At}q(t)] = e^{-At}Bx(t). \tag{13.30}$$

Integrating both sides of (13.30) gives

$$e^{-At}q(t) = q(0) + \int_0^t e^{-A\lambda} Bx(\lambda)\, d\lambda.$$

Finally, multiplying both sides on the left by e^{At}, we obtain

$$q(t) = e^{At}q(0) + \int_0^t e^{A(t-\lambda)} Bx(\lambda)\, d\lambda. \tag{13.31}$$

Equation (13.31) is the complete solution to (13.21) resulting from initial state $q(0)$ and input $x(t)$ applied for $t \geq 0$. This equation is a generalization of the expression we gave for the solution of a first-order scalar differential equation [see (2.5)].

OUTPUT RESPONSE Inserting (13.31) into (13.22), we have that the complete output response $y(t)$ resulting from initial state $q(0)$ and input $x(t)$ is given by

$$y(t) = Ce^{At}q(0) + \int_0^t C e^{A(t-\lambda)} Bx(\lambda)\, d\lambda + Dx(t), \qquad t \geq 0. \tag{13.32}$$

Using the definition of the unit impulse $\delta(t)$, we can rewrite (13.32) in the form

$$y(t) = Ce^{At}q(0) + \int_0^t \{C e^{A(t-\lambda)} Bx(\lambda)$$
$$+ D\delta(t - \lambda)x(\lambda)\}\, d\lambda, \qquad t \geq 0. \tag{13.33}$$

Defining

$$y_{zi}(t) = Ce^{At}q(0) \tag{13.34}$$

and

$$y_{zs}(t) = \int_0^t \{C \, e^{A(t-\lambda)} Bx(\lambda) + D\delta(t - \lambda)x(\lambda)\} \, d\lambda, \tag{13.35}$$

we have

$$y(t) = y_{zi}(t) + y_{zs}(t).$$

The term $y_{zi}(t)$ is called the *zero-input response* since it is the complete output response when the input $x(t)$ is zero. The term $y_{zs}(t)$ is called the *zero-state response* since it is the complete output response when the initial state $q(0)$ is zero.

The zero-state response $y_{zs}(t)$ is the same as the response to input $x(t)$ with no initial energy in the system at time $t = 0$. In the single-input single-output case, from the results in Chapter 4 we know that

$$y_{zs}(t) = (h * x)(t) = \int_0^t h(t - \lambda)x(\lambda) \, d\lambda, \qquad t \geq 0, \tag{13.36}$$

where $h(t)$ is the impulse response of the system. Equating the right sides of (13.35) and (13.36), we have

$$\int_0^t \{C \, e^{A(t-\lambda)} Bx(\lambda) + D\delta(t - \lambda)x(\lambda)\} \, d\lambda = \int_0^t h(t - \lambda)x(\lambda) \, d\lambda. \tag{13.37}$$

For (13.37) to hold for all inputs $x(t)$, it must be true that

$$h(t - \lambda) = C \, e^{A(t-\lambda)} B + D\delta(t - \lambda), \qquad t \geq 0$$

or

$$h(t) = Ce^{At}B + D\delta(t), \qquad t \geq 0. \tag{13.38}$$

Using the relationship (13.38), we can compute the impulse response directly from the coefficient matrices of the state model of the system.

SOLUTION VIA THE LAPLACE TRANSFORM

Again consider a p-input r-output n-dimensional system given by the state model (13.21) and (13.22). To compute the state and output responses resulting from initial state $q(0)$ and input $x(t)$ applied for $t \geq 0$, we can use the expressions (13.31) and (13.32).

To avoid having to evaluate the integrals in (13.31) and (13.32), we can use the Laplace transform to compute the state and output responses of the system. The transform approach is a matrix version of the procedure we gave for solving a scalar first-order differential equation. The steps are as follows.

Taking the Laplace transform of the state equation (13.21), we obtain

$$sQ(s) - q(0) = AQ(s) + BX(s), \tag{13.39}$$

where $Q(s)$ is the Laplace transform of the state vector $q(t)$ with the transform taken component by component; that is,

$$Q(s) = \begin{bmatrix} Q_1(s) \\ Q_2(s) \\ \cdot \\ \cdot \\ \cdot \\ Q_n(s) \end{bmatrix},$$

where $Q_i(s)$ is the Laplace transform of $q_i(t)$. The term $X(s)$ in (13.39) is the Laplace transform of the input $x(t)$, where again the transform is taken component by component.

Now we can rewrite (13.39) in the form

$$(sI - A)Q(s) = q(0) + BX(s), \tag{13.40}$$

where I is the $n \times n$ identity matrix. Note that in factoring out s from $sQ(s)$, we multiplied s by I. The reason for this is that we cannot subtract A from the scalar s since A is a $n \times n$ matrix. However, we can subtract A from the diagonal matrix sI. By definition of the identity matrix, the product $(sI - A)Q(s)$ is equal to $sQ(s) - AQ(s)$.

It turns out that the matrix $sI - A$ in (13.40) always has an inverse $(sI - A)^{-1}$. (We will see why this is true later.) Thus we can multiply both sides of (13.40) on the left by $(sI - A)^{-1}$. This gives

$$Q(s) = (sI - A)^{-1}q(0) + (sI - A)^{-1}BX(s). \tag{13.41}$$

The right side of (13.41) is the Laplace transform of the state response resulting from initial state $q(0)$ and input $x(t)$ applied for $t \geq 0$. The state response $q(t)$ can then be computed by taking the inverse Laplace transform of the right side of (13.41).

Comparing (13.41) with (13.31), we see that $(sI - A)^{-1}$ is the Laplace transform of the state-transition matrix e^{At}. Since e^{At} is a well-defined function of t, this shows that the inverse $(sI - A)^{-1}$ must exist. Also note that

$$e^{At} = \text{inverse Laplace transform of } (sI - A)^{-1}. \tag{13.42}$$

The relationship (13.42) is very useful for computing e^{At}. We shall illustrate this in an example given below.

Taking the Laplace transform of the output equation (13.22), we have

$$Y(s) = CQ(s) + DX(s). \tag{13.43}$$

Inserting (13.41) into (13.43), we get

$$Y(s) = C(sI - A)^{-1}q(0) + [C(sI - A)^{-1}B + D]X(s). \tag{13.44}$$

The right side of (13.44) is the Laplace transform of the complete output response resulting from initial state $q(0)$ and input $x(t)$.

If $q(0) = 0$ (no initial energy in the system), from the results in Section 6.7, we know that

$$Y(s) = Y_{zs}(s) = H(s)X(s), \tag{13.45}$$

where $H(s)$ is the transfer function matrix of the system. Setting $q(0) = 0$ in (13.44) and comparing (13.44) and (13.45), we see that

$$H(s) = C(sI - A)^{-1}B + D. \tag{13.46}$$

So the transfer function matrix can be computed directly from the coefficient matrices of the state model of the system. Note that $H(s)$ is the Laplace transform of the impulse-response function matrix $H(t)$ given by

$$H(t) = \begin{cases} 0, & t < 0 \\ Ce^{At}B + D\delta(t), & t \geq 0. \end{cases}$$

EXAMPLE 13.5. Consider the two-input three-output two-dimensional system with state model $\dot{q}(t) = Aq(t) + Bx(t)$, $y(t) = Cq(t)$, where

$$A = \begin{bmatrix} -3 & -1 \\ -2 & -1 \end{bmatrix}, \quad B = \begin{bmatrix} 3 & 2 \\ 2 & 1 \end{bmatrix}, \quad C = \begin{bmatrix} 1 & 2 \\ -2 & 2 \\ 1 & -1 \end{bmatrix}.$$

A realization of this system is given in Figure 13.5.

We shall first compute the state-transition matrix e^{At} of the system. We have that

$$(sI - A)^{-1} = \begin{bmatrix} s + 3 & 1 \\ 2 & s + 1 \end{bmatrix}^{-1} = \frac{1}{s^2 + 4s + 5} \begin{bmatrix} s + 1 & -1 \\ -2 & s + 3 \end{bmatrix}$$

$$= \frac{1}{(s + 2)^2 + 1} \begin{bmatrix} s + 1 & -1 \\ -2 & s + 3 \end{bmatrix}.$$

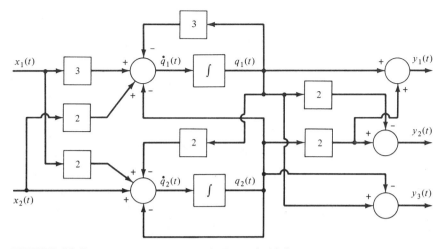

FIGURE 13.5. Realization of system in Example 13.5.

By (13.42) e^{At} is equal to the inverse Laplace transform of $(sI - A)^{-1}$. Using table lookup, we get

$$e^{At} = e^{-2t} \begin{bmatrix} \cos t - \sin t & -\sin t \\ -2 \sin t & \cos t + \sin t \end{bmatrix}.$$

Now the state response $q(t)$ resulting from any initial state $q(0)$ with zero input is given by

$$q(t) = e^{At}q(0).$$

For example, if

$$q(0) = \begin{bmatrix} 1 \\ -1 \end{bmatrix},$$

then

$$q(t) = e^{-2t} \begin{bmatrix} \cos t - \sin t & -\sin t \\ -2 \sin t & \cos t + \sin t \end{bmatrix} \begin{bmatrix} 1 \\ -1 \end{bmatrix}$$

$$= e^{-2t} \begin{bmatrix} \cos t \\ -\cos t - 3 \sin t \end{bmatrix}.$$

Now let us compute the state response $q(t)$ resulting from the input

$$x(t) = \begin{bmatrix} u(t) \\ e^{-t} \end{bmatrix}, \qquad t \geq 0.$$

Then

$$X(s) = \begin{bmatrix} \dfrac{1}{s} \\ \dfrac{1}{s+1} \end{bmatrix}.$$

Using (13.41) with $q(0) = 0$, we have

$$Q(s) = \begin{bmatrix} s+3 & 1 \\ 2 & s+1 \end{bmatrix}^{-1} \begin{bmatrix} 3 & 2 \\ 2 & 1 \end{bmatrix} \begin{bmatrix} \dfrac{1}{s} \\ \dfrac{1}{s+1} \end{bmatrix}$$

$$= \frac{1}{s^2 + 4s + 5} \begin{bmatrix} s+1 & -1 \\ -2 & s+3 \end{bmatrix} \begin{bmatrix} \dfrac{5s+3}{s(s+1)} \\ \dfrac{3s+2}{s(s+1)} \end{bmatrix}$$

$$= \frac{1}{[(s+2)^2 + 1]s(s+1)} \begin{bmatrix} 5s^2 + 5s + 1 \\ 3s^2 + s \end{bmatrix}.$$

Expanding the components of $Q(s)$, we have

$$Q(s) = \begin{bmatrix} \dfrac{0.3s + 5.7}{(s+2)^2 + 1} + \dfrac{0.2}{s} - \dfrac{0.5}{s+1} \\ \dfrac{s+6}{(s+2)^2 + 1} - \dfrac{1}{s+1} \end{bmatrix}.$$

Taking the inverse Laplace transform of $Q(s)$, we get

$$q(t) = \begin{bmatrix} e^{-2t}(0.3 \cos t + 5.1 \sin t) + 0.2 - 0.5e^{-t} \\ e^{-2t}(\cos t + 4 \sin t) - e^{-t} \end{bmatrix}.$$

The output response is

$$y(t) = Cq(t) = \begin{bmatrix} e^{-2t}(2.3 \cos t + 13.1 \sin t) + 0.2 - 2.5e^{-t} \\ e^{-2t}(1.4 \cos t - 2.2 \sin t) - 0.4 - e^{-t} \\ e^{-2t}(-0.7 \cos t + 1.1 \sin t) + 0.2 + 0.5e^{-t} \end{bmatrix}.$$

Since $q(0) = 0$, we could have computed $y(t)$ by using the transfer function representation $Y(s) = H(s)X(s)$. In this example, the transfer function is

$$H(s) = C(sI - A)^{-1}B$$

$$= \begin{bmatrix} 1 & 2 \\ -2 & 2 \\ 1 & -1 \end{bmatrix} \begin{bmatrix} s + 3 & 1 \\ 2 & s + 1 \end{bmatrix}^{-1} \begin{bmatrix} 3 & 2 \\ 2 & 1 \end{bmatrix}$$

$$= \frac{1}{s^2 + 4s + 5} \begin{bmatrix} 1 & 2 \\ -2 & 2 \\ 1 & -1 \end{bmatrix} \begin{bmatrix} s + 1 & -1 \\ -2 & s + 3 \end{bmatrix} \begin{bmatrix} 3 & 2 \\ 2 & 1 \end{bmatrix}$$

$$= \frac{1}{s^2 + 4s + 5} \begin{bmatrix} s - 3 & 2s + 5 \\ -2s - 6 & 2s + 8 \\ s + 3 & -s - 4 \end{bmatrix} \begin{bmatrix} 3 & 2 \\ 2 & 1 \end{bmatrix}$$

$$= \frac{1}{s^2 + 4s + 5} \begin{bmatrix} 7s + 1 & 4s - 1 \\ 4s - 10 & -2s - 4 \\ s + 1 & s + 2 \end{bmatrix}.$$

13.4

DISCRETE-TIME SYSTEMS

A p-input r-output finite-dimensional linear time-invariant discrete-time system can be modeled by the state equations

$$q(kT + T) = Aq(kT) + Bx(kT) \qquad (13.47)$$

$$y(kT) = Cq(kT) + Dx(kT). \qquad (13.48)$$

The state vector $q(kT)$ is the n-element column vector

$$q(kT) = \begin{bmatrix} q_1(kT) \\ q_2(kT) \\ \cdot \\ \cdot \\ \cdot \\ q_n(kT) \end{bmatrix}.$$

As in the continuous-time case, the state $q(kT)$ at time kT represents the past history (before time kT) of the system.

The input $x(kT)$ and output $y(kT)$ are the column vectors

$$x(kT) = \begin{bmatrix} x_1(kT) \\ x_2(kT) \\ \cdot \\ \cdot \\ \cdot \\ x_p(kT) \end{bmatrix}, \qquad y(kT) = \begin{bmatrix} y_1(kT) \\ y_2(kT) \\ \cdot \\ \cdot \\ \cdot \\ y_r(kT) \end{bmatrix}.$$

The matrices A, B, C, and D in (13.47) and (13.48) are $n \times n$, $n \times p$, $r \times n$, and $r \times p$, respectively. Equation (13.47) is a first-order vector difference equation. Equation (13.48) is the output equation of the system.

CONSTRUCTION OF STATE MODELS   Suppose that we have a single-input single-output linear time-invariant discrete-time system with the input/output difference equation

$$y(kT + nT) + \sum_{i=0}^{n-1} a_i y(kT + iT) = bx(kT). \qquad (13.49)$$

By defining the states

$$q_i(kT) = y(kT + iT), \qquad 0, 1, 2, \ldots, n - 1, \qquad (13.50)$$

we obtain a state model of the form (13.47) and (13.48) with

$$A = \begin{bmatrix} 0 & 1 & 0 & \cdots & 0 \\ 0 & 0 & 1 & & 0 \\ \cdot & \cdot & & \cdot & \cdot \\ \cdot & \cdot & & \cdot & \cdot \\ 0 & 0 & 0 & \cdots & 1 \\ -a_0 & -a_1 & -a_2 & \cdots & -a_{n-1} \end{bmatrix}, \quad B = \begin{bmatrix} 0 \\ 0 \\ \cdot \\ \cdot \\ 0 \\ b \end{bmatrix},$$

$$C = [1 \ 0 \ 0 \ \cdots \ 0], \qquad D = 0.$$

If the right side of (13.49) is modified so that it is in the more general form

$$\sum_{i=0}^{n-1} b_i x(kT + iT),$$

the state model above is still a state model, except that B and C must be modified so that

$$B = \begin{bmatrix} 0 \\ 0 \\ \cdot \\ \cdot \\ \cdot \\ 0 \\ 1 \end{bmatrix}$$

$$C = [b_0 \ b_1 \ \cdots \ b_{n-1}].$$

In this case, the states $q_i(kT)$ are functions of $y(kT)$, $x(kT)$ and left shifts of $y(kT)$ and $x(kT)$; that is, $q_i(kT)$ is no longer given by (13.50). This state model is the discrete-time counterpart of the state model we generated from a linear constant-coefficient input/output differential equation in the continuous-time case (see Section 13.2).

UNIT-DELAYER REALIZATIONS Given a discrete-time system with the n-dimensional state model (13.47) and (13.48), we can construct a realization of the system consisting of an interconnection of n unit delayers and combinations of adders, subtracters, and scalar multipliers. Conversely, if a discrete-time system is specified by an interconnection of unit delayers, adders, subtracters, and scalar multipliers, we can generate a state model of the form (13.47) and (13.48) directly from the interconnection. The procedure for going from unit-delayer realizations to state models (and conversely) is analogous to that given in the continuous-time case.

EXAMPLE 13.6. Consider the three-input two-output three-dimensional discrete-time system given by the interconnection in Figure 13.6. From the diagram we have

$$q_1(kT + T) = -q_2(kT) + x_1(kT) + x_3(kT)$$

$$q_2(kT + T) = q_1(kT) + x_2(kT)$$

$$q_3(kT + T) = q_2(kT) + x_3(kT)$$

$$y_1(kT) = q_2(kT)$$

$$y_2(kT) = q_1(kT) + q_3(kT) + x_2(kT).$$

Writing these equations in matrix form, we get the state model

$$\begin{bmatrix} q_1(kT + T) \\ q_2(kT + T) \\ q_3(kT + T) \end{bmatrix} = \begin{bmatrix} 0 & -1 & 0 \\ 1 & 0 & 0 \\ 0 & 1 & 0 \end{bmatrix} \begin{bmatrix} q_1(kT) \\ q_2(kT) \\ q_3(kT) \end{bmatrix} + \begin{bmatrix} 1 & 0 & 1 \\ 0 & 1 & 0 \\ 0 & 0 & 1 \end{bmatrix} \begin{bmatrix} x_1(kT) \\ x_2(kT) \\ x_3(kT) \end{bmatrix}$$

$$\begin{bmatrix} y_1(kT) \\ y_2(kT) \end{bmatrix} = \begin{bmatrix} 0 & 1 & 0 \\ 1 & 0 & 1 \end{bmatrix} \begin{bmatrix} q_1(kT) \\ q_2(kT) \\ q_3(kT) \end{bmatrix} + \begin{bmatrix} 0 & 0 & 0 \\ 0 & 1 & 0 \end{bmatrix} \begin{bmatrix} x_1(kT) \\ x_2(kT) \\ x_3(kT) \end{bmatrix}.$$

SOLUTION OF STATE EQUATIONS Again consider the p-input r-output discrete-time system with the state model

$$q(kT + T) = Aq(kT) + Bx(kT) \tag{13.51}$$

$$y(kT) = Cq(kT) + Dx(kT). \tag{13.52}$$

We can solve the vector difference equation (13.51) by using a matrix version of recursion. The process is a straightforward generalization of the recursive procedure for solving a first-order scalar difference equation (see Chapter 3). The steps are as follows.

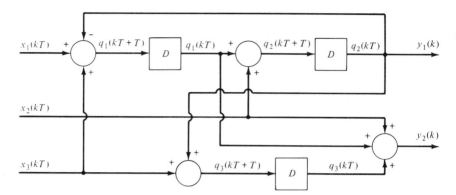

FIGURE 13.6. System in Example 13.6.

We assume that the initial state of the system is the state $q(0)$ at initial time $kT = 0$. Then setting $k = 0$ in (13.51), we obtain

$$q(T) = Aq(0) + Bx(0). \tag{13.53}$$

Setting $k = 1$ in (13.51) and using (13.53), we have

$$\begin{aligned} q(2T) &= Aq(T) + Bx(T) \\ &= A[Aq(0) + Bx(0)] + Bx(T) \\ &= A^2q(0) + ABx(0) + Bx(T). \end{aligned}$$

If we continue this process, for any integer value of $k \geq 1$, we have

$$q(kT) = A^kq(0) + \sum_{i=0}^{k-1} A^{k-i-1}Bx(iT), \qquad k \geq 1. \tag{13.54}$$

The right side of (13.54) is the state response resulting from initial state $q(0)$ and input $x(kT)$ applied for $k \geq 0$. Note that if $x(kT) = 0$ for $k \geq 0$, then

$$q(kT) = A^kq(0). \tag{13.55}$$

From (13.55) we see that the state transition from initial state $q(0)$ to state $q(kT)$ at time kT (with no input applied) is equal to $q(0)$ times the matrix A^k. Therefore, in the discrete-time case the state-transition matrix is the matrix A^k.

Inserting (13.54) into the output equation (13.52), we obtain

$$y(kT) = CA^kq(0) + \sum_{i=0}^{k-1} CA^{k-i-1}Bx(iT) + Dx(kT), \qquad k \geq 1. \tag{13.56}$$

The right side of (13.56) is the complete output response resulting from initial state $q(0)$ and input $x(kT)$. The term

$$y_{zi}(kT) = CA^kq(0), \qquad k \geq 0,$$

is the zero-input response, and the term

$$y_{zs}(kT) = \sum_{i=0}^{k-1} CA^{k-i-1}Bx(iT) + Dx(kT), \qquad k \geq 1$$

$$= \sum_{i=0}^{k} [CA^{k-i-1}Bu(k-i-1) + \Delta(k-i)D]x(iT), \qquad k \geq 1$$

(13.57)

is the zero-state response [where $\Delta(k)$ = unit pulse located at $k = 0$].
In the single-input single-output case,

$$y_{zs}(kT) = \sum_{i=0}^{k} h(kT - iT)x(iT),$$

(13.58)

where $h(kT)$ is the unit-pulse response of the system. Comparing (13.57) and
(13.58), we have that

$$h(kT - iT) = CA^{k-i-1}Bu(k-i-1) + \Delta(k-i)D$$

$$h(kT) = CA^{k-1}u(k-1) + \Delta(k)D$$

or

$$h(kT) = \begin{cases} D, & k = 0 \\ CA^{k-1}B, & k \geq 1. \end{cases}$$

SOLUTION VIA THE z-TRANSFORM Taking the z-transform of the vector difference equation (13.51), we get

$$zQ(z) - zq(0) = AQ(z) + BX(z),$$

(13.59)

where $Q(z)$ and $X(z)$ are the z-transforms of $q(kT)$ and $x(kT)$, respectively, with
transforms taken component by component. Solving (13.59) for $Q(z)$ gives

$$Q(z) = (zI - A)^{-1}zq(0) + (zI - A)^{-1}BX(z).$$

(13.60)

The right side of (13.60) is the z-transform of the state response $q(kT)$ resulting
from initial state $q(0)$ and input $x(kT)$.

Comparing (13.60) and (13.54), we see that $(zI - A)^{-1}$ is the z-transform
of the state-transition matrix A^k. Thus

$$A^k = \text{inverse } z\text{-transform of } (zI - A)^{-1}.$$

(13.61)

Taking the z-transform of (13.52) and using (13.60), we have

$$Y(z) = C(zI - A)^{-1}zq(0) + [C(zI - A)^{-1}B + D]X(z).$$

(13.62)

The right side of (13.62) is the z-transform of the complete output response resulting from initial state $q(0)$ and input $x(kT)$.

If $q(0) = 0$,

$$Y(z) = Y_{zs}(z) = [C(zI - A)^{-1}B + D]X(z). \tag{13.63}$$

Since

$$Y_{zs}(z) = H(z)X(z),$$

where $H(z)$ is the transfer function matrix, by (13.63)

$$H(z) = C(zI - A)^{-1}B + D. \tag{13.64}$$

EXAMPLE 13.7. Again consider the system in Example 13.6. We shall first compute the state-transition matrix A^k for this system. We have that

$$(zI - A)^{-1} = \begin{bmatrix} z & 1 & 0 \\ -1 & z & 0 \\ 0 & -1 & z \end{bmatrix}^{-1} = \frac{1}{(z^2 + 1)z} \begin{bmatrix} z^2 & -z & 0 \\ z & z^2 & 0 \\ -1 & -z & 1 \end{bmatrix}.$$

Expanding the components of $(zI - A)^{-1}$ and using table lookup, we get

$$A^k = \begin{bmatrix} \sin \dfrac{\pi}{2} k & -\left(\cos \dfrac{\pi}{2} k\right) u(k - 1) & 0 \\ \left(\cos \dfrac{\pi}{2} k\right) u(k - 1) & \sin \dfrac{\pi}{2} k & 0 \\ -\Delta(k - 1) + \sin \dfrac{\pi}{2} k & -\left(\cos \dfrac{\pi}{2} k\right) u(k - 1) & \Delta(k - 1) - \sin \dfrac{\pi}{2} \end{bmatrix}$$

We can then determine the state response $q(t)$ resulting from any initial state $q(0)$ (with no input applied) by using (13.55). For example, if

$$q(0) = \begin{bmatrix} 1 \\ 0 \\ 1 \end{bmatrix},$$

then

$$q(kT) = A^k q(0) = \begin{bmatrix} \sin \dfrac{\pi}{2} k \\ \left(\cos \dfrac{\pi}{2} k\right) u(k - 1) \\ 0 \end{bmatrix}, \quad k \geq 0.$$

We conclude the example by computing the transfer-function matrix of the system. From (13.64)

$$H(z) = C(zI - A)^{-1}B + D$$

$$= \frac{1}{(z^2 + 1)z} \begin{bmatrix} 0 & 1 & 0 \\ 1 & 0 & 1 \end{bmatrix} \begin{bmatrix} z^2 & -z & 0 \\ z & z^2 & 0 \\ -1 & -z & 1 \end{bmatrix} \begin{bmatrix} 1 & 0 & 1 \\ 0 & 1 & 0 \\ 0 & 0 & 1 \end{bmatrix} + \begin{bmatrix} 0 & 0 & 0 \\ 0 & 1 & 0 \end{bmatrix}$$

$$= \frac{1}{(z^2 + 1)z} \begin{bmatrix} z & z^2 & z \\ z^2 - 1 & z(z^2 - 1) & z^2 \end{bmatrix}.$$

13.5

EQUIVALENT STATE REPRESENTATIONS

Unlike the transfer function model, the state model of a system is not unique. However, all n-dimensional state models of a given system are "essentially equivalent." We consider the relationship between state models in this section. The following analysis is developed in terms of continuous-time systems. The theory in the discrete-time case is very similar, so we will restrict our attention to the continuous-time case.

Consider a p-input r-output n-dimensional linear time-invariant continuous-time system given by the state model

$$\dot{q}(t) = Aq(t) + Bx(t) \tag{13.65}$$

$$y(t) = Cq(t) + Dx(t). \tag{13.66}$$

Let P denote a fixed $n \times n$ matrix with entries that are real numbers. We require that P be invertible, so the determinant $|P|$ of P is nonzero (see Appendix C). The inverse of P is denoted by P^{-1}.

In terms of the matrix P, we can define a new state vector $\bar{q}(t)$ for the given system, where

$$\bar{q}(t) = Pq(t). \tag{13.67}$$

The relationship (13.67) is called a *coordinate transformation* since it takes us from the original state coordinates to the new state coordinates.

Multiplying both sides of (13.67) on the left by the inverse P^{-1} of P, we get

$$P^{-1}\bar{q}(t) = P^{-1}Pq(t).$$

By definition of the inverse, $P^{-1}P = I$, where I is the $n \times n$ identity matrix. Hence

$$P^{-1}\bar{q}(t) = Iq(t) = q(t)$$

or

$$q(t) = P^{-1}\overline{q}(t). \tag{13.68}$$

Using (13.68), we can go from the new state vector $\overline{q}(t)$ back to the original state vector $q(t)$. Note that if P were not invertible, we would not be able to go back to $q(t)$ from $\overline{q}(t)$. This is the reason P must be invertible.

In terms of the new state vector $\overline{q}(t)$, we can generate a new state-equation model of the given system. The steps are as follows.

Taking the derivative of both sides of (13.67) and using (13.65), we have

$$\dot{\overline{q}}(t) = P\dot{q}(t) = P[Aq(t) + Bx(t)]$$
$$= PAq(t) + PBx(t). \tag{13.69}$$

Inserting the expression (13.68) for $q(t)$ into (13.69) and (13.66), we get

$$\dot{\overline{q}}(t) = PA(P^{-1})\overline{q}(t) + PBx(t) \tag{13.70}$$

$$y(t) = C(P^{-1})\overline{q}(t) + Dx(t). \tag{13.71}$$

Defining the matrices

$$\overline{A} = PA(P^{-1}), \quad \overline{B} = PB, \quad \overline{C} = C(P^{-1}), \quad \overline{D} = D, \tag{13.72}$$

we can write (13.70) and (13.71) in the form

$$\dot{\overline{q}}(t) = \overline{A}\overline{q}(t) + \overline{B}x(t) \tag{13.73}$$

$$y(t) = \overline{C}\overline{q}(t) + \overline{D}x(t). \tag{13.74}$$

Equations (13.73) and (13.74) are the state equations of the given system in terms of the new state vector $\overline{q}(t)$. So we have been able to generate a new n-dimensional state model from the original n-dimensional state model. Since the above construction can be carried out for any invertible $n \times n$ matrix P, and there are an infinite number of such matrices, we can generate an infinite number of new state models from a given state model.

Let us denote the original state model by the quadruple (A, B, C, D) and the new state model by the quadruple $(\overline{A}, \overline{B}, \overline{C}, \overline{D})$. The state models (A, B, C, D) and $(\overline{A}, \overline{B}, \overline{C}, \overline{D})$ are said to be related by the coordinate transformation P since the state vector $\overline{q}(t)$ of the latter is related to the state vector $q(t)$ of the former by the relationship $\overline{q}(t) = Pq(t)$. Any two such state models are said to be *equivalent*. The only difference between two equivalent state models is in the labeling of states. More precisely, the states of $(\overline{A}, \overline{B}, \overline{C}, \overline{D})$ are linear combinations [given by $\overline{q}(t) = Pq(t)$] of the states of (A, B, C, D).

It turns out that any two n-dimensional state models of the same system are equivalent. In other words, if one has two n-dimensional state models of a system, there must be a coordinate transformation that relates the state vectors of the two models. The proof of the existence of the coordinate transformation will not be given.

It should be stressed that the notion of equivalent state models applies only to state models having the same dimension. State models with different dimensions cannot be related by a coordinate transformation.

Any two equivalent state models have exactly the same input/output relationship. In particular, the transfer function matrices corresponding to any two equivalent models are the same. To prove this, let $H(s)$ and $\overline{H}(s)$ denote the transfer function matrices associated with (A, B, C, D) and $(\overline{A}, \overline{B}, \overline{C}, \overline{D})$, respectively; that is,

$$H(s) = C(sI - A)^{-1}B + D \tag{13.75}$$

$$\overline{H}(s) = \overline{C}(sI - \overline{A})^{-1}\overline{B} + \overline{D} \tag{13.76}$$

Inserting the expressions (13.72) for $\overline{A}, \overline{B}, \overline{C}, \overline{D}$ into (13.76), we have

$$\overline{H}(s) = C(P^{-1})[sI - PA(P^{-1})]^{-1}PB + D.$$

Now

$$sI - PA(P^{-1}) = P(sI - A)(P^{-1}).$$

In addition, for any $n \times n$ invertible matrices X, Y, W,

$$(WXY)^{-1} = (Y^{-1})(X^{-1})(W^{-1}).$$

Thus

$$\begin{aligned}
\overline{H}(s) &= C(P^{-1})P(sI - A)^{-1}(P^{-1})PB + D \\
&= C(sI - A)^{-1}B + D \\
&= H(s).
\end{aligned}$$

So the transfer function matrices are the same.

By considering a coordinate transformation, it is sometimes possible to go from one state model (A, B, C, D) to another state model $(\overline{A}, \overline{B}, \overline{C}, \overline{D})$ for which one or more of the coefficients matrices $\overline{A}, \overline{B}, \overline{C}$ have a special form. Such models are called canonical models or canonical forms. Examples are the diagonal form, control-canonical form, and the observer-canonical form. Due to the special structure of these canonical forms, they can result in a significant simplication in the solution to certain classes of problems. For example, the control-canonical form is very useful in the study of state feedback. For an in-depth development of the various canonical forms, we refer the reader to Chen [1984], Kailath [1980], or Brogan [1985].

EXAMPLE OF THE DIAGONAL FORM Consider the *RLC* series circuit shown in Figure 13.7. We shall first determine a state model of the circuit. As noted in Section 13.1, we can define the states to be the current $i(t)$ in the

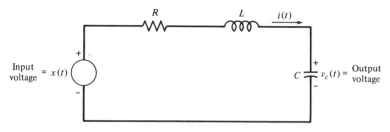

FIGURE 13.7.   Series *RLC* circuit.

inductor and the voltage $v_c(t)$ across the capacitor. Thus we can set

$$q_1(t) = i(t)$$

$$q_2(t) = v_c(t).$$

Summing the voltages around the loop, we have

$$Ri(t) + L \frac{di(t)}{dt} + v_c(t) = x(t).$$

Hence

$$\dot{q}_1(t) = -\frac{R}{L} q_1(t) - \frac{1}{L} q_2(t) + \frac{1}{L} x(t).$$

Also,

$$\dot{q}_2(t) = \frac{1}{C} q_1(t)$$

$$y(t) = q_2(t).$$

So the circuit has the state model $\dot{q}(t) = Aq(t) + Bx(t)$, $y(t) = Cq(t)$, where

$$A = \begin{bmatrix} -\dfrac{R}{L} & -\dfrac{1}{L} \\[2ex] \dfrac{1}{C} & 0 \end{bmatrix}, \qquad B = \begin{bmatrix} \dfrac{1}{L} \\[2ex] 0 \end{bmatrix}, \qquad C = [0 \quad 1].$$

Now we would like to know if there is a coordinate transformation $\overline{q}(t) = Pq(t)$ such that $\overline{A} = PA(P^{-1})$ is in the diagonal form

$$\overline{A} = \begin{bmatrix} \overline{a}_1 & 0 \\ 0 & \overline{a}_2 \end{bmatrix}.$$

Part of the interest in the diagonal form is the simplification that results from this form. For instance, if $\overline{A}$ is in the diagonal form given above, the state transition matrix has the simple form

$$e^{\overline{A}t} = \begin{bmatrix} e^{\overline{a}_1 t} & 0 \\ 0 & e^{\overline{a}_2 t} \end{bmatrix}.$$

Not every matrix A can be put into a diagonal form $\overline{A}$ by a coordinate transformation $\overline{q}(t) = Pq(t)$. There are systematic methods for studying the existence and computation of diagonal forms. In the following development we consider a direct procedure for determining the existence of a diagonal form for the series *RLC* circuit.

First, letting det $(sI - A)$ denote the determinant of $sI - A$ (see Appendix C), we have

$$\det (sI - A) = \det \begin{bmatrix} s + \dfrac{R}{L} & \dfrac{1}{L} \\ -\dfrac{1}{C} & s \end{bmatrix} = s^2 + \dfrac{R}{L} s + \dfrac{1}{LC}. \quad (13.77)$$

With $\overline{A}$ equal to the diagonal form given above, we have

$$\det (sI - \overline{A}) = \det \begin{bmatrix} s - \overline{a}_1 & 0 \\ 0 & s - \overline{a}_2 \end{bmatrix} = (s - \overline{a}_1)(s - \overline{a}_2). \quad (13.78)$$

From results in matrix algebra, it can be shown that

$$\det (sI - A) = \det (sI - \overline{A}).$$

Then equating (13.77) and (13.78), we see that $\overline{a}_1$ and $\overline{a}_2$ must be the zeros of $s^2 + (R/L)s + (1/LC)$. Thus

$$\overline{a}_1^2 + \dfrac{R}{L} \overline{a}_1 + \dfrac{1}{LC} = 0$$

$$\overline{a}_2^2 + \dfrac{R}{L} \overline{a}_2 + \dfrac{1}{LC} = 0.$$

Now since $\overline{A} = PA(P^{-1})$, we have that $\overline{A}P = PA$. Setting $\overline{A}P = PA$ gives

$$\begin{bmatrix} \overline{a}_1 & 0 \\ 0 & \overline{a}_2 \end{bmatrix} \begin{bmatrix} p_1 & p_2 \\ p_3 & p_4 \end{bmatrix} = \begin{bmatrix} p_1 & p_2 \\ p_3 & p_4 \end{bmatrix} \begin{bmatrix} -\dfrac{R}{L} & -\dfrac{1}{L} \\ \dfrac{1}{C} & 0 \end{bmatrix}.$$

Equating the entries of $\overline{A}P$ and PA, we obtain the equations

$$\overline{a}_1 p_1 = -\frac{Rp_1}{L} + \frac{p_2}{C} \qquad (13.79)$$

$$\overline{a}_1 p_2 = -\frac{p_1}{L} \qquad (13.80)$$

$$\overline{a}_2 p_3 = -\frac{Rp_3}{L} + \frac{p_4}{C} \qquad (13.81)$$

$$\overline{a}_2 p_4 = -\frac{p_3}{L}. \qquad (13.82)$$

Equations (13.79) and (13.80) reduce to the single constraint

$$p_2 = -\frac{p_1}{L\overline{a}_1}.$$

Equations (13.81) and (13.82) reduce to the single constraint

$$p_4 = -\frac{p_3}{L\overline{a}_2}.$$

Therefore,

$$P = \begin{bmatrix} p_1 & -\dfrac{p_1}{L\overline{a}_1} \\ p_3 & -\dfrac{p_3}{L\overline{a}_2} \end{bmatrix}.$$

Finally, since P must be invertible, $\det P \neq 0$, and thus

$$-\frac{p_1 p_3}{L\overline{a}_2} + \frac{p_1 p_3}{L\overline{a}_1} \neq 0. \qquad (13.83)$$

Equation (13.83) is satisfied if and only if $\overline{a}_1 \neq \overline{a}_2$. So we have proved that the diagonal form exists if and only if the zeros of $s^2 + (R/L)s + (1/LC)$ are distinct. Note that the transformation matrix P given above is not unique.

13.6

DISCRETIZATION OF STATE MODEL

Again consider a p-input r-output n-dimensional continuous-time system given by the state model

$$\dot{q}(t) = Aq(t) + Bx(t) \qquad (13.84)$$

$$y(t) = Cq(t) + Dx(t). \qquad (13.85)$$

In this section we show that the state representation can be discretized in time, which results in a discrete-time simulation of the given continuous-time system.

From the results in Section 13.3, we know that the state response $q(t)$ resulting from initial state $q(0)$ and input $x(t)$ is given by

$$q(t) = e^{At}q(0) + \int_0^t e^{A(t-\lambda)}Bx(\lambda)\, d\lambda, \qquad t > 0.$$

Now suppose that we change the initial time from $t = 0$ to $t = \tau$, where τ is any real number. Solving the vector difference equation (13.84) with the initial time $t = \tau$, we get

$$q(t) = e^{A(t-\tau)}q(\tau) + \int_\tau^t e^{A(t-\lambda)}Bx(\lambda)\, d\lambda, \qquad t > \tau. \qquad (13.86)$$

Let T be a fixed positive number. Then setting $\tau = kT$ and $t = kT + T$ in (13.86), we obtain

$$q(kT + T) = e^{AT}q(kT) + \int_{kT}^{kT+T} e^{A(kT+T-\lambda)}Bx(\lambda)\, d\lambda. \qquad (13.87)$$

Equation (13.87) looks like a state equation for a discrete-time system except that the second term on the right side of (13.87) is not in the form of a matrix times $x(kT)$. We can express this term in such a form if the input $x(t)$ is constant over the T-second intervals $kT \le t < kT + T$; that is,

$$x(t) = x(kT), \qquad kT \le t < kT + T. \qquad (13.88)$$

If $x(t)$ satisfies (13.88), we can write (13.87) in the form

$$q(kT + T) = e^{AT}q(kT) + \left\{ \int_{kT}^{kT+T} e^{A(kT+T-\lambda)}B\, d\lambda \right\} x(kT).$$

$$(13.89)$$

Let B_d denote the $n \times p$ matrix defined by

$$B_d = \int_{kT}^{kT+T} e^{A(kT+T-\lambda)}B\, d\lambda. \qquad (13.90)$$

If we define the change of variables $\bar\lambda = kT + T - \lambda$ in the integral in (13.90), we have

$$B_d = \int_T^0 e^{A\bar\lambda}B(-d\bar\lambda) = \int_0^T e^{A\lambda} B\, d\lambda.$$

From this expression we see that B_d is independent of the time index k.

Now let A_d denote the $n \times n$ matrix defined by

$$A_d = e^{AT}.$$

Then in terms of A_d and B_d, we can write the difference equation (13.89) in the form

$$q(kT + T) = A_d q(kT) + B_d x(kT). \tag{13.91}$$

Setting $t = kT$ in both sides of (13.85), we obtain the discretized output equation

$$y(kT) = Cq(kT) + Dx(kT). \tag{13.92}$$

Equations (13.91) and (13.92) are the state equations of a linear time-invariant n-dimensional discrete-time system. This discrete-time system is a discretization in time of the given continuous-time system. If the input $x(t)$ is constant over the T-second intervals $kT \le t < kT + T$, the values of the state response $q(t)$ and output response $y(t)$ for $t = kT$ can be computed exactly by solving the state equations (13.91) and (13.92). Since (13.91) can be solved recursively, the discretization process yields a numerical method for solving the state equation of a linear time-invariant continuous-time system.

If $x(t)$ is not constant over the T-second intervals $kT \le t < kT + T$, the solution of (13.91) and (13.92) will yield approximate values for $q(kT)$ and $y(kT)$. In general, the accuracy of the approximate values will improve as T is made smaller. So even if $x(t)$ is not piecewise constant, the representation (13.91) and (13.92) serves as a discrete-time simulation of the given continuous-time system.

Since the step function $u(t)$ is constant for all $t > 0$, in the single-input single-output case the step response of the discretization (13.91) and (13.92) will match the values of the step response of the given continuous-time system. Hence the above discretization process is simply a state version of step-response matching (see Section 12.3).

The discretization in time given above can be used to discretize any system given by a nth-order linear constant-coefficient input/output differential equation. In particular, we first construct a state model of the system by using the realization given in Section 13.3. We can then discretize the coefficient matrices of this state model, which yields a discretization in time of the given continuous-time system.

EXAMPLE 13.8. Consider the automobile on a level surface given by the input/output differential equation

$$\ddot{y}(t) + \frac{k_f}{M} \dot{y}(t) = \frac{1}{M} x(t),$$

where $y(t)$ is the position of the car at time t and $x(t)$ is the force applied to the car at time t. We can construct a state model for the car by defining the states

$$q_1(t) = y(t), \qquad q_2(t) = \dot{y}(t).$$

With respect to these states, the state model is

$$
\begin{bmatrix} \dot{q}_1(t) \\ \dot{q}_2(t) \end{bmatrix} = \begin{bmatrix} 0 & 1 \\ 0 & \dfrac{-k_f}{M} \end{bmatrix} \begin{bmatrix} q_1(t) \\ q_2(t) \end{bmatrix} + \begin{bmatrix} 0 \\ \dfrac{1}{M} \end{bmatrix} x(t)
$$

$$
y(t) = \begin{bmatrix} 1 & 0 \end{bmatrix} \begin{bmatrix} q_1(t) \\ q_2(t) \end{bmatrix}.
$$

To compute the discretized matrices A_d and B_d for this system, we first need to calculate the state transition matrix e^{At}. We have

$$
(sI - A)^{-1} = \begin{bmatrix} s & -1 \\ 0 & s + \dfrac{k_f}{M} \end{bmatrix}^{-1} = \frac{1}{s(s + k_f/M)} \begin{bmatrix} s + \dfrac{k_f}{M} & 1 \\ 0 & s \end{bmatrix}.
$$

Thus

$$
e^{At} = \begin{bmatrix} 1 & \dfrac{M}{k_f}\left[1 - \exp\left(-\dfrac{k_f}{M} t \right) \right] \\ 0 & \exp\left(-\dfrac{k_f}{M} t \right) \end{bmatrix}
$$

and

$$
e^{AT} = \begin{bmatrix} 1 & \dfrac{M}{k_f}\left[1 - \exp\left(-\dfrac{k_f T}{M} \right) \right] \\ 0 & \exp\left(-\dfrac{k_f T}{M} \right) \end{bmatrix}
$$

$$
B_d = \int_0^T e^{A\lambda} B \, d\lambda = \begin{bmatrix} \displaystyle\int_0^T \dfrac{1}{k_f}\left[1 - \exp\left(-\dfrac{k_f}{M} \lambda \right) \right] d\lambda \\ \displaystyle\int_0^T \dfrac{1}{M} \exp\left(-\dfrac{k_f}{M} \lambda \right) d\lambda \end{bmatrix}
$$

$$
= \begin{bmatrix} \dfrac{T}{k_f} - \dfrac{M}{k_f^2}\left[1 - \exp\left(-\dfrac{k_f T}{M} \right) \right] \\ \dfrac{1}{k_f}\left[1 - \exp\left(-\dfrac{k_f T}{M} \right) \right] \end{bmatrix}.
$$

When $M = 1$, $k_f = 0.1$ and $T = 0.1$, the discretized matrices are

$$A_d = \begin{bmatrix} 1 & 0.09950166 \\ 0 & 0.99004983 \end{bmatrix}, \qquad B_d = \begin{bmatrix} 0.00498344 \\ 0.09950166 \end{bmatrix}.$$

The state model of the resulting discrete-time simulation is

$$\begin{bmatrix} q_1(0.1k + 0.1) \\ q_2(0.1k + 0.1) \end{bmatrix} = \begin{bmatrix} 1 & 0.09950166 \\ 0 & 0.99004983 \end{bmatrix} \begin{bmatrix} q_1(0.1k) \\ q_2(0.1k) \end{bmatrix}$$

$$+ \begin{bmatrix} 0.00498344 \\ 0.09950166 \end{bmatrix} x(0.1k),$$

where

$$q_1(0.1k) = y(0.1k) \qquad \text{and} \qquad q_2(0.1k) = \dot{y}(0.1k).$$

A computer program for computing the discretized matrices A_d and B_d is given in Figure A.15. Running this program with

$$A = \begin{bmatrix} 0 & 1 \\ 0 & -0.1 \end{bmatrix}, \qquad B = \begin{bmatrix} 0 \\ 1 \end{bmatrix},$$

we get

$$A_d = \begin{bmatrix} 1 & 0.09950166 \\ 0 & 0.9900498 \end{bmatrix}, \qquad B_d = \begin{bmatrix} 0.004983375 \\ 0.09950166 \end{bmatrix}.$$

This result is consistent with the result obtained in Example 13.8.

PROBLEMS

Chapter 13

13.1 For the circuit in Figure P13.1, find the state model with the state variables defined to be $q_1(t) = i_L(t)$, $q_2(t) = v_C(t)$, and with the output defined as $y(t) = i_L(t) + v_C(t)$.

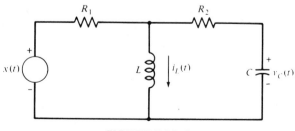

FIGURE P13.1

13.2 When the input $x(t) = \cos t$, $t \geq 0$, is applied to a linear time-invariant continuous-time system, the resulting output response (with no initial energy in the system) is

$$y(t) = 2 - e^{-5t} + 3 \cos t, \qquad t \geq 0.$$

Find a state model of the system with the smallest possible number of state variables.

13.3 A linear time-invariant continuous-time system has transfer function

$$H(s) = \frac{s^2 - 2s + 2}{s^2 + 3s + 1}.$$

Find a state model of the system with the smallest possible number of state variables.

13.4 Consider the two-input two-output linear time-invariant continuous-time system shown in Figure P13.4.
(a) Find the state model of the system with the state variables defined to be $q_1(t) = v(t)$, $q_2(t) = y_1(t) - v(t) - x_2(t)$, and $q_3(t) = y_2(t)$.
(b) Find the state model of the system with the state variables defined to be $q_1(t) = y_1(t) - x_2(t)$, $q_2(t) = y_2(t)$, and $q_3(t) = y_1(t) - v(t) - x_2(t)$.

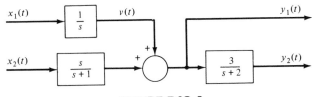

FIGURE P13.4

13.5 For the system shown in Figure P13.5, find a state model with the number of state variables equal to two.

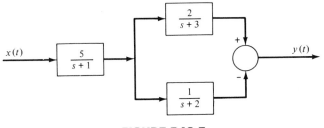

FIGURE P13.5

13.6 A linear time-invariant continuous-time system has transfer function

$$H(s) = \frac{bs}{s^2 + a_1 s + a_0}.$$

Find the state model of the system with the state variables defined to be

$$q_1(t) = y(t) \quad \text{and} \quad q_2(t) = \int_{-\infty}^{t} y(\lambda)\, d\lambda.$$

13.7   For the two-input two-output system shown in Figure P13.7, find a state model with the smallest possible number of state variables.

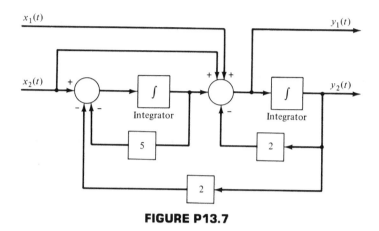

FIGURE P13.7

13.8   The input $x(k) = -2 + 2^k$, $k = 0, 1, 2, \ldots$, is applied to a linear time-invariant discrete-time system. The resulting response is $y(k) = 3^k - 4(2^k)$, $k = 0, 1, 2, \ldots$, with no initial energy in the system. Find a state model of the system with the smallest possible number of state variables.

13.9   A linear time-invariant continuous-time system has the state model

$$\begin{bmatrix} \dot{q}_1(t) \\ \dot{q}_2(t) \end{bmatrix} = \begin{bmatrix} 0 & 1 \\ -1 & 2 \end{bmatrix} \begin{bmatrix} q_1(t) \\ q_2(t) \end{bmatrix} + \begin{bmatrix} 0 \\ 1 \end{bmatrix} x(t)$$

$$y(t) = \begin{bmatrix} 1 & 2 \end{bmatrix} \begin{bmatrix} q_1(t) \\ q_2(t) \end{bmatrix}.$$

Derive an expression for $q_1(t)$ and $q_2(t)$ in terms of $x(t)$, $y(t)$, and (if necessary) the derivatives of $x(t)$ and $y(t)$.

13.10 A two-input two-output linear time-invariant continuous-time system has the transfer function matrix

$$H(s) = \begin{bmatrix} \dfrac{1}{s+1} & 0 \\ \dfrac{1}{s+!} & \dfrac{1}{s+2} \end{bmatrix}.$$

Find the state model of the system with the state variables defined to be

$q_1(t) = y_1(t)$, $q_2(t) = y_2(t)$, where $y_1(t)$ is the first system output and $y_2(t)$ is the second system output.

13.11 A linear time-invariant continuous-time system is given by the state model $\dot{q}(t) = Aq(t) + Bx(t)$, $y(t) = Cq(t)$, where

$$A = \begin{bmatrix} 0 & 2 \\ 0 & -1 \end{bmatrix}, \qquad B = \begin{bmatrix} 1 & 0 \\ 1 & -1 \end{bmatrix}, \qquad C = [1 \quad 3].$$

(a) Compute the state transition matrix e^{At}.
(b) Compute the transfer function matrix $H(s)$.
(c) Compute the impulse response function matrix $H(t)$.
(d) Compute the state response $q(t)$ for $t > 0$ resulting from initial state $q(0) = [-1 \quad 1]'$ (prime denotes transpose) and input $x(t) = [u(t) \quad u(t)]'$.
(e) If possible, find a nonimpulsive input $x(t)$ with $x(t) = 0$ for $t < 0$ such that the state response $q(t)$ resulting from initial state $q(0) = [1 \quad -1]'$ and input $x(t)$ is given by $q(t) = [u(t) \quad -u(t)]'$.

13.12 Consider a single car moving on a level surface given by the input/output differential equation

$$\frac{d^2y(t)}{dt^2} + \frac{k_f}{M}\frac{dy(t)}{dt} = \frac{1}{M}x(t),$$

where $y(t)$ is the position of the car at time t. With states $q_1(t) = y(t)$, $q_2(t) = \dot{y}(t)$, the car has the state model $\dot{q}(t) = Aq(t) + bx(t)$, $y(t) = cq(t)$, where

$$A = \begin{bmatrix} 0 & 1 \\ 0 & \dfrac{-k_f}{M} \end{bmatrix}, \qquad b = \begin{bmatrix} 0 \\ \dfrac{1}{M} \end{bmatrix}, \qquad c = [1 \quad 0].$$

In the following independent parts, take $M = 1$, and $k_f = 0.1$.
(a) Using the state model, derive an expression for the state response $q(t)$ resulting from initial conditions $y(0) = y_0$, $\dot{y}(0) = v_0$, with $x(t) = 0$, all t.
(b) With $x(t) = 0$ for $0 \leq t \leq 10$, it is known that $y(10) = 0$, $\dot{y}(10) = 55$. Compute $y(0)$ and $\dot{y}(0)$.
(c) The force $x(t) = 1$ is applied to the car for $0 \leq t \leq 10$. The state $q(5)$ at time $t = 5$ is known to be $q(5) = [50 \quad 20]'$. Compute the initial state $q(0)$ at time $t = 0$.
(d) Now suppose that $x(t) = 0$ for $10 \leq t \leq 20$, and $y(10) = 5$, $y(20) = 50$. Compute the state $q(10)$ at time $t = 10$.

13.13 A two-input two-output linear time-invariant continuous-time system is given by the state model $\dot{q}(t) = Aq(t) + Bx(t)$, $y(t) = Cq(t)$, where

$$A = \begin{bmatrix} 1 & -1 \\ 0 & 2 \end{bmatrix}, \qquad B = \begin{bmatrix} 1 & 1 \\ 0 & 1 \end{bmatrix}, \qquad C = \begin{bmatrix} 1 & -1 \\ 1 & -1 \end{bmatrix}.$$

(a) The output response $y(t)$ resulting from some initial state $q(0)$ with $x(t) = 0$ for all $t \geq 0$ is given by

$$y(t) = \begin{bmatrix} 2e^{2t} \\ 2e^{2t} \end{bmatrix}, \qquad t \geq 0.$$

Compute $q(0)$.

(b) The output response $y(t)$ resulting from some initial state $q(0)$ and input $x(t) = [u(t) \quad u(t)]'$ is given by

$$y(t) = \begin{bmatrix} 4e^t - 2e^{2t} - 2 \\ 4e^t - 2e^{2t} - 2 \end{bmatrix}, \qquad t \geq 0.$$

Compute $q(0)$.

13.14 A linear time-invariant continuous-time system has state model $\dot{q}(t) = Aq(t) + Bx(t)$, $y(t) = Cq(t)$, where

$$A = \begin{bmatrix} -8 & -4 \\ 12 & 6 \end{bmatrix}, \qquad B = \begin{bmatrix} 1 & 1 \\ 2 & 2 \end{bmatrix}, \qquad C = [1 \quad -2].$$

The following parts are independent.

(a) Suppose that $y(2) = 3$ and $\dot{y}(2) = 5$. Compute $q(2)$.

(b) Suppose that $x(t) = 0$ for $0 \leq t \leq 1$ and that $q(1) = [1 \quad 1]'$. Compute $q(0)$.

(c) Suppose that $q(0) = [1 \quad 1]'$. If possible, find an input $x(t)$ such that the output response resulting from $q(0)$ and $x(t)$ is zero, that is, $y(t) = 0$, $t > 0$.

13.15 Consider the inverted pendulum given by the linearized equations

$$(J + mL^2)\ddot{\theta}(t) - mgL\theta(t) + mL\ddot{d}(t) = 0$$

$$(M + m)\ddot{d}(t) + mL\ddot{\theta}(t) = x(t).$$

Recall that $\theta(t)$ is the angle of the pendulum from the vertical position, $d(t)$ is the position of the cart at time t, $x(t)$ is the force applied to the cart, M is the mass of the cart, and m is the mass of the pendulum (see Problem 6.19). In the following parts, take $J = 1$, $L = 1$, $g = 9.8$, $M = 1$, and $m = 0.1$.

(a) With the states defined to be $q_1(t) = \theta(t)$, $q_2(t) = \dot{\theta}(t)$, $q_3(t) = d(t)$, and $q_4(t) = \dot{d}(t)$, find the state model of the inverted pendulum.

(b) With A equal to the system matrix found in part (a), compute the state transition matrix e^{At}.

(c) Compute the inverse $(e^{At})^{-1}$ of the state transition matrix.

(d) Using your answer in part (c), compute the state $q(5)$ at time $t = 5$ assuming that $q(10) = [10° \quad 0 \quad 5 \quad 2]'$ and $x(t) = 0$ for $5 \leq t \leq 10$.

(e) Using the state model, compute the state response $q(t)$ for all $t > 0$ when $\theta(0) = 10°$, $d(0) = 0$, $\dot{\theta}(0) = 0$, $\dot{d}(0) = 0$, and $x(t) = 0$ for $t \geq 0$.

(f) Repeat part (e) with $\theta(0) = 0$, $\dot{\theta}(0) = 1$, $d(0) = 0$, $\dot{d}(0) = 0$, and $x(t) = 0$ for $t \geq 0$.

(g) Repeat part (e) with $\theta(0) = 0$, $\dot{\theta}(0) = 0$, $d(0) = 0$, $\dot{d}(0) = 1$, and $x(t) = 0$ for $t \geq 0$.

13.16 A linear time-invariant discrete-time system is given by the state model $q(k + 1) = Aq(k) + Bx(k)$, $y(k) = Cq(k)$, where

$$A = \begin{bmatrix} -1 & 1 \\ -1 & -2 \end{bmatrix}, \quad B = \begin{bmatrix} 0.5 & 1 \\ -1 & -0.5 \end{bmatrix}, \quad C = \begin{bmatrix} 2 & 1 \\ -1 & -2 \end{bmatrix}.$$

(a) Compute $q(1)$, $q(2)$, and $q(3)$ when $q(0) = [1 \quad 1]'$ and $x(k) = [k \quad k]'$.

(b) Compute the transfer function matrix $H(z)$.

(c) Suppose that $q(0) = [0 \quad 0]'$. Find an input $x(k)$ that sets up the state $q(2) = [-1 \quad 2]'$; that is, the state $q(2)$ of the system at time $k = 2$ resulting from input $x(k)$ is equal to $[-1 \quad 2]'$.

(d) Now suppose that $q(0) = [1 \quad -2]'$. Find an input $x(k)$ that drives the system to the zero state at time $k = 2$; that is, $q(2) = [0 \quad 0]'$.

13.17 Consider the discrete-time system with state model $q(k + 1) = Aq(k) + Bx(k)$, $y(k) = Cq(k)$, where

$$A = \begin{bmatrix} 1 & 0 \\ 0.5 & 1 \end{bmatrix}, \quad B = \begin{bmatrix} 2 \\ 1 \end{bmatrix}, \quad C = \begin{bmatrix} 2 & -1 \\ 1 & 0 \\ 1 & -1 \end{bmatrix}.$$

The following parts are independent.

(a) Compute $y(0)$, $y(1)$ and $y(2)$ when $q(0) = [-1 \quad 2]'$ and $x(k) = \sin{(\pi/2)k}$.

(b) Suppose that $q(3) = [1 \quad -1]$. Compute $q(0)$ assuming that $x(k) = 0$ for $k = 0, 1, 2, \ldots$.

(c) Suppose that $y(3) = [1 \quad 2 \quad -1]'$. Compute $q(3)$.

13.18 A discrete-time system has the state model $q(k + 1) = Aq(k) + Bx(k)$, $y(k) = Cq(k)$, where

$$CB = \begin{bmatrix} 6 \\ 3 \end{bmatrix} \quad \text{and} \quad CAB = \begin{bmatrix} 22 \\ 11 \end{bmatrix}.$$

When $q(0) = 0$, it is known that $y(1) = [6 \quad 3]'$ and $y(2) = [4 \quad 2]'$. Compute $x(0)$ and $x(1)$.

13.19 A continuous-time system has state model $\dot{q}(t) = Aq(t) + bx(t)$, $y(t) = Cq(t)$, where

$$A = \begin{bmatrix} 3 & -2 \\ 9 & -6 \end{bmatrix}, \quad b = \begin{bmatrix} 1 \\ 2 \end{bmatrix}.$$

Determine if there is a coordinate transformation $\bar{q}(t) = Pq(t)$ such that $\bar{A}$ is in diagonal form. If such a transformation exists, give P and $\bar{A}$.

13.20 We are given two continuous-time systems with state models

$$\dot{q}(t) = A_1 q(t) + b_1 x(t), \quad y(t) = C_1 q(t)$$

$$\dot{\bar{q}}(t) = A_2 \bar{q}(t) + b_2 x(t), \quad y(t) = C_2 \bar{q}(t),$$

where

$$A_1 = \begin{bmatrix} 1 & 1 \\ 0 & 2 \end{bmatrix}, \quad A_2 = \begin{bmatrix} 4 & 2 \\ -3 & -1 \end{bmatrix}, \quad b_1 = \begin{bmatrix} 1 \\ 1 \end{bmatrix},$$

$$b_2 = \begin{bmatrix} 5 \\ 3 \end{bmatrix}, \quad C_1 = \begin{bmatrix} 1 & 2 \\ 1 & 2 \end{bmatrix}, \quad C_2 = \begin{bmatrix} 0 & 1 \\ 0 & 1 \end{bmatrix}.$$

Determine if there is a coordinate transformation $\bar{q}(t) = Pq(t)$ between the two systems. Determine P if it exists.

13.21 A linear time-invariant continuous-time system has state model $\dot{q}(t) = Aq(t) + bx(t)$, $y(t) = cq(t)$. It is known that there is a coordinate transformation $\bar{q}(t) = P_1 q(t)$ such that

$$\bar{A} = \begin{bmatrix} -2 & 0 & 0 \\ 0 & -1 & 0 \\ 0 & 0 & 1 \end{bmatrix}, \quad \bar{b} = \begin{bmatrix} 1 \\ 1 \\ 1 \end{bmatrix}, \quad \bar{c} = \begin{bmatrix} -1 & 1 & 1 \end{bmatrix}.$$

It is also known that there is a second transformation $\bar{\bar{q}}(t) = P_2 q(t)$ such that

$$\bar{\bar{A}} = \begin{bmatrix} 0 & 1 & 0 \\ 0 & 0 & 1 \\ -a_0 & -a_1 & -a_2 \end{bmatrix}, \quad \bar{\bar{b}} = \begin{bmatrix} 0 \\ 0 \\ 1 \end{bmatrix}.$$

(a) Determine a_0, a_1, and a_2.
(b) Compute the transfer function $H(s)$ of the system.
(c) Compute $\bar{\bar{c}}$.

13.22 A discretized continuous-time system is given by the state model $q(kT + T) = A_d q(kT) + B_d x(kT)$, $y(kT) = Cq(kT)$, where

$$A_d = \begin{bmatrix} 1 & 2 \\ -2 & 4 \end{bmatrix}, \quad B_d = \begin{bmatrix} 1 \\ 1 \end{bmatrix}, \quad C = \begin{bmatrix} 1 & 1 \end{bmatrix}.$$

(a) It is known that $y(0) = 1$, $y(T) = -2$ when $x(0) = 2$, $x(T) = 4$. Compute $q(0)$.
(b) It is known that $q(0) = \begin{bmatrix} 2 & -3 \end{bmatrix}'$, $y(T) = -1$. Compute $x(0)$.

13.23 Determine the discrete-time simulation (with $T = 1$) for the continuous-time system $\dot{q}(t) = Aq(t) + bx(t)$, where

$$A = \begin{bmatrix} 1 & 1 \\ -1 & 1 \end{bmatrix}, \quad b = \begin{bmatrix} 0 \\ 1 \end{bmatrix}.$$

13.24. Repeat Problem 13.23 for the system with

$$A = \begin{bmatrix} 0 & 1 \\ 0 & -1 \end{bmatrix}, b = \begin{bmatrix} 0 \\ 1 \end{bmatrix}.$$

13.25 The state model that interrelates $\theta(t)$ and $x(t)$ in the inverted pendulum (see Problem 13.15) is given by

$$\begin{bmatrix} \dot{\theta}(t) \\ \ddot{\theta}(t) \end{bmatrix} = \begin{bmatrix} 0 & 1 \\ a & 0 \end{bmatrix} \begin{bmatrix} \theta(t) \\ \dot{\theta}(t) \end{bmatrix} + \begin{bmatrix} 0 \\ b \end{bmatrix} x(t),$$

where

$$a = \frac{(M + m)mgL}{(M + m)J + Mm(L^2)} \quad \text{and} \quad b = \frac{-mL}{(M + m)J + Mm(L^2)}.$$

Taking $g = 9.8$, $L = 1$, $J = 1$, $M = 1$, $m = 0.1$, and $T = 0.1$, compute the discretized state model.

13.26 Consider the two-car system given by

$$\dot{v}_1(t) + \frac{k_f}{M} v_1(t) = \frac{1}{M} x_1(t)$$

$$\dot{v}_2(t) + \frac{k_f}{M} v_2(t) = \frac{1}{M} x_2(t)$$

$$\dot{w}(t) = v_2(t) - v_1(t).$$

With the states $q_1(t) = v_1(t)$, $q_2(t) = v_2(t)$, $q_3(t) = w(t)$, and the outputs $y_1(t) = q_1(t)$, $y_2(t) = q_2(t)$, $y_3(t) = q_3(t)$, and with $M = 1$, $k_f = 0.1$, and $T = 0.1$, compute the discretized state model.

Computer Programs

The computer programs mentioned in the text are listed in this appendix. A summary of the programs is given in Table A.1. The programs are written in the most elementary form of Basic. They were written for the IBM Personal Computers; however, with minor changes the programs should run on any of the standard personal computers. It is emphasized that the programs are primarily for instructional purposes rather than for serious applications. Students should be able to follow the sequence of steps in the programs. It is expected that some people will discover ways of improving or expanding the programs given here.

By first loading the plotting subroutine, most of the programs can be typed into a computer in a reasonable period of time. To avoid having to type in the programs, a $5\frac{1}{4}$-inch diskette containing all the programs is available by sending $ 5.00 to

> Text Software
> P.O. Box 8301
> Pittsburgh, PA 15218-0301

The programs will run on any machine that will read a MS-DOS diskette and run Basic.

TABLE A.1

Program	Location	Program Function
PLOTSUB	Figure A.1	Plotting subroutine
RECUR	Figure A.3	Solves nth-order recursion
CONV	Figure A.4	Performs discrete-time convolution
NUMCONV	Figure A.5	Performs numerical convolution
ROOTS	Figure A.6	Solves for roots of a polynomial
INVLT	Figure A.7	Computes inverse Laplace transform
INVZT	Figure A.8	Computes inverse z-transform
ROUTH	Figure A.9	Routh stability test
CONTFREQ	Figure A.10	Continuous-time frequency response
DISCFREQ	Figure A.11	Discrete-time frequency response
DFT	Figure A.12	Computes DFT
FFT	Figure A.13	Computes DFT and inverse DFT via FFT
CTFT	Figure A.14	Computes continuous-time Fourier transform
STATEDIS	Figure A.15	Discretizes state equations

The first program, called PLOTSUB, is listed in Figure A.1. This simple program plots the values of a function $y(k)$ for $k = 0, 1, 2, \ldots, Q$. The plot is displayed with k on the vertical axis and $y(k)$ on the horizontal axis; that is, the plot is rotated $90°$ from the standard format. PLOTSUB also displays the

```
1000 YH=Y(0):YL=Y(0)
1010 FOR K=1 TO Q
1020 IF Y(K)>=YH THEN YH=Y(K):GOTO 1050
1030 IF Y(K)>=YL THEN YH=YH:GOTO 1050
1040 YL=Y(K)
1050 NEXT K
1060 FOR K=0 TO Q
1070 Y(K)=1+(Y(K)-YL)*29/(YH-YL):Y(K)=INT(Y(K)+.5)
1080 NEXT K
1090 Z=1+INT(-YL*29/(YH-YL)+.5):YH=.01*INT(100*YH+.5)
1100 YL=.01*INT(100*YL+.5):INPUT "Plot on screen (y/n)";A$
1110 IF A$="n" THEN GOTO 1190
1120 PRINT "min=";YL;TAB(25);"max=";YH
1130 FOR K=0 TO Q
1140 IF Y(K)=Z THEN PRINT TAB(Z);"*":GOTO 1180
1150 IF Z<=30 AND Y(K)<Z THEN PRINT TAB(Y(K));"*";TAB(Z);"0":GOTO 1180
1160 IF Z>=1 AND Y(K)>Z THEN PRINT TAB(Z);"0";TAB(Y(K));"*":GOTO 1180
1170 PRINT TAB(Y(K));"*"
1180 NEXT K
1190 INPUT "Plot on printer (y/n)";A$
1200 IF A$="n" THEN END
1210 LPRINT "min=";YL;TAB(25);"max=";YH
1220 FOR K=0 TO Q
1230 IF Y(K)=Z THEN LPRINT TAB(Z);"*":GOTO 1270
1240 IF Z<=30 AND Y(K)<Z THEN LPRINT TAB(Y(K));"*";TAB(Z);"0":GOTO 1270
1250 IF Z>=1 AND Y(K)>Z THEN LPRINT TAB(Z);"0";TAB(Y(K));"*":GOTO 1270
1260 LPRINT TAB(Y(K));"*"
1270 NEXT K
1280 END
```

FIGURE A.1. Plotting subroutine.

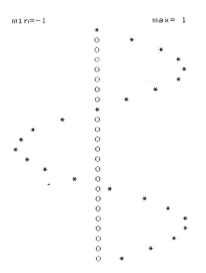

FIGURE A.2.   Output of PLOTSUB.

minimum and maximum values of $y(k)$. If the plot crosses the zero axis ($y = 0$), PLOTSUB will print a line of zeros showing the location of the value $y = 0$. The plotting routine operates in the text mode, so that the number of points that can be displayed on a screen depends on the number of lines the screen can display in the text mode. However, PLOTSUB will plot any number of desired points on a printer. A special version of the programs in this book is available that supports high-resolution plots directly to any printer equivalent to the IBM graphics printer (e.g., Epson MX80 or OkiData with plug and play) or to the screen if a graphics card is available. A diskette with modified text programs and the stand-alone plotting program is available for $15 from Text Software at the above address.

To illustrate the use of PLOTSUB, the subroutine was loaded into a computer and then the following statements were added:

```
10 INPUT "Q";Q
20 DIM Y(Q)
30 FOR K = 0 TO Q
40 Y(K) = SIN(.4*K)
50 NEXT K
```

The resulting output was sent to a printer. A copy of the printer output is shown in Figure A.2.

Listed in Figures A.3 to A.5 are the next three programs, called RECUR, CONV, AND NUMCONV. To input these programs on a computer, one should first load PLOTSUB and then type in the other statements. If the plotting routine is not desired, do not type in lines 1000–1280. Instructions on how to use these programs are given in the REM statements (also see the appropriate parts of the text).

```
10 REM  Solution of nth-order difference equation y(kT+nT) +
20 REM  a(n-1)y(kT+nT-T) + ... + a(1)y(kT+T) + a(0)y(kT) =
30 REM  b(n)x(kT+nT) + ... + b(1)x(kT+T) + b(0)x(kT).
40 REM  Program by E. W. Kamen
50 REM  Program computes y(0),y(T),...,y(QT).
60 REM  If x(kT) is given in function form for k>0, define
70 REM  x(kT) on line 330 using function statement DEF FNX(K).
80 CLS
90 INPUT "Order n of difference equation";N
100 PRINT:INPUT "No Q of time points";Q
110 DIM X(Q+N),Y(Q+N),A(N),B(N)
120 FOR I=0 TO N-1
130 PRINT:PRINT "Enter a(i) for i =";I:INPUT A(I)
140 NEXT I
150 A(N) = 0
160 FOR I=0 TO N
170 NEXT I
180 FOR I=0 TO N
190 PRINT:PRINT "Enter b(i) for i =";I:INPUT B(I)
200 NEXT I
210 FOR K=1 TO N
220 PRINT
230 PRINT "Enter y(kT),x(kT) for k =";(-1)*K:INPUT Y(N-K),X(N-K)
240 NEXT K
250 PRINT:INPUT "Is x(kT) in function form (y/n)";A$
260 IF A$="y" THEN 310
270 FOR K=0 TO Q-1
280 PRINT:PRINT "Enter x(kT) for k =";K:INPUT X(K+N)
290 NEXT K
300 GOTO 370
310 PRINT:INPUT "Is x(kT) defined on line 330 (y/n)";A$
320 IF A$="n" THEN 460
330 DEF FNX(K) =
340 FOR K=0 TO Q
350 X(K+N) = FNX(K)
360 NEXT K
370 PRINT:PRINT "k             y(kT)":PRINT
380 FOR K=0 TO Q
390 Y(K+N) = 0
400 FOR I=0 TO N
410 Y(K+N) = Y(K+N)-A(I)*Y(K+I)+B(I)*X(K+I)
420 NEXT I
430 Y(K)=Y(K+N):PRINT K; TAB(10);Y(K)
440 NEXT K
450 GOTO 480
460 PRINT:PRINT "Enter function form of x(kT) on line 330"
470 PRINT "and rerun program":END
480 PRINT:INPUT"Want a plot of y(k) (y/n)";A$
490 PRINT:PRINT:IF A$="y" THEN 1000
500 END
1000 YH=Y(0):YL=Y(0)
1010 FOR K=1 TO Q
1020 IF Y(K)>=YH THEN YH=Y(K):GOTO 1050
1030 IF Y(K)>=YL THEN YH=YH:GOTO 1050
1040 YL=Y(K)
1050 NEXT K
1060 FOR K=0 TO Q
1070 Y(K)=1+(Y(K)-YL)*29/(YH-YL):Y(K)=INT(Y(K)+.5)
1080 NEXT K
1090 Z=1+INT(-YL*29/(YH-YL)+.5):YH=.01*INT(100*YH+.5)
1100 YL=.01*INT(100*YL+.5):INPUT "Plot on screen (y/n)";A$
1110 IF A$="n" THEN GOTO 1190
1120 PRINT "min=";YL;TAB(25);"max=";YH
1130 FOR K=0 TO Q
1140 IF Y(K)=Z THEN PRINT TAB(Z);"*":GOTO 1180
1150 IF Z<=30 AND Y(K)<Z THEN PRINT TAB(Y(K));"*";TAB(Z);"0":GOTO 1180
1160 IF Z>=1 AND Y(K)>Z THEN PRINT TAB(Z);"0";TAB(Y(K));"*":GOTO 1180
1170 PRINT TAB(Y(K));"*"
1180 NEXT K
1190 INPUT "Plot on printer (y/n)";A$
1200 IF A$="n" THEN END
```

FIGURE A.3. Program RECUR.

```
1210 LPRINT "min=";YL;TAB(25);"max=";YH
1220 FOR K=0 TO Q
1230 IF Y(K)=Z THEN LPRINT TAB(Z);"*":GOTO 1270
1240 IF Z<=30 AND Y(K)<Z THEN LPRINT TAB(Y(K));"*";TAB(Z);"0":GOTO 1270
1250 IF Z>=1 AND Y(K)>Z THEN LPRINT TAB(Z);"0";TAB(Y(K));"*":GOTO 1270
1260 LPRINT TAB(Y(K));"*"
1270 NEXT K
1280 END
```

FIGURE A.3.   (Continuation of RECUR.)

```
10 REM   Discrete-time convolution by E. W. Kamen
20 REM   Program computes (x*v)(kT) for k=0,1,2,...,Q.
30 REM   If x(kT) and v(kT) are given in function form,
40 REM   define x(kT) on line 210 and v(kT) on line 280.
50 CLS
60 INPUT "No Q of time points";Q
70 DIM X(Q),V(Q),Y(Q)
80 PRINT:INPUT "Is x(kT) in function form (y/n)"; A$
90 IF A$ = "y" THEN 190
100 FOR K=0 TO Q
110 PRINT:PRINT "Enter x(kT) for k =";K:INPUT X(K)
120 NEXT K
130 PRINT:INPUT "Is v(kT) in function form (y/n)";A$
140 IF A$ = "y" THEN 260
150 FOR K=0 TO Q
160 PRINT:PRINT "Enter v(kT) for k =";K:INPUT V(K)
170 NEXT K
180 GOTO 320
190 PRINT:INPUT "Is x(kT) defined on line 210 (y/n)";A$
200 IF A$ = "n" THEN 410
210 DEF FNX(K) =
220 FOR K = 0 TO Q
230 X(K) = FNX(K)
240 NEXT K
250 GOTO 130
260 PRINT:INPUT "Is v(kT) defined on line 280 (y/n)";A$
270 IF A$ = "n" THEN 430
280 DEF FNV(K) =
290 FOR K=0 TO Q
300 V(K) = FNV(K)
310 NEXT K
320 PRINT:PRINT " k          (x*v)(kT)":PRINT
330 FOR K=0 TO Q
340 Y(K) = 0
350 FOR I = 0 TO K
360 Y(K) = Y(K) + X(I)*V(K-I)
370 NEXT I
380 PRINT K;TAB(10);Y(K)
390 NEXT K
400 GOTO 450
410 PRINT:PRINT "Enter function form of x(kT) on line 210"
420 PRINT "and rerun program":END
430 PRINT:PRINT "Enter function form of v(kT) on line 280"
440 PRINT "and rerun program"
450 PRINT:INPUT"Want a plot of (x*v)(kT) (y/n)";A$
460 PRINT:PRINT:IF A$="y" THEN 1000
470 END
1000 -1280 PLOTSUB
```

FIGURE A.4.   Program CONV.

```
10 REM   Numerical convolution by E. W. Kamen
20 REM   Program computes discretized output y(kT) =
30 REM   T[h(0)x(kT) + h(T)x(kT-T) + h(kT-T)x(T) + h(kT)x(0)]
40 REM   for k=0,1,2,...,Q, where
50 REM   h(kT) = discretized impulse response,
60 REM   x(kT) = discretized input,
70 REM   and T = discretization interval.
80 REM   If h(kT) and x(kT) are given in function form,
90 REM   define h(kT) on line 270 and x(kT) on line 340.
100 CLS
110 INPUT "No Q of time points";Q
120 DIM H(Q),X(Q),Y(Q)
130 PRINT:INPUT "Discretization interval T";T
140 PRINT:INPUT "Is h(kT) in function form (y/n)";A$
150 IF A$ = "y" THEN 250
160 FOR K = 0 TO Q
170 PRINT:PRINT "Enter h(kT) for k =";K:INPUT H(K)
180 NEXT K
190 PRINT:INPUT "Is x(kT) in function form (y/n)";A$
200 IF A$ = "y" THEN 320
210 FOR K = 0 TO Q
220 PRINT:PRINT "Enter x(kT) for k = ";K:INPUT X(K)
230 NEXT K
240 GOTO 380
250 PRINT:INPUT "Is h(kT) defined on line 270 (y/n)";A$
260 IF A$ = "n" THEN 470
270 DEF FNH(K) =
280 FOR K = 0 TO Q
290 H(K) = FNH(K)
300 NEXT K
310 GOTO 190
320 PRINT:INPUT "Is x(kT) defined on line 340 (y/n)";A$
330 IF A$ = "n" THEN 490
340 DEF FNX(K) =
350 FOR K = 0 TO Q
360 X(K) = FNX(K)
370 NEXT K
380 PRINT:PRINT " k                y(kT)":PRINT
390 FOR K = 0 TO Q
400 Y(K) = 0
410 FOR I = 0 TO K
420 Y(K) = Y(K) + T*H(I)*X(K-I)
430 NEXT I
440 PRINT K; TAB(10); Y(K)
450 NEXT K
460 GOTO 510
470 PRINT:PRINT "Enter function form of h(kT) on line 270"
480 PRINT "and rerun program";END
490 PRINT:PRINT "Enter function form of x(kT) on line 340"
500 PRINT "And rerun program":END
510 PRINT:INPUT"Want a plot of y(k) (y/n)";A$
520 PRINT:PRINT:IF A$="y" THEN GOTO 1000
530 END
1000 -1280 PLOTSUB
```

FIGURE A.5. Program NUMCONV.

The roots of a polynomial with real coefficients can be computed using the program ROOTS listed in Figure A.6. The ROOTS program is used in the program INVLT given in Figure A.7. In fact, note that lines 10000–10580 are identical in both programs. These lines should be typed in first as a subroutine and then the other statements in ROOTS and INVLT can be added to the subroutine. The program INVLT computes the inverse Laplace transform of a rational function $X(s)$ whose poles are distinct. We invite the interested reader to consider expanding this program to include the case of repeated poles.

```
10 REM   Roots by S. Svoronos
20 REM   Program computes roots of
30 REM   a(n)s^n+a(n-1)s^(n-1)+...+a(1)s+a(0) = 0.
40 CLS
50 INPUT "Degree n of polynomial";N:PRINT
60 N9=N+2:IR=1
70 DIM A(N9),U(N9),V(N9),H(N9),B(N9),C(N9)
80 FOR I = 0 TO N
90 PRINT"Enter a(i) for i=";I:INPUT A(I+1)
100 NEXT I
110 GOSUB 10000
120 CLS
130 PRINT"  Root No     Real Part       Imaginary Part"
140 FOR I=1 TO N
150 PRINT:PRINT TAB(5);I;TAB(15);U(I);TAB(30);V(I)
160 NEXT I
170 END
10000 IV=IR
10010 NC=N+1
10020 FOR I=1 TO NC: H(I)=A(I): NEXT
10030 P=0: R=0: Q=0
10040 IF H(1)<>0 THEN 10080
10050 NC=NC-1: U(NC)=0: V(NC)=0
10060 FOR I=1 TO NC: H(I)=H(I+1): NEXT
10070 GOTO 10040
10080 IF NC=1 THEN RETURN
10090 IF NC=2 THEN R=-H(1)/H(2): GOTO 10390
10100 IF NC=3 THEN P=H(2)/H(3): Q=H(1)/H(3): GOTO 10460
10110 IF ABS(H(NC-1)/H(NC))>=ABS(H(2)/H(1)) THEN GOTO 10180
10120 IV=-IV
10130 M=INT(NC/2+.1)
10140 FOR I=1 TO M:NL=NC+1-I:F=H(NL):H(NL)=H(I):H(I)=F:NEXT
10150 IF Q=0 THEN P=0: GOTO 10170
10160 P=P/Q: Q=1/Q
10170 IF R<>0 THEN R=1/R
10180 E0=1E-09: REM sets accuracy
10190 B(NC)=H(NC):C(NC)=H(NC):B(NC+1)=0:C(NC+1)=0:NP=NC-1
10200 FOR J=1 TO 1000
10210 FOR I1=1 TO NP:I=NC-I1:B(I)=H(I)+R*B(I+1):C(I)=B(I)+R*C(I+1)
10220 NEXT:IF ABS(B(1)/H(1))<=E0 THEN 10390
10230 IF C(2)=0 THEN R=R+1: GOTO 10250
10240 R=R-B(1)/C(2)
10250 FOR I1=1 TO NP
10260 I=NC-I1:B(I)=H(I)-P*B(I+1)-Q*B(I+2)
10270 C(I)=B(I)-P*C(I+1)-Q*C(I+2):NEXT I1
10280 IF H(2)<>0 THEN 10310
10290 IF ABS(B(2)/H(1))<=E0 THEN 10320
10300 IF ABS(B(2)/H(1))>E0 THEN 10330
10310 IF ABS(B(2)/H(2))>E0 THEN 10330
10320 IF ABS(B(1)/H(1))<=E0 THEN 10460
10330 CB=C(2)-B(2): D=C(3)*C(3)-CB*C(4): IF D<>0 THEN 10350
10340 P=P-2: Q=Q*(Q+1): GOTO 10370
10350 P=P+(B(2)*C(3)-B(1)*C(4))/D
10360 Q=Q+(-B(2)*CB+B(1)*C(3))/D
10370 NEXT J
10380 E0=E0*10: GOTO 10200
10390 NC=NC-1
10400 V(NC)=0
10410 IF IV>=0 THEN 10430
10420 U(NC)=1/R: GOTO 10440
10430 U(NC)=R
10440 FOR I=1 TO NC: H(I)=B(I+1): NEXT
10450 GOTO 10080
10460 NC=NC-2
10470 IF IV>=0 THEN 10490
10480 QP=1/Q: PP=P/(Q*2): GOTO 10500
10490 QP=Q: PP=P/2
10500 F=PP*PP-QP: IF F>=0 THEN 10520
10510 U(NC+1)=-PP:U(NC)=-PP:V(NC+1)=SQR(-F):V(NC)=-V(NC+1):GOTO 10570
10520 IF PP<>0 THEN 10540
10530 U(NC+1)=-SQR(F): GOTO 10550
```

FIGURE A.6. Program ROOTS.

```
10540 U(NC+1)=-(PP/ABS(PP))*(ABS(PP)+SQR(F))
10550 V(NC+1)=0
10560 U(NC)=QP/U(NC+1): V(NC)=0
10570 FOR I=1 TO NC: H(I)=B(I+2): NEXT
10580 GOTO 10080
```

FIGURE A.6. (Continuation of ROOTS.)

```
10 '    The inverse Laplace transform of a rational function
20 '    X(s)=N(s)/D(s) where degree of N(s) < degree of D(s).
30 '    Input the polynomials in the form
40 '             p(s)=a(n)*s^n+a(n-1)*s^(n-1)+...+a(1)s+a(0).
50 '
60 '
70 '  Program by Robert S. Cooper -- University of Florida
80 '
90 '
100 CLS
110 IR=1  ' Dictates the order in which coefficients are read
120 PRINT: INPUT" Degree of N(s)=";N1
130 CLS
140 N=N1: PRINT: PRINT"Enter coefficients of N(s)":PRINT
150 '
160 ' Read in the numerator coefficients
170 '
180 GOSUB 1350
190 TOPCONST=A(N+1)
200 '
210 ' Find the roots of the numerator -- label them U(i) and V(i),
220 ' the real and imaginary parts respectively.
230 '
240 GOSUB 10000
250 CLS
260 FOR I=1 TO N1: U1(I)=U(I): V1(I)=V(I):NEXT
270 PRINT: INPUT" Degree of D(s)=";N2
280 N=N2: PRINT: PRINT"Enter coefficients of D(s)":PRINT
290 '
300 ' Read in the denominator coefficients
310 '
320 GOSUB 1350
330 BOTCONST=A(N+1)
340 '
350 ' Find the roots of the denominator -- label them U(i) and V(i),
360 ' the real and imaginary parts respectively.
370 '
380 GOSUB 10000
390 '
400 ' TOTCONST is the ratio of leading coefficients.
410 '
420 TOTCONST=TOPCONST/BOTCONST
430 CLS
440 FOR J=1 TO N2: U2(J)=U(J): V2(J)=V(J):NEXT
450 '
460 ' Outer loop determines the coefficient (magnitude and angle)
470 ' of each partial fraction expansion term.  First nested
480 ' loop substitutes the Ith root into the transfer function
490 ' numerator.  The second nested loop substitutes the root
500 ' into the rest of the transfer function's denominator.
510 ' Then the two are divided, yielding the coefficient.
520 '
530 FOR I=1 TO N2
540 TOPMAG=1:TOPANGLE=0
550      FOR J=1 TO N1
560      TOPREAL=U2(I) - U1(J)
570      TOPIMAG = V2(I) - V1(J)
580      TOPMAG = SQR(TOPREAL^2 + TOPIMAG^2) * TOPMAG
```

FIGURE A.7. Program INVLT.

```
590      IF TOPREAL=0 AND SGN(TOPIMAG)=1  THEN ANGLE=90:GOTO 640
600      IF TOPREAL=0 AND SGN(TOPIMAG)=0  THEN ANGLE=0 :GOTO 640
610      IF TOPREAL=0 AND SGN(TOPIMAG)=-1 THEN ANGLE=-90:GOTO 640
620      ANGLE = ATN(TOPIMAG/TOPREAL)*57.29577951#
630      IF SGN(TOPREAL) = -1 THEN ANGLE=ANGLE + 180
640      TOPANGLE = TOPANGLE + ANGLE
650      NEXT J
660 BOTMAG = 1:BOTANGLE = 0
670      FOR J=1 TO N2
680      IF J=I THEN GOTO 780
690      BOTREAL=U2(I) - U2(J)
700      BOTIMAG = V2(I) - V2(J)
710      BOTMAG = SQR(BOTREAL^2 + BOTIMAG^2) * BOTMAG
720      IF BOTREAL=0 AND SGN(BOTIMAG)=1  THEN ANGLE=90:GOTO 770
730      IF BOTREAL=0 AND SGN(BOTIMAG)=0  THEN ANGLE=0 :GOTO 770
740      IF BOTREAL=0 AND SGN(BOTIMAG)=-1 THEN ANGLE=-90:GOTO 770
750      ANGLE = ATN(BOTIMAG/BOTREAL)*57.29577951#
760      IF SGN(BOTREAL) = -1 THEN ANGLE=ANGLE + 180
770      BOTANGLE = BOTANGLE + ANGLE
780      NEXT J
790 TOTALMAG(I) = TOPMAG/BOTMAG
800 TOTANGLE(I) = TOPANGLE - BOTANGLE
810 NEXT I
820      FOR I=1 TO N2
830      TOTANGLE(I)=(TOTANGLE(I)/360 - FIX(TOTANGLE(I)/360))*360
840      IF TOTANGLE(I)>180 THEN TOTANGLE(I)=TOTANGLE(I)-360
850      IF TOTANGLE(I)<-180 THEN TOTANGLE(I)=360+TOTANGLE(I)
860      DUMMY1=ABS(ABS(TOTANGLE(I))-180)
870      IF DUMMY1 < .00001 THEN TOTANGLE(I)=0:TOTALMAG(I)=-TOTALMAG(I)
880      NEXT I
890 '
900 ' This is simply a reordering routine.  The loop
910 ' reorders complex conjugate pairs so that the terms
920 ' with positive imaginary parts come first.
930 '
940 FOR I=1 TO N2
950      FOR J=(I+1) TO N2
960      IF V2(I) >= V2(J) THEN GOTO 1090
970      DUMMY1 = TOTANGLE(J)
980      DUMMY2 = TOTALMAG(J)
990      DUMMY3 = U2(J)
1000     DUMMY4 = V2(J)
1010     TOTANGLE(J) = TOTANGLE(I)
1020     TOTALMAG(J) = TOTALMAG(I)
1030     U2(J) = U2(I)
1040     V2(J) = V2(I)
1050     TOTANGLE(I) = DUMMY1
1060     TOTALMAG(I) = DUMMY2
1070     U2(I) = DUMMY3
1080     V2(I) = DUMMY4
1090     NEXT J
1100 NEXT I
1110 '
1120 ' This loop COUNTs the number of complex conjugate pairs.
1130 '
1140 COUNT = 0
1150 FOR I=1 TO N2
1160 IF (V2(I))>.00001 THEN COUNT=COUNT+1
1170 NEXT I
1180 CLS
1190 '
1200 ' Un-normalize each partial fraction term. Print the result.
1210 '
1220 FOR I=1 TO N2: TOTALMAG(I)=TOTALMAG(I)*TOTCONST:NEXT I
1230 PRINT "The inverse Laplace transform of X(s) = "
1240     FOR I=1 TO COUNT
1250     DUMMY2 = 2*TOTALMAG(I)
1260     PRINT USING "+####.####exp(####.##t)";DUMMY2;U2(I);
1265     PRINT USING "cos(####.##t";V2(I);
1270     PRINT USING " +###.##) ";TOTANGLE(I);
1280     NEXT I
```

FIGURE A.7. (Continuation of INVLT.)

```
1290 DUMMY1 = N2-COUNT
1300 FOR I=(COUNT+1) TO DUMMY1
1310 PRINT USING "+####.####exp(####.##t) ";TOTALMAG(I);U2(I);
1320 NEXT I
1330 END
1340 '
1350 '    Subroutine for entering polynomial coefficient array
1360 '
1370 FOR I=0 TO N
1380 PRINT" for i="I: INPUT"a(i)=";A(I+1)
1390 NEXT I
1400 RETURN
10000 IV=IR
10010 NC=N+1
10020 FOR I=1 TO NC: H(I)=A(I): NEXT
10030 P=0: R=0: Q=0
10040 IF H(1)<>0 THEN 10080
10050 NC=NC-1: U(NC)=0: V(NC)=0
10060 FOR I=1 TO NC: H(I)=H(I+1): NEXT
10070 GOTO 10040
10080 IF NC=1 THEN RETURN
10090 IF NC=2 THEN R=-H(1)/H(2): GOTO 10390
10100 IF NC=3 THEN P=H(2)/H(3): Q=H(1)/H(3): GOTO 10460
10110 IF ABS(H(NC-1)/H(NC))>=ABS(H(2)/H(1)) THEN GOTO 10180
10120 IV=-IV
10130 M=INT(NC/2+.1)
10140 FOR I=1 TO M:NL=NC+1-I:F=H(NL):H(NL)=H(I):H(I)=F:NEXT
10150 IF Q=0 THEN P=0: GOTO 10170
10160 P=P/Q: Q=1/Q
10170 IF R<>0 THEN R=1/R
10180 E0=1E-09: REM sets accuracy
10190 B(NC)=H(NC):C(NC)=H(NC):B(NC+1)=0:C(NC+1)=0:NP=NC-1
10200 FOR J=1 TO 1000
10210 FOR I1=1 TO NP:I=NC-I1:B(I)=H(I)+R*B(I+1):C(I)=B(I)+R*C(I+1)
10220 NEXT:IF ABS(B(1)/H(1))<=E0 THEN 10390
10230 IF C(2)=0 THEN R=R+1: GOTO 10250
10240 R=R-B(1)/C(2)
10250 FOR I1=1 TO NP
10260 I=NC-I1:B(I)=H(I)-P*B(I+1)-Q*B(I+2)
10270 C(I)=B(I)-P*C(I+1)-Q*C(I+2):NEXT I1
10280 IF H(2)<>0 THEN 10310
10290 IF ABS(B(2)/H(1))<=E0 THEN 10320
10300 IF ABS(B(2)/H(1))>E0 THEN 10330
10310 IF ABS(B(2)/H(2))>E0 THEN 10330
10320 IF ABS(B(1)/H(1))<=E0 THEN 10460
10330 CB=C(2)-B(2): D=C(3)*C(3)-CB*C(4): IF D<>0 THEN 10350
10340 P=P-2: Q=Q*(Q+1): GOTO 10370
10350 P=P+(B(2)*C(3)-B(1)*C(4))/D
10360 Q=Q+(-B(2)*CB+B(1)*C(3))/D
10370 NEXT J
10380 E0=E0*10: GOTO 10200
10390 NC=NC-1
10400 V(NC)=0
10410 IF IV>=0 THEN 10430
10420 U(NC)=1/R: GOTO 10440
10430 U(NC)=R
10440 FOR I=1 TO NC: H(I)=B(I+1): NEXT
10450 GOTO 10080
10460 NC=NC-2
10470 IF IV>=0 THEN 10490
10480 QP=1/Q: PP=P/(Q*2): GOTO 10500
10490 QP=Q: PP=P/2
10500 F=PP*PP-QP: IF F>=0 THEN 10520
10510 U(NC+1)=-PP:U(NC)=-PP:V(NC+1)=SQR(-F):V(NC)=-V(NC+1):GOTO 10570
10520 IF PP<>0 THEN 10540
10530 U(NC+1)=-SQR(F): GOTO 10550
10540 U(NC+1)=-(PP/ABS(PP))*(ABS(PP)+SQR(F))
10550 V(NC+1)=0
10560 U(NC)=QP/U(NC+1): V(NC)=0
10570 FOR I=1 TO NC: H(I)=B(I+2): NEXT
10580 GOTO 10080
```

FIGURE A.7.   (Continuation of INVLT.)

```
10 REM   Inverse z-transform by E. W. Kamen
20 REM   Program computes first Q + 1 values
30 REM   of the inverse z-transform of X(z) =
40 REM   b(n)z^n + b(n-1)z^n-1 + ... b(0) divided by
50 REM   a(n)z^n + a(n-1)z^n-1 ... + a(0),
60 REM   where n is the order of X(z).    |
70 CLS
80 INPUT "Order n of rational function";N
90 PRINT:INPUT "No Q of time points";Q
100 DIM X(Q+N),A(N),B(N),V(Q+N),Y(Q)
110 FOR I = 0 TO N
120 PRINT:PRINT "Enter b(i) for i =";I:INPUT B(I)
130 NEXT I
140 FOR I = 0 TO N
150 PRINT:PRINT "Enter a(i) for i =";I:INPUT A(I)
160 NEXT I
170 FOR K = 0 TO N-1
180 X(K)=0:V(K)=0:A(K)=A(K)/A(N):B(K)=B(N)/A(N)
190 NEXT K
200 B(N) = B(N)/A(N):A(N)=0
210 PRINT:PRINT " k           x(kT)":PRINT
220 FOR K=0 TO Q
230 V(K+N) = 1-SGN(K)
240 NEXT K
250 FOR K=0 TO Q
260 X(K+N) = 0
270 FOR I = 0 TO N
280 X(K+N)=X(K+N)-A(I)*X(K+I)+B(I)*V(K+I)  |
290 NEXT I
300 Y(K)=X(K+N):PRINT K;TAB(10);Y(K)
310 NEXT K
320 PRINT:INPUT"Want a plot of x(kT) (y/n)";A$
330 PRINT:PRINT:IF A$="y" THEN 1000
340 END
1000 -1280 PLOTSUB
```

FIGURE A.8.   Program INVZT.

The inverse z-transform of a rational function can be calculated using the program INVZT in Figure A.8. The Routh stability test is implemented in the program ROUTH listed in Figure A.9. The frequency response functions of continuous-time and discrete-time systems can be determined using the programs CONTFREQ and DISCFREQ listed in Figures A.10 and A.11. Note that lines 1000–1270 of PLOTSUB are reproduced in both CONTFREQ and DISCFREQ. Therefore, in putting CONTFREQ and DISCFREQ on a computer, remember to load PLOTSUB first.

```
10 REM   Routh stability test
20 REM   Program by E. W. Kamen
30 REM   Coefficients of polynomial are given by
40 REM   a(n)s^n + a(n-1)s^(n-1) + ... +a(1)s + a(0),
45 REM   where a(n) > 0.
50 CLS
60 INPUT "Number n of poles";N
70 DIM A(N),B(N,N)
80 FOR I = 0 TO N
90 PRINT:PRINT "Enter a(i) for i =";I:INPUT A(I)
100 NEXT I
```

FIGURE A.9.   Program ROUTH.

```
110 FOR I = 0 TO INT(N/2)
120 B(N,N-I) = A(N-2*I)
130 NEXT I
140 FOR I = 0 TO INT((N-1)/2)
150 B(N-1,N-I) = A(N-2*I-1)
160 NEXT I
165 IF N < 3 GOTO 230
170 FOR I = 2 TO N-1
180 FOR K = 0 TO N-I
185 IF B(N-I+1,N) = 0 THEN B(N-I+1,N) = .00001
190 B(N-I,N-K) = -B(N-I+2,N)*B(N-I+1,N-K-1)/B(N-I+1,N)
200 B(N-I,N-K) = B(N-I,N-K) + B(N-I+2,N-K-1)
210 NEXT K
220 NEXT I
230 B(0,N) = A(0)
240 PRINT "First column of Routh array is":PRINT
250 FOR I = 0 TO N
260 PRINT B(N-I,N)
270 NEXT I
280 P = 0
285 FOR I = N-1 TO 0 STEP -1
290 IF B(I,N) < 0 THEN 320
300 IF B(I+1,N) < 0 THEN P = P + 1
310 GOTO 340
320 IF B(I+1,N) > 0 THEN P = P + 1
330 P = P
340 NEXT I
350 PRINT:PRINT:PRINT
360 PRINT "        There are";P;"poles not in the"
370 PRINT:PRINT "            open left-half plane."
380 END
```

FIGURE A.9. (Continuation of ROUTH.)

```
10 REM  Frequency response curves of system
20 REM  given by transfer function H(s) =
30 REM  b(n)s^n+b(n-1)s^(n-1)+...+b(1)s+b(0) divided by
40 REM  a(n)s^n+a(n-1)s^(n-1)+...+a(1)s+a(0).
50 REM  Program by E. W. Kamen
60 CLS
70 INPUT "Order n of transfer function";N
80 PRINT:INPUT "Minimum frequency in rad/sec";WL
90 PRINT:INPUT "Maximum frequency in rad/sec";WH
100 PRINT:INPUT "Number of points";Q
110 DIM A(N),B(N),W(Q),RN(Q),IN(Q),RD(Q),ID(Q)
120 DIM MH(Q),PH(Q),X(Q),Y(Q),LH(Q)
130 FOR I = O TO N
140 PRINT:PRINT "Enter b(i) for i =";I:INPUT B(I)
150 NEXT I
160 FOR I = O TO N
170 PRINT:PRINT "Enter a(i) for i =";I:INPUT A(I)
180 NEXT I:PRINT
190 PRINT "Freq        Mag        Mag in dB        Phase"
200 PRINT:FOR K = 0 TO Q
210 W(K) = WL+K*(WH-WL)/Q
220 FOR I = 0 TO INT(N/2)
230 RN(K)=RN(K)+(-1)^I*B(2*I)*(W(K)^(2*I))
240 RD(K)=RD(K)+(-1)^I*A(2*I)*(W(K)^(2*I))
250 NEXT I
260 FOR I = 1 TO INT(N/2+.5)
270 IN(K)=IN(K)+(-1)^(I+1)*B(2*I-1)*(W(K)^(2*I-1))
280 ID(K)=ID(K)+(-1)^(I+1)*A(2*I-1)*(W(K)^(2*I-1))
290 NEXT I
300 PI = 3.1415927#
310 MH(K)=SQR((RN(K)^2+IN(K)^2)/(RD(K)^2+ID(K)^2))
320 LH(K)=20*LOG(MH(K)+.00001)/LOG(10)
330 IF RN(K)>0 THEN X(K)=ATN(IN(K)/RN(K))*180/PI:GOTO 360
340 IF RN(K)=0 THEN X(K)=SGN(IN(K))*90:GOTO 360
350 X(K)=180+ATN(IN(K)/RN(K))*180/PI
360 IF RD(K)>0 THEN Y(K)=ATN(ID(K)/RD(K))*180/PI:GOTO 390
```

FIGURE A.10. Program CONTFREQ.

```
370 IF RD(K)=0 THEN Y(K)=SGN(ID(K))*90:GOTO 390
380 Y(K)=180+ATN(ID(K)/RD(K))*180/PI
390 IF K=0 THEN 440
400 IF X(K)>X(K-1)+180 THEN X(K)=X(K)-360
410 IF X(K)<X(K-1)-180 THEN X(K)=X(K)+360
420 IF Y(K)>Y(K-1)+180 THEN Y(K)=Y(K)-360
430 IF Y(K)<Y(K-1)-180 THEN Y(K)=Y(K)+360
440 PH(K)=X(K)-Y(K)
450 W(K)=.001*INT(1000*W(K)+.5)
460 MH(K)=.001*INT(1000*MH(K)+.5)
470 LH(K)=.001*INT(1000*LH(K)+.5)
480 PH(K)=.001*INT(1000*PH(K)+.5)
490 PRINT W(K);TAB(13);MH(K);TAB(29);LH(K);TAB(48);PH(K)
500 NEXT K
510 PRINT:INPUT"Want a plot of the mag function (y/n)";D$
520 IF D$="y" THEN FOR K=0 TO Q:Y(K)=MH(K):NEXT:GOTO 1000
530 PRINT:INPUT"Want a plot of the log mag function (y/n)";B$
540 IF B$="y" THEN FOR K=0 TO Q:Y(K)=LH(K):NEXT:GOTO 1000
550 PRINT:INPUT"Want a phase plot (y/n)";C$
560 IF C$="y" THEN FOR K=0 TO Q:Y(K)=PH(K):NEXT:GOTO 1000
570 END
1000 YH=Y(0):YL=Y(0)
1010 FOR K=1 TO Q
1020 IF Y(K)>=YH THEN YH=Y(K):GOTO 1050
1030 IF Y(K)>=YL THEN YH=YH:GOTO 1050
1040 YL=Y(K)
1050 NEXT K
1060 FOR K=0 TO Q
1070 Y(K)=1+(Y(K)-YL)*29/(YH-YL):Y(K)=INT(Y(K)+.5)
1080 NEXT K
1090 Z=1+INT(-YL*29/(YH-YL)+.5):YH=.01*INT(100*YH+.5)
1100 YL=.01*INT(100*YL+.5):INPUT "Plot on screen (y/n)";A$
1110 IF A$="n" THEN GOTO 1190
1120 PRINT "min=";YL;TAB(25);"max=";YH
1130 FOR K=0 TO Q
1140 IF Y(K)=Z THEN PRINT TAB(Z);"*":GOTO 1180
1150 IF Z<=30 AND Y(K)< Z THEN PRINT TAB(Y(K));"*";TAB(Z);"0":GOTO 1180
1160 IF Z>=1 AND Y(K)>Z THEN PRINT TAB(Z);"0";TAB(Y(K));"*":GOTO 1180
1170 PRINT TAB(Y(K));"*"
1180 NEXT K
1190 INPUT "Plot on printer (y/n)";A$
1200 IF A$="n" THEN 1280
1210 LPRINT "min=";YL;TAB(25);"max=";YH
1220 FOR K=0 TO Q
1230 IF Y(K)=Z THEN LPRINT TAB(Z);"*":GOTO 1270
1240 IF Z<=30 AND Y(K)<Z THEN LPRINT TAB(Y(K));"*";TAB(Z);"0":GOTO 1270
1250 IF Z>=1 AND Y(K)>Z THEN LPRINT TAB(Z);"0";TAB(Y(K));"*":GOTO 1270
1260 LPRINT TAB(Y(K));"*"
1270 NEXT K
1280 IF B$<>"y" AND C$<>"y" THEN 530
1290 IF B$="y" AND C$<>"y" THEN 550
1300 END
```

FIGURE A.10. (Continuation of CONTFREQ.)

```
10 REM  Frequency response curves of discrete-time system
20 REM  given by transfer function H(z) =
30 REM  b(n)z^n+b(n-1)z^(n-1)+...+b(1)z+b(0) divided by
40 REM  a(n)z^n+a(n-1)z^(n-1)+...+a(1)z+a(0).
50 REM  Program by E. W. Kamen
60 CLS
70 INPUT "Order n of transfer function";N
80 PRINT:INPUT "Number of points";Q
90 DIM A(N),B(N),W(Q),RN(Q),IN(Q),RD(Q),ID(Q)
100 DIM MH(Q),PH(Q),X(Q),Y(Q),LH(Q)
110 FOR I = 0 TO N
120 PRINT:PRINT "Enter b(i) for i =";I:INPUT B(I)
130 NEXT I
140 FOR I = 0 TO N
150 PRINT:PRINT "Enter a(i) for i =";I:INPUT A(I)
```

FIGURE A.11. Program DISCFREQ.

```
160 NEXT I:PRINT
170 PRINT "Freq          Mag            Mag in dB          Phase"
180 PRINT:FOR K = 0 TO Q
190 LET PI = 3.1415927#
200 W(K) = K*(2*PI/Q)
210 FOR I = 0 TO N
220 RN(K) = RN(K) + B(I)*COS(I*W(K))
230 RD(K) = RD(K) + A(I)*COS(I*W(K))
240 IN(K) = IN(K) + B(I)*SIN(I*W(K))
250 ID(K) = ID(K) + A(I)*SIN(I*W(K))
260 NEXT I
270 MH(K)=SQR((RN(K)^2+IN(K)^2)/(RD(K)^2+ID(K)^2))
280 LH(K)=20*LOG(MH(K)+.00001)/LOG(10)
290 IF RN(K)>0 THEN X(K)=ATN(IN(K)/RN(K))*180/PI:GOTO 320
300 IF RN(K)=0 THEN X(K)=SGN(IN(K))*90:GOTO 320
310 X(K)=180+ATN(IN(K)/RN(K))*180/PI
320 IF RD(K)>0 THEN Y(K)=ATN(ID(K)/RD(K))*180/PI:GOTO 350
330 IF RD(K)=0 THEN Y(K)=SGN(ID(K))*90:GOTO 350
340 Y(K)=180+ATN(ID(K)/RD(K))*180/PI
350 IF K=0 THEN 400
360 IF X(K)>X(K-1)+180 THEN X(K)=X(K)-360
370 IF X(K)<X(K-1)-180 THEN X(K)=X(K)+360
380 IF Y(K)>Y(K-1)+180 THEN Y(K)=Y(K)-360
390 IF Y(K)<Y(K-1)-180 THEN Y(K)=Y(K)+360
400 PH(K)=X(K)-Y(K)
410 W(K)=.001*INT(1000*W(K)+.5)
420 MH(K)=.001*INT(1000*MH(K)+.5)
430 LH(K)=.001*INT(1000*LH(K)+.5)
440 PH(K)=.001*INT(1000*PH(K)+.5)
450 PRINT W(K);TAB(13);MH(K);TAB(29);LH(K);TAB(48);PH(K)
460 NEXT K
470 PRINT:INPUT"Want a plot of the mag function (y/n)";D$
480 IF D$="y" THEN FOR K=0 TO Q:Y(K)=MH(K):NEXT:GOTO 1000
490 PRINT:INPUT"Want a plot of the log mag function (y/n)";B$
500 IF B$="y" THEN FOR K=0 TO Q:Y(K)=LH(K):NEXT:GOTO 1000
510 PRINT:INPUT"Want a phase plot (y/n)";C$
520 IF C$="y" THEN FOR K=0 TO Q:Y(K)=PH(K):NEXT K:GOTO 1000
530 END
1000 YH=Y(0):YL=Y(0)
1010 FOR K=1 TO Q
1020 IF Y(K)>=YH THEN YH=Y(K):GOTO 1050
1030 IF Y(K)>=YL THEN YH=YH:GOTO 1050
1040 YL = Y(K)
1050 NEXT K
1060 FOR K=0 TO Q
1070 Y(K)=1+(Y(K)-YL)*29/(YH-YL):Y(K)=INT(Y(K)+.5)
1080 NEXT K
1090 Z=1+INT(-YL*29/(YH-YL)+.5):YH=.01*INT(100*YH+.5)
1100 YL=.01*INT(100*YL+.5):INPUT "Plot on screen (y/n)";A$
1110 IF A$="n" THEN 1190
1120 PRINT "min=";YL;TAB(25);"max=";YH
1130 FOR K=0 TO Q
1140 IF Y(K)=Z THEN PRINT TAB(Z);"*":GOTO 1180
1150 IF Z<=30 AND Y(K)<Z THEN PRINT TAB(Y(K));"*";TAB(Z);"0":GOTO 1180
1160 IF Z>=1 AND Y(K)>Z THEN PRINT TAB(Z);"0";TAB(Y(K));"*":GOTO 1180
1170 PRINT TAB(Y(K));"*"
1180 NEXT K
1190 INPUT "Plot on printer (y/n)";A$
1200 IF A$="n" THEN 1280
1210 LPRINT "min=";YL;TAB(25);"max=";YH
1220 FOR K=0 TO Q
1230 IF Y(K)=Z THEN LPRINT TAB(Z);"*":GOTO 1270
1240 IF Z<=30 AND Y(K)<Z THEN LPRINT TAB(Y(K));"*";TAB(Z);"0":GOTO 1270
1250 IF Z>=1 AND Y(K)>Z THEN LPRINT TAB(Z);"0";TAB(Y(K));"*":GOTO 1270
1260 LPRINT TAB(Y(K));"*"
1270 NEXT K
1280 IF B$<>"y" AND C$<>"y" THEN 490
1290 IF B$="y" AND C$<>"y" THEN 510
1300 END
```

FIGURE A.11.   (Continuation of DISCFREQ.)

```
10 REM   Program computes N-point DFT.
20 REM   Program by E. W. Kamen
30 REM   If signal x(kT) is given in function form,
40 REM   define x(kT) on line 140.
50 PI = 3.14159265#
60 CLS
70 INPUT "No N of points";N
80 DIM X(N),R(N),I(N),M(N),P(N),Y(N)
90 PRINT:INPUT"Is x(kT) in function form (y/n)";A$
100 PRINT:IF A$="y" THEN 140
110 FOR K=0 TO N-1
120 PRINT:PRINT"Enter x(kT) for k =";K:INPUT X(K)
130 NEXT K:GOTO 160
140 DEF FNX(K) =
150 FOR K=1 TO N-1:X(K)=FNX(K):NEXT
160 PRINT "Point #        Mag           Phase"
170 PRINT:FOR K=0 TO N-1
180 R(K)=0:I(K)=0
190 FOR I=0 TO N-1
200 R(K)=R(K)+X(I)*COS(2*PI*I*K/N)
210 I(K)=I(K)-X(I)*SIN(2*PI*I*K/N)
220 NEXT I
230 M(K) = SQR(R(K)^2+I(K)^2)
240 IF R(K)>0 THEN P(K)=ATN(I(K)/R(K))*180/PI:GOTO 270
250 IF R(K)=0 THEN P(K)=SGN(I(K))*90:GOTO 270
260 P(K)=180+ATN(I(K)/R(K))*180/PI
270 IF K=0 THEN 300
280 IF P(K)>P(K-1)+180 THEN P(K)=P(K)-360
290 IF P(K)<P(K-1)-180 THEN P(K)=P(K)+360
300 M(K) = .001*INT(1000*M(K)+.5)
310 P(K)=.001*INT(1000*P(K)+.5)
320 PRINT K;TAB(13);M(K);TAB(29);P(K)
330 NEXT K
340 Q=N-1
350 PRINT:INPUT"Want a plot of mag spectrum (y/n)";B$
360 IF B$="y" THEN FOR K=0 TO Q:Y(K)=M(K):NEXT:GOTO 1000
370 PRINT:INPUT"Want a plot of phase spectrum (y/n)";C$
380 IF C$="y" THEN FOR K=0 TO Q:Y(K)=P(K):NEXT:GOTO 1000
390 END
1000 YH=Y(0):YL=Y(0)
1010 FOR K=1 TO Q
1020 IF Y(K)>=YH THEN YH=Y(K):GOTO 1050
1030 IF Y(K)>=YL THEN YH=YH:GOTO 1050
1040 YL = Y(K)
1050 NEXT K
1060 FOR K=0 TO Q
1070 Y(K)=1+(Y(K)-YL)*29/(YH-YL):Y(K)=INT(Y(K)+.5)
1080 NEXT K
1090 Z=1+INT(-YL*29/(YH-YL)+.5):YH=.01*INT(100*YH+.5)
1100 YL=.01*INT(100*YL+.5):INPUT "Plot on screen (y/n)";A$
1110 IF A$="n" THEN GOTO 1190
1120 PRINT "min=";YL;TAB(25);"max=";YH
1130 FOR K=0 TO Q
1140 IF Y(K)=Z THEN PRINT TAB(Z);"*":GOTO 1180
1150 IF Z<=30 AND Y(K)<Z THEN PRINT TAB(Y(K));"*";TAB(Z);"0":GOTO 1180
1160 IF Z>=1 AND Y(K)>Z THEN PRINT TAB(Z);"0";TAB(Y(K));"*":GOTO 1180
1170 PRINT TAB(Y(K));"*"
1180 NEXT K
1190 INPUT "Plot on printer (y/n)";A$
1200 IF A$="n" THEN 1280
1210 LPRINT "min=";YL;TAB(25);"max=";YH
1220 FOR K=0 TO Q
1230 IF Y(K)=Z THEN LPRINT TAB(Z);"*":GOTO 1270
1240 IF Z<=30 AND Y(K)<Z THEN LPRINT TAB(Y(K));"*";TAB(Z);"0":GOTO 1270
1250 IF Z>=1 AND Y(K)>Z THEN LPRINT TAB(Z);"0";TAB(Y(K));"*":GOTO 1270
1260 LPRINT TAB(Y(K));"*"
1270 NEXT K
1280 IF B$="y" AND C$<>"y" THEN 370
1290 END
```

FIGURE A.12. Program DFT.

```
10 REM    FFT Algorithm
20 REM    Program computes DFT or inverse DFT by FFT method.
30 REM    Program by Hiep Vu and R. S. Cooper
40 REM    If signal or transform is given in function form,
50 REM    define real part A(k) on line 220 and
60 REM    imaginary part B(k) on line 230.
70 CLS
80 PRINT"If you want to calculate the DFT, set FFT = -1"
90 PRINT
100 PRINT"If you want to calculate the inverse DFT, set FFT=1"
110 PRINT:INPUT "FFT =  ";FFT:PRINT
120 INPUT "Give M where N = 2^M = no of points, M=  ";M
130 N=2^M
140 DIM A(N),B(N),M(N),UA(N),UB(N),P(N),Y(N)
150 PRINT:INPUT "Is the signal in function form (y/n)";F$
160 IF F$="y" THEN 220
170 FOR I=1 TO N
180 PRINT:PRINT "Enter real part A(k), imag part B(k) ";
190 PRINT"for k = ";I-1:INPUT A(I),B(I)
200 NEXT I
210 GOTO 270
220 DEF FNA(K) =
230 DEF FNB(K) =
240 FOR J=1 TO N
250 K=J-1:A(J)=FNA(K):B(J)=FNB(K)
260 NEXT J
270 N2=INT(N/2)
280 N1=N-1
290 J=1
300 FOR I=1 TO N1
310      IF I>=J THEN 380
320      TA=A(J)
330      TB=B(J)
340      A(J)=A(I)
350      B(J)=B(I)
360      A(I)=TA
370      B(I)=TB
380      K=N2
390      IF K>=J GOTO 430
400      J=J-K
410      K=INT(K/2)
420      GOTO 390
430      J=J+K
440 NEXT I
450 REM
460 REM    THE FOLLOWING IS THE ACTUAL CALCULATION
470 REM
480 PRINT:PRINT "point #          Mag              Angle"
490 PRINT:PI=3.141592653#
500 FOR L=1 TO M
510 LE=2^L
520 L2=INT(LE/2)
530 UA(1)=1:UB(1)=0
540 WA=COS(PI/L2):IF ABS(WA)<1E-08 THEN WA=0
550 WB=FFT*SIN(PI/L2):IF ABS(WB)<1E-08 THEN WB=0
560 FOR G=1 TO L2
570   FOR H=G TO N STEP LE
580   IP=H+L2
590   VA=A(IP)*UA(G) - B(IP)*UB(G)
600   VB=B(IP)*UA(G) + A(IP)*UB(G)
610   A(IP)=A(H)-VA
620   B(IP)=B(H)-VB
630   A(H)=A(H)+VA
640   B(H)=B(H)+VB
650   NEXT H
660 UA(G+1)=UA(G)*WA - UB(G)*WB
670 UB(G+1)=UB(G)*WA + UA(G)*WB
680 NEXT G
690 NEXT L
700 IF FFT=-1 GOTO 750
```

FIGURE A.13. Program FFT.

```
710 FOR I=1 TO N
720 A(I)=A(I)/N
730 B(I)=B(I)/N
740 NEXT I
750 FOR J=1 TO N
760 M(J)=(SQR(A(J)^2  +  B(J)^2))
770 IF A(J)>0 THEN P(J)=ATN(B(J)/A(J))*180/PI:GOTO 800
780 IF A(J)=0 THEN P(J)=SGN(B(J))*90:GOTO 800
790 P(J)=180+ATN(B(J)/A(J))*180/PI
800 IF J=1 THEN 830
810 IF P(J)>P(J-1)+180 THEN P(J)=P(J)-360
820 IF P(J)<P(J-1)-180 THEN P(J)=P(J)+360
830 PRINT J-1;TAB(13);M(J);TAB(29);P(J)
840 NEXT J
850 Q=N-1
860 PRINT:INPUT"Want a plot of mag function (y/n)";B$
870 IF B$="y" THEN FOR K=0 TO Q:Y(K)=M(K+1):NEXT:GOTO 1000
880 PRINT:INPUT"Want a plot of angle function (y/n)";C$
890 IF C$="y" THEN FOR K=0 TO Q:Y(K)=P(K+1):NEXT:GOTO 1000
900 END
1000 YH=Y(0):YL=Y(0)
1010 FOR K=1 TO Q
1020 IF Y(K)>=YH THEN YH=Y(K):GOTO 1050
1030 IF Y(K)>=YL THEN YH=YH:GOTO 1050
1040 YL = Y(K)
1050 NEXT K
1060 FOR K=0 TO Q
1070 Y(K)=1+(Y(K)-YL)*29/(YH-YL):Y(K)=INT(Y(K)+.5)
1080 NEXT K
1090 Z=1+INT(-YL*29/(YH-YL)+.5):YH=.01*INT(100*YH+.5)
1100 YL=.01*INT(100*YL+.5):INPUT "Plot on screen (y/n)";A$
1110 IF A$="n" THEN GOTO 1190
1120 PRINT "min=";YL;TAB(25);"max=";YH
1130 FOR K=0 TO Q
1140 IF Y(K)=Z THEN PRINT TAB(Z);"*":GOTO 1180
1150 IF Z<=30 AND Y(K)<Z THEN PRINT TAB(Y(K));"*";TAB(Z);"0":GOTO 1180
1160 IF Z>=1 AND Y(K)>Z THEN PRINT TAB(Z);"0";TAB(Y(K));"*":GOTO 1180
1170 PRINT TAB(Y(K));"*"
1180 NEXT K
1190 INPUT "Plot on printer (y/n)";A$
1200 IF A$="n" THEN 1280
1210 LPRINT "min=";YL;TAB(25);"max=";YH
1220 FOR K=0 TO Q
1230 IF Y(K)=Z THEN LPRINT TAB(Z);"*":GOTO 1270
1240 IF Z<=30 AND Y(K)<Z THEN LPRINT TAB(Y(K));"*";TAB(Z);"0":GOTO 1270
1250 IF Z>=1 AND Y(K)>Z THEN LPRINT TAB(Z);"0";TAB(Y(K));"*":GOTO 1270
1260 LPRINT TAB(Y(K));"*"
1270 NEXT K
1280 IF B$="y" AND C$<>"y" THEN 880
1290 END
```

FIGURE A.13. (Continuation of FFT.)

Listed in Figures A.12 to A.15 are the last four programs, DFT, FFT, CTFT, and STATEDIS. Note that lines 1000–1270 of PLOTSUB are reproduced in DFT, FFT, and CTFT. Thus PLOTSUB should be loaded first before typing in these programs. It should also be noted that since the program FFT computes the discrete Fourier transform, it is not necessary to implement the program DFT. However, the program DFT is interesting from an instructional standpoint. In particular, it is interesting to compare the time of computation between the two programs DFT and FFT.

```
10 REM   Continuous-time Fourier transform
20 REM   This program computes the continuous-time
30 REM   Fourier transform by using the FFT algorithm.
40 REM   Program by R. S. Cooper and E. W. Kamen
50 REM   If the signal x(t) is given in function form,
60 REM   define x(kT) on line 210.
70 CLS
80 PI = 3.14159265#
90 FFT = -1
100 INPUT "Give M where N = 2^M = no of points, M =  ";M
110 N = 2^M
120 DIM A(N),B(N),M(N),P(N),UA(N),UB(N),Y(N)
130 PRINT:INPUT "Sampling interval T = ";T
140 FOR K=1 TO N:B(K)=0:NEXT
150 PRINT:INPUT "Is the signal in function form (y/n)";A$
160 PRINT:IF A$="y" THEN 210
170 FOR I=1 TO N
180 PRINT:PRINT "Enter x(kT) for k = ";I-1:INPUT A(I)
190 NEXT I
200 GOTO 230
210 DEF FNA(K) =
220 FOR J=1 TO N:K=J-1:A(J)=FNA(K):NEXT
230 PRINT "Freq            Mag             Phase"
240 N2=INT(N/2)
250 N1=N-1
260 J=1
270 FOR I=1 TO N1
280     IF I>=J THEN 350
290     TA=A(J)
300     TB=B(J)
310     A(J)=A(I)
320     B(J)=B(I)
330     A(I)=TA
340     B(I)=TB
350     K=N2
360     IF K>=J GOTO 400
370     J=J-K
380     K=INT(K/2)
390     GOTO 360
400     J=J+K
410 NEXT I
420 REM
430 REM    THE FOLLOWING IS THE ACTUAL CALCULATION
440 REM
450 FOR L=1 TO M
460 LE=2^L
470 L2=INT(LE/2)
480 UA(1)=1:UB(1)=0
490 WA=COS(PI/L2):IF ABS(WA)<1E-08 THEN WA=0
500 WB=FFT*SIN(PI/L2):IF ABS(WB)<1E-08 THEN WB=0
510 FOR G=1 TO L2
520  FOR H=G TO N STEP LE
530  IP=H+L2
540  VA=A(IP)*UA(G) - B(IP)*UB(G)
550  VB=B(IP)*UA(G) + A(IP)*UB(G)
560 A(IP)=A(H)-VA
570 B(IP)=B(H)-VB
580 A(H)=A(H)+VA
590 B(H)=B(H)+VB
```

FIGURE A.14. Program CTFT.

```
600 NEXT H
610 UA(G+1)=UA(G)*WA - UB(G)*WB
620 UB(G+1)=UB(G)*WA + UA(G)*WB
630 NEXT G
640 NEXT L
650 W=2*PI/(N*T)
660 M(0)=T*SQR(A(1)^2+B(1)^2):IF A(1)>=0 THEN P(0)= 0
670 IF A(1)<0 THEN P(0)= 180
680 FOR I=1 TO N-1
690 R = SIN(I*W*T)/(I*W)
700 IM = (COS(I*W*T)-1)/(I*W)
710 X = A(I+1)*R -B(I+1)*IM
720 IX = A(I+1)*IM + B(I+1)*R
730 M(I) = SQR(X^2+IX^2)
740 IF X>0 THEN P(I)=ATN(IX/X)*180/PI:GOTO 770
750 IF X=0 THEN P(I)=SGN(IX)*90:GOTO 770
760 P(I)=180+ATN(IX/X)*180/PI
770 IF P(I)>P(I-1)+180 THEN P(I)=P(I)-360
780 IF P(I)<P(I-1)-180 THEN P(I)=P(I)+360
790 NEXT I
800 FOR I=0 TO N-1
810 PRINT .001*INT(1000*W*I);TAB(13);M(I);TAB(29);P(I)
820 NEXT I
830 Q=N-1
840 PRINT:INPUT "Want a plot of mag spectrum (y/n)";B$
850 IF B$="y" THEN FOR K=0 TO Q:Y(K)=M(K+1):NEXT:GOTO 1000
860 PRINT:INPUT "Want a plot of the phase spectrum (y/n)";C$
870 IF C$ = "y" THEN FOR K=0 TO Q:Y(K)=P(K+1):NEXT:GOTO 1000
880 END
1000 YH=Y(0):YL=Y(0)
1010 FOR K=1 TO Q
1020 IF Y(K)>=YH THEN YH=Y(K):GOTO 1050
1030 IF Y(K)>=YL THEN YH=YH:GOTO 1050
1040 YL = Y(K)
1050 NEXT K
1060 FOR K=0 TO Q
1070 Y(K)=1+(Y(K)-YL)*29/(YH-YL):Y(K)=INT(Y(K)+.5)
1080 NEXT K
1090 Z=1+INT(-YL*29/(YH-YL)+.5):YH=.01*INT(100*YH+.5)
1100 YL=.01*INT(100*YL+.5):INPUT "Plot on screen (y/n)";A$
1110 IF A$="n" THEN GOTO 1190
1120 PRINT "min=";YL;TAB(25);"max=";YH
1130 FOR K=0 TO Q
1140 IF Y(K)=Z THEN PRINT TAB(Z);"*":GOTO 1180
1150 IF Z<=30 AND Y(K)<Z THEN PRINT TAB(Y(K));"*";TAB(Z);"0":GOTO 1180
1160 IF Z>=1 AND Y(K)>Z THEN PRINT TAB(Z);"0";TAB(Y(K));"*":GOTO 1180
1170 PRINT TAB(Y(K));"*"
1180 NEXT K
1190 INPUT "Plot on printer (y/n)";A$
1200 IF A$="n" THEN 1280
1210 LPRINT "min=";YL;TAB(25);"max=";YH
1220 FOR K=0 TO Q
1230 IF Y(K)=Z THEN LPRINT TAB(Z);"*":GOTO 1270
1240 IF Z<=30 AND Y(K)<Z THEN LPRINT TAB(Y(K));"*";TAB(Z);"0":GOTO 1270
1250 IF Z>=1 AND Y(K)>Z THEN LPRINT TAB(Z);"0";TAB(Y(K));"*":GOTO 1270
1260 LPRINT TAB(Y(K));"*"
1270 NEXT K
1280 IF B$="y" AND C$<>"y" THEN 860
1290 END
```

FIGURE A.14. (Continuation of CTFT.)

```
10 REM   Discretization of state model by Robert Cooper
20 CLS
30 INPUT "Size of B matrix (row,column)";R,C
40 AF=0
50 DIM A(R,R), AT(R,R), AP(R,R), AD(R,R), BT(R,C)
60 DIM B(R,C), BP(R,C), BD(R,C)
70 AC=1
80 BC=2
90 PRINT:INPUT "Discretization interval = ";T
100 CY = .000001
110 FOR I = 1 TO R
120 FOR J = 1 TO R
130 AT(I,J) = 0
140 AD(I,J) = 0
150 NEXT J,I
160 FOR I = 1 TO R
170 AT(I,I) = 1
180 FOR J = 1 TO C
190 BD(I,J) = 0
200 NEXT J,I
210 CLS
220 GOSUB 310
230 GOSUB 440
240 GOSUB 490
250 GOSUB 590
260 IF AF = 0 GOTO 230
270 GOSUB 810
280 GOSUB 860
290 GOSUB 960
300 GOTO 270
310 FOR I = 1 TO R
320 FOR J = 1 TO R
330 PRINT "A("I","J") = ":INPUT A(I,J)
340 NEXT J,I
350 CLS
360 FOR I = 1 TO R
370 FOR J = 1 TO C
380 PRINT "B(";I;",";J;") = ":INPUT B(I,J)
390 BT(I,J) = B(I,J)*T
400 NEXT J,I
410 CLS
420 PRINT "Computing discretized matrices..."
430 RETURN
440 FOR I = 1 TO R
450 FOR J = 1 TO R
460 AD(I,J) = AD(I,J) + AT(I,J)
470 NEXT J,I
480 RETURN
490 FOR I = 1 TO R
500 FOR J = 1 TO R
510 DU = 0
520 FOR K = 1 TO R
530 DU = DU + AT(I,K)*A(K,J)
540 NEXT K
550 AP(I,J) = DU*T/AC
560 NEXT J,I
570 AC = AC + 1
```

FIGURE A.15.   Program STATEDIS.

```
 580 RETURN
 590 FOR I = 1 TO R
 600 FOR J = 1 TO R
 610 DE = AT(I,J) - AP(I,J)
 620 IF ABS(DE) > CY GOTO 760
 630 NEXT J,I
 640 CLS
 650 PRINT "The discretized system matrix:"
 660 PRINT:PRINT:PRINT
 670 FOR I = 1 TO R
 680 FOR J = 1 TO R
 690 Y = .00001*INT(100000!*AD(I,J))
 700 PRINT Y;"        ";
 710 NEXT J
 720 PRINT
 730 NEXT I
 740 AF = 1
 750 RETURN
 760 FOR I = 1 TO R
 770 FOR J = 1 TO R
 780 AT(I,J) = AP(I,J)
 790 NEXT J,I
 800 RETURN
 810 FOR I = 1 TO R
 820 FOR J = 1 TO C
 830 BD(I,J) = BD(I,J) + BT(I,J)
 840 NEXT J,I
 850 RETURN
 860 FOR I = 1 TO R
 870 FOR J = 1 TO C
 880 DU = 0
 890 FOR K = 1 TO R
 900 DU = DU + A(I,K)*BT(K,J)
 910 NEXT K
 920 BP(I,J) = DU*T/BC
 930 NEXT J,I
 940 BC = BC + 1
 950 RETURN
 960 FOR I = 1 TO R
 970 FOR J = 1 TO C
 980 DE = BT(I,J) - BP(I,J)
 990 IF ABS(DE) > CY GOTO 1100
1000 NEXT J,I
1010 PRINT:PRINT:PRINT:PRINT:PRINT:PRINT:PRINT:PRINT
1020 PRINT "The discretized input matrix:":PRINT:PRINT
1030 FOR I = 1 TO R
1040 FOR J = 1 TO C
1050 PRINT .00001*INT(100000!*BD(I,J));"        ";
1060 NEXT J
1070 PRINT
1080 NEXT I
1090 END
1100 FOR I = 1 TO R
1110 FOR J = 1 TO C
1120 BT(I,J) = BP(I,J)
1130 NEXT J,I
1140 RETURN
```

FIGURE A.15. (Continuation of STATEDIS.)

Brief Review
of Complex Variables

A very brief and elementary review of complex variables is given in this appendix. For an in-depth treatment of complex variables, the reader may consult Churchill et al. [1976].

Let s be a complex number. The rectangular form of s is given by

$$s = a + jb, \tag{B.1}$$

where a and b are real numbers and where $j = \sqrt{-1}$. The real number a is the real part of s and we write $a = \text{Re } s$. The real number b is the imaginary part of s and we write $b = \text{Im } s$.

The polar form of a complex number s is given by

$$s = \rho e^{j\theta}, \tag{B.2}$$

where ρ is a nonnegative real number and θ is a real number. The number ρ is the magnitude of s and we write $\rho = |s|$. The real number θ is the angle of s and we write $\theta = \sphericalangle s$.

As illustrated in Figure B.1, complex numbers may be represented as points in the complex plane. The horizontal axis of the complex plane is the *real axis*, and the vertical axis is the *imaginary axis*. The complex number $s = a + jb$

611

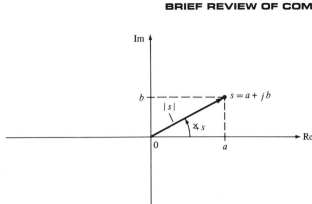

FIGURE B.1.   Point s in the complex plane.

with both a and b positive is the point in the complex plane shown in Figure B.1. The *vector representation* of s is the vector beginning at the origin of the complex plane and ending at the point $s = a + jb$. The direction of the vector representation is indicated by the arrow shown in Figure B.1. Note that the magnitude $|s|$ of s is the length of the vector representation of s and the angle $\sphericalangle s$ of s is the angle between the vector representation of s and the real axis of the complex plane.

Two complex numbers s_1 and s_2 are equal if and only if their real parts are equal and their imaginary parts are equal; that is,

$$\operatorname{Re} s_1 = \operatorname{Re} s_2 \quad \text{and} \quad \operatorname{Im} s_1 = \operatorname{Im} s_2.$$

With respect to the polar representation, two complex numbers s_1 and s_2 are equal if and only if their magnitudes are equal and their angles are equal; that is,

$$|s_1| = |s_2| \quad \text{and} \quad \sphericalangle s_1 = \sphericalangle s_2.$$

It is often necessary to convert from the rectangular form of s to the polar form of s, and vice versa. This can be accomplished as follows. Suppose that we have a complex number s with the rectangular form (B.1) and the polar form (B.2). *Euler's formula* states that

$$e^{j\theta} = \cos \theta + j \sin \theta. \tag{B.3}$$

Inserting (B.3) into the polar representation (B.2), we have

$$s = \rho(\cos \theta + j \sin \theta). \tag{B.4}$$

Equating the expression (B.1) for the rectangular form of s with the expression (B.4), we have

$$a = \rho \cos \theta \quad \text{and} \quad b = \rho \sin \theta. \tag{B.5}$$

Using (B.5), we can compute the rectangular form of s given the polar form of s.

Now from (B.5) we have

$$a^2 + b^2 = \rho^2(\cos^2\theta + \sin^2\theta) = \rho^2 \tag{B.6}$$

and

$$\frac{b}{a} = \frac{\sin\theta}{\cos\theta} = \tan\theta. \tag{B.7}$$

Solving (B.6) for ρ and (B.7) for θ, we have

$$\rho = \sqrt{a^2 + b^2} \tag{B.8}$$

and

$$\theta = \begin{cases} \tan^{-1}\left(\dfrac{b}{a}\right) & \text{when } a > 0 \\[3mm] 180° + \tan^{-1}\left(\dfrac{b}{a}\right) & \text{when } a < 0. \end{cases} \tag{B.9}$$

Using (B.8) and (B.9), we can compute the polar form of s given the rectangular form of s.

The *complex conjugate* of $s = a + jb$, denoted by $\bar{s}$, is given by

$$\bar{s} = a - jb.$$

Thus conjugation changes the sign of the imaginary part of a complex number. If s is given in the polar form (B.2), the complex conjugate $\bar{s}$ is defined by

$$\bar{s} = \rho e^{-j\theta}.$$

So conjugation changes the sign of the angle of a complex number.

Given two complex numbers $s_1 = a_1 + jb_1$ and $s_2 = a_2 + jb_2$, the sum $s_1 + s_2$ is defined by

$$s_1 + s_2 = (a_1 + a_2) + j(b_1 + b_2). \tag{B.10}$$

From (B.10) we see that the real part of the sum of two complex numbers is equal to the sum of the real parts of the two numbers, and the imaginary part of the sum is equal to the sum of the imaginary parts.

The product $s_1 s_2$ is defined by

$$s_1 s_2 = (a_1 a_2 - b_1 b_2) + j(a_1 b_2 + b_1 a_2).$$

The division s_1/s_2 is given by

$$\frac{s_1}{s_2} = \frac{a_1 + jb_1}{a_2 + jb_2}$$

$$= \frac{(a_1 + jb_1)(a_2 - jb_2)}{(a_2 + jb_2)(a_2 - jb_2)}$$

$$= \frac{(a_1a_2 + b_1b_2) + j(-a_1b_2 + b_1a_2)}{a_2^2 + b_2^2}.$$

If s_1 and s_2 are written in polar form,

$$s_1 = \rho_1 e^{j\theta_1} \quad \text{and} \quad s_2 = \rho_2 e^{j\theta_2},$$

the product s_1s_2 is given by

$$s_1s_2 = (\rho_1\rho_2) \exp[j(\theta_1 + \theta_2)], \tag{B.11}$$

and the division s_1/s_2 is given by

$$\frac{s_1}{s_2} = \frac{\rho_1}{\rho_2} \exp[j(\theta_1 - \theta_2)]. \tag{B.12}$$

From (B.11) we see that the magnitude $|s_1s_2|$ of the product s_1s_2 is equal to the product of the magnitude $|s_1|$ of s_1 and the magnitude $|s_2|$ of s_2, and the angle $\angle(s_1s_2)$ of the product is the sum $\theta_1 + \theta_2$ of the angles of s_1 and s_2; that is,

$$|s_1s_2| = |s_1||s_2| \quad \text{and} \quad \angle s_1s_2 = \angle s_1 + \angle s_2.$$

From (B.12) we have that the magnitude $|s_1/s_2|$ of the division s_1/s_2 is equal to the magnitude $|s_1|$ of s_1 divided by the magnitude $|s_2|$ of s_2, and the angle $\angle(s_1/s_2)$ is equal to the angle θ_1 of s_1 minus the angle θ_2 of s_2; that is,

$$\left|\frac{s_1}{s_2}\right| = \frac{|s_1|}{|s_2|} \quad \text{and} \quad \angle\left(\frac{s_1}{s_2}\right) = \angle s_1 - \angle s_2.$$

The above results show that if we want to multiply or divide complex numbers, it is easier to use the polar form (rather than the rectangular form). On the other hand, if we want to add or subtract complex numbers, it is easier to use the rectangular form.

Brief Review
of Matrix Algebra

This appendix contains a very brief and elementary review of matrix algebra. For a thorough treatment of matrix algebra, see Ayres [1962].

Let n be a positive integer. An n-element *row vector* q is a row of elements defined by

$$q = [q_1 \quad q_2 \quad \cdots \quad q_n].$$

An n-element *column vector* q is a column of elements defined by

$$q = \begin{bmatrix} q_1 \\ q_2 \\ \cdot \\ \cdot \\ \cdot \\ q_n \end{bmatrix}.$$

The elements or components of the (row or column) vector q are the quantities $q_1, q_2, \ldots, q_n$. These components may be real or complex numbers. They may also be functions of time or functions of a complex variable s.

615

Given two positive integers m and n, a $m \times n$ (m by n) *matrix A* is an array of elements given by

$$A = \begin{bmatrix} a_{11} & a_{12} & \cdots & a_{1n} \\ a_{21} & a_{22} & \cdots & a_{2n} \\ \cdot & \cdot & & \cdot \\ \cdot & \cdot & & \cdot \\ \cdot & \cdot & & \cdot \\ a_{m1} & a_{m2} & \cdots & a_{mn} \end{bmatrix}.$$

This matrix has m rows and n columns. For $i = 1, 2, \ldots, m$ and $j = 1, 2, \ldots, n$, the element a_{ij} in A is the i, j entry or component of the matrix A. As in the case of vectors, the components of a matrix may be numbers or functions.

To simplify the notation, we shall sometimes denote a matrix by $A = (a_{ij})$, where a_{ij} is the i, j entry of A.

The *diagonal components* of the matrix $A = (a_{ij})$ are the elements $a_{11}, a_{22}, \ldots, a_{qq}$, where $q = m$ if $m \leq n$, $q = n$ if $n \leq m$. A matrix is said to be a diagonal matrix if all the nondiagonal components are zero.

Note that if $n = 1$, a $m \times 1$ matrix is a m-element column vector and if $m = 1$, a $1 \times n$ matrix is a n-element column vector. Thus row vectors and column vectors can be viewed as special cases of matrices.

If $n = m$, the $m \times n$ matrix A is said to be a *square matrix*. In other words, a matrix is square if it has the same number of rows and columns. An example of a square matrix is the *identity matrix* I_n, defined by

$$I_n = \begin{bmatrix} 1 & 0 & 0 & & \cdots & 0 \\ 0 & 1 & 0 & & \cdots & 0 \\ \cdot & & & \cdot & & \cdot \\ \cdot & & & & \cdot & \cdot \\ \cdot & & & & & \cdot \\ 0 & 0 & 0 & & \cdots & 1 \end{bmatrix}.$$

So the identity matrix I_n is a diagonal matrix with 1's on the diagonal.

Given the $m \times n$ matrix $A = (a_{ij})$, the *transpose* of A is the $n \times m$ matrix A' defined by

$$A' = \begin{bmatrix} a_{11} & a_{21} & \cdots & a_{m1} \\ a_{12} & a_{22} & \cdots & a_{m2} \\ \cdot & \cdot & & \cdot \\ \cdot & \cdot & & \cdot \\ \cdot & \cdot & & \cdot \\ a_{1n} & a_{2n} & \cdots & a_{mn} \end{bmatrix}.$$

Using our shorthand notation, we have that $A' = (a_{ji})$.

From the definition of A', we see that the rows of the transpose are the same as the columns of A. In particular, the transpose of a row vector is equal to a column vector, and the transpose of a column vector is equal to a row vector.

MATRIX OPERATIONS Two matrices can be added if and only if they have the same size, that is, they have the same number of rows and the same number of columns. In particular, suppose that A and B are two $m \times n$ matrices given by $A = (a_{ij})$, $B = (b_{ij})$. Then the sum $A + B$ is defined by

$$A + B = (a_{ij} + b_{ij}).$$

So the i, j entry of the sum $A + B$ is equal to the sum of the i, j entries of A and B. Note that if $m = 1$, A and B are row vectors and thus $A + B$ is the addition of two row vectors. If $n = 1$, A and B are column vectors and $A + B$ is the addition of two column vectors.

Given a $m \times n$ matrix A and a $p \times r$ matrix B, the product AB can be defined if and only if $n = p$; that is, the number of columns of A must be equal to the number of rows of B. If this is the case, the product AB is defined by

$$AB = (c_{ij}), \qquad \text{where } c_{ij} = \sum_{k=1}^{n} a_{ik}b_{kj}.$$

EXAMPLE C.1. Suppose that

$$A = \begin{bmatrix} -1 & 2 \\ 3 & 1 \\ 2 & 4 \end{bmatrix} \quad \text{and} \quad B = \begin{bmatrix} -3 & 2 \\ 2 & 1 \end{bmatrix}.$$

Then writing AB in the form $AB = (c_{ij})$, we have

$$c_{11} = (-1)(-3) + (2)(2) = 7$$

$$c_{12} = (-1)(2) + (2)(1) = 0$$

$$c_{21} = (3)(-3) + (1)(2) = -7$$

$$c_{22} = (3)(2) + (1)(1) = 7$$

$$c_{31} = (2)(-3) + (4)(2) = 2$$

$$c_{32} = (2)(2) + (4)(1) = 8.$$

Thus

$$AB = \begin{bmatrix} 7 & 0 \\ -7 & 7 \\ 2 & 8 \end{bmatrix}.$$

It is important to observe that matrix multiplication is *not* commutative in general. In other words, $AB \neq BA$. For instance, in the above example the product BA cannot be defined since the number of columns of B is not equal to the number of rows of A. Even if A and B are square matrices of the same size, in general $AB \neq BA$. For example, suppose that

$$A = \begin{bmatrix} 1 & 1 \\ 0 & 0 \end{bmatrix} \quad \text{and} \quad B = \begin{bmatrix} 0 & 0 \\ 1 & 1 \end{bmatrix}.$$

Then

$$AB = \begin{bmatrix} 1 & 1 \\ 0 & 0 \end{bmatrix} = A \quad \text{and} \quad BA = \begin{bmatrix} 0 & 0 \\ 1 & 1 \end{bmatrix} = B.$$

Thus $AB \neq BA$.

Given any $n \times n$ matrix A, we have that

$$A(I_n) = I_n A = A,$$

where I_n is the $n \times n$ identity matrix. Hence multiplication of a square matrix by the identity matrix reproduces the matrix. This property implies that I_n is the identity element of matrix multiplication.

Given the $m \times n$ matrix $A = (a_{ij})$ and a scalar b, the scalar multiplication bA is defined by

$$bA = (ba_{ij}).$$

In other words, the multiplication of a matrix by a scalar is defined by multiplying each component of the matrix by the scalar.

DETERMINANT OF A SQUARE MATRIX   Let $A = (a_{ij})$ be a square matrix of size n. The *determinant* of A is a scalar denoted by $|A|$. The reader should note that our symbol for the determinant is the same as our symbol for the magnitude of a complex number. It will always be clear from the context whether we are considering the determinant of a square matrix or the magnitude of a complex number. Thus this dual use of the symbol $|\ |$ should not be confusing.

When $n = 1$, so that A is a scalar, the determinant of A is equal to A. Now suppose that $n = 2$, so that

$$A = \begin{bmatrix} a_{11} & a_{12} \\ a_{21} & a_{22} \end{bmatrix}.$$

In this case, the determinant of A is

$$|A| = a_{11}a_{22} - a_{12}a_{21}. \tag{C.1}$$

Now let $n = 3$, so that

$$A = \begin{bmatrix} a_{11} & a_{12} & a_{13} \\ a_{21} & a_{22} & a_{23} \\ a_{31} & a_{32} & a_{33} \end{bmatrix}.$$

The determinant of A is given by

$$|A| = a_{11} \begin{vmatrix} a_{22} & a_{23} \\ a_{32} & a_{33} \end{vmatrix} - a_{21} \begin{vmatrix} a_{12} & a_{13} \\ a_{32} & a_{33} \end{vmatrix} + a_{31} \begin{vmatrix} a_{12} & a_{13} \\ a_{22} & a_{23} \end{vmatrix}.$$

$$= a_{11}(a_{22}a_{33} - a_{32}a_{23}) - a_{21}(a_{12}a_{33} - a_{32}a_{13}) \qquad \text{(C.2)}$$
$$+ a_{31}(a_{12}a_{23} - a_{22}a_{13}).$$

The expression (C.2) for the determinant was computed by expanding with respect to the first column of A. We can calculate $|A|$ by expanding with respect to any row or column of A. For instance, if we expand with respect to the first row of A, we obtain

$$|A| = a_{11} \begin{vmatrix} a_{22} & a_{23} \\ a_{32} & a_{33} \end{vmatrix} - a_{12} \begin{vmatrix} a_{21} & a_{23} \\ a_{31} & a_{33} \end{vmatrix} + a_{13} \begin{vmatrix} a_{21} & a_{22} \\ a_{31} & a_{32} \end{vmatrix}. \qquad \text{(C.3)}$$

From the pattern in (C.2) and (C.3), it should be clear how to expand with respect to any row or column of A. To minimize the number of computations, one should expand with respect to the row or column containing the most zeros. In particular, note that if A has a row or column consisting of all zeros, then $|A| = 0$.

When $n = 4$, if we compute $|A|$ by expanding with respect to the first column of A, we obtain

$$|A| = a_{11} \begin{vmatrix} a_{22} & a_{23} & a_{24} \\ a_{32} & a_{33} & a_{34} \\ a_{42} & a_{43} & a_{44} \end{vmatrix} - a_{21} \begin{vmatrix} a_{12} & a_{13} & a_{14} \\ a_{32} & a_{33} & a_{34} \\ a_{42} & a_{43} & a_{44} \end{vmatrix}$$

$$+ a_{31} \begin{vmatrix} a_{12} & a_{13} & a_{14} \\ a_{22} & a_{23} & a_{24} \\ a_{42} & a_{43} & a_{44} \end{vmatrix} - a_{41} \begin{vmatrix} a_{12} & a_{13} & a_{14} \\ a_{22} & a_{23} & a_{24} \\ a_{32} & a_{33} & a_{34} \end{vmatrix}. \qquad \text{(C.4)}$$

The pattern in (C.2) and (C.4) extends to the computation of $|A|$ when $n > 4$. We omit a detailed development of the general case (see Ayres [1962]).

It is worth noting that if A is a diagonal matrix with values a_{ii} on the diagonal, the determinant of A is

$$|A| = a_{11}a_{22} \cdots a_{nn}.$$

MATRIX INVERSION   Let A be a $n \times n$ matrix. The matrix A has an *inverse* denoted by A^{-1} if and only if

$$A(A^{-1}) = A^{-1}A = I_n.$$

The inverse A^{-1} (if it exists) is also a $n \times n$ matrix.

It is a fundamental result in matrix algebra that A has an inverse A^{-1} if and only if the determinant $|A|$ is nonzero. If $|A| \neq 0$, the inverse is given by

$$A^{-1} = \frac{1}{|A|} \operatorname{cof}(A)',$$

where $\operatorname{cof}(A)'$ is the transpose of the *cofactor matrix* $\operatorname{cof}(A)$. When $n = 2$, $\operatorname{cof}(A)$ is given by

$$\operatorname{cof}(A) = \begin{bmatrix} a_{22} & -a_{12} \\ -a_{21} & a_{11} \end{bmatrix}.$$

When $n = 3$,

$$\operatorname{cof}(A) = \begin{bmatrix} b_{11} & b_{12} & b_{13} \\ b_{21} & b_{22} & b_{23} \\ b_{31} & b_{32} & b_{33} \end{bmatrix},$$

where

$$b_{11} = \begin{vmatrix} a_{22} & a_{23} \\ a_{32} & a_{33} \end{vmatrix}, \qquad b_{12} = -\begin{vmatrix} a_{21} & a_{23} \\ a_{31} & a_{33} \end{vmatrix}, \qquad b_{13} = \begin{vmatrix} a_{21} & a_{22} \\ a_{31} & a_{32} \end{vmatrix}.$$

$$b_{21} = -\begin{vmatrix} a_{12} & a_{13} \\ a_{32} & a_{33} \end{vmatrix}, \qquad b_{22} = \begin{vmatrix} a_{11} & a_{13} \\ a_{31} & a_{33} \end{vmatrix}, \qquad b_{23} = -\begin{vmatrix} a_{11} & a_{12} \\ a_{31} & a_{32} \end{vmatrix}.$$

$$b_{31} = \begin{vmatrix} a_{12} & a_{13} \\ a_{22} & a_{23} \end{vmatrix}, \qquad b_{32} = -\begin{vmatrix} a_{11} & a_{13} \\ a_{21} & a_{23} \end{vmatrix}, \qquad b_{33} = \begin{vmatrix} a_{11} & a_{12} \\ a_{21} & a_{22} \end{vmatrix}.$$

Again, there is a pattern in the computation of $\operatorname{cof}(A)$ that extends to $n > 3$.

It is easy to see that if A is a diagonal matrix with nonzero values a_{ii} on the diagonal, the inverse A^{-1} exists and is also diagonal with the values $1/a_{ii}$ on the diagonal. To verify this, observe that

$$\begin{bmatrix} a_{11} & 0 & \cdots & 0 \\ 0 & a_{22} & \cdots & 0 \\ \cdot & & & \cdot \\ \cdot & & & \cdot \\ \cdot & & & \cdot \\ 0 & 0 & \cdots & a_{nn} \end{bmatrix} \begin{bmatrix} \dfrac{1}{a_{11}} & 0 & \cdots & 0 \\ 0 & \dfrac{1}{a_{22}} & & 0 \\ \cdot & & & \cdot \\ \cdot & & & \cdot \\ \cdot & & & \cdot \\ 0 & 0 & \cdots & \dfrac{1}{a_{nn}} \end{bmatrix} = \begin{bmatrix} 1 & 0 & \cdots & 0 \\ 0 & 1 & & 0 \\ \cdot & & & \cdot \\ \cdot & & & \cdot \\ \cdot & & & \cdot \\ 0 & 0 & \cdots & 1 \end{bmatrix}$$

Bibliography

The references mentioned in this book are listed here according to subject area. In addition to these references, we have included additional textbooks on signals and systems and related areas.

Circuits

HAYT, W. H., JR., and KEMMERLY, J. E., *Engineering Circuit Analysis,* 3rd ed., McGraw-Hill, New York, 1978.

HUFAULT, J. R., *Op Amp Network Design,* Wiley, New York, 1985.

O'MALLEY, J. R., *Basic Circuit Analysis,* Schaum's Outline Series, McGraw-Hill, New York, 1981.

NILSSON, J. W., *Electrical Circuits,* Addison-Wesley, Reading, Mass., 1983.

VAN VALKENBURG, M. E., *Network Analysis,* 3rd ed., Prentice-Hall, Englewood Cliffs, N.J., 1974.

VAN VALKENBURG, M. E., and KINARIWALA, B. K., *Linear Circuits,* Prentice-Hall, Englewood Cliffs, N.J., 1982.

Communications

CARLSON, A. B., *Communication Systems,* 3rd ed., McGraw-Hill, New York, 1986.

COUCH, L., *Analog and Digital Communications,* Macmillan, New York, 1983.

GALLAGER, R. G., *Information Theory and Reliable Communication,* Wiley, New York, 1968.

KANEFSKY, M., *Communication Techniques for Digital and Analog Signals,* Harper & Row, New York, 1985.

SCHWARTZ, M., *Information, Transmission, Modulation, and Noise,* 3rd ed., McGraw-Hill, New York 1980.

Controls

BROGAN, W. L., *Modern Control Theory,* 2nd ed., Prentice-Hall, Englewood Cliffs, N.J., 1985.

CHEN, C. T., *Linear System Theory and Design,* Holt, Rinehart and Winston, New York, 1984.

DORF, R. C., *Modern Control Systems,* 3rd ed., Addison-Wesley, Reading, Mass., 1980.

FRANKLIN, G. F., and POWELL, J. D., *Digital Control,* Addison-Wesley, Reading, Mass., 1980.

FRIEDLAND, B., *Control System Design,* McGraw-Hill, New York, 1986.

KAILATH, T., *Linear Systems,* Prentice-Hall, Englewood Cliffs, N.J., 1980.

KUO, B. C., *Digital Control Systems,* Holt, Rinehart and Winston, New York, 1980.

KUO, B. C., *Automatic Control Systems,* 4th ed., Prentice-Hall, Englewood Cliffs, N.J., 1982.

OGATA, K., *Modern Control Engineering,* Prentice-Hall, Englewood Cliffs, N.J., 1970.

VAN DE VEGTE, J., *Feedback Control Systems,* Prentice-Hall, Englewood Cliffs, N.J., 1986.

VANLANDINGHAM, H. F., *Introduction to Digital Control Systems,* Macmillan, New York, 1985.

Digital Signal Processing/Digital Filtering

BOSE, N. K., *Digital Filters,* Elsevier North-Holland, New York, 1985.

BRIGHAM, E. O., *The Fast Fourier Transform,* Prentice-Hall, Englewood Cliffs, N.J., 1974.

BURRUS, C. S., and PARKS, T. W., *DFT/FFT and Convolution Algorithms,* Wiley, New York, 1984.

LUDEMAN, L. C., *Fundamentals of Digital Signal Processing,* Harper & Row, New York, 1986.

OPPENHEIM, A. V., and SCHAFER, R. W., *Digital Signal Processing,* Wiley, New York, 1975.

RABINER, L. R., and GOLD, B., *Theory and Application of Digital Signal Processing,* Prentice-Hall, Englewood Cliffs, N.J., 1975.

SCHWARTZ, M., AND SHAW, L., *Signal Processing,* McGraw-Hill, New York, 1975.

WILLIAMS, L. S., *Designing Digital Filters,* Prentice-Hall, Englewood Cliffs, N.J., 1986.

Mathematics

AYRES, F., JR., *Matrices,* Schaum Publishing Co., New York, 1962.

CHURCHILL, R. V., BROWN, J. W., and VERHEY, R. F., *Complex Variables and Applications,* 3rd ed., McGraw-Hill, New York, 1976.

JAMES, M. L., SMITH, G. M., and WOLFORD, J. C., *Applied Numerical Methods for Digital Computation,* Harper & Row, New York, 1985.

Microprocessors

ANDREWS, M., *Programming Microprocessor Interfaces for Control and Instrumentation,* Prentice-Hall, Englewood Cliffs, N.J., 1982.

BANSAL, V. K., *Design of Microprocessor-Based Systems,* Wiley, New York, 1985.

Optical Communications

BARNOSKI, M. K., ed., *Fundamentals of Optical Fiber Communications,* Academic Press, New York, 1976.

GOWAR, J., *Optical Communication Systems,* Prentice-Hall, Hemel Hempstead, Hertfordshire, England, 1984.

KEISER, G., *Optical Fiber Communications,* McGraw-Hill, New York, 1983.

PALAIS, J. C., *Fiber Optic Communications,* Prentice-Hall, Englewood Cliffs, N.J., 1984.

SENIOR, J. M., *Optical Fiber Communications, Principles and Practice,* Prentice-Hall, Hemel Hempstead, Hertfordshire, England, 1985.

WILSON, J., and HAWKES, J. F. B., *Optoelectronics: An Introduction,* Prentice-Hall, Hemel Hempstead, Hertfordshire, England, 1983.

Signals and Systems

GLISSON, T. H., *Introduction to System Analysis,* McGraw-Hill, New York, 1985.

LUENBERGER, D. G., *Introduction to Dynamic Systems,* Wiley, New York, 1979.

MAYHAN, R. J., *Discrete-Time and Continuous-Time Linear Systems,* Addison-Wesley, Reading, Mass., 1984.

McGILLEM, C. D., and COOPER, G. R., *Continuous and Discrete Signal and System Analysis,* 2nd ed., Holt, Rinehart and Winston, New York, 1984.

NEFF, H. P., JR., *Continuous and Discrete Linear Systems,* Harper & Row, New York, 1984.

OPPENHEIM, A. V., and WILLSKY, A. S., *Signals and Systems,* Prentice-Hall, Englewood Cliffs, N.J., 1983.

PAPOULIS, A., *Circuits and Systems,* Holt, Rinehart and Winston, New York, 1980.

POULARIKAS, A. D., and SEELY, S., *Signals and Systems,* PWS Publishers, Boston, Mass., 1985.

TRUXAL, J. G., *Introductory System Engineering,* McGraw-Hill, New York, 1972.

ZIEMER, R. E., TRANTER, W. H., and FANNIN, D. R., *Signals and Systems,* Macmillan, New York, 1983.

INDEX

Properties of the z-Transform

Property	Transform Pair/Property
Linearity	$ax(kT) + bv(kT) \leftrightarrow aX(z) + bV(z)$
Right shift in time	$x(kT - qT)u(kT - qT) \leftrightarrow z^{-q}X(z)$
Second version of right shift	$x(kT - T) \leftrightarrow z^{-1}X(z) + x(-T)$ $x(kT - 2T) \leftrightarrow z^{-2}X(z) + x(-2T) + z^{-1}x(-T)$ $\vdots$ $x(kT - qT) \leftrightarrow z^{-q}X(z) + x(-qT) + z^{-1}x(-qT + T)$ $\qquad\qquad\qquad + \cdots + z^{-q+1}x(-T)$
Left shift in time	$x(kT + T) \leftrightarrow zX(z) - x(0)z$ $x(kT + 2T) \leftrightarrow z^2X(z) - x(0)z^2 - x(T)z$ $\vdots$ $x(kT + qT) \leftrightarrow z^qX(z) - x(0)z^q - x(T)z^{q-1} - \cdots$ $\qquad\qquad\qquad - x(qT - T)z$
Multiplication by k	$kx(kT) \leftrightarrow -z\dfrac{d}{dz}X(z)$
Multiplication by k^2	$k^2x(kT) \leftrightarrow z\dfrac{d}{dz}X(z) + z^2\dfrac{d^2}{dz^2}X(z)$
Multiplication by a^k	$a^kx(kT) \leftrightarrow X\left(\dfrac{z}{a}\right)$
Multiplication by $\cos \omega kT$	$\cos(\omega kT)\,x(kT) \leftrightarrow \frac{1}{2}[X(e^{j\omega T}z) + X(e^{-j\omega T}z)]$
Multiplication by $\sin \omega kT$	$\sin(\omega kT)x(kT) \leftrightarrow \dfrac{j}{2}[X(e^{j\omega T}z) - X(e^{-j\omega T}z)]$
Convolution	$(x * v)(kT) \leftrightarrow X(z)V(z)$
Summation	$v(kT) \leftrightarrow \dfrac{z}{z-1}X(z), \qquad \text{where } v(kT) = \displaystyle\sum_{i=0}^{k} x(iT)$
Initial-value theorem	$x(0) = \lim_{z\to\infty} X(z)$ $x(T) = \lim_{z\to\infty} [zX(z) - zx(0)]$ $\vdots$ $x(qT) = \lim_{z\to\infty} [z^qX(z) - z^qx(0) - z^{q-1}x(T)$ $\qquad\qquad\qquad\qquad - \cdots - zx(qT - T)]$
Final-value theorem	If $\lim_{k\to\infty} x(kT)$ exists, then $\lim_{k\to\infty} x(kT) = \lim_{z\to 1} \left[\dfrac{z-1}{z}X(z)\right]$